AN APPROACH
TO THE HISTORICAL GEOGRAPHY
OF THE
SOUTHERN BLACK SEA LITTORAL
(FIRST MILLENNIUM BC)

COLLOQUIA ANTIQUA

Supplements to the Journal ANCIENT WEST & EAST

Colloquia Antiqua is a refereed publication

For proposals and editorial and other matters, please contact the Series Editor:

The Gallery
Spa Road
Llandrindod Wells
Powys LD1 5ER
UK
E-mail: g.tsetskhladze63@gmail.com

COLLOQUIA ANTIQUA
— 36 —

AN APPROACH TO THE HISTORICAL GEOGRAPHY OF THE SOUTHERN BLACK SEA LITTORAL (FIRST MILLENNIUM BC)

By

MANOLIS MANOLEDAKIS

PEETERS
LEUVEN – PARIS – BRISTOL, CT
2022

A catalogue record for this book is available from the Library of Congress.

ISBN 978-90-429-4906-5
eISBN 978-90-429-4907-2
D/2021/0602/129

To Maria,
To my mother,
To the memory of my father.

TABLE OF CONTENTS

SERIES EDITOR'S PREFACE

It is a pleasure to include in this series a volume on the southern Black Sea, a region to which the author, Manolis Manoledakis, has devoted many years of research, some of the fruits of which are presented here. He is also very active in Black Sea Studies more broadly at the International Hellenic University outside Thessaloniki, his base since 2008, where he immediately started to organise an MA in the subject, launched in 2010 (now the MA in Black Sea and Eastern Mediterranean Studies), then holding a series of international workshops on the Black Sea in antiquity (2012, 2015, 2018, with more to come) and bringing the proceedings of each to publication, and finally playing host to the (seventh) International Congress on Black Sea Antiquities, postponed from September 2021 to September 2022.

In antiquity, the southern coast of the Black Sea and its hinterland provided a home to many people(s), the bulk of them unknown, and to many places mentioned in ancient written sources – about 80 – but again with locations largely unknown. The region has attracted the attention of such noted scholars as M.I. Maksimova (*Antichnye goroda Yugo-Vostochnogo Prichernomor'ya: Sinopa, Amis, Trapezunt*, Moscow/Leningrad 1956), D.R. Wilson (*The Historical Geography of Bithynia, Paphlagonia and Pontus in the Greek and Roman Periods*, BLitt., Oxford 1960), J.G.F. Hind (*Greek Colonisation in the Black Sea Area in the Archaic and Classical Periods*, PhD, Cambridge 1971), S.M. Burstein (*Outpost of Hellenism: The Emergence of Heraclea on the Black Sea*, Berkeley 1976), and A. Bryer and D. Winfield (*The Byzantine Monuments and Topography of the Pontos*, Washington, DC 1985). A few articles by the late D. Asheri on Heraclea Pontica and the Mariandynoi made an impact on scholarship. At a glance it may appear that much has been done and this is true up to a point. However, each of these authors concentrated on some particular part of the southern Black Sea, not the whole. The area attracted my own attention as well. In the summer of 2003, thanks to the help of Prof. S. Atasoy, I was able to take a trip along the Turkish northern seaboard and its hinterland, visiting sites and museums. The results of my observations have been published as: 'Greeks and Locals in the Southern Black Sea Littoral: A Re-Examination'. In G. Herman and I. Shatzman (eds.), *Greeks between East and West. Essays in Greek Literature and History in Memory of David Asheri* (Jerusalem 2007), 160–95; and 'The Southern Black Sea Coast and its Hinterland: an Ethno-Cultural Perspective'. In G.R. Tsetskhladze (ed.), *The Black Sea, Paphlagonia, Pontus and Phrygia in Antiquity: Aspects of Archaeology and Ancient History* (Oxford 2012), 235–41. I have continued to devote much time to study of the Greek colonies of this coast as well. A year on year increase in new excavation and survey projects along the southern Black Sea, conducted by Turkish and overseas colleagues, means that we shall benefit in the future from much new information about this region's historical geography: see, for example, my 'Greeks, locals and others around the Black Sea and its hinterland: recent developments'. In G.R Tsetskhladze *et al.* (eds.), *The Danubian Lands between the Black, Aegean and Adriatic Seas (7th Century BC–10th Century AD)* (Oxford 2015), 16–21.

This book's focus is on the historical geography, broadly conceived, of the southern Black Sea littoral in the 1st millennium BC. Although the investigation of the ancient Black Sea has surged ahead, not least thanks to the demise of the Communist regimes that dominated three of its four shores, the southern littoral remains the poor relation: systematic archaeological investigation is in its infancy. The present volume, the first recent attempt to do so, combines all the available written and archaeological evidence to consider the littoral's natural environment, the development of local peoples (of whom little is known), cultures and economies, and, of course, the Greek presence and Greek colonisation, better (but not well) known. Other peoples are considered, neither indigenes nor colonists, yet with a stake here, be they Cimmerian invaders or Persian suzerains. Urban development is also examined: what kind, where, when, etc.

I should like to thank Bert Verrept and his colleagues at Peeters for their exemplary help in bringing the volume to press, and James Hargrave for assistance with copy-editing. May I also thank the author for bringing this fascinating and encompassing volume to *Colloquia Antiqua*.

Gocha R. Tsetskhladze
Series Editor

FOREWORD

The writing of a book is usually a long and exciting journey. I cannot say that what matters in the end is only the journey itself and not the destination. I have always believed that both are equally important; and indeed, especially when it comes to a book, it is probably only the latter that matters. Nevertheless, in this brief foreword I grasp the opportunity to refer to a significant aspect of my journey: no one can deny that good travel companions always make the journey more enjoyable. I was lucky to have such good travel companions, and I wish to thank them all here.

Firstly, I am deeply indebted to Sümer Atasoy and Şahin Yıldırım, respectively former and current directors of the excavation at Tieion. From the very first day we met, they have always been willing to help me in any way: they generously discussed anything about their research with me, supplied me with their papers, and informed me without the slightest hesitation of their always impressive excavation results, even before their publication. Moreover, they warmly hosted and guided me around several Black Sea sites during a journey of mine in Turkey. I am most grateful for their honest interest, their valuable help (also to my post-graduate students), and their unforgettable hospitality.

Equally valuable, generous and diverse has always been the help I received from Gocha Tsetskhladze, again from the first day of our meeting. He is always willing to share views, give ideas, warn against risks and slips, and transmit much of his vast knowledge on the Black Sea in antiquity. I am most grateful to him as well.

Owen Doonan, the director of the Sinop Regional Archaeological Project, has shared with me both information about his excavation results and his thoughts about the ancient history of the settlements in Sinope and its wider region. I thank him for this. Several other colleagues and friends assisted me in various ways, either by enlightening me as to several matters arising in this book, or by sending me relevant papers of theirs before or just after publication, or simply by offering me some ideas during a discussion. These are the late Alexandru Avram, Eleni Manakidou, Eleni Mentesidou, Michalis Tiverios and Despoina Tsiafaki, whom I wish to thank.

My decision to write a book on the historical geography of the southern Black Sea in antiquity has always been followed by the desire to create new maps of the region that would display settlements, geographical features, peoples' territories, and also the natural resources of the littoral, always based on the available evidence of the ancient sources; maps that would be topographically accurate and reliable as well as informative. The creation of such maps was achieved thanks to Jamieson Donati, who prepared the topographical maps (in Figs. 8, 10, 13, 14) in GIS using a digital elevation model from the Shuttle Radar Topography Mission. I am most grateful to him.

Many warm thanks go also to Maria Levanti, who helped with the copy-editing and the creation of the index, as well as to Janet Koniordos, for improving my English. I also thank the referee of my manuscript for *Colloquia Antiqua* and James Hargrave for the final copy-editing.

Finally, I warmly thank *Colloquia Antiqua* and the series editor, Gocha Tsetskhladze, for including my book in their monograph series.

All translations, unless otherwise attributed, are from the Loeb edition.

The first figure of the Section numbers mentioned in the text refers to the respective Part of the book.

Manolis Manoledakis
Thessaloniki, July 2021

LIST OF ILLUSTRATIONS

LIST OF ABBREVIATIONS

ACSS	*Ancient Civilizations from Scythia to Siberia.*
AJA	*American Journal of Archaeology.*
AS	*Anatolian Studies.*
ANES	*Ancient Near Eastern Studies.*
AWE	*Ancient West and East.*
CAH	*Cambridge Ancient History.*
CIG	*Corpus Inscriptionum Graecarum.*
EGF	*Epicorum Graecorum Fragmenta.*
FGrH	F. Jacoby, *Die Fragmente der griechischen Historiker.*
GGM	K. Müller, *Geographi Graeci minores.*
IG	*Inscriptiones Graecae* (Berlin 1903–).
IK Sinope	D.H. French, *The Inscriptions of Sinope* (Inschriften Griechischer Städte aus Kleinasien 64) (Bonn 2004).
IstMitt	*Mitteilungen des Deutschen Archäologischen Instituts. Abteilung Istanbul.*
JHS	*Journal of Hellenic Studies.*
JNES	*Journal of Near Eastern Studies.*
RE	*Paulys Realencyclopädie der classischen Altertumswissenschaft.*
TAPA	*Transactions of the American Philological Association.*

INTRODUCTION

1. Why the Southern Black Sea?

The southern Black Sea littoral is a region with significant particularities, all the more so when it comes to the 1st millennium BC. These particularities pertain to several aspects. One of them is that the inland (southern) boundary of the littoral is remarkably well defined. An intrinsic difficulty when dealing with a littoral is how far from the sea to place its inland boundaries, but in our case the natural relief solves the problem. The huge Pontides, widely known as the Paphlagonian mountains, extend parallel to the sea along almost the entire Black Sea shore, and in fact very close to it, cutting it off from the rest of Anatolia (Fig. 1). They thus effectively determine the southern boundaries of the littoral, which is, accordingly, held to be the region between the northern slopes of the Pontides and the Black Sea, a ribbon of land roughly 1000 km in length as the crow flies, and 30–70 km wide.

This particularity has in its turn given rise to another, relating to the geographical distribution of peoples in Asia Minor in the 1st millennium BC, and indeed earlier. While the Phrygians, and before them the Hittites, managed to dominate the greater part of the Anatolian hinterland, neither of them succeeded in gaining control over the littoral. This, as we shall see, was inhabited by many peoples during the 1st millennium BC, but with few exceptions they remained essentially confined to that coastal strip. This image finds no parallels on other Black Sea coasts, nor does it in any way recall the rest of Anatolia, whose northern coast forms our littoral.

And so we come to the next and very significant particularity of the southern Black Sea littoral, which concerns its cultural identity. Would we say that our area of study is more 'Black Sea' or more 'Anatolian'? In the middle of the previous century M.I. Maksimova pointed out that the geographical isolation of the southern Black Sea shore from the inner regions of Asia Minor 'determined to a considerable extent the course of the historical development of its inhabitants'.[1] Going a step further, G.I. Bratianu stressed the role that the north–south axis of the Black Sea played in the unfolding of its history.[2]

[1] Maksimova 1951, 80.
[2] Bratianu 1969.

The end of the Cold War signalled the outburst of a rapid and unprecedented worldwide interest in the study of the Black Sea region[3] – not only in antiquity. This interest also engendered an eagerness, not unjustified, to emphasise the features that are common all along the Black Sea through time, and thus treat the latter as a unit. For example, Y.E. Özveren, strongly influenced by F. Braudel's celebrated work on the Mediterranean in the Age of Philip II (16th century), attempted to consider the Black Sea an economic unit by applying to it the Braudelian *longue durée* approach. Following Bratianu's theory, Özveren reiterated the importance of the north–south Pontic corridor in the long-term development of the Black Sea, but focusing rather on modern periods, mainly the 19th century, by examining relations between the Russian (north) and Ottoman (south) empires.[4]

More recently, in the light of some results of an archaeological investigation at and around Sinope,[5] A.A. Bauer and O.P. Doonan further emphasised the unity of the Black Sea, sparked by the similarity of pottery in their area of research with pottery from the northern and western Black Sea shores, unlike that of the Anatolian hinterland, which looks different. They proposed evidence for patterns of maritime exchange in the Black Sea and the significant role of trade from as early as the Early Bronze Age, perhaps even the Chalcolithic period, and confirmed Özveren's theory regarding the historical unity of the Black Sea.[6]

In sharp contrast, other scholars overemphasised the presence or influence of Anatolian cultural elements in the southern Black Sea, often almost concealing any impact coming from the sea. Ş. Dönmez, for instance, referring to the same place and time as Bauer and Doonan, noted that finds from the Bronze Age 'point to the fact that the Sinop area was one of the northern extremities of the commercial network of the Assyrian Trade Colonies Period, centred at Kültepe/Kaneš'.[7] Dönmez insisted on the dominant role of the cultures of central Anatolia in the littoral's cultural landscape in many of his publications.[8] He even ascribed the numerous decorative architectural elements, particularly terracotta revetments on buildings in local settlements all

[3] See most recently Tsetskhladze 2021b.
[4] Özveren 1997; 2001.
[5] On which, see section 2.1 below.
[6] Bauer and Doonan 2012. See also Bauer 2006; Doonan 2006; 2007; Doonan *et al.* 2016; 2017; Rempel and Doonan 2020.
[7] Dönmez 2012, 154.
[8] See most recently Dönmez 2012; 2019a; 2019b.

along the Halys valley,[9] to the cultures of the central Anatolia region with no Greek influence.[10]

Both theories have the advantage of being based on archaeological evidence, among other things. The trouble with them, however, is that they are sometimes formulated in terms sufficiently absolute to suggest a refusal to examine any contrary evidence: there remains little doubt, for example, that the afore-mentioned terracotta revetments display a clear Greek influence and that their moulds must have come from the Greek colonies on the coast.[11] On the other hand, while it is certainly true that water, especially in closed bodies, facilitates communication and contacts, we can hardly equate this axiom with the exclusion of any contacts between a coast and its hinterland, such as that expressed in the following example: it has been argued that 'the Black Sea region's orientation toward fishing and the sea has *always* made the Black Sea groups separate from inland economies and more "cosmopolitan", or accessible to foreign influences from their neighbours'.[12] *Often*, maybe yes, but not *always*. If we examine the evidence regarding the indigenous peoples that inhabited the southern Black Sea littoral in the 1st millennium BC,[13] we shall very rarely notice either fishing playing a dominant economic role or a clear separation from inland economies (especially in the mountainous far east), and even less a 'cosmopolitanism'.[14]

If we overestimate the natural isolation of the littoral, we risk associating ourselves with an exaggerated environmental determinism, not to mention ignoring facts like the indisputable influence of the Phrygian culture on the littoral,[15] which has sometimes been identified even in Sinope, one of the most physically isolated spots on the coast. Similarly, if we close our eyes to the maritime experiences and activities of communities living on a coast, we simply limit ourselves to a naively one-sided, and thus false, examination of the region. As we shall see in this book, the most remarkable particularity of

[9] On which, see Section 4.4.2.

[10] Dönmez 2006, 13, 22–29, 38–39; 2007b, 1208; 2010, 517–22.

[11] Summerer 2005a, 148; 2005b; 2007, 31–34; 2008, 267–78.

[12] Bauer and Doonan 2012, 22.

[13] On the evidence regarding every single of these peoples, see Part 2 in detail.

[14] We should also note that Maksimova was not exactly speaking of cultural affinities of the littoral with the rest of the Black Sea shores, as mentioned by Bauer and Doonan (2012, 23), but only with the eastern Black Sea and its hinterland as far as the Caspian (Maksimova 1951, 80: 'The culture of these regions was more closely connected with Trans-Caucasia and Caucasia than with the inner regions of Anatolia'). It is a remark that has also been made by other scholars and concerns mainly the eastern part of the littoral and its peoples. See the bibliography in Sections 2.13.2, 2.15, 2.16.3.

[15] See Section 3.4 in detail.

the southern Black Sea littoral is that it belongs to two worlds: the Black Sea and Anatolia. Ruling out one of them distorts its real identity.

The Pontides indeed shape a wall that seems to separate the littoral from the rest of Asia Minor and necessarily orientate it towards the sea; but the valleys of the Halys and the Iris are there to remind us that it also belongs to Anatolia, the cradle of so many cultures, being its access to the Black Sea. We can note this combination by looking both at the indigenous peoples (Part 2), and the peoples that came to invade, occupy or colonise the littoral (Parts 3 and 4). And actually the best period to note this is the middle of the 1st millennium BC, when the coast was at the same time colonised by the Greeks, who came from the sea, and politically subject to the Persians of Asia. One would expect that the naturally isolated southern Black Sea littoral would be culturally isolated as well, but judging by the number of different peoples who came and inhabited it – from all four cardinal directions, from both the sea and the hinterland – it was exactly the opposite that happened. This might be, *inter alia*, a result of the littoral's rich natural resources (see Part 1). No other side of the Black Sea displays such a cultural and ethnic mosaic, or offers such difficulty in recognising an ethnic, tribal, or cultural unit there.

For these reasons I believe that it is worth examining the historical geography of the southern Black Sea littoral, and especially in the 1st millennium BC, when the first written sources broadly discussing the area and the tremendous mass population movements there appeared, and when the foundations for several of the cultural features that would prevail in the region over the next two millennia were laid. And all the more so since more than 65 years have passed since the publication of M.I. Maksimova's still influential book,[16] and 35 since that of A. Bryer and D. Winfield,[17] which offers more than its title describes,[18] during which time most of the relevant archaeological evidence has come to light.

The first part of the book is dedicated to the natural environment of the southern Black Sea littoral, its particularities, and the natural resources it was blessed with; in other words, the geographic stage on which the numerous peoples that inhabited it developed their civilisations and economies, which were to a large degree defined by it. The indigenous peoples of the littoral are

[16] Maksimova 1956.

[17] Bryer and Winfield 1985.

[18] Both books deal only with the eastern half of the littoral, which would later become more widely known as 'the Pontus'. Unfortunately, it has been impossible for me to consult D.R. Wilson's *The Historical Geography of Bithynia, Paphlagonia, and Pontus in the Greek and Roman Periods: A New Survey with Particular Reference to Surface Remains Still Visible* (Oxford 1960).

examined in the second part.[19] The reader will realise that this part is by far the largest, constituting almost half the total. There are reasons for this. First of all is the startling number of these peoples: the written sources mention about twenty different peoples living along the coast in the 1st millennium BC. Even if not all of them occupied parts of the littoral synchronously, and even if few of them may have shared kinship with others or even been fictional, the number is still large for such a narrow strip of land.

Most importantly, however, the reason for such an extensive treatment is the fact that, save for some few exceptions, these peoples have never been given as detailed an examination as, for example, the Greek colonists there, and our knowledge of them is still scanty. This is also justified: with one possible exception none of them has left any identifiable product of material culture and – for want of a script in their languages – none of them has left any written text, while the Greek (and Roman) sources that speak of them deal with them much more summarily than with the Greeks. Nevertheless, they were still the indigenous peoples of the littoral, who were already protagonists in the region when the Greeks appeared on the scene. Therefore, I thought that in this monograph on the historical geography of the littoral I should give them the space they deserve.

The third part is devoted to peoples that cannot be considered indigenous in the same sense but still had a presence in the littoral, whether as invaders, as was the case of the Cimmerians, or as political overlords, as happened with the Persians and the Romans. The latter, however, are not treated here, for they appeared very late in our period of study and thus had not yet managed to leave their imprint on the littoral by the end of the 1st millennium BC. The subject of Part Four is the Greek presence on the coast, culminating with the celebrated Greek colonisation, a particular and unique phenomenon with a tremendous, long-lasting impact on the area. Finally, the fifth part examines the urban development along the littoral over the 1st millennium BC: how all those peoples surveyed in the previous parts intervened in the natural environment described in Part 1, what kinds of settlements and installations they created, in which places, and with what frequency, over the last thousand years before Christ. Inevitably, owing to the littoral's elongated shape, the detailed maps created for this book have been segmented.

[19] The definition of 'indigenous' is provided in Section 2.1.

2. The Available Evidence

To examine the historical geography of a place in a period that ended two millennia ago is obviously a rather difficult task, since it depends largely on fragmentary evidence, which might, moreover, often be of debatable validity. The two pillars of evidence that are available for our study are archaeological data and written sources. This will become apparent in every single part of the book. We shall, therefore, devote a few lines to presenting both an outline of the archaeological research in the southern Black Sea littoral and the most important written sources that tell us of peoples, places and events there. The more these two complement each other on any specific matter, the better that matter can be examined.[20]

We shall start with the archaeological evidence. This has the inherent advantage of potentially going much further back in time than the written sources; besides, it is considered to be more objective and consequently more reliable than the accounts of ancient authors, as the latter might sometimes be mistaken, misinformed or merely subjective, which would also be misleading for the modern scholar. And in fact, there are several occasions where each of the above seems to be the case, as we shall see. Nevertheless, while it is true that archaeological evidence does not lie, it frequently confuses or deludes. Even in the latter case, though, it is not the archaeological material itself, however fragmentary, that may lead to erroneous conclusions, but rather the often overhasty desire of archaeologists (and others) to reach a definitive interpretation. Examples of such cases will be encountered as we progress.

2.1. The Archaeological Evidence

Unfortunately, when it comes to the southern Black Sea, archaeological evidence is scanty, sometimes disappointingly so, especially in comparison to other parts of both the Black Sea and Anatolia. The reasons for our area of study being so poorly archaeologically investigated are numerous.[21] First of all is construction, especially the construction of modern cities upon the ruins of ancient ones, often those of the greatest significance. In some cases military installations or prisons have been built over the hearts of ancient settlements, such as, for example, in Amisos and Sinope respectively, while the modern

[20] The epigraphic evidence from the littoral is of no particular help to matters of historical geography. *Cf.* Braund 2021, 133. Some exceptions are mentioned in the appropriate sections below. For the inscriptions from the southern Black Sea, see the bibliography in Avram 2021, 10–11.

[21] See generally Tsetskhladze 2007, 176–80; Manoledakis 2018b, 181.

coastal highway in the western half of the littoral cuts through areas where ancient settlements once stood, destroying all trace of them forever. Furthermore, we should certainly mention the rise in sea level since antiquity,[22] which has submerged the ruins of many ancient settlements.

A few years ago, a Turkish archaeologist reiterated her complaint about the negligence surrounding archaeological studies in southern Pontus, contrary to the situation in other countries bordering the Black Sea.[23] The absence of foreign archaeological school missions during the last century, unlike Turkey's Aegean coast, in what is in any case a politically awkward region is also worth noting. It is only in the last couple of decades that some significant efforts to investigate the southern Black Sea littoral archaeologically have begun to appear.

It should be stressed, however, that overbuilding relates mainly to the ancient Greek colonies; other sites, most of which have been inhabited since the Stone or Bronze Age, have escaped this fate. Excavations at prehistoric sites along the southern Black Sea and its hinterland started as early as the late 19th century and still continue, albeit concentrated mainly on the central part of the littoral, around and along the Halys (Kızılırmak) and the Iris (Yeşilırmak) and the hinterland of Sinop.[24] One of the best investigated sites in this category is Ikiztepe, very close to the Halys and northwest of Bafra.[25] But all this is outside the time period examined in this book.

Since most of the important Greek colonies were overbuilt, archaeological investigations there have been extremely difficult to conduct. What archaeological evidence survived was usually the architectural remains of some large ancient building constructions and a few portable finds collected from individual spots on various occasions, which were either brought to local (or other) museums or, in the case of coins or gems, even sold abroad, and these have inevitably given rise to fragmentary assumptions about the ancient history of these colonies. An indicative example is Heraclea (at modern Ereğli),[26] a colony which, apart from its inherent significance, is dogged by the ambiguity concerning the identity of its founders, an uncertainty created by the very

[22] See Part 1 in detail.

[23] Erciyas 2007, 1195.

[24] See, indicatively, Burney 1956; Dönmez 2002; 2012; Doonan 2004; 2019; Doonan *et al.* 2017, all with the previous bibliography. Most recently Çalışkan Akgül 2019. See also Özdoğan 2007.

[25] Bilgi 1999; Alkım *et al.* 2003.

[26] Hoepfner 1972; Atasoy 2018, 111–12.

ancient sources.[27] In 2011, however, five shipwrecks from the Roman and the Byzantine periods were identified off the city's coast.[28]

Amisos (at modern Samsun) also suffered from being overbuilt as well as from plundering of archaeological sites in the wider region,[29] although the city has always been famous for the manufacture of terracottas, which have also been the subject of a study.[30] Nevertheless, limited rescue excavations have been successfully undertaken[31] in some areas within the territory of the ancient city and especially its necropolis, while in the city itself a short excavation took place in 1996.[32] Special interest attaches to the mainly rescue excavations in a wider area around Samsun, which have led to the location of several ancient sites.[33] In one of these, at Kurupelit, about 9 km from Amisos, several multi-level structures, rooms and roads were brought to light in 2009, within a large area that may belong to a sanctuary.[34]

The earliest and most important colony of all, Sinope (at modern Sinop), is almost entirely overbuilt, making it impossible for archaeologists to conduct research to the desired extent, which could lead to safe and incontestable conclusions concerning the ancient history of the city and, consequently, Greek colonisation in the southern Black Sea. Nevertheless, archaeological investigations took place in several areas within and without the ancient city in 1951–53,[35] which resulted in important finds, both architectural and mobile, that are still the subject of debate among scholars, such as the well-known temple initially identified as the Sarapeion.[36] Among the mobile finds the trade amphorae and the architectural terracottas occupy an exceptional position, providing valuable data about their manufacture and the intense trade activity of this major Greek colony.[37] In 1987–90 field surveys were conducted over a very wide area around Sinope and its hinterland.[38] Finally, since 1996, the Sinop Regional Archaeological Project, an international, interdisciplinary archaeological research project under the direction of Owen Doonan, has been

[27] See more in Section 4.4.
[28] Brennan *et al.* 2013, 96–97.
[29] Atasoy 2018, 134.
[30] Summerer 1999.
[31] Atasoy 2018; 2019.
[32] Atasoy and Ertuğrul 1998.
[33] Yiğitpaşa 2018, with bibliography.
[34] See more in Section 3.4.
[35] Akurgal 1955b; 1956; Budde 1956; Akurgal and Budde 1956.
[36] Akurgal and Budde 1956, 27–32. See most recently Summerer 2014, 193–96, with the previous bibliography.
[37] Indicative is the volume dedicated to 15 years of research on Sinope between the end of the 20th and the beginning of the 21st centuries: Kassab Tezgör 2012.
[38] Işın 1998.

researching and conducting test excavations in the entire Sinop Promontory. Within its framework, (currently) limited excavations have taken place close to the castle (*kalesi*) of Sinop (Fig. 2), in 2000 and since 2015, while surveys have also been conducted in the valleys of Demirci and Karasu, to the west and south of Sinope.[39] At the same time, other research teams have brought to light building constructions and graves in several of the city's districts, including Zeytinlik, Gelincik and Balatlar.[40] In addition, the sea around Sinop has attracted the interest of marine archaeologists, and the four shipwrecks discovered to date have provided valuable evidence relating, among other things, to Sinope's sea trade up to Late Antiquity.[41]

Despite the desire, since the middle of the previous century, to excavate as many areas of Sinope and its territory as possible, a desire impeded so dramatically by the modern settlements standing on the ancient ruins, nothing can be compared to the archaeological status of another ancient city, Tieion. Although among the earliest primary Greek colonies, thus in the same category as the three cities mentioned above,[42] Tieion was until very recently by far the least known of the four, historically as well as archaeologically.[43] This is utterly astonishing, considering that it is the only significant Greek colony in the whole of the southern Black Sea littoral that has never been overbuilt after its abandonment – the modern village-port of Filyos just touches the south-southwest fringes of the ancient city (Fig. 3). Nonetheless, it was only in 2006 that systematic excavations began in the ancient Greek settlement, under the direction of Sümer Atasoy, which have continued since 2012 under the direction of Şahin Yıldırım.

The on-going excavations at Tieion have changed the archaeological landscape of the southern Black Sea once and for all, together with our knowledge of the latter's ancient history. Within less than ten years, many parts of both the Lower and the Upper City (Acropolis), as well as the necropolis, have been revealed and are being excavated, while surface surveys – as far as the banks of the Billaeus (Fig. 4) – and geomagnetic studies have also been carried out. The most important buildings on the acropolis (Fig. 5) are Archaic subterranean constructions, a Roman temple, a Byzantine church and parts of the mediaeval fortification, and in the Lower City parts of its walls, the aqueduct, the Roman baths, a cistern and, above all, the 5000-seat theatre (Fig. 6). Along

[39] Doonan 2004; 2006; 2007; 2009; 2012; 2016, 217; Doonan *et al.* 2015; 2016; 2017. The reader may also visit https://srap.newmedialab.cuny.edu and https://www.sinopexcavations.org/.

[40] See the bibliography on them in Atasoy 2018, 127–30.

[41] Ballard *et al.* 2001.

[42] See in Section 4.4.1.

[43] See the relevant short entry in Avram *et al.* 2004, 963–64.

with these, finds from an amazing variety of categories and from the Archaic to the Byzantine periods are shedding light not only on the history of the city but also on the history of Greek colonisation in the Black Sea in general.[44] Furthermore, three shipwrecks of the Roman period were discovered by fishermen in 2013 about 100 m off the Tieion coast, but underwater archaeological investigation still remains to be done.[45]

As regards other settlements, a very important archaeological expedition was initiated in 2017 in the ancient city of Amastris (modern Amasra) by scholars from many institutions in an interdisciplinary context designed to provide the infrastructure for systematic excavation. The team's goal is to shed light on the history of the city and at the same time to preserve antiquities in the region before they disappear.[46] Limited rescue excavations in the 1970s and 1990s had resulted *inter alia* in the discovery of some marble statues.[47] Far less explored – indeed, almost not at all – is Trapezous (at modern Trabzon), where so far only a brief rescue excavation in 1997 has taken place.[48]

Recently, there have been efforts to reconstruct the port of Calpe (Kerpe),[49] while an impressive international survey project, interdisciplinary and with a large number of participants representing several specialisations and many institutions, was carried out between 2009 and 2011 in the area of Cide, seeking answers to several questions relating to the area's past from as early as the Neolithic period.[50] The region was the site of the ancient settlements of Cytoron and Aegialos.

Several archaeological excavations have been conducted in the wider areas around Fatsa and Ordu,[51] but not in the cities themselves, which are identified with, respectively, the ancient settlements of Phadisane/Phabda and Cotyora. As for the area of Giresun, identified with another colony of Sinope, Kerasous, very limited investigations in its mountainous hinterland have been made so far, leading to identifications of some local cult areas, but no specific publication on the matter has appeared yet.[52] Although very small, the Island of Ares

[44] See, merely indicatively, Atasoy and Erpehlivan 2012; Öztürk 2013; Atasoy and Yıldırım 2015a; 2015b; Atasoy 2016; 2018, 112–22; Aksoy and Yıldırım 2017; Yıldırım 2021, with further bibliography.

[45] Atasoy 2016, 212; 2018, 117.

[46] Çam *et al.* 2019.

[47] Ateşoğulları and Şimşek 1995; Atasoy 2018, 124–26.

[48] Kaya 1999.

[49] Aslan 2014.

[50] Düring and Glatz 2015a.

[51] Şenyurt 2015; Erol 2015; 2018; Erol and Yıldırım 2019.

[52] The information is derived from the abstract delivered by E.M. Doksanaltı and M. Tekocak for the Fifth International Congress on Black Sea Antiquities (Belgrade 17–21 September 2013),

(Giresun Adasi) lying opposite has been excavated since 2011 following two years of surveys, revealing part of its history, mainly of the Byzantine period,[53] as is the case with Thynias Island, in Bithynia.[54]

In view of the above, it is apparent that, notwithstanding the situation up to the end of the previous century as well as the difficulties in excavating the ancient coastal sites of the southern Black Sea, the 21st century seems to have signalled a new era in the archaeological investigation of the thus far so inadequately explored littoral. The excavations in Tieion and Sinope, but also the new international and interdisciplinary programmes like the Amastris project, indicate a significant change in direction as regards both the interest of Turkish archaeologists in the region and the participation of foreign scholars in its investigation. In addition, they have also sparked studies focusing on specific aspects or find categories, such as, for example, the terracotta production of Sinope,[55] the Attic red-figure pottery from the same city,[56] or the Hellenistic relief bowls from the areas of Ordu and Fatsa.[57] All this betokens a highly promising future in the archaeology of the southern Black Sea.

2.2. *The Written Sources*

The most valuable sources in the examination of historical geography are not necessarily those that are traditionally considered the main sources for the study of pure history. The latter may describe historical events, in greater or lesser detail, and such events – at least in the ancient literature that is available to us – are usually connected with the Greeks, paying little attention to, for example, indigenous peoples or local settlements that did not play an important role in Greek affairs. What we need here, however, is first of all names, names of peoples and settlements; then, information – of any kind – about these indigenous peoples; and finally, information about the natural environment. Since our area of study is coastal, we have the advantage of being able to use the *Periploi*, which may make no mention of historical events but do provide us with the names of a number of settlements, smaller and larger, most

but apparently in the end the authors neither participated nor submitted a paper. See Tsetskhladze *et al.* 2015, 529.

53 Doksanaltı and Mimiroğlu 2010; Doksanaltı 2012; Doksanaltı and Ekici 2017; Temür and Iltar 2018.

54 Tülek 2018.

55 Erdem 2021. For other finds from Sinope, see Baş 2018.

56 Sağlan and Korkmaz 2021.

57 Erol and Tamer 2021.

of which, especially in the Roman period, we would be totally unaware of, were it not for this kind of text.[58]

There are, moreover, other sources that provide important information of ethnographic as well as geographic content concerning the southern Black Sea region, as well as on the natural environment and the history of specific areas. Xenophon's *Anabasis* and Strabo's *Geography* are certainly the most important of these. Here, we shall refer generally to all of them as 'geographical sources', of which we may define two categories: those that are purely geographic, i.e. that give strictly geographical information, such as place names, distances between places, or even co-ordinates (e.g. the *Periploi*, Ptolemy's *Geography*, etc.), and those whose main purpose was the narration of an historic (e.g. Xenophon's *Anabasis*) or mythical event (e.g. Apollonius' *Argonautica*) but that also provide geographical and ethnographic information along the way. Strabo's *Geography* is a good combination of both.

We shall now briefly present the literary sources for the southern Black Sea of the 1st millennium BC.[59] The purpose of this chapter is not to elaborate on each of these (although an indicative bibliography is given), but merely to present them together with a short assessment of their treatment of the southern Black Sea littoral. However, for many of the authors mentioned here, specific comments regarding our area of study, which may refer to dates, influences from or on other authors, including ones whose works are now lost (e.g. Ephorus), etc., are made in subsequent parts of the book, when this becomes necessary.[60]

The earliest of our 'geographical sources' is the work of the geographer and historian Hecataeus of Miletus (*ca.* 560–*ca.* 480 BC) called *Periegesis*. It was written *circa* 500 BC, in two books, the first on Europe, the second on Asia, including Africa, and was most probably followed by a map, the second world map known to us after the one supposedly created by Hecataeus' fellow citizen and almost certainly teacher, Anaximander. The map must have been designed according to the Archaic geographic perception of the world as a disc surrounded by Ocean. Following in general lines the tradition of the *Periploi*, Hecataeus mentioned places and peoples in the coastal regions of the then known world, but without always strictly confining himself to settlements on the seaboard, as the authors of the *Periploi* did. Which of all these places

[58] The value of these sources in this respect will be further explained in Part 5.

[59] In the introduction to Table 3 in Section 5.1 the reader can see which 'geographical sources' from the 1st millennium AD are used in this book and which not, and why.

[60] See Sections 2.9, 2.10, 2.11, 2.15.

Hecataeus travelled to himself is not known,[61] but he must have relied on the records of several navigators and geographers.

Nothing has survived from the work itself, which is known to us only fragmentarily, through entries in Aelius Herodianus and Stephanus of Byzantium – more than 300 fragments about the whole known world, twelve in the southern Black Sea. Hecataeus merely gives the names of places and peoples, usually with the spatial relations among the latter; he also locates Greek cities in the territories of indigenous peoples. Since the extracts from his work are found alphabetically by city name in Aelius Herodianus and Stephanus of Byzantium, we do not know the geographical order in which Hecataeus mentioned the cities.[62]

Another important *Periplus* from roughly the same period, written probably just a few years earlier, is that by Scylax of Caryanda, an explorer famous in his day whom – according to Herodotus (4. 44) – Darius I charged with investigating the route of the River Indus. His *Periplus* was most likely one of the works Hecataeus made use of.[63] The *Periplus* that has survived under his name, however, is actually a much later work, composed in the third quarter of the 4th century BC in Athens by a person who did not leave his own name and is thus conventionally called Pseudo-Scylax. Unlike Scylax himself, there is no indication that this unknown author ever set foot in any of the places described in the work,[64] which is rather a compilation from several sources written for purposes other than those normally guiding the writer of a pure *Periplus*.[65] The southern Black Sea region occupies eleven of the 114 chapters of Pseudo-Scylax's *Periplus* (82–92), and is described from east to west. Each chapter is headed by the name of a local people (e.g. Mossynoikoi, Tibarenoi) or land (e.g. Assyria, Paphlagonia) and merely lists significant settlements,

[61] See also Section 2.15.1.

[62] Jacoby (1957) has tried to put them in geographical order. On Hecataeus, see Jacoby 1912; 1957, with the relevant fragments on pp. 15–16, 30–31; Dorati 1999–2000. On the specific fragments referring to the southern Black Sea, see Part 2 and Section 5.2.3.

[63] For the work of Scylax and the authors that were influenced by it, see Peretti 1979, and briefly also Counillon 2004 and Shipley 2011. For the use of Scylax's work by Hecataeus, see Peretti 1979, 149.

[64] Some scholars doubt even whether Scylax himself had ever travelled around the Black Sea or the Mediterranean: Shipley 2011, 6, *contra* Parker 2008, 14–18.

[65] For as full an evaluation as possible of Pseudo-Scylax's *Periplus* and its legacy, see most recently Counillon 2004; Shipley 2011, with the earlier bibliography. The degree to which Pseudo-Scylax's work was based on Scylax's one is still a matter of debate. Shipley (2011, 11, 15–16, 21) seems to exclude any connection of the former's work with that of the latter, though admitting the inability to name any of Pseudo-Scylax's possible sources. See more views in Counillon 2004, 24–27, who, on the other hand, does not see in the work just a simple compilation of purely geographical works.

rivers or islands in the particular territory, without the slightest further description or information of any kind. The only exception is a single sentence stating that the Mossynoikoi lived in the mountains (*Periplus* 86).[66]

Next in the *Periploi* series comes the periegetic account entitled *Periodos to Nicomedes*, addressed to a Bithynian king of that name and written in iambic trimeter. It was once erroneously attributed to Scymnus of Chios (early 2nd century BC), who was known in later sources as the author of a *Periegesis* but whose work appears to have been written in prose. As a result the name Pseudo-Scymnus has been attached to the author of our work, which is dated to the end of the 2nd century BC. The author describes the southern Black Sea region from east to west, mentions names of peoples and some important cities, but also provides some anthropological information as well as significant data about the founding of certain Greek colonies, Sinope included.[67]

Less than a century later, *ca.* 23 BC, a *Periplus maris interni* was written by Menippus of Pergamon. This work has not survived, but in the 4th century the geographer Marcian of Heraclea produced an edition of it, with some improvements, which are thought to be neither numerous nor significant.[68] This means that most of what we read in Marcian's epitome belongs actually to Menippus.[69] The work is written in prose. The southern Black Sea is described from west to east and it is the first time in the preserved ancient literature that we have such a rich and precise description of the littoral. In total, 55 coastal settlements and 40 rivers and other geographical features are mentioned,[70] with the distances between them given in stades. It is a purely geographical work, following the tradition of the *Periploi*, without providing any further information.

The next preserved work of this kind is the *Periplus Ponti Euxini* written by Arrian of Nicomedia in *ca.* AD 130–132. Arrian was among other things governor of Cappadocia in the time of Hadrian, and it was probably in this period of his life that he wrote the *Periplus*, addressing it to his emperor. Composed in epistolary form, Arrian's *Periplus* is more easily read and contains more

[66] For how much of the part dedicated to the southern Black Sea might belong to the original *Periplus* of Scylax, see Peretti 1979, 456–69. For the scant reliability of Pseudo-Scylax's *Periplus* as regards chapter 89, see Flensted-Jensen and Hansen 1996, 152. The work appears not to have been used by Pseudo-Scymnus, Mela or Pliny: Shipley 2011, 21–22.

[67] On Pseudo-Scymnus, see Diller 1952, 165–76; Marcotte 2000; Boshnakov 2004, who suggests that the author was actually Semos of Delos, and cites the earlier identification proposals in the previous bibliography. Asheri (1998, 267) has suggested that an important source of Pseudo-Scymnus must have been Ephorus.

[68] Marcian was the only one of the geographers known to have used the works of others who gives both his own name and the names of the author and the work that he summarised. Neither Pseudo-Skylax, Pseudo-Scymnus nor Anonymous (see below) felt the need to do this.

[69] Diller 1952, 147–64.

[70] See in Part 1, Table 1 and Section 5.2.2, Table 6.

than strictly geographical data, which for the rest are as rich as those of Menippus.[71] Nevertheless, the great value of this work derives from the fact that Arrian actually visited the places he described, having a personal interest in conveying reliable information to his emperor. This is clearly stated in his text, as is also the fact that he had studied the works of other eye-witnesses describing the littoral, including Xenophon's *Anabasis*, whose information Arrian sometimes tries to confirm.

These four *Periploi* were used by an unknown author in the 6th century to compile a *Periplus Ponti Euxini*, adding some information from his own era, such as names of districts or the fact that some of the indigenous peoples (e.g. Chalybes, Tibarenoi, Mossynoikoi, etc.) no longer existed. This work, which we call the *Periplus* of the Anonymous, may even provide lost parts of some of the four earlier *Periploi*, while inevitably it mentions more place names than any other. Determining just what the Anonymous took from each of his four sources would assuredly be a noteworthy endeavour.[72]

Many more geographical texts were produced in the Roman era, which indeed was the peak period for this kind of work. Strabo's celebrated *Geography* is not merely a geographical work, thus will be treated separately. About AD 44, the Spaniard Pomponius Mela from Tingentera wrote his *De chorographia*, in which we read about the most important – not all – settlements and peoples (a general characteristic of his work) from west to east. His text is quite balanced as regards the combination of purely geographical with other information.[73] However, the author not only appears not to have travelled to the places he describes, but also to have selected them based on odd and vague criteria.[74] In fact, in the very first sentence of his book Mela himself admits that his work is puzzling and time-consuming, and modern scholars generally do not speak of it in flattering terms.[75]

A few decades later, towards the end of his life, Gaius Plinius Secundus or Pliny the Elder (AD 23–79), author of the Natural History (*Historiae Naturalis*, hereafter *NH*), dedicated Books 3–6 of the 37 comprising his encyclopaedic work to the geography of the world. In the sixth, he dealt *inter alia* with our area of study, again with an eastward progression. A bit richer in place names than Mela's, this work was also based mostly on known data collected by other sources, not on personal experience. Like Mela's, Pliny's work

[71] One settlement more (56) and one geographical feature less (39). The direction is again from west to east and the distances are given in stades.

[72] Diller 1952 is still the best study of the Anonymous *Periplus*.

[73] On Mela, see Romer 2001.

[74] *Cf.* Romer 2001, 6, 21.

[75] Tozer 1971, 13, 262–63; Romer 2001, 23.

cannot be considered among our most reliable 'geographical sources', the author apparently having drawn his information from sources of very different dates, resulting in anachronisms.[76]

In *ca.* 120 Dionysius Periegetes (the Traveller), a poet from Alexandria, wrote a descriptive survey of the world in Greek hexameter verse, the *Oikoumenes periegesis*, widely known as *Orbis terrae descriptio*. As regards the southern Black Sea, only Sinope is mentioned, but it lists twelve of the indigenous peoples, from east to west. His work appears to have been fairly popular; indeed Eustathius of Thessalonica wrote a commentary on it, which will be frequently cited in this book, certainly more often than Dionysius' work itself.[77] That opus is considered as having the same defect as Mela's, namely relying on authorities widely differing in date.[78]

Geographical research undoubtedly culminated with Ptolemy (Claudius Ptolemaeus) and his *Geographike Hyphegesis* or, more simply *Geography*. Following a long tradition of great geographers, which goes back as at least as far as Eratosthenes, and living in Alexandria, where he had access to a huge variety of works, Ptolemy (AD 90–168) became one of the most important geographers of antiquity, and by far the most influential one in the world. His *Geography* is actually nothing more than a list of about 8000 places (settlements as well as geographical features) covering the entire known world, with their co-ordinates. What deserve special attention are, firstly, the fact that it happens to be the earliest preserved work of ancient literature providing co-ordinates, though several earlier ones were of this kind, and secondly, that the first of its eight books is a detailed introduction to the methodology used by the author. This notwithstanding, the possibility of errors in a work dealing with the whole world cannot be avoided, and the southern Black Sea, whose places are listed from west to east, is indeed one of the regions where a few errors appear.[79]

[76] For example, one cannot fail to observe his remark that the towns Carambis and Harmene no longer exist, despite the fact that Harmene especially is mentioned by almost all his colleagues, both earlier and later. *Cf.* Tozer 1971, 13, 264. More recently on Pliny: Brodersen 2015.

[77] On Dionysius Periegetes, see now Lightfoot 2014.

[78] Tozer 1971, 286.

[79] For example, he puts Eusene and Cordyle inland, and Hyssos limen west of Trapezous. *Cf.* also Bekker-Nielsen 2014, 74. However, we should bear in mind that Ptolemy's co-ordinates are given with an intrinsic approximation of five spherical minutes of arc, which means *ca.* 10 kilometres on an actual terrestrial sphere. This means that a place mentioned by Ptolemy with specific co-ordinates could in reality lie anywhere within a circle of uncertainty of 10 km radius around the point indicated by those co-ordinates, and thus possibly, in the case of the aforementioned 'exceptions', in the correct (according to other geographers) location. But perhaps this is not enough to justify such mistakes. For details about this characteristic of Ptolemy's work, see

Without doubt, special mention must be made of Strabo. His celebrated *Geography*, written probably between AD 18 and 24 and comprising 17 books, is the only geographical work that is not limited to a list of place names but is a prose composition in which less weight is given to the completeness of a geographic catalogue and more to the presentation of a comprehensive history of the regions under study, including geographic, historical, ethnographic and environmental data. Book 12 especially, which is devoted to the area that we are examining here, has the additional value of being written by an author speaking of his own homeland, a land where, despite a long period of absence, he had lived, thus what he writes is based on first-hand knowledge and observation as well as on personal access to archives and other sources of information, which is evident from some of his references.[80]

Of particular value are Strabo's references to the natural environment, a subject to which ancient authors generally paid little attention. He is the only author of the Roman period to speak of the environment and natural resources of the southern Black Sea as it was in his day; the only possible comparison can be with the much briefer references of Xenophon (see below). Strabo attaches much importance to the natural environment and especially to the goods that the land of each region offers its inhabitants, as we shall see in the first part of the book.

With his reports he becomes the only geographer to provide extensive and reliable information on the economy of the communities living along the southern coast of the Black Sea during the period of the Julio-Claudian dynasty. Overall, it appears that Strabo had read several earlier texts, and we can detect the influence of Xenophon's *Anabasis* (although not cited by name) in the description of several peoples of the eastern part of the coast (e.g. in 12. 3. 18). One can also see the influence of the *Periploi* in the reference to adjacent coastal localities in 12. 3. 17, where we also find 'Trapezous, polis Hellenis', a term that Strabo uses nowhere else in reference to this area and which clearly recalls Pseudo-Scylax.[81]

Strabo's work is unique among the geographic sources, because it offers what in the other texts we must try to detect: the linking of geographical information with historical fact. Living the last part of his life (probably) in his native Amaseia,[82] Strabo dedicated many lines to the recent history of the

Manoledakis and Livieratos 2006. For a general evaluation of Ptolemy's work, see most recently Aujac 2015.

[80] *Cf.* Tozer 1971, 258; Dueck 2000, 4–5, 185.

[81] For the influence of the *Periploi* on Strabo's work, see Dueck 2000, 40–45.

[82] On the question of whether Strabo ended both his work and life (and when the latter occurred) in Amaseia or in Rome there are differing views, although Amaseia seems more likely.

southern Black Sea hinterland to which this city belongs. We could say that Strabo's work goes a stage further than other geographical texts,[83] for example the *Periploi*, marking the passage from geography to historical geography according to the model of Polybius.[84] However, Strabo's work apparently became known only long after his death, as indicated by the fact that neither Pliny nor Mela seem to have used or even been aware of it. It is in Dionysius Periegetes that we first see the influence of Strabo's *Geography*, or, according to other views, at least a century-and-a-half after Strabo's death or not even until Marcian.[85]

*

* *

Let us now move on to the second category of our 'geographical sources', namely those whose main purpose was the narration of an historic or mythical event but which also provide geographical and ethnographic information along the way. One work clearly stands out in this category, and this is Xenophon's *Anabasis*. Its significance is so great that it could stand alone in this category (as in a way it does) and still be as important as all the works of the first category put together. Born in Athens at the outset of the Peloponnesian War, Xenophon was about 30 years old when he assumed command of one of the biggest Greek mercenary armies, the famous 'Ten Thousand', which marched to fight alongside Cyrus in his unsuccessful attempt to oust his brother Artaxerxes II from the Persian throne. Returning to Greece after the lost Battle of Cunaxa (401 BC), Xenophon and his roughly 8600 surviving men reached the Black Sea at Trapezous in 400 BC and, partly on foot and partly by ship, traversed the littoral until they reached Bithynia and passed into Greece.

Several decades later, between 370 and 360 BC, therefore not long before his death (in 354), Xenophon wrote a description of this expedition in the

Cf. Tozer 1971, 241–44, 258; Diller 1975, 3–7; Dueck 2000, 1–5, 145, with previous bibliography.

[83] For all these purely geographical works, apart from Strabo's one, see also Müller 1855–1861.

[84] For a comparison between Strabo and Polybius, see Dueck 2000, 46–52. For a general evaluation of Strabo's *Geography*, see Dueck 2000, 145–87. For a good guide to Strabo's work, see recently Roller 2018.

[85] Former: Diller 1975, 7–8; Dueck 2000, 151–52. Latter: Tozer 1971, 248–49. Scholars seem to disagree also on whether Ptolemy was influenced by Strabo or not. For an evaluation of Strabo's *Geography*, see most recently Prontera 2015. For the sources of the Roman era specifically, including those of the post-Christian period (e.g. *Itineraria Romana*, *Tabula Peutingeriana*), see most recently Manoledakis 2022a. Olshausen (2015) offers an overview (very brief, though) of their possible sources for their description of the Black Sea.

seven books of the *Anabasis*.[86] The march along the southern Black Sea coast from east to west is described in the last part of Book 4 and Books 5–6. The narrative is full of details – names, distances, dialogues, numbers, etc. – that would have been impossible to recall had Xenophon, by then more than 60 years of age and the author of many additional works, not kept notes during the expedition. The value of the *Anabasis* for the examination of the historical geography of the southern Black Sea in the 1st millennium BC is incomparable. The reader will realise that the whole second part of this book (actually almost half of it) as well as much of the last part owe most of their material to the *Anabasis*. Xenophon and his soldiers met many of the indigenous peoples of the littoral, sojourned in their territories, fought or co-operated with them, feasted with members of their communities and were guided by them to their settlements, learned about their names, occupations, social and economic life, manners and customs.

All these things are described in fairly extensive detail in the work, together with information about the peoples' houses, distances between villages, relations between each other and the Greek colonists, etc. With Part 2 and Section 5.3 dedicated to the indigenous peoples of the littoral and their settlements, Xenophon appears as the sole voice of the indigenous peoples in our whole group of sources, which is otherwise entirely Hellenocentric, presenting everything from the Greek viewpoint: only Strabo gives some few – general, unclear and sometimes of debatable credibility – data about certain, always 'barbarian', locals. Xenophon had in any case a sincere interest in the peoples and places of the Persian Empire, which is evident throughout his work,[87] and has therefore been justifiably described as the 'Greek intellectual who knew this land and its people far better than most'.[88]

Unfortunately, we cannot rely in the same way on the other great historian of the 5th century BC, Herodotus. In contrast to his extensive preoccupation with the west and north sides of the Black Sea, Herodotus did not deal with the south coast, and his scarce references to the area concern mainly an enumeration of the Persian satrapies under Darius and the peoples who joined Xerxes in the Persian Wars (*ca.* 481/80 BC) and their equipment. For the rest, he refers briefly to the Cimmerian and Scythian presence in Anatolia, but does deal with the Colchians. Nonetheless, Herodotus' references raise doubts as to

[86] From the vast bibliography on the *Anabasis*, see indicatively Lane Fox 2004b. See also Briant 1995, more generally, not that much for our area of study. Lendle (1995) has offered a valuable commentary on the work.

[87] On which, see Hobden and Tuplin 2012.

[88] Gruen 2011, 65. *Cf.* Hirsch 1985; Tuplin 2004; 2007a, 7.

whether he really visited all the Black Sea areas he talks about; for the southern littoral in particular, this is highly improbable.[89]

Finally, we have the poetic works that recount mythical events and specifically the Argonautic expedition. Chief among them is undoubtedly the celebrated four-book Hellenistic epic entitled *Argonautica*, composed by Apollonius of Rhodes in about 250–230 BC.[90] Based on a very long tradition of poems dedicated to this myth, which was well known already in Homer's time, Apollonius's work happens to be the first wholly preserved one narrating it. The description of the southern Black Sea coast, which the Argonauts followed on their way to Colchis, includes references to many peoples and places, most of which were borrowed from some of the 'geographical works' mentioned above.[91] Apollonius' poem was widely used by Valerius Flaccus in his homonymous work of the 1st century AD (who even omits some data), as well as by the *Orphica Argonautica* (probably 6th century AD), although sometimes with errors in the geographical order of the local peoples. In any case, these works add nothing of importance to our study.

Of course, there were many other texts dealing with the southern Black Sea which are now lost but may have been significant sources for some of the works presented above.[92] On the other hand, several preserved sources later than ours were based on them and on some of the lost ones, directly or otherwise, one example being Diodorus' *Historical Library*, its relation to Ephorus and the latter's relation to Xenophon.[93]

Apart from all these primary sources, one should not fail to mention the works of several scholars of Late Antiquity and the Byzantine period, who commented on the ancient texts or have just transmitted to us passages from otherwise lost ones. Characteristic are the cases of Aelius Herodianus (2nd century) and Stephanus of Byzantium (6th century), thanks to whom we are aware of some parts of Hecataeus' work. And of course, there are the scholiasts, who, commenting on texts like the Homeric epics,[94] Apollonius' *Argonautica*,

[89] See, for example, Kimball Armayor 1978a; Asheri 2007, 2–6; Corcella 2007, 548–63, 561. *Cf.* Ivantchik 1999.

[90] For the dating, see Manoledakis 2013a.

[91] For the possible sources of Apollonius, see Hoefer 1904; Stylianou 2004; and Section 2.11.2 here. *Cf.* also Delage 1930, 14–19, 277–81. For the characterisation of the *Argonautica* as a geographic epic, see Delage 1930, 9. For the southern Black Sea in the poem, see Delage 1930, 135–90.

[92] Such are, for example, the works of Ephorus, Nymphodorus, Sophaenetus and many others. We know about them through much later works of lexicographers and other grammarians. Their information about the southern Black Sea is presented in several parts of the book.

[93] See more in Section 2.11.2.

[94] On the short passage of the *Iliad*'s Trojan Catalogue referring to the southern Black Sea, part of which must have been a 3rd-century BC interpolation, see Manoledakis 2013a.

or Dionysius Periegetes' *Orbis terrae descriptio*, imbue their studies with much information derived from many other sources. These scholars, among them the 12th-century Eustathius of Thessalonica,[95] span a huge period of time from the Roman to the Late Byzantine era. Nevertheless, these works, although usually treated as belonging to the same category as the sources presented above, should not actually be considered as such. For their chronological distance from the events they deal with in their commentary upon the ancient texts is huge – it may be as much as 1200 to 2000 years. That is like a scholar of today writing about Strabo or Stephanus of Byzantium. Therefore, all these sources must certainly be numbered among the secondary ones; such is also the Anonymous' compilation of the four *Periploi*. Most dictionaries (e.g. the *Suda*, the *Etymologicum Magnum*, etc.), too, are rarely reliable as regards the geographical position of places and peoples.

To conclude, the evidence originating from the written sources with regard to our area of interest would be highly fragmentary (except for place names), were it not for two wholly-preserved works, Xenophon's *Anabasis* and Strabo's *Geography*. Both offer a great amount of knowledge, but the pre-eminence of the *Anabasis* is indisputable. It is certainly by far the most reliable work we have for the study of the historical geography of the southern Black Sea in the 1st millennium BC. What does reliable mean? Certainly not infallible. But Xenophon is the only eye-witness who deals with so many aspects of the situation in the area.[96] For Strabo we cannot be absolutely sure that he visited the whole littoral himself, and even if he did, his personal experience concerns a period after the time we are dealing with here, while his historical information is derived from earlier sources.

[95] On these scholars, see Dickey 2007. Here always abbreviated as Schol.

[96] Xenophon's reliability is, of course, here confined to the specific books of the *Anabasis* mentioned above and regards the specific place at the specific time. One should not be left with the impression that all Xenophon's works (e.g. the *Cyropaedia* – see Hirsch 1985; Kurht 2007, 19) are totally reliable in everything they narrate; probably not even the *Anabasis* itself as regards, for example, the relation of the expedition's generals with one another. But the place with its people was just the background to the events narrated, and Xenophon had no reason to distort what he saw.

PART 1

THE NATURAL ENVIRONMENT

1. A History of a Sea

A seafarer approaching the Anatolian coast of Pontus faces a huge, imposing mountainous curtain extending along its whole length, which gives the impression of almost rising from the sea; indeed, in some parts of the littoral this is actually the case. Only two places are different: the western part of Bithynia, and the plains shaped by the estuaries of the Halys (Kızılırmak) and the Iris (Yeşilırmak) on both sides of Amisos (Samsun), almost in the middle of the coast (Fig. 1). One might reasonably wonder how much arable land is left in such a narrow strip between the sea and the rugged mountain ranges hugging the coast. And whether agriculture was one of the pillars of the economy of the several peoples that inhabited the coast or if they were forced to rely on fishing, cattle-breeding and trade. Reading the ancient sources, however, one would be surprised to realise not only how little concern the indigenous peoples had with maritime trade, at least before the Greek colonisation, but also how much local economies depended on agriculture. For that very same mountainous terrain, extending far inland into Anatolia, together with the endless plateau to its south, gives birth to countless rivers and streams flowing into the Black Sea, many of which create fertile valleys along their way, supplementing the few existing plains.

The subject of this initial part of the book is the natural environment that hosted the southern Black Sea littoral's peoples in the 1st millennium BC; the environment which these peoples lived and acted in, and interacted with; and where they created their urban environment, which will be examined in the last part. Comprehension of the natural environment is essential when attempting to study the culture, the society and the different kinds of economy of those who lived in it, as well as the unfolding of the most important historical events that marked them, which are the aspects that concern the next parts of the book.

Without doubt, the Black Sea is one of the earth's most singular basins. Its formation is considered to have begun in the Cretaceous Period (145–66 million years ago), at the end of the Mesozoic Era. It is probably a remnant of the Paleo-Tethys Ocean, which closed during the Late Triassic Period (237–201 million years ago).[1] Towards the end of the Late Pleistocene, at the Last Glacial Maximum, namely about 22,000–18,000 years ago, the Black Sea was a lake (the Neoeuxinian Lake), isolated from both the Sea of Marmara (the

[1] Yılmaz 2007, 540, with bibliography.

Propontis) and the Caspian Sea. As a result of the warming climate about 17,000–10,000 years ago, a massive discharge of water from the Caspian through the Manych outlet (the channel shaped by the rivers Volga and Don) increased the level of the Black Sea, so that it spilled over into the Sea of Marmara and from there into the Mediterranean.[2] Despite the intensive research that has been conducted over the last decades on the matter, there are still differences among scholars on several 'details' of this event. For example, it is not unanimously agreed whether the Neoeuxinian lake was a freshwater lake,[3] or semi-fresh to brackish;[4] whether its initial level was *ca.* 140[5] or *ca.* 100 m below the present one;[6] whether it was the Sea of Marmara and the Mediterranean that first spilled their water into the Black Sea,[7] or *vice versa*;[8] whether this exchange of waters occurred in the Early Holocene[9] or much earlier, in the Late Pleistocene;[10] and whether this transformation of the Black Sea was gradual, catastrophic, or fluctuating.[11]

Numerous as these disagreements as they may be,[12] the one certain thing is that the Black Sea experienced a dramatic, even if gradual, transformation, which literarily left the people inhabiting its coast to the mercy of the elements. Most probably, especially if we accept the scenario of a catastrophic

[2] See, indicatively, Jablonka 2003; Yanko-Hombach 2007a; 2007b; Chepalyga 2007.

[3] E.g. Ryan *et al.* 2003. The bibliography for all the theories given here is only indicative. The reader can find a detailed bibliography on all these matters in the collective volume published by Yanko-Homach *et al.* (2007).

[4] E.g. Yanko-Hombach 2007a; 2007b.

[5] Ryan and Pitman 1998; Ryan 2007.

[6] Yanko-Hombach 2007a; 2007b. For theories about the southern coast in particular, see below. According to Yanko-Hombach (2007a, 14), during the last 10,000 years the level of the Black Sea has never again exhibited a maximum amplitude of fluctuation greater than 20 m. During this period, the sea level has risen gradually, but in an oscillating manner, to the present level and perhaps slightly higher.

[7] Ryan and Pitman 1998; Ryan 2007.

[8] Yanko-Hombach 2007a; 2007b.

[9] Ryan and Pitman 1998: 7200 years ago (= about 5200 BC); Ryan *et al.* 2003; Ryan 2007: 8400 years ago. See also Coleman and Ballard 2007.

[10] About 17,000 years ago: Yanko-Hombach 2007a; 2007b. *Cf.* Chepalyga 2007. More theories in Yanko-Hombach 2007a.

[11] For the three theories, see Yanko-Hombach 2007a, 7–8, who supports the third one. *Cf.* Yanko-Hombach *et al.* 2011; Düring and Glatz 2015b, 13–15.

[12] About a quarter of a century ago, W. Ryan and W. Pitman (1998) published their theory in a book, the very title of which attracted the lights of publicity and with it a big wave of reaction among scholars challenged by the theory itself, as well as the connection of the events in question with the mythical traditions of several Eurasian cultures about a catastrophic flood that had happened in the distant past. Much of this reaction is encapsulated in Yanko-Hombach *et al.* 2007, though also with opposite views. For a response to this reaction, see Ryan 2007.

event, foragers and farmers of the Stone Age[13] were forced to leave the once fertile and now flooded areas of the former lake, and migrate massively to Europe[14] or change their whole way of life.[15]

But the particularities of the Black Sea, the world's largest semi-enclosed marginal sea,[16] do not end here. As a meromictic basin – in fact, the largest that exists – it has three water layers, which do not mix with each other. Beneath the surface layer (0 to 50 m deep), which is well oxygenated, is a suboxic one (50–100 m); the deepest layer, from 100 m to the sea floor, is anoxic. The surface layer receives water mainly from the large rivers flowing into the Black Sea (mostly in its northern half), which thus results in lower salinity[17] and temperature. Conversely, the deep layer receives water from the Mediterranean via the Sea of Marmara and the Bosporus, which is warmer and saltier. Deprived of atmospheric oxygen due to its inability to mix with the surface layer, the deep layer of the Black Sea is an anoxic environment.[18] As a result, organic material, including anthropogenic artefacts such as boat hulls, is well preserved, since there are no species to destroy it.[19] Indeed, archaeological evidence and geological samples below today's sea level has been invoked by the supporters of the theory of a catastrophic spill-over from the Bosporus into the Black Sea in the Neolithic Period (see above),[20] although other scholars strongly disagree that there is any archaeological evidence for such a theory.[21]

Notwithstanding the many disputes regarding the Black Sea's distant geological past, the natural events that created the Pontus as we know it, with all its particularities, are undoubtedly remarkable. Equally impressive is how early ancient scholars began to be concerned with how these events took place and how the Black Sea was created, not to mention their first environmental

[13] Whether of the Paleolithic (e.g. Chepalyga 2007; Yanko-Hombach 2007a, 10) or the Neolithic period (e.g. Ryan and Pitman 1998), according to the theory put forward (see above).

[14] Ryan and Pitman 1998.

[15] Chepalyga 2007, especially 141–44, who discusses even the possibility of the 'appearance of new ethnic communities'.

[16] Murray *et al.* 2007, 1; Algan *et al.* 2007, 604.

[17] About 1.8%, while the average in the open seas is 3.5%.

[18] Murray *et al.* 2007; Algan *et al.* 2007, both with bibliography; Yanko-Hombach *et al.* 2011. According to Murray *et al.* (2007, 3–4), the Black Sea is 'the classic anoxic marine ocean basin and is considered a prototype for the earth's ancient ocean'.

[19] In fact, several well-preserved ships, dating from the Roman and the Byzantine periods, have been spotted on the sea floor, for example during expeditions conducted between 1998 and 2003: Ballard *et al.* 2001; Coleman and Ballard 2007. For recently spotted shipwrecks close to Tieion, see Atasoy and Yıldırım 2015b, 441.

[20] Ballard *et al.* 2001; Ryan *et al.* 2003; Coleman and Ballard 2007.

[21] Jablonka 2003: 'contrary to what Ballard *et al.* 2000 suggest'; Özdoğan 2003; 2007; Yanko-Hombach 2007a; 2007b; Yanko-Hombach *et al.* 2011.

worries about the dangers and the future of the sea. Thanks to Strabo's detailed examination of and commentaries on earlier scientists' essays, and to the fortunate preservation of his whole work, we happen to know that as early as the 5th century BC Xanthus of Lydia had reported that there had been so great a drought in his time that the rivers, lakes and wells dried up; and that he himself had often seen fossil shells a long way from the sea, some like cockles, others resembling scallop shells, also salt lakes in the land of the Armenians and that of the Matienoi,[22] which induced him to believe that sea had formerly been where the land now was (Strabo 1. 3. 4).[23] As regards the Black Sea specifically, Diodorus, writing in the 1st century BC, invoked a legend narrated by the people of Samothrace about a great flood that befell their ancestors: at some indefinite time in the distant past, the Black Sea was a lake, which received such a large volume of water from the rivers that flow into it that it burst forth violently into the Hellespont and flooded a large part of the coast of Asia Minor, even turning part of the land of Samothrace into a sea; because of this, 'fishermen have now and then brought up in their nets the stone capitals of columns, since even cities were covered by the inundation' (Diodorus 5. 47. 3–5). Even if the last part of the story reveals its improbability, it is still worth noting the perception of the Black Sea as a former lake, which in some age became connected with the Aegean.

By the Hellenistic period research on the shaping of the Black Sea had moved to a different level, as one can see by reading the assessments of Strato of Lampsacus (4th–3rd centuries BC) and Polybius (2nd century BC). Both of them attributed the outrush of the Black Sea over the Propontis to the result of silting-up from alluvial matter deposited by the numerous rivers flowing into it,[24] and both of them expressed concern that this being a constant process, the Black Sea would in course of time be entirely filled, Polybius (being the more pessimist of the two) stressing the navigating difficulties in Lake Maeotis already observed in his day (Strato in Strabo 1. 3. 4; Polybius 4. 39–42).

The two scholars' argumentations may have some significant differences, but the overall perception is similar. For example, Strato implied that the Black Sea was initially a closed lake, until silting raised its level to the point where it broke through into the Hellespont. Since, he said, the Black Sea has a higher floor than the Mediterranean, its shallowest parts would be soon left dry, due to the constant outflow of its waters (Strabo 1. 3. 4). Polybius, on the other

[22] For the latter, see Section 2.6.1, n. 198.

[23] In the same passage of Strabo, we read that Eratosthenes had also dealt with the issue and studied Xanthus' work.

[24] They both use the verb (ἐγ)χώννυμι. Similar is the case with Lake Maeotis (the Sea of Azov) in relation to the Black Sea through the Cimmerian Bosporus (Kerch Straits).

hand, stated that the rivers flowing into the Black Sea bring alluvial deposits, which cause it to silt up. Such a constant process would turn the Black Sea into a shallow fresh-water lake, which would gradually be entirely filled (Polybius 4. 40, 42).[25]

Research on the particularities of the Black Sea continued in the Roman period. Strabo criticised Strato, claiming that the causes he propounded for the rising and the falling of the sea were not correct: in his opinion, the cause was not to be sought for in the different levels of the sea beds, but in the fact that the sea floors themselves sometimes rise and sometimes are depressed, causing the sea to rise or subside with them. Therefore, the explanation has to be sought in the sea floor, either that which underlies the sea or that which is temporarily flooded, but more likely the submarine floor (Strabo 1. 3. 5–6). Later, Arrian (*Periplus Ponti Euxini* 8) reported that the water of the Phasis:

> if you take it from the surface, is fresh; but if any one lets down a jar deep into the stream, he finds the water brackish. However, the water of the Pontus is much fresher (i.e. less salty) than the water of the outer sea. And this is a result of the rivers which flow into the Pontus, the number and size of which are beyond computation. We may bring as a proof of its freshness (if we need any proof for something that we can feel), that all the people who live close to this sea lead out their cattle to drink of its water, which the latter willingly do; and it is said that they thrive better with this than with fresh water.

The low salinity of the Black Sea, as well as the even lower of the Maeotis, had been also mentioned by Polybius (4. 42; *cf.* also Strabo 7. 3. 18, for the northern Black Sea).

Finally, the currents. The surface current of the Bosporus straits runs from the Black Sea to the Aegean; it is stronger in summer, when the volume of fresh water entering the sea is greater than during the rest of the year. A counter-current of heavier salt water and lower velocity flows from the Mediterranean to the Black Sea along the bottom.[26] This had already been observed in antiquity, as is evident from the description provided by Procopius in the 6th century (*de Bellis* 8. 6. 26–28).

[25] One should note that Polybius laid weight also on presenting his data in a scientific manner. He says: 'I must leave no point unelaborated and barely stated, as is the habit of most writers, but must rather give a description of the facts supported by proofs, so that no doubts may be left in the reader's mind. For this is the characteristic of our times, in which, all parts of the world being accessible by land or sea, it is no longer proper to cite the testimony of poets and mythographers regarding matters of which we are ignorant, […] but we should aim at laying before our readers a credible narrative' (4. 40). It is not impossible that Polybius was referring here also to the legends presented by Diodorus (see above).

[26] West 2003, 152–53. For the currents in the southern Black Sea, see also Murray *et al.* 2007, 4–12; Coleman and Ballard 2007. *Cf.* Doonan 2006, 49.

2. The Southern Black Sea

Let us now focus on the southern shore of Pontus, this 'well worth seeing sea' (Herodotus 4. 85. 1). Its most striking characteristic, in comparison with the northern coast,[27] is the sharp contrast between the two in terms of geology and climate. As regards the former – which affects the latter – the dissimilarity is apparent from a single glance at a physical map (Fig. 1). Whereas vast steppe plains run back from the north shore, in the south everything is different: parallel to the coast and suffocatingly close to it, from the Sakarya river to the easternmost end of the littoral runs a long mountain range, the Pontides, widely known as the Pontic Mountains or Pontic Alps (in Turkish: Kuzey Anadolu Dağları). They are young and uplifting mountains that reach an elevation of 3937 m in their eastern part, not far from the borders between Turkey and Georgia. The Black Sea shelf in front of this range is relatively narrow, varying from 1 to 10 km, beyond which a steep continental slope descends to a basin more than 2000 m deep.[28] To the south of the Pontides lies the North Anatolian Fault, which is one of the factors responsible for their formation. Known for strong and frequent earthquakes, this fault is the transform boundary between the Eurasian and the Anatolian Plates. Similarly, the cold winters and hot dry summers of the north are in stark contrast to the subtropical climate of the south, which provides relatively mild winters and humid summers with high precipitation.[29]

The southern Black Sea coast is about 1625 km long.[30] This figure, so much larger than the direct line between the coast's western and eastern ends (1000 km), is due to the numerous capes and bays that shape the shoreline, formed by the many rivers flowing into the sea and the several mountainous areas extending to it. The actual length of such a tortuous coast has always been difficult to estimate, as is apparent from some of the figures provided by ancient authors. According to Herodotus, the length of the Black Sea, which corresponded to a voyage of nine days and eight nights (4. 86. 2),[31] is 11,100 stades (4. 85. 2), or about 2000 km – much more than the actual length. Nevertheless, the measurement given by Procopius (*de Bellis* 8. 5. 33) about

[27] The western and eastern shores generally resemble the northern and southern ones respectively, as regards the aspects examined here.

[28] Yılmaz 2007, 539; Algan *et al.* 2007, 608–11.

[29] *Cf.* Rostovtzeff and Ormerod 1932, 212; Burney 1956, 179; Maksimova 1956, 13, 17; Bryer and Winfield 1985, 3–4, 7.

[30] Algan *et al.* 2007, 609. Not far off this was the estimation of certain ancient authors (e.g. about 1200 km in Menippus – *cf.* Marek 1993, 7).

[31] Xenophon mentioned that it was a long day's journey for a trireme to row from Byzantium to Heraclea (Xenophon *Anabasis* 6. 4. 2).

ten centuries later is strikingly similar: 'the right side of the Euxine Sea, namely from Calchedon to the Phasis river, is a journey of 52 days for an unencumbered traveller'. The Byzantine historian had earlier (3. 1. 17) calculated that '210 stades are covered in a day'. Thus, in 52 days one could traverse 10,920 stades, which again means a bit more than 2000 km.

The natural relief of the coast determined the existence of routes and therefore communications. A crucial feature of the littoral as regards communication between settlements is the rivers. The widest ones especially, with the valleys they created, provided people with the otherwise very rare possibility of breaking through the chain of mountains and heading south.[32] However, large rivers like the Halys and the Iris were a double-edged sword, since they also impeded east–west traffic. Addressing the Ten Thousand, who wished to continue their homeward journey from Cotyora along the southern Black Sea coast, the Sinopean envoy Hecatonymus, fully conversant with the north Anatolian land, warned them of the large rivers they would encounter:

> First is the Thermodon, three plethra in width [92 m], which I think would be difficult to cross, especially with great numbers of the enemy in front and great numbers following behind; second, the Iris, likewise three plethra wide; third, the Halys, not less than two stadia in width [370 m], which you could not cross without boats – and who will be there to supply you with boats? – and similarly impassable is also the Parthenius, to which you would come if you should get across the Halys (Xenophon *Anabasis* 5. 6. 9).

An additional inhibitory factor to east–west movement close to the coast was the number of rushing streams and steep crags tumbling down to the sea. Therefore, communication between different coastal places had to be done mainly by sea.[33] But even this was not as easy as it might seem. First of all, the Black Sea has always been an especially windy sea, with the prevailing wind in the south being the north-westerly. Therefore, sheltered harbours would have to be situated on the south-east sides of headlands, but not, of course, craggy ones, which were quite frequent along the southern Black Sea coast, with the mountains coming so close to the sea. In addition, simply providing shelter was not enough:[34] a good harbour also needed access to the hinterland. Places combining these features are few.[35] The Sinop peninsula was certainly one of those providing excellent shelter from any westerly and also any northerly wind, which was an intrinsic difficulty on a southern coast

[32] See more in Section 5.5.

[33] Maksimova 1956, 16; Bryer and Winfield 1985, 18.

[34] That is, in cases when more was needed than just an installation to provide shelter, small scale ship repairs and generally brief mooring.

[35] *Cf.* Bryer and Winfield 1985, 7.

like ours. This is where one of the earliest and most important Greek colonies would be founded. However, it lacked good access to the hinterland; the champion in this respect was Amisos, built at one end of the only wide natural north–south route (see above), but with a rather inadequate harbour. Other places providing good shelter for ships were Cape Iasonion (Yason Burnu)[36] and the nearby Cape Genetes (Çamburnu), which is considered by several scholars as the best winter harbour on the southern Black Sea littoral,[37] a claim that would be difficult to believe were it not for the greater depth of water than in Sinope.[38]

Be that as it may, the sea route was still a good solution, given the difficulty of the terrain, especially for those coming from afar. Its use from very early on is reflected in the Argonautic myth. Jason and the Argonauts sailed along the southern Black Sea coast in order to reach Colchis and obtain the Golden Fleece.[39] It is the same route that served the Greek colonists who dotted the coast with settlements in the 1st millennium BC, and the unnumbered subsequent journeys of any kind between these settlements and Greece. It was also the one used in part by Xenophon and the Ten Thousand in 400 BC to return to Greece after fighting alongside Cyrus in the Battle of Cunaxa (401 BC). Some centuries later, the Romans created several outposts along the coast.[40]

3. In the 1st Millennium BC

We cannot know what exactly the natural environment of the southern Black Sea littoral looked like in the 1st millennium BC; we can, however, try to shape an image as close to reality as possible by examining relevant evidence provided through historical, archaeological or geological studies. Our area of interest has not yet become the subject of an aggregated examination, as has, for example, a part of the Paphlagonian hinterland, to the south of the Pontides.[41] Archaeological investigation in the littoral is quite insufficient, as has been mentioned in the Introduction Section 2, and is moreover largely confined to the sphere of human activity, void of any, for example, palaeo-environmental dimension.

[36] *Cf.* Bryer and Winfield 1985, 119.

[37] Hamilton 1842, 269; Bryer and Winfield 1985, 120, with more references; Lendle 1995, 332.

[38] For other places on the coast offering an eastern side of a headland as a protection, see Winfield 1977, 155–57. *Cf.* also Bryer and Winfield 1985, 7–12.

[39] Hence Cape Iasonion; see no. 49 in Table 1, below.

[40] For the settlements and installations on the coast, see Part 5.

[41] Marsh *et al.* 2009. For other studies, with results concerning the 3rd and 2nd millennia BC, see Dönmez 2006.

The only such aspect that has been examined to any extent so far is sea-level change, but even this has not resulted in unanimously accepted conclusions, nor do the suggested figures for the several parts of the coast display any homogeneity. It has been suggested that the level of the sea, including the southern shore, rose at several times in antiquity and has risen by a further 3–4 m since the 1st century AD.[42] More recently, research focused on the bay of Gideros indicated that the ancient Greek port of Cytoron, which was established there, would not have existed if the sea had been even 10 m lower than at present, for the harbour is very shallow.[43] On the other hand, there are also converse data: for example, geological investigation has shown that the settlement at Ikiztepe, near Bafra,[44] which is now about 7 km inland, was coastal in antiquity.[45] While this seems to contradict the theory of a rise in sea level, the site happens to be extremely close to the Halys and its estuary, which means that in this case we have rather an example of the silting-up caused by alluvial deposits brought down by the river.

All this is indicative of the divergences that can be observed between different parts of the coast, or that might even just be the result of different research projects. Divergences which, naturally, increase as we go back in time. It has been pointed out that in the Neolithic period the sea level in the area of Sinop was 155 m lower, and near the Sakarya estuary 90 m lower, than today.[46] Another study on the former place suggests that at the start of the Holocene the sea would have been about 40 m below current levels, would then have stabilised at -20 m for some time around 6000 BC, and would finally have risen to more or less current levels at around 3000 BC.[47] Nevertheless, such differences are characteristic not only of the south shore but of the whole Black Sea,[48] where again little research has been conducted yet.[49]

[42] Tsetskhladze 2007, 177–80, especially 178 and n. 38 on the southern Black Sea.

[43] Düring and Glatz 2015b, 14–15.

[44] The latter could be identified with Strabo's (12. 3. 14) Gadelon. See Section 5.1, Table 3, no. 105 with Comment 21.

[45] See Tsetskhladze 2007, 178.

[46] Ballard *et al.* 2000; 2001; Coleman and Ballard 2007; Algan *et al.* 2007.

[47] Düring and Glatz 2015b, 14.

[48] *Cf.* Düring and Glatz 2015b, 14.

[49] For the northern Black Sea, see Porotov 2007, according to whom a review of existing data from various sites shows the presence of submerged cultural layers that did not exceed 2.5–3.5 m below present levels. For the western Black Sea, see Coleman 2000, 109, who suggests that 'the considerable rise of the Black Sea sea-level must have taken place after *ca.* 2700 BC, since remains of that date have been found along the western coast of the Black Sea submerged to a depth of 8–10 metres'. Conversely, Peev (2016) claims that archaeological data show that the relative sea level was around -4 m below present in the 6th century BC, rising to the present mean sea level around the 6th century AD. For recent research in the Black Sea and the relevant methodological issues, see Baralis *et al.* 2016. For the geological, palaeontological and

Further evidence could be provided by the submerged breakwater off the coast of Filyos (Fig. 7), not investigated yet, which once provided shelter from northerly winds.[50] If it belonged to the harbour of the Greek colony of Tieion,[51] which is highly probable, then we can estimate quite precisely the rise in sea level in the specific area from the 1st millennium BC.[52]

On the other hand, trying to draw conclusions from comparisons with the situation in neighbouring regions or in the same region in later periods might be risky, since the environment may change considerably in space and time.[53] For example, there are some very detailed reports by travellers who visited or described our area in the 18th and 19th centuries,[54] not to mention contemporary scholarship.[55] The 18 or 20 centuries that have passed since the (end of the) 1st millennium BC may seem a short span compared to the millennia that preceded the formation of the Black Sea, but are still a considerable period during which many environmental changes can take place.[56] Apart from changes in sea level, one might mention by way of example that the landscape of Sinope described by D.M. Robinson[57] in the early 20th century differs from the account in Strabo, who lived twenty centuries earlier, while the plain of Themiscyra, the largest in the eastern half of the littoral, which was particularly fertile in antiquity (see below), was described by M.I. Maksimova in the middle of the previous century as being swamp.[58] One might even consider the fact that, although Kerasous (Giresun) seems to have been famous in antiquity for cherry trees, hence its name (see below), in the modern period it is rather the hazelnut that is widely produced and traded there.[59]

It seems, then, that our best means of approaching the natural environment of the southern Black Sea littoral in the 1st millennium BC is to study the texts written in the period. All the more so when some of the texts in question are

palynological methods used to study Late Glacial to Holocene sea level as well as climate changes in the Black Sea, see Yanko-Hombach *et al.* 2011.

[50] Maybe two of them very close to each other: Atasoy and Yıldırım 2015b, 441; Aksoy and Yıldırım 2017; Atasoy 2016, 212; 2018, 117.

[51] See Part 4.

[52] Ş. Yıldırım has informed me that the breakwaters are now 5–6 m below sea level.

[53] For a brief overview of the southern Black Sea's geology and climate, as well as the present situation as regards resources, economy and pollution, see Kerestecioğlu *et al.* 1998.

[54] E.g. Cramer 1832; Hamilton 1842; Ramsay 1890. For the several descriptions of the region of Cide (Aegialos) specifically through the ages, see Karasalihoğlu and Düring 2015.

[55] The most noteworthy is the one by Bryer and Winfield (1985), unfortunately limited to the eastern half of the littoral. More descriptions of the littoral's landscape by modern travellers and contemporary scholars in Marek 1993, 7, n. 50.

[56] An example is the later cultivation of tea: Dönmez 2006, 31.

[57] Robinson 1906, 127–29.

[58] Maksimova 1956, 16.

[59] Magie 1950, 1073–74; Bryer and Winfiled 1985, 128.

among the most reliable written sources we possess for our area of interest, like Xenophon's *Anabasis* and Strabo's *Geography*.[60]

In the following pages we shall therefore be examining the natural environment of the littoral and the opportunities it afforded its inhabitants in the 1st millennium BC, through the descriptions of people who had been there in this period of time, acknowledging that our data will be necessarily fragmentary and limited only to the areas they described. Inevitably, our long area of study will be divided into separate parts following the division that may result from both our sources' relevant passages and the landscape characteristics of the several parts of the littoral. Table 1 and the maps of Fig. 13 present all the geographical features of the littoral that have been mentioned and named by ancient authors, listed in geographical order from west to east. Unlike the western boundary of the littoral, which is formed by the Bosporus straits, the eastern one cannot be easily defined, since the coastline describes a smooth curve to the north. But since an eastern boundary there must be, I adopt for our purposes the modern border between Turkey and Georgia.

The first column of the Table gives the names of the features, and the second its kind. The third column cites the sources that mention each of the features. Only what I term 'geographical sources' are listed.[61] The sources appear in abbreviated form.[62] If a name is rendered somewhat differently in any source, this is recorded in a parenthesis after that source.

In the last column the features are located on the modern map and, where possible, identified with modern ones. For this identification I had recourse to several modern sources, chiefly the *Barrington Atlas*,[63] the *Pleiades* project,[64] the *Digital Atlas of the Roman Empire* by the Centre for Digital Humanities, University of Gothenburg,[65] as well as Bryer and Winfield (1985). In several cases there are disagreements among them; in the few for which no identification is provided, I give those I consider most probable.[66] Some of the modern names may have changed in recent years or decades, and thus do not appear in the earlier modern bibliography; here I use the names that are currently valid and appear on Google maps today (2021).

[60] For which, see Section 2 in the Introduction.
[61] For the term see in the Introduction Section 2.
[62] See all the abbreviations in the introduction to Table 3, Section 5.1, n. 4.
[63] Talbert 2000.
[64] https://pleiades.stoa.org.
[65] https://dh.gu.se/dare/.
[66] In a few other cases, the attributions of these sources are considered erroneous; for example, the Rhebas was only a river and no ancient source mentions it as a settlement as well, as we see in Talbert 2000, map 52, and vol. II, 791. Only Aelius Herodianus and Stephanus of Byzantium (*s.v.* Ῥήβας) mention it also as a *chora*.

Roman geographical sources of the 2nd century AD – notably Ptolemy and Arrian – are included, although later than the period covered by this book, since they belong to the same geographical tradition as Menippus, Strabo, Mela and Pliny, and for other reasons that are mentioned in Section 5.1. The much later anonymous *Periplus* is also taken into account, because it is essentially a compilation of four earlier works and may even provide lost parts of some of them. On the contrary, much later works with no clear influences from our 'geographical' sources, are not taken into consideration.[67]

Table 1. Geographical features of the southern Black Sea littoral (from west to east).

No.	Name	Category	Sources	Location/Identification
1	Rhebas	River	Pseudo-Scylax 92; Apollonius 349, 650; Men./Marc. 5701; Pliny 6. 4; Dionysius P. 794–796; Arrian 12; Anon. 3.	Riva/Çayağzı creek
2	Colone	Peak	Apollonius 650.	unknown, probably fictional
3	Melaena/Cale Akre	Cape	Apollonius 349, 651; Men./Marc. 5701; Ptolemy 5. 1. 5 (Bithynias akra); Arrian 12; Anon. 3.	Karaburun
4	Artanes	River	Ps-Scylax 92; Men./Marc. 5701; Arrian 12; Anon. 3.	Türknil river
5	Psillis	River	Apollonius 652 (Phylleis); Men./Marc. 5701; Strabo 12. 3. 7; Ptolemy 5. 1. 5; Arrian 12; Anon. 4.	Ağva/Koca creek
6	Calpe	River	Apollonius 659; Men./Marc. 5701; Strabo 12. 3. 7; Ptolemy 5. 1. 5; Anon. 5.	Sarısu stream
7	Thynias/Apollonia/Daphne/Daphnousia	Island	Pseudo-Scylax 92; Apollonius 350, 673; Pseudo-Scymnus 1025; Men./Marc. 5701; Strabo 12. 3. 7; Mela 2. 98; Ptolemy 5. 1. 15; Arrian 13; Anon. 6.	Kefken island

[67] For example, the Cytoris mentioned by Ps.-Zonaras is not included.

No.	Name	Category	Sources	Location/Identification
8	Sangarius	River	Pseudo-Scylax 92; Apollonius 722; Pseudo-Scymnus 1022; Men./Marc. 5710; Strabo 12. 3. 7; Pliny 6. 4; Ptolemy 5. 1. 6; Arrian 13; Anon. 7.	Sakarya
9	Hyrius/Hypius	River	Pseudo-Scylax 91; Pseudo-Scymnus 1020; Men./ Marc. 5710; Ptolemy 5. 1. 7; Arrian 13; Anon. 8.	Büyükmelen stream
10	Elaeus	River	Men./Marc. 5710; Ptolemy 5. 1. 7 (Elata); Anon. 9.	Kocaman stream
11	Cales	River	Men./Marc. 5710; Anon. 9.	Alaplı stream
12	Lycus	River	Xenophon 6. 2. 3; Pseudo-Scylax 91; Apollonius 724; Pliny 6. 4; Arrian 13; Anon. 9.	Aydınlar stream
13	Anthemoeisis	Lake	Apollonius 724.	unknown
14	Acherousias	Cape	Xenophon 6. 2. 2; Apollonius 354, 728.	promontory of Ereğli (Baba Burnu)
15	Acherousia	Cave	Mela 1. 103; Pliny 6. 4.	Cehennemağzı Caves?
16	Paedopides	River	Pliny 6. 4.	unknown
17	Acheron/Soonautes	River	Apollonius 355, 743–746; Pliny 6. 4.	Limanbaşı creek
18	Oxinas/Callichorus (not surely the same)	River	Pseudo-Scylax 90 (Callichorus); Apollonius 904 (Callichorus); Men./Marc. 5710 (Oxinas); Pliny 6. 4; Arrian 13; Anon. 12.	Ilıksu
19	Aulion	Cave	Apollonius 910.	unknown, probably fictional
20	Billaeus	River	Men./Marc. 5801; Pliny 6. 4; Arrian 13; Anon. 13.	Yenice
21	Parthenius	River	Xenophon 5. 6. 9; 6. 2. 1; Pseudo-Scylax 90; Apollonius 936; Pseudo-Scymnus 1012; Men./Marc. 5801; Strabo 12. 3. 5, 8; Mela 1. 104; Pliny 6. 5; Ptolemy 5. 1. 7; Arrian 13; Anon. 13.	Bartın
22	Amastris, former Sesamos	River	Men./Marc. 5801; Anon. 15.	an unknown creek close to Amasra

No.	Name	Category	Sources	Location/Identification
23	Erythinoi	Reefs	Strabo 12. 3. 10; Ptolemy 5. 1. 15.	off the coast of Çakraz
24	Cytorus	Mount	Apollonius 942(?); Pliny 6. 5.	to the south of Gideros Bay
25	Aegialos	Seacoast	Apollonius 945; Strabo 12. 3. 10.	to the west of Cape Carambis
26	Carambis	Cape	Apollonius 361, 943; Men./Marc. 5801; Strabo 12. 3. 10; Mela 1. 104; Pliny 6. 6; Dionysius P. 785; Ptolemy 5. 4. 2; Arrian 14; Anon. 17.	Kerempe
27	Aeginetes	River	Men./Marc. 5905; Anon. 20.	İlişi stream
28	Kinolis	River	Men./Marc. 5905.	Kirazlı stream
29	Lepte/Syrias Akre	Cape	Men./Marc. 5905; Arrian 14; Anon. 20.	Inceburun
30	Ochosbanes/ Ochthomanes	River	Pseudo-Scylax 89 (Ocherainos); Men./Marc. 5905; Anon. 20.	Karasu river
31	Skopelos	Mount on Cape	Men./Marc. 5905; Anon. 21.	Boztepe Cape (of Sinop)
32	Euarchus/Euechus	River	Men./Marc. 6001; Pliny 6. 6; Anon. 23.	Kabalı stream
33	Zali(s)cus	River	Men./Marc. 6001; Ptolemy 5. 4. 3; Anon. 24.	Uluçay
34	Halys	River	Xenophon 5. 6. 9; 6. 2. 1; Pseudo-Scylax 89; Apollonius 366, 963; Pseudo-Scymnus 982; Men./Marc. 6001; Strabo 12, *passim*; Mela 1. 105; Pliny 6. 6; Dionysius P. 784; Ptolemy 5. 4. 3; Arrian 14; Anon. 24.	Kızılırmak
35	'Lake close to Naustathmos'	Lake	Men./Marc. 6001; Arrian 15; Anon. 26(?).	Balık lake
36	Conopeion	Lake	Men./Marc. 6001; Arrian 15; Anon. 26.	Gernek lake
37	Lycastus	River	Pseudo-Scylax 89; Men./Marc. 6020; Anon. 28.	Mert river
38	Chadisius	River	Men./Marc. 6020; Anon. 28.	Abdal stream

No.	Name	Category	Sources	Location/Identification
39	Iris	River	Xenophon 5. 6. 9; 6. 2. 1; Apollonius 367, 963; Men./Marc. 9r9; Strabo 12. 3. 15, 30; Pliny 6. 10; Dionysius P. 783; Ptolemy 5. 6. 2; Arrian 15; Anon. 29.	Yeşilırmak
40	Heracleios	Cape	Apollonius 371 (Themiscyreion), 965 (Amazonidon akre); Men./Marc. 9r9; Strabo 12. 3. 17; Ptolemy 5. 6. 3; Arrian 15; Anon. 29.	close to Amazon Tabiat Parkı
41	Thermodon	River	Xenophon 5. 6. 9; 6. 2. 1; Pseudo-Scylax 89; Apollonius 370, 970; Men./Marc. 9r9; Strabo 12. 3. 15; Mela 1. 105; Pliny 6. 10; Dionysius P. 774; Ptolemy 5. 6. 4; Arrian 15; Anon. 29.	Terme
42	Themiscyra	Plain	Apollonius 373 (Doiandos pedion), 988; Strabo 12. 3. 15; Mela 1. 105.	plain around the Terme
43	Beris	River	Arrian 16; Anon. 29.	under the Milic bridge
44	Thoaris	River	Arrian 16; Anon. 29.	Say creek
45	Oenius	River	Men./Marc. 9r9; Anon. 30.	River through Ünye
46	Phigamous	River	Arrian 16; Anon. 30.	Kurna creek
47	Sidene	Plain	Strabo 12. 3. 16.	to the west of Yason Burnu
48	Sidenos	River	Pliny 6. 11.	Boloman creek
49	Iasonion	Cape	Xenophon 6. 2. 1; Pseudo-Scylax 88; Strabo 12. 3. 17; Ptolemy 5. 6. 4; Arrian 16; Anon. 32.	Yason Burnu
50	Kilikon	Island	Arrian 16; Anon. 32.	Hoynat Adasi
51	Genetes	Cape	Pseudo-Scylax 88; Apollonius 378, 1009; Strabo 12. 3. 17 (or the town?).	Çamburnu
52	Genepus/Genetus	River	Men./Marc. 9r9; Anon. 32.	under the bridge Kacalı Köprüsü
53	Boon	Cape	Men./Marc. 9r9; Anon. 32.	at the estuary of the Genepus

No.	Name	Category	Sources	Location/Identification
54	Melanthius	River	Men./Marc. 9r9; Pliny 6. 11; Arrian 16; Anon. 34.	Melet river
55	Pharmatenus	River	Arrian 16; Anon. 34.	Aksu stream
56	Island of Ares/Aretias	Island	Pseudo-Scylax 86; Apollonius 382–387, 1031–1230; Pseudo-Scymnus 952; Men./Marc. 9r9; Mela 2. 98; Pliny 6. 32 (Chalceritis); Arrian 16; Anon. 34–35.	Giresun islet
57	Philyreis	Island	Apollonius 1231.	probably fictional
58	Zephyrion	Cape	Arrian 16.	the cape from Uluburnu to Çamburnu
59	Tripolis	River	Pliny 6. 11.	Doğankent stream
60	Kerasous	River	Men./Marc. 9r9(?); Anon. 36.	Fol creek or Kireson stream
61	Hieron Oros	Cape/ Mount	Men./Marc. 9r9.	Cape near Yoroz
62	Hyssus	River	Arrian 7.	Karadere stream
63	Ophius	River	Men./Marc. 9r9; Arrian 7; Anon. 38.	Çataklı stream
64	Psychros	River	Arrian 7; Anon. 38.	Yeniköy creek
65	Calos/Cale Parembole	River	Arrian 7; Anon. 39.	Ikizdere stream
66	Rhizius	River	Men./Marc. 9r9; Arrian 7; Anon. 39.	Taşlı creek
67	Ascurus	River	Arrian 7; Anon. 39.	a small stream north-east of Rize
68	Adienus	River	Men./Marc. 9r9; Arrian 7; Anon. 39.	Büyük stream
69	Athenon or Adienon Akron	Cape	Ptolemy 5. 6. 6.	close to Pazar
70	Zagatis	River	Arrian 7; Anon. 40 (Zaggales).	Hemşin creek
71	Prytanis	River	Pseudo-Scylax 83 (Pordanis); Arrian 7; Anon. 40.	Fırtına creek
72	Pyxites	River	Pliny 6. 12; Arrian 7; Anon. 40.	Tahiroğlu river[68]

[68] The identification of Pyxites with Değirmendere, close to Trapezous, which is much further west than the place defined by the ancient sources, is an error that has for some reason appeared in modern scholarship (e.g. Bryer and Winfield 1985, 48–49, 251–54). *Cf.* also n. 199 in Part 5 and n. 626 in Part 2.

No.	Name	Category	Sources	Location/Identification
73	Archabis	River	Pseudo-Scylax 83 (Arabis); Ptolemy 5. 6. 6; Arrian 7; Anon. 40.	Orçi stream
74	Kissa	River	Ptolemy 5. 6. 6.	Hopa stream
75	Arion	River	Pseudo-Scylax 82.	Çam creek, or Karaosmaniye creek, or Hopa
76	Daraanon	River	Pseudo-Scylax 82.	Çamlıl, or Karaosmaniye Creek, or Hopa

As a first comment, one might remark on the large number of rivers – 49 out of 76 features, or about two-thirds of them. The fact that people in antiquity not only mentioned so many rivers in this narrow strip of land – on average one every 20 km – but also named them is indicative of their significance for the landscape and the human activities developed within it. This is natural, considering the relief of the littoral. With the mountains coming so close to the coast, the rivers, especially the larger ones, were a significant feature that offered the possibility of some arable land as well as access to the hinterland (see above). Next come the capes, one of the striking characteristics of the mountainous coastline; twelve of them are known by their ancient names, far fewer than their actual number.

Having once been a lake, the Black Sea lacks islands, apart from a few very small ones close to the coast. The biggest near the southern Black Sea coast is Thynias (Kefken), in the western part, called also Apollonia, Daphne or Daphnousia, with an area of only 11 ha. It is noteworthy that this small island was already inhabited in the Classical period.[69] For the rest, another two uninhabited islets are mentioned in the eastern part of the littoral: Kilikon Nesos (*ca.* 1 ha) and the famous Island of Ares (*ca.* 4 ha). The Philyreis Island mentioned only by Apollonius in the *Argonautica* is probably fictional.[70] Finally, there are the reefs called Erythinoi, another ambiguous case.[71]

Lakes were another uncommon feature. Two have been mentioned (nos. 35, 36), both to the east of the Halys estuary and still existing, while the third, Anthemoeisis (no. 13), again known only to Apollonius, is obscure, as is the cave called Acherousia (no. 15) in the same general area, which is mentioned in a mythical context (see below). Probably also fictional is Apollonius' Aulion cave (no. 19). There is also the case of Aegialos (no. 25), mentioned apparently

[69] See Sections 4.4.1 and 5.1, and Comment 1 to Table 3.
[70] See Section 2.14.1.
[71] See Section 5.2.1.

as a seacoast, but also with a known homonymous town in the region.[72] Plains were, of course, also very rare, and always around large rivers; the only ones named are Themiscyra (no. 42), close to the homonymous city,[73] and Sidene (no. 47), upon which stood, among other places, Side.[74]

Some other plains are mentioned in the ancient literature as affecting the littoral's economy, even though they were a bit farther from the coast: for example, Strabo's Phanaroia (12. 3. 15, 30), traversed by the Iris, which should probably be connected with Ptolemy's plain of Phanagoria (5. 6. 3). Lying to the south of the plain called Themiscyra, Phanaroia was an extremely fertile plain, which Strabo called the best place in Pontus, probably with a certain degree of subjectivity since it bordered his home city of Amaseia. Not far from there must have been Phanoria, Pliny's fortress (no. 67 in Table 3). Ammianus Marcellinus' statement that 'beyond the Chalybes are open plains (*terras patentes*), inhabited by the Byzares, Sapeires, Tibarenoi, Mossynoikoi, Macrones and Philyres' (*Rerum Gestarum* 22. 8. 21) is certainly the result of exaggeration or confusion, since the part of the littoral indicated by the reference to all these peoples[75] is almost its entire eastern half and most mountainous part.[76]

And finally, there are the mountains. The Pontides are the southern boundary of the littoral and often even part of it, especially in the east. One might therefore consider it strange that we encounter so few mountain names in the ancient literature and particularly in the Roman period. There are only Cytorus (no. 24) and Hieron Oros (no. 61), plus Skopelos (no. 31) on the Sinop peninsula. The peak called Colone by Apollonius (no. 2) is probably fictional, like other features mentioned by the poet (see above). Nevertheless, several mountains are mentioned that are not on the littoral but clearly in the Pontides, such as Olgassys in Paphlagonia, which was full of temples erected by the Paphlagonians (Strabo 12. 3. 40);[77] Amazonius, to the south of the Themiscyra plain (Pliny *NH* 6. 10; Apollonius *Argonautica* 2. 979); Theches, south of Trapezous, from where the Ten Thousand on their way back to Greece could finally see the sea (Xenophon *Anabasis* 4. 7. 20–21);[78] Paryadres and Scydises, which joined the Moschian Mountains that extended to the Caucasus (Strabo 12. 3. 18; 11. 2. 1, 15). Let us now examine the littoral in its separate parts (see also Fig. 8.1–3 and Table 2).

[72] No. 30 in Table 3.

[73] No. 66 in Table 3.

[74] No. 72 in Table 3.

[75] For these peoples, see Part 2.

[76] The very part of the sentence that follows ('peoples not known to us through any intercourse') may strengthen this view.

[77] See Sections 2.6.3 and 2.6.4A.

[78] See Sections 2.9.3 and 2.13.1.

3.1. From the Bosporus to the Promontory of Ereğli (Baba Burnu) (Fig. 8.1)

The western part of the southern Black Sea littoral, from the straits of Bosporus to the promontory of Ereğli and generally known in antiquity as Bithynia,[79] is quite an interesting region. It is the only part with such a broad lowland, before the endless Pontides begin hugging the coast. On the other hand, it is not one of the littoral's best described regions in ancient literature; were it not for Xenophon's detailed narrative, our knowledge of its natural environment in antiquity would be quite meagre. Xenophon's description of it is so rare in its detail, in fact perhaps his most detailed description of a place in the whole littoral and indeed in the *Anabasis*, that it is worth quoting:

> As for Calpe Harbour, it lies midway of the voyage between Heraclea and Byzantium and is a bit of land jutting out into the sea, the part of it which extends seaward being a precipitous mass of rock, not less than twenty fathoms high at its lowest point, and the isthmus which connects this head with the mainland being about four plethra in width; and the space to the seaward of the isthmus is large enough for ten thousand people to dwell in. At the very foot of the rock there is a harbour whose beach faces toward the west, and an abundantly flowing spring of fresh water close to the shore of the sea and commanded by the headland. There is also a great deal of timber of various sorts, but an especially large amount of fine ship-timber, on the very shore of the sea. The ridge extends back into the interior for about twenty stadia, and this stretch is deep-soiled and free from stones, while the land bordering the coast is thickly covered for a distance of more than twenty stadia with an abundance of heavy timber of all sorts.[80] The rest of the region is fair and extensive, and contains many inhabited villages; for the land produces barley, wheat, beans of all kinds, millet and sesame, a sufficient quantity of figs, an abundance of grapes which yield a good sweet wine, and in fact everything except olives' (*Anabasis* 6. 4. 3–7).[81]

Xenophon's praise of the area was not without reason. What he actually aimed at was to persuade his readers that the specific place was indeed a good choice for establishing a new Greek colony; something he had not managed to convince his soldiers of, when they were there.[82] Bithynia seems to have been a region abundant in trees, since we read about ravines (Xenophon *Anabasis* 6. 5. 12–23, 31), as well as thickly overgrown areas (6. 4. 27). Unlike the region of Calpe, more to the east, close to Cape Acherousias (the promontory of Ereğli or Baba Burnu), wild olive trees flourished (Apollonius *Argonautica* 2. 842–850). Our knowledge of the trees producing edible fruit in Bithynia is

[79] For the names of the several parts of the littoral in antiquity, see the next part of the book.

[80] For the availability of timber in Bithynia, see also Roebuck 1984, 123.

[81] Xenophon mentions again the abundance in wheat and barley, wine, beans, millet and figs, as well as the lack of olive oil (6. 6. 1).

[82] *Cf.* Lendle 1995, 385–87; Tsetskhladze 2007, 162–64. See more in Section 4.4.3.

increased thanks to a brief anthology by Athenaeus (*Deipnosophistae* 2. 35) of several passages from authors from the 5th to the 1st centuries BC:

> Asclepiades of Myrlea, mentioning a kind of ground-cherry, spoke of it thus: 'In the country of the Bithynians grows the bush-cherry, the root of which is not large, nor, for that matter, is the tree, but equal in size to the rose-bush; its fruit, in all other respects, resembles the cherry, but it causes drowsiness, as of wine, to those who eat too much, and makes the headache.' These are Asclepiades' words, and I think that he is speaking of the arbutus. For not only does the tree bearing this fruit correspond to this description, but it is also true that whoever eats more than seven berries of it gets a headache. Aristophanes said: 'On the mountains, without cultivation, the arbutus-trees used to grow in plenty for their enjoyment.' Theopompus said: 'They eat myrtle-berries and ripe fruit of the arbutus-tree.' Crates said: 'The ripe loveliness of her breasts is as the apple or the arbutus-berry.' Amphis said: 'The mulberry-tree, you see, bears mulberries, the ilex acorns, the strawberry-tree arbutus.' And Theophrastus spoke: 'The strawberry-tree, which bears the edible arbutus-berry'.

Nonetheless, Bithynia seems to have been more famous for its cheese, although this was not actually produced right in its coastal part, but a bit more inland, to the south of Heraclea. It was in the area called Salona, close to Bithynion, where 'the best pasturage for cattle' produced the Salonian cheese, as reported by both Strabo (12. 4. 7) and Pliny (*NH* 11. 97). The latter commented that pastures there contained salt, so that as the cheese aged it turned salty.

Strabo tells us (12. 3. 7) that the Sangarius river was navigable in his day, which earlier it was not. The geographer did not, like many authors before him, fail to mention aconitum, a plant which grows in the territory of Heraclea. According to a Greek myth (first in Herodorus, according to the Scholiast of Apollonius Rhodius 2. 354), aconitum was the poisonous herb that spewed from the slavering mouth of Cerberus when the latter was dragged from the Underworld by Heracles. The supposed entrance to the Underworld where this happened is described by Apollonius in his *Argonautica*, since the Argonauts were supposed to have passed by there.[83] Apparently the description better

[83] 'Here is a downward path to the abode of Hades, and the headland of Acherousias (no. 14 on Table 1) stretches aloft, and eddying Acheron cleaves its way at the bottom, even through the headland, and sends its waters forth from a huge ravine' (*Argonautica* 2. 353–356). And then again: 'It rises aloft with steep cliffs, looking towards the Bithynian sea; and beneath it smooth rocks, ever washed by the sea, stand rooted firm; and round them the wave rolls and thunders loud, but above, wide-spreading plane trees grow on the topmost point. And from it towards the land a hollow glen slopes gradually away, where there is a cave of Hades overarched by wood and rocks. From here an icy breath, unceasingly issuing from the chill recess, ever forms a glistening rime which melts again beneath the midday sun. And never does silence hold that grim headland, but there is a continual murmur from the sounding sea and the leaves that quiver in the winds from the cave. And here is the outfall of the river Acheron which bursts its way through

befitted the dreadful image appropriate to the entrance to the Underworld than the reality, although some elements, like the existence of a cave and the cool breeze coming out of it, might well be true. It has therefore been proposed that the cave that inspired the creation of the specific mythical version could be the Cehennemağzı Caves at Ereğli (ancient Heraclea). In fact, not far from the area lay Aconae, a settlement probably named after the plant in question.[84]

Unfortunately, Strabo did not lay more stress on describing the landscape when dealing with Bithynia (12. 4), let alone its Black Sea coastal region. But later authors also praised Bithynia's fertility, its abundance of trees and the wealth of quarries in its mountains (Dionysius Periegetes *Orbis terrae description* 793; Eustathius *Comm. in Dion. Per.* 793; *Paraphrases in Dion. Per.* 793; Nicephorus Blemmydes *Conspectus geographiae* 464. 2. 36), while in the *Geoponica* (20. 46. 3) we read how the Bithynians prepared liquamen, a kind of garum. Taking all these references together, it is readily apparent that Bithynia was a blessed region, able to provide its inhabitants with plenty of opportunity to develop a variety of economic activities; these would include agriculture, carpentry (see also Pliny *NH* 16. 197; *Epistulae ad Traianum* 41 [50]), viniculture and also livestock rearing, since Xenophon several times refers to the many sheep in the area (*Anabasis* 6. 3. 3, 22; 6. 5).

Athenaeus also mentions the so-called fossil fish, 'which occur in Heraclea and around the Milesian colony of Tieion in Pontus'. According to Theophrastus

> these fish are frozen in the winter's ice, so that they have no feeling and cannot move until they are put into the saucepans and cooked. And these fish have this especial peculiarity, which also belongs in some degree to the fish which are called fossil fish in Paphlagonia. For they say that ditches are dug in those places to an exceeding depth, where no overflow of rivers ever reaches, nor of any other waters whatever; and yet living fishes are found in these ditches' (*Deipnosophistae* 8. 331; *cf.* Pseudo-Aristotle *de Mirabilibus Auscultationibus* 835B).

For the rest, we have also some data provided by Procopius in the 6th century AD, which refer not to the part of Bithynia bordering the Black Sea but to 'a certain road leading into the Phrygian territory', so an area more to the south, but still interesting: several times countless men and beasts perished on this road during winter. The soil of this region is exceedingly deep, and after deluges of rain or the melting of heavy snows, but even after occasional showers, it turns into a deep and impassable marsh, making the roads quagmires (Procopius *De aedificiis* 5. 3. 12–13). Such a harsh climate might not suit the

the headland and falls into the Eastern sea, and a hollow ravine brings it down from above' (*Argonautica* 2. 729–745).

[84] See also Sections 4.4.1, n. 153, and 5.1, no. 13 on Table 3.

above-cited descriptions, nonetheless the famous 2nd-century physician Galen attributed the fatness of the Bithynians to the cold climates they were living in (*De simplicium medicamentorum* 11. 513–14).[85]

3.2. From the Promontory of Ereğli to the Thermodon (Terme) (Fig. 8.2)

Heraclea was built on a bay ending in a promontory which ascends to the north, giving the impression of a barrier separating the lands to the west from the rest of the southern Black Sea coast. The main change in the terrain to the east of this promontory is that from here on the elevation easily reaches three digits within 2 km of the coast. This means that the mountains almost touch the sea, which is actually one of the overall characteristics of the littoral. Another feature of this stretch, as far as the Sinop promontory, is the lack of notable bays or promontories, unlike the picture presented by the coast from that point eastwards. Cape Carambis (Kerempe), somewhere in the middle of this region, is merely a gentle blip in the monotony of an otherwise almost straight coastline on either side. The northernmost point of the littoral is not here,[86] but at Lepte or Syrias Akre (Inceburun);[87] it is, however, the closest point to the northern Black Sea littoral's southernmost one, Kriou Metopon (Cape Sarych) in the Crimea, as observed already in antiquity (Pseudo-Scymnus 998–1000; Strabo 2. 5. 22; 7. 4. 3, 27. 6; 11. 2. 14; 12. 3. 10). Therefore, the reference to a long seacoast to the west of Cape Carambis is not at all unexpected. This is Aegialos (see above), 'a long shore of more than a hundred stadia' (Strabo 12. 3. 10; *cf.* Apollonius *Argonautica* 2. 365, 944–945), so about 18.5 km.

Notwithstanding the length of this part of the littoral, but also the importance of the Sinop peninsula, where one of the earliest and most significant Greek colonies of the whole ancient world would one day be established – not to mention the several earlier local settlements there, from prehistoric times[88] – this part of the coast, generally defined as Paphlagonia, is in terms of environment the least described of the whole shore in ancient literature;[89] in sharp contrast to its hinterland.

[85] For freezing episodes in the Black Sea during the last two millennia, see Yavuz *et al.* 2007.

[86] As Bryer and Winfiled (1985, 2) claimed.

[87] Kerempe: 42° 01′ 11″ N; Inceburun: 42° 05′ 53″ N.

[88] For Sinope, see Section 4.3.2. For the prehistoric settlements in the region, see Section 5.2.2, n. 89.

[89] By contrast, it is one of best described ones in modern archaeological literature: Düring and Glatz 2015b.

Starting from the west, the Bartın river seems to have acquired its ancient name (Parthenius) from the homonymous plant and its general ability to create flowery areas along its path (Strabo 12. 3. 8). The peninsula to its east provides harbours on either side of the isthmus – a miniature of the Sinop peninsula – and this is where the city of Amastris was founded (Strabo 12. 3. 10). Unlike the area of Calpe, where Xenophon noted the lack of olive trees (see above), it seems that olive oil was produced in the wider region of the Parthenius, in the territory of Amastris (Amasra).[90] More to the east is the area of Cytoros, at the foot of the homonymous mountain, where the most and best box wood grew (Strabo 12. 3. 10; Theophrastus *Historia Plantarum* 3. 15. 5). The mountain range facing the sea all through Paphlagonia, from Bithynia to Sinope, had good shipbuilding timber, which was easy to transport (Strabo 12. 3. 12).

As far as fish is concerned, thanks to Aelian (*De natura animalium* 15. 5) we know that the whole coast from Heraclea to Tieion and Amastris was abundant in tunny and that the inhabitants of that land knew exactly when this fish would come in the summer and how best to catch it. This is almost everything we know about the landscape of Paphlagonia as far as Sinope.[91] The words of an envoy of the Paphlagonians addressed to Xenophon and the Ten Thousand in 400 BC condense in a phrase the diversity of the landscape: 'the Paphlagonian land possesses both the fairest plains and the loftiest mountains' (*Anabasis* 5. 6. 6).

The imposing peninsula of Sinop, providing excellent shelter for ships from both north and south winds, stands out from the littoral like a stronghold, jutting into the Black Sea almost at the mid-point of the coast.[92] Naturally, it could not escape the authors' attention, and the several surface surveys and excavations that been conducted during the last decades have supplemented our knowledge of the area and its economic opportunities. An overview is provided by Strabo (12. 3. 11):

> Sinope is beautifully equipped both by nature and by human foresight, for it is situated on the neck of a peninsula, and has on either side of the isthmus harbours and roadsteads and wonderful pelamydes-fisheries, of which I have already made mention, saying that the Sinopeans get the second catch and the Byzantians the third. Furthermore, the peninsula is protected all round by ridgy shores, which have hollowed-out places in them, rock-cavities, as it were, which the people call 'choenicides'; these are filled with water when the sea rises, and therefore the

[90] Mitchell 2005, 88; 2010, 92. Lipka (1995, 72) seems to ignore the distinction between the two regions made by Xenophon.

[91] See also Marek 1993, 10–11. On the wild mules that were said to originate from the area of the Enetoi, somewhere in Paphlagonia (Homer *Iliad* 2. 852; Strabo 12. 3. 8), see Section 2.7.

[92] One of the best modern descriptions of the Sinop peninsula is still Robinson's (1906, 126–35). Among other things, he highlights the volcanic nature of the peninsula.

> place is hard to approach, not only because of this, but also because the whole surface of the rock is prickly and impassable for bare feet. Higher up, however, and above the city, the ground is fertile and adorned with diversified gardens; and especially the suburbs of the city.[93]

Fishing appears to have been a vital activity of Sinope (*cf.* also Strabo 7. 6. 2; Athenaeus *Deipnosophistae* 7. 307B, for mullet). Pliny (*NH* 9. 47–49) described in detail when and how male tunnies entered the Black Sea from the Mediterranean in springtime, and why they were given the name pelamyda. All this has sometimes led to the theory that even the initial purpose of the founding of Sinope was probably 'to secure fish', especially tuna fish, rather than to open up trade with the interior.[94] However, such a theory sounds rather exaggerated. The importance of fishing for the inhabitants of a coastal area with such limited access to the hinterland is indisputable,[95] but I cannot see it as a decisive reason for people living in areas equally rich in fish to leave the latter and colonise Sinope.[96]

Moreover, there has often been a simplistic generalisation of data provided by ancient sources regarding fish and fisheries in the Black Sea, in ascribing them specifically to Sinope.[97] For example, when Pliny talks about the variety of fish species in the Black Sea (*NH* 9. 47–52), he is referring to the whole Black Sea and not only to Sinope. Similar is the case with Aelian's detailed description of the fishing methods used by fishermen in the southern Black Sea (*De natura animalium* 15. 5–6), as well as with Diodorus' information (37. 3. 5) that a jar of salted fish from the Black Sea was sold for 400 drachmae.[98]

The region of Sinope also produced also maple and mountain-nut trees, which yield a wood suitable for tables (Strabo 12. 3. 12). The development of carpentry in the city is further attested by Polyaenus (7. 21. 2), who mentions the city's great number of builders, especially ship-builders, and carpenters in the 4th century BC. Apparently, Sinope was also overgrown with olive trees, which offered its people the opportunity to develop olive oil production. Apart from reports in several authors (Strabo 12. 3. 12; Athenaeus *Deipnosophistae* 2. 54D), there is also a gravestone commemorating Manes the oil-merchant

[93] See also Polybius 4. 56.

[94] Roebuck 1984, 47, 116, 122–23. Curtis (2005, 31) also suggested that the development of the fisheries resources of the Black Sea (in general) may have been a prime motivation behind the Greek colonisation of the region, perhaps as early as, or before, the 7th century BC. *Contra* Rostovtzeff and Ormerod 1932, who stress the role of the area's minerals.

[95] Maksimova 1956, 68; Doonan 2002, 188–89; 2004, 95–96; Doonan *et al.* 2016.

[96] For the main reasons for Greek colonisation in the southern Black Sea, see Part 4.

[97] E.g. Doonan 2004, 95.

[98] For fish salting in the Black Sea, see Curtis 2005, 37–38.

(ἐλαιοπώλης), dating from the Classical period.[99] This has led to a justifiable linking of Sinope's oil production with the city's great production of amphorae, which is well attested by archaeological evidence.[100]

Finally, Sinope was also renowned for its exports of Cappadocian *miltos* (red earth), which was considered among the best in the world (Strabo 12. 2. 10; Theophrastus *De Lapidibus* 52–53; Pliny *NH* 35. 12–14; Vitruvius *De architectura* 7. 7).[101] It was appreciated for its multiple medical qualities, including the treatment of wounds and diseases in many parts of the human body (Pliny *NH* 32; 34). Furthermore, *miltos* was especially valuable in painting, since it was a natural red pigment, one of the four main ones (Pliny *NH* 30). By the Roman period Cappadocian *miltos* had become so famous that it was called *sinopis*, form the place where it was first discovered. Lumps of it were red, but speckled on the outside. Pliny mentions that painters in earlier times employed it to give a glow. There were three kinds of *sinopis*, the red, the faintly red and the intermediate. The price of the best sort in the Roman period was two dinars per pound and it was used for painting with a brush or for colouring wood (Pliny *NH* 31). It is worth noting that, according to Theophrastus (*De Lapidibus* 52), one of the several kinds of *miltos* comes from mines, since iron mines also contain red earth. Since a large part of the littoral to the east of Sinope was known for its iron mines (see below), one cannot help but suspect that the Cappadocian *miltos* of Sinope was of this sort.

After a large bay to the east of the Sinop peninsula, which curiously enough seems to have been quite difficult to traverse by land until recently,[102] the next place that attracts one's attention is undeniably the Halys (Kızılırmak), the largest of all the rivers crossing the littoral. The Halys, which took its name from the salt pans along it (Strabo 12. 3. 12), is frequently presented in ancient Greek literature as a political boundary (usually Paphlagonia's eastern border),[103] which sometimes becomes a symbolic one, when authors refer to peoples, areas or events on its near or far side.[104] This is easily understandable, considering that its broad valley is the only wide natural corridor between the Black Sea and the Anatolian hinterland – one might say the only escape route through

[99] French 2004 = *IK Sinope* 28.

[100] Maksimova 1956, 35, 67, 69; Doonan 2002, 192–94; 2004, 95; Braund 2005c, 122; Mitchell 2005, 85, 90; 2010, 92, with more bibliography. See also Rempel and Doonan 2020.

[101] Roebuck 1984, 122. For the variety of *miltos'* uses, see Robinson 1906, 142–43. For its composition, see Maksimova 1956, 79–80.

[102] Bryer and Winfield 1985, 89.

[103] See Section 2.6.1.

[104] For example, Herodotus 1. 6. 1, 28. 3; 5. 102. 4; Thucydides 1. 16; Strabo 2. 5. 24; 6. 4. 2; 11. 1. 7; 12. 1. 3, 3. 1, etc.; and many other authors. See more recently Bekker-Nielsen 2014, with bibliography.

the endless Pontides and the only break in their monotony. Herodotus says (1. 72. 3), tellingly, that 'the Halys river cuts off nearly the whole of the lower part of Asia, from Cyprus to the Black Sea. This is the neck of all this land; an unencumbered man needs five days to traverse its length.' As will frequently be demonstrated in the next parts of the book, this natural north–south route, which leads into the heart of Anatolia, became from very early on a cultural route, and the many settlements that were created along it have provided archaeological evidence linking them with several civilisations. Thus the Halys valley became a melting pot of local Anatolian, Greek, Phrygian and Persian cultural elements.

The estuary of the Halys creates a fairly large peninsula, with several lakes on both sides. Its eastern side marks the beginning of the next bay, upon which the city of Amisos would be founded. To its east, the same landscape is repeated, thanks to the estuary of the littoral's second largest river, the Iris (Yeşilırmak). In fact, from here on the alternation of smaller bays, headlands and river mouths will be the main characteristic of the coast as far as its eastern end.

Between the mouth of the Halys and Amisos was Gazelonitis, a fertile plain – rare in the eastern part of the littoral – where everything was grown. It also had a sheep-industry, raising flocks clothed in skins and yielding soft wool, of which there was a great scarcity throughout the whole of Cappadocia and Pontus (Strabo 12. 3. 13).[105] An equally large and fertile plain surrounded the mouth of the Iris, to the east of Amisos; its name was Themiscyra and it was traversed by the Thermodon. This was the place where the mythical Amazons lived. Describing this plain, Strabo notes that

> it is always moist and covered with grass and can support herds of cattle and horses alike and admits of the sowing of millet- and sorghum seeds in very great, or rather unlimited, quantities. Indeed, their plenty of water offsets any drought, so that no famine comes down on these people, never once; and the country along the mountain yields so much fruit, self-grown and wild, I mean grapes and pears and apples and nuts, that those who go out to the forest at any time in the year get an abundant supply – the fruits at one time still hanging on the trees and at another lying on the fallen leaves or beneath them, which are shed deep and in great quantities. And numerous, also, are the catches of all kinds of wild animals, because of the good yield of food (12. 3. 15).[106]

Let us dwell a moment on the nuts. The word used by Strabo is καρυώδη, which literally means 'walnut-like'. There has always been a debate as to

[105] It has been proposed that Lycastus, the name of the river and the city just to the east of Amisos, attest to the frequency of wolves (*λύκοι*) in the region (Nollé 2005, 63).

[106] See also Apollonius *Argonautica* 2. 970–84. *Cf.* Magie 1950, 182.

precisely what kind of nut is meant, especially since several ancient authors have talked about the 'Pontic nuts'. These authors usually mean the hazelnut (e.g. Pliny *NH* 15. 88), though this is not always absolutely clear. For example, we read in Athenaeus (*Deipnosophistae* 2. 41) that

> Nicander mentions also the nuts of Pontus, which some writers call λόπιμα ('peelable', i.e. which have a skin and not a shell); while Hermonax and Timachidas, in the *Dictionary*, say that the acorn (βελανίδι) of Jupiter is what is called the nut of Pontus.

Apparently, the 'Pontic nuts' were quite well-known in antiquity, but it was not always certain what kind of nuts they were. Hazelnuts and acorns are fairly similar, and these authors were not necessarily deep connoisseurs of everything they wrote about. Walnuts, however, are quite different. Besides, any reference to nuts in Pontus, like Strabo's specific one, does not necessarily pertain to the so-called 'Pontic nuts', the exact origin of which is never actually given. Hazel trees were and indeed are abundant in the area around Kerasous (Giresun),[107] but so too are walnuts.[108] We make no claim to be sure about the kind of the nuts grown in the Themiscyra plain.

3.3. To the East of the Thermodon (Fig. 8.3)

In most cases, the descriptions of the natural landscape of this long stretch of our littoral in ancient literature followed the authors' reports on specific local peoples and their territories.[109] Since there are several discrepancies regarding the exact part of the littoral that each of these peoples possessed, and a consequent vagueness about the position of many of them, one cannot be always sure where exactly several specific landscape descriptions encountered in the sources should be placed on the map, unless the description is accompanied by reference to a settlement. For the rest, it is only the wider area that can be assumed, which is of little use, given the general uniformity of the landscape the east of Cape Heracleios.

The last plain to the east was Sidene, fertile, but not well-watered like Themiscyra (Strabo 12. 3. 16). It is followed immediately by the large 'double' peninsula of Iasonion (Yason Burnu) and Genetes. The landscape from this point eastwards to the border with Georgia is fairly undifferentiated. There are many rivers but no notable plains, since the rivers are small and the steep

[107] See the sources in Magie 1950, 1073–74. *Cf.* Bryer and Winflied 1985, 128.

[108] Magie (1950, 1074) talks about hazelnuts *and* walnuts, not just the former, as Bryer and Winflied (1985, 128), who invoke him.

[109] All the indigenous peoples that are mentioned here are examined separately and in detail in the next part of the book.

mountains hug the coast, so that not much of the land is tilled (Strabo 12. 3. 19). These mountains are the main feature of the eastern part of the littoral. Here, carpentry appears to have been one of the most important occupation of the local peoples.[110] Indeed, one of them, the Mossynoikoi, were named after the wooden houses and towers they lived in. Wood was also used by these peoples in the making of several kinds of weapons, as well as ships (Xenophon *Anabasis* 5. 4. 10–12; Strabo 12. 3. 18). The villages of the indigenous peoples were built on deep ravines; in some cases 'the inhabitants could hear one another shouting from one town to the next, such heights and valleys there were in the country' (Xenophon *Anabasis* 4. 8. 2, 10; 5. 2. 4–8, 5. 31).

The mountains provided the locals with more than just timber, for the land between Capes Iasonion and Hieron Oros (nos. 49 and 61) was fabled for its mineral wealth.[111] Indeed, the very name of one of the indigenous peoples, the Chalybes, was used by the Greeks to denote steel, and the important city of Amisos was known for the export of the famed Chalybian steel.[112] A description of the qualities of the country's iron is found in Pseudo-Aristotle's *De mirabilibus* (833b. 48):

> It is said that the origin of Chalybian and Amisenian iron is most extraordinary. For it grows, so they say, from the sand which is borne down by the rivers. Some say that they simply wash this and heat it in a furnace; others say that they repeatedly wash the residue which is left after the first washing and heat it, and that they put into it a stone which is called fire-proof; and there is much of this in the district. This iron is much superior to all other kinds. If it were not burned in a furnace, it would not apparently be very different from silver. They say that it alone is not liable to rust, but that there is not much of it.

In the 19th century W.J. Hamilton visited the region, seeking landscapes that would corroborate the ancient sources' reports about the mines of the Chalybes. He was surprised by 'the remarkable change in the geological structure of the country' to the west of Cape Iasonion, around the ancient town of Oenoe (modern Ünye), in the mountainous hinterland of which he finally found what he thought he could match to the iron mines of the Chalybes.[113]

According to Strabo (12. 3. 19), in earlier times there were also silver-mines in this land. Actually, more to the east, near Tripolis, Arrian (*Periplus* 16; *cf.* Anonymous *Periplus* 36) mentions a settlement called Argyria, which most

[110] *Cf.* Mehl 1987, 116, 123; Lipka 1995, 72.

[111] Mines were also found in the area around Mt Olgassys (see above), in the Paphlagonian hinterland (Strabo 12. 3. 40).

[112] Roebuck 1984, 103, 122; Lipka 1995, 72.

[113] Hamilton 1842, 271–78. For the area's mineral wealth, see also Rostovtzeff and Ormerod 1932, 212; Magie 1950 II, 1068–70, with more bibliography; Maksimova 1956, 69–70; Roebuck 1959, 103; Herzfeld 1968, 128; Degnate 1973, 245–46; Bryer 1982, 135; Kaymakçi 2021.

probably denotes silver-mining activities.[114] Furthermore, Pliny noted that the stones of the wider region were most highly esteemed (like those of India), provided they were not of a mottled hue (*NH* 37. 126).

A bit more to the east, iron was succeeded by copper. We read in Pseudo-Aristotle's *De mirabilibus* (835a. 62) that the copper of the Mossynoikoi's land was said to be very shiny and white:

> not because there is tin mixed with it, but because some earth is combined and molten with it. But they say that the man who discovered the mixture never taught anyone; so the copper vessels which were made in earlier days have this distinction, but subsequent ones have not.[115]

Finally, according to Pliny (*NH* 37. 115), a deep blue variety of jasper was found near the Thermodon, a variety similar to the Indian one (which resembled emerald) in Amisos and a purplish blue one in Cappadocia, without further specification; he also failed to specify where exactly in Pontus alum was mined (*NH* 37. 52).[116]

Despite the limited arable land, the area did not lack crops and viticulture. Xenophon reports that in 400 BC, when the Greeks were plundering the strongholds of the Mossynoikoi, they found in the houses ancestral stores, as the Mossynoikoi described them, of heaped-up loaves, while the new corn was laid away with the straw, most of it being spelt. They also found large quantities of flat nuts without any divisions. From these nuts, by boiling them and baking them with a lot of grain, they made the bread which they used most.[117] Finally, they also found wine, 'which by reason of its harshness appeared to be sharp when taken unmixed, but mixed with water was fragrant and delicious' (*Anabasis* 5. 4. 27–29). When the Ten Thousand reached Trapezous, the city's inhabitants gave them hospitality gifts, among them wine (*Anabasis* 4. 8. 23). As the name of Kerasous, an important colony of Sinope, indicates,[118] the area abounded with cherry trees (Pliny *NH* 15. 102; Athenaeus *Deipnosophistae* 2. 51A; Ammianus Marcellinus 22. 8. 16, etc.).[119]

In all this region, its westernmost part, the country of the Tibarenoi, was said to have been flatter and less rugged, providing pasture for many flocks and herds (Xenophon *Anabasis* 5. 5. 2; Apollonius *Argonautica* 2. 377–378; Dionysius Periegetes *Orbis terrae descriptio* 767; Eustathius *Comm. in Dion.*

114 Hamilton 1842, 259; Bryer and Winfield 1985, 3, 139.

115 *Cf.* de Jesus 1978, 98, with previous bibliography.

116 Perhaps at Şebinkarahisar, according to Bryer and Winfield (1985, 3, 148), which is inland.

117 For this bread, see Section 2.11.3; Manoledakis 2021a, 58.

118 See more recently Manoledakis 2010a, 147; Düring and Glatz 2015b, 23. There was also a second city named Kerasous to the west of Trapezous. See Section 5.1, nos. 79 and 88 on Table 3 and Comment 14.

119 See also Baumann 2011, 98.

Per. 767. 1–3; *Paraphrases in Dion. Per.* 767. *Cf.* Nicephorus Blemmydes *Conspectus geographiae* 464. 2. 15–16). Indeed, A. Bryer and D. Winfield refer to the advantages of the country east of Cotyora (Ordu), observing that it provides plenty of rich arable and pasture land for the support of the city.[120] Oxen, as well as barley-meal, were also mentioned in the area of Trapezous (Xenophon *Anabasis* 4. 8. 23–24).[121]

Moving to the easternmost part, especially noteworthy is the production of a special kind of honey, which was allegedly poisonous. Just before reaching Trapezous, Xenophon's soldiers had a very unpleasant experience:

> The swarms of bees in the neighbourhood were numerous, and the soldiers who ate of the honey all went off their heads, and suffered from vomiting and diarrhoea, and not one of them could stand up, but those who had eaten a little were like people exceedingly drunk, while those who had eaten a great deal seemed like crazy, or even, in some cases, dying men. So they lay there in great numbers as though the army had suffered a defeat, and great despondency prevailed. On the next day, however, no one had died, and at approximately the same hour as they had eaten the honey they began to come to their senses; and on the third or fourth day they got up, as if from a drugging (*Anabasis* 4. 8. 20–21).

It seems that the honey of this general area and the symptoms it caused in those who ate it were well known in the region down the ages. For Strabo (12. 3. 18) did not omit to mention that thanks to it the Mossynoikoi, who lived in the neighbourhood, cut down three maniples of Pompey's army when they were passing through their mountainous country: they mixed bowls of this honey, which is yielded by the tree-twigs, and placed them in the roads, and when the soldiers drank the mixture and lost their senses, they attacked them and easily disposed of them. According to Pliny (*NH* 21. 77), this honey was called *maenomenon* (crazing) from its effects. Its poison was supposed to have been extracted from the flowers of the oleander, which grew plentifully in the area.[122]

Contrary to what one might assume, people in this part of the littoral, the most mountainous of all, did not ignore fishing. In fact, it was in the coast around Pharnakia (Giresun)[123] and Trapezous that the pelamydes were first caught, according to Strabo (7. 6. 2; 12. 3. 19). The geographer then added

[120] Bryer and Winfield 1985, 8–9. *Cf.* the green lakes mentioned by Valerius Flaccus (*Argonautica* 5. 147).

[121] *Cf.* Mehl 1987, 124; Lipka 1995, 71.

[122] On the 'mad honey' and the efforts that have been made to find its sources (e.g. chestnut trees or tobacco fields) and side effects, see Lehmann-Haupt 1931, 816–22; Bryer and Winfield 1985, 4; Lendle 1995, 286–87; Mayor 1995; Lane Fox 2004, 36–46; Rood 2005, 213–14; Baumann 2011, 117. Generally for honey in the Black Sea, see also Lipka 1995, 72.

[123] For the identification of Pharnakia, see Section 5.1, Table 3, no. 80 and Comment 15.

that the pelamydes were hatched in the marshes of Lake Maeotis, and after they had gained a little strength they rushed out through the mouth of the lake in schools and moved as far as Trapezous and Pharnakia. However, the catch there was not considerable, because the fish had not yet grown to their normal size. But when they reached Sinope, they were mature enough for catching and salting, hence Sinope's wonderful pelamyda fisheries (see above).

Furthermore, the ancient sources inform us of the abundance of dolphins in the Black Sea. Xenophon reports that in the houses of the Mossynoikoi he and his soldiers found slices of dolphin salted away in jars, and in other vessels dolphin blubber, which 'the Mossynoikoi used in the same way as the Greeks use olive oil' (*Anabasis* 5. 4. 28). Several centuries later, Strabo described how the dolphins were caught: they pursued the schools of fish, mainly cordyle,[124] tunny-fish and pelamydes (see above), and grew fat on them. Thus they become easy to catch, also because they tend to approach the land. According to the geographer, the people there are the only ones who cut up the dolphins, which are caught with bait, and used their abundance of fat for all purposes (12. 3. 19). It seems that the dolphins were the only large marine species in the Black Sea (Aristotle *Historia animal*ium 8. 13; Pliny *NH* 9. 50).

Table 2. The natural resources of the southern Black Sea littoral and their spatial distribution. In the first column appear all those that are mentioned in ancient literature, by category: bovidae and their products, other animals, agricultural products, starting with cereals, trees with fruit or appropriate for carpentry, plants, fish, mines and minerals. The last row gives the places where each of them appeared in the 1st millennium BC. In some cases the sources mention a wider area or people's country, in others they are more specific.

	Bithynia	**Calpe**	**Heraclea**	**Salona**	**Tieion**	**Cytoron**	**Amastris**	**Paphlagonia**	**Sinope**	**Halys**	**Gazelonitis**	**Amisos**	**Themiscyra**	**Kerasous**	**Chalybes**	**Argyria**	**Tibarenoi**	**Mossynoikoi**	**Trapezous**
Oxen													×						×
Sheep		×									×						×		×
Gazelles											×								
Cheese				×															
Horses													×						
Wild animals													×					×	

[124] Which was also the name of a town to the west of Trapezous (no. 90 on Table 3 in Section 5.1).

	Bithynia	Calpe	Heraclea	Salona	Tieion	Cytoron	Amastris	Paphlagonia	Sinope	Halys	Gazelonitis	Amisos	Themiscyra	Kerasous	Chalybes	Argyria	Tibarenoi	Mossynoikoi	Trapezous
Flour									×										×
Wheat		×																×	
Barley		×																	
Sorghum													×						
Millet		×											×						
Sesame		×																	
Legumes		×																	
Vegetables									×										
Honey																		×	×
Vines		×							×				×					×	×
Olives			×				×		×										
Cherries, berries			×											×					
Figs		×																	
Apples													×						
Pears													×						
Nuts									×				×					×	
Timber	×	×						×	×									×	
Plane trees			×																
Box trees						×													
Maples									×										
Aconitum			×																
Fish (edible)			×		×		×		×					×	×				×
Dolphins															×			×	
Salt pans										×									
Mines	×														×	×			
Stones												×	×				×		
Miltos									×										

3.4. Summing Up

Taking all of the above into account, one can easily comprehend that the southern Black Sea littoral was a strip of land appreciated by people in the 1st millennium BC for the many natural gifts it offered, despite its straightness and the many difficulties in communication resulting from the terrain. Apparently, the huge mountains that embrace the littoral did not oppress the communities that happened to live there, but rather landscaped a long and narrow land of opportunity,[125] as can be deduced with a look at Table 2 and Figs. 8.1–3, but also from the export activities of cities like Heraclea and Sinope over a. long period of time. The words of the 2nd-century BC historian Polybius, that the Pontus furnished all that people require for the support of life (4. 38. 3), although referring to the whole of the Black Sea could easily describe the situation of its southern side in particular.

The mountainous slopes, apart from being so rich in timber and minerals, support many fruit-bearing trees, while in summer they are overgrown with verdure, offering fine pasturage for cattle (Fig. 9).[126] And we should not forget the flora, which, as Bryer and Winfield have pointed out, 'is much too rich and varied to be described in detail'.[127] The volcanic period of Pontic geological history produced quantities of mineral-bearing ores of different kinds.[128] Thus north Anatolia became a quite significant area for metallurgy from the Neolithic period, as indicated by several relevant products that have been excavated.[129]

Among the several agricultural opportunities and the subsequent trade benefits, one should certainly stress the olives and the vines. Finally, there was the abundance of fish, which was always appreciated in the literature.[130] Apart from the importance of fishing for Greek colonisation (see above), it has even been suggested that the appearance of some fish species on the coins of cities like Heraclea and Sinope implies their dependency on marine resources.[131] Besides, the littoral's wealth of fish could have resulted in the creation of several installations for fish processing, such as watchtowers.[132]

[125] *Cf.* Rostovtzeff and Ormerod 1932, 212.
[126] Hewsen 2009, 38.
[127] Bryer and Winfield 1985, 5–6. *Cf.* Hamilton 1842, 250–51.
[128] Bryer and Winfield 1985, 3.
[129] Bilgi 2001; Dönmez 2006, 33–38; Yalçın 2008.
[130] For Galen's references to fishing in the Black Sea and especially in the southern littoral, see Wilkins 2005, 26.
[131] Stolba 2005.
[132] Højte 2005, 138, and 156 on a possible tank at Harmene, close to Sinope.

The production and export of all these products inevitably benefited the manufacture of another category of product, namely pottery. Indeed, a huge number of trade amphorae from the most significant export cities of the littoral, like Heraclea and Sinope, have been discovered at ancient sites on and around the Black Sea.[133] Their content, which has frequently been the subject of research, could certainly have been olive oil or wine, but it could also have been fish.[134]

[133] See more on the production of amphorae in Section 4.4.3. For the export potential of the littoral, see Braund 2005c.

[134] Lund and Gabrielsen 2005.

PART 2

THE INDIGENOUS PEOPLES

1. Introduction

In this part of the book we will be dealing with the second protagonist of historical geography: the people who inhabited the natural environment outlined in the previous part and created the history of the southern Black Sea littoral in the 1st millennium BC; the people whose activity was determined by the opportunities provided by the environment and the limitations imposed by it, but who at the same time also exploited the environment and affected it, especially through their urban activities.

More specifically, we will be dealing with the indigenous peoples of the southern Black Sea littoral. But who is considered indigenous? Since no people has always lived in the same place and, conversely, very few peoples totally cease to exist in an area, it is difficult to define the term. One of the commonly accepted objective criteria for considering a people indigenous is 'the habitation of the country, or a geographical region to which the country belongs, at the time of conquest or colonisation or the establishment of present state boundaries'.[1] This definition is based on a comparison, which is apparently marked by the arrival of a new people via conquest or colonisation. The latter term is decisive when speaking of the southern Black Sea in the 1st millennium BC, because Greek colonisation in the Black Sea region – part of the huge movement designated the Second Greek Colonisation, which lasted for almost four centuries – was a determinative event that not only changed the anthropogeography and the history of the southern Black Sea littoral once and for all, but also became the inevitable occasion of most of our knowledge of the area, since the vast majority of our written sources, as indicated in the Introduction Section 2.2, were Greek.

So despite the fact that most of the peoples examined in the following sections are not among the earliest known inhabitants of the littoral, it is these that will be presented here as the indigenous peoples of the area in the 1st millennium BC, when the Greeks came and colonised the region – and gradually became locals themselves. The Greek presence will be examined separately, as will also the presence, more or less probable, of other peoples (e.g. the Cimmerians, the Scythians, the Persians and the Phrygians), which, however, was in any case different from the Greek case.

[1] See Article 1 of the Indigenous and Tribal Peoples Convention, 1989 (No. 169) of the United Nations' International Labour Organization. See more at www.un.org and www.ilo.org.

One of the particularities of the southern Black Sea littoral is that, despite its narrow breadth, it was inhabited by many different peoples throughout the 1st millennium BC. The exact time of their appearance is in most cases unknown, but it is usually connected to the general causes and events that led, among other things, to the fall of the Hittite Empire. In one of the classic chapters of modern literature on the beginning of the Iron Age in Anatolia, R.D. Barnett writes:

> The Hittite Empire collapsed in ruins in about 1200 BC [...] and a horde or series of hordes flooded over the land [...]. When the curtain rises again, central Anatolia is ruled by an invading people, a horse-rearing military aristocracy called the Phrygians, or the Mushki and Tabal.[2]

Barnett may be referring to central Anatolia and not our area of interest, but his phrase 'when the curtain rises again' perfectly matches the southern Black Sea littoral and many other areas as well, in fact even better than central Anatolia itself, where, as Barnett himself says, the Phrygians seem to have arrived not long after than the Hittite collapse.[3] For the phrase indicates a complete and lengthy gap in our historical knowledge,[4] which in our case may extend as far as the 8th century BC. An additional problem with our area, which may be due to its natural isolation from the rest of Anatolia, is that it was not previously possessed by the Hittites, but by other peoples, among which the Kaska seem to have played an important role. Thus, things are much more complex here than the, even if simplistic,[5] replacement of the Hittites by the Phrygians.

Our knowledge of the Kaska is desperately scant. Modern scholarship is not unanimous on even such basic elements as whether they were Indo-Europeans[6] or not,[7] or whether they came from the east[8] or the north-west.[9] Similar uncertainty attaches to the degree of their contribution to the collapse of the Hittite

[2] Barnett 1975, 417. The relation between the Phrygians and the Mushki and the Tabal is disputable. Nonetheless, the latter have sometimes been associated with some of the peoples we are going to examine. Barnett 1975, 420–21. See also Section 2.10.3.

[3] Barnett 1975, 418.

[4] Which has resulted in the use of the disputable term 'Dark Ages' (e.g. Dönmez 2006, 15; Tsetskhladze 2012b, 236). *Cf.* Goetze 1957, 186; Genz 2011, 332. For the transition from the Bronze to the Iron Age in central Anatolia, see Genz 2003.

[5] Goetze 1957, 186.

[6] Mellaart 1978, 62.

[7] Luraghi 1998, 172.

[8] Luraghi 1998, 172.

[9] Mellaart 1978, 62. It has also been suggested that they were autochthonous people (Dönmez 2019, 453–54).

Empire[10] and the borders between them and the Hittites.[11] What is beyond doubt is the considerable trouble they caused to the Hittites,[12] as is visible through the Hittite records, the only sources we have on these people, which consequently give us a one-sided image of them.[13]

Another discrete area of the littoral in the 2nd millennium BC appears to be the land called Pala. Geographically, this land roughly coincides with the territory of the later Paphlagonia.[14] The language that was spoken by the people of Pala, termed Palaic, was a 2nd-millennium BC Anatolian language related to the Hittite and Luwian tongues, but practically nothing is known about it. Relations between the people of Pala and the Hittites are supposed to have been uninterrupted until the Kaska people invaded and occupied the southern Black Sea littoral, where the Palaic people lived; the latter were thus displaced by the Kaska and actually disappear from the historical sources.[15] According to one theory, the people who became known as Paphlagonians in the 1st millennium BC sprang from the Palaic populations that survived the Kaska occupation and were mixed with other local elements of northern Anatolia, and the very name of the Paphlagonians evolved from the word *pala*.[16]

How and when most of the peoples examined in elow emerged in the area is vague. Was it a result of the Hittite collapse or of later migrations, in the late second or early 1st millennium BC? Whatever the case may be, since this book focuses on the 1st millennium BC, we shall examine the peoples dwelling on the southern Black Sea littoral in that period of time, who thus were

[10] Mellaart (1978, 72) thinks that they were the main reason for it. *Cf.* Beal 2011, 596; *contra* Moreu (2003, 122), who suggests that the advance of the Kaska was helped by the defeat of the Hittites by the Mushki.

[11] See, for example, an indicative map in Mellaart 1978, 58–59. *Cf.* Glatz and Matthews 2005; Dönmez 2002, 275–76; 2006, 14. For the Hittites, see also Seeher 2011; Beal 2011. Maksimova (1951) has argued that Hittite objects had reached the Black Sea shore.

[12] Glatz and Matthews 2005, 47; Matthews 2009a, 10; Glatz, Matthews and Schachner 2009, 112; Beal 2011.

[13] As has correctly been pointed out, things would have been presented differently if there were written sources from the Kaska as well. But nothing is preserved from these people, neither texts nor identifiable products of material culture. See Glatz and Matthews 2005, 49–63. The authors suggest that the Kaska-Hittite relations were much more complex than a mere war; they talk about efforts to make peace, even about close political, commercial and social dealings between them, as well as intermarriages, and provide earlier bibliography on the matter. More recently on the Kaska, see also de Boer 2015, 73–74. For the one-sided information that we have on the Kaska, see also Dönmez 2019, 449.

[14] See Section 2.6.1.

[15] Forrer 1921, 24; Ruge 1949, 2511, 2521; Maksimova 1956, 19; Mellaart 1978, 53; Melchert 1994, 10; Finkelberg 1997, 5; Luraghi 1998, 169–72; Popko 2008, 60–62; Taracha 2009, 58–59; Bryce 2006, 140; 2009, 552, who claims that the land of Pala was subject to the Hittites; *contra* Ruge 1949, 2515.

[16] See more in Section 2.6.2.

indigenous when the Greek colonists arrived. As we shall see, the curtain might have risen at different moments of the millennium in different parts of the littoral, but in most cases it is ancient Greek literature that raised it for us.

The privileged but at the same time isolated southern Black Sea coast appears to have been inhabited by about 20 peoples; in fact, the number refers to the names of peoples mentioned in the written sources, some of which may, however, have shared kinship with others. None of them has left any identifiable product of material culture, with the possible exception of the Paphlagonians and some burial monuments in their and neighbouring areas, where, however, speculation prevails over certainty.[17] Thus, ethnicity is not archaeologically identified in what is in any case an archaeologically unexplored region.[18] What we are left with, therefore, given the additional absence of local texts or inscriptions, and most probably of local languages with a script, is ancient Greek and Latin literature.[19] That is why for most peoples we do not even know what they called themselves.[20]

Consequently, one cannot claim to be able to compile the history of each people. What is examined in this part is, rather, the data, however minimal, we have for their identities – who they were, where they came from, which area of the coast they possessed – and some of their anthropological features. And since our sources for all these aspects are essentially Greek or Greek-influenced, this examination is hampered by the significant drawback of a one-sided, subjective image. Thus, one reason why, where possible, some sections contain material on the manners and customs of the peoples presented or the Greeks' perception of them is to remind the reader that our knowledge of these peoples comes from texts written by Greeks, in particular by highly educated Greeks; texts, moreover, which were written after the Persian Wars, which according to a widely accepted theory played a crucial role in the Greeks' conception of their ethnic identity as well as in the decisive change in the meaning of the word *barbarian*.[21]

[17] See Section 2.6.2.

[18] For the problems and the risks in trying to trace ethnicity through archaeological data, see Antonaccio 2001, 124–33. *Cf.* Jones 1997, 124. See more recently Tsetskhladze 2012a; 2012b; 2014, 316, with a discussion on the relevant problem 'How do we excavate ethnicity?' For the general problem of the ethnic identification of the Paphlagonians, see also Maksimova 1956, 19; Manoledakis 2021c.

[19] In the vast majority of cases the latter repeats data already provided by the former.

[20] The only certain exception is the Macrones, who are reported by Xenophon to have told him what their name was (see Section 2.13.3).

[21] See, for example, bibliography in Hall 2002, 175. Also, Hornblower 1991, 11. For a different approach, see Mitchell 2007, 77–112; Hornblower 2008, 38–39; Vlassopoulos 2013, 35–36.

But this is not the only difficulty we encounter in attempting to examine the indigenous southern Black Sea peoples. In fact, this is at least a difficulty of which we are conscious, and we can proceed accordingly. The problems begin with the most basic elements, such as defining the territory of each people. Discrepancies regarding the borders between two peoples as given by different sources can be frustrating. And while it is already far from unnatural to read about one river as a border between the territories of two peoples in a 6th century BC source and a different one in a 3rd-century BC text, it is when the discrepancies appear between sources of the same period and relate not only to where a border between two peoples lies but also to their spatial relation to each other that things get complicated.

The most striking example is the geographic relation between the Tibarenoi and the Chalybes. In his *Anabasis*, written probably sometime around 370–360 BC, Xenophon clearly mentions the Tibarenoi as dwelling in the land to the west of the Chalybes, while the *Periplus* compiled by Pseudo-Scylax only a few decades later has the Chalybes west of the Tibarenoi.[22] In other cases, we find the Colchians, for example, possessing the eastern part of the coast as far to the west as the Mossynoikoi in the *Anabasis*, while in other sources several other peoples are mentioned between them.[23] For some of these discrepancies more or less plausible solutions may be proposed;[24] for others, the mystery remains.

One of the factors that could serve to explain some issues of this kind is that in terms of geographic spread the names of the peoples do not always coincide with the names of the territories that were called after them. Paphlagonia is a case in point. For reasons which are examined in the homonymous section,[25] the name Paphlagonia frequently ended up denoting a geographical area larger than the one inhabited strictly by the Paphlagonians – at least according to how both names are presented in the written sources. Consequently, important cities and rivers, like Sinope and the Halys, are placed by different writers in the territory of different peoples: for example, in one case in the country of the Paphlagonians and in another in the land of the Leucosyroi. This might sometimes have to do with the influential culture of a specific people, or with administrative territories, as these were formed especially in the Roman period, inhabited by more than one people. There are instances, especially in later sources, where a distinction is made between the geographical and national

[22] For the dating of the two works, see in the Introduction 2.2. For the issue of the geographic relation between the Chalybes and the Tibarenoi, see Sections 2.9.1 and 2.10.1.

[23] See Sections 2.14 and 2.15.1.

[24] As in the latter case; see Section 2.15.1.

[25] Section 2.6.

meanings of names.[26] But we should always keep in mind that most of the later authors were often confused, attempting to reconcile all the different versions of the earlier sources they were aware of.

The examination of a people's territory, which is usually treated first in every section in this part of the book, includes information about its natural environment and resources, as well as the types of settlements and the general urban environment, when this is possible. Thus, on the eastern and central parts of the coast we encounter similarities in local urban environments,[27] such as, for example, the existence of wooden houses, towers and acropoleis, which do not appear to be characteristic of the settlement model in the western part.

Stress is laid also on the origin of the several peoples. This is another difficult matter, since we often have to rely on mythical traditions, not necessarily created by the peoples in question themselves. The emergence of most peoples goes back to this early part of history, which is totally shrouded by dense mist as regards any kind of evidence. Nonetheless, for a rough picture one could divide the littoral into three main parts, describing the western one, as far as the River Billaeus or Parthenius, as having been overrun by peoples of Thracian origin and therefore coming from the west-northwest. The central one, roughly as far as the Iris or the Thermodon, seems to have been occupied mainly by peoples formed from surviving populations of local Bronze Age tribes mixed with peoples moving in from south-east Anatolia; a determinative factor here must have been the access provided by the Halys valley, a natural corridor between north and south. Finally, the eastern part, where almost two thirds of the littoral's peoples lived, is usually presented as inhabited by nations connected to peoples arriving from the north-east, such as the Scythians and the Colchians (Fig. 10.1–3). Unlike the western and central parts of the littoral, where the state of affairs was more or less established before the middle of the 1st millennium BC, it seems that in the eastern part things continued to be fluid even until the last century of the millennium, when a significant migration of tribes from the Caucasian coast is supposed to have come to further change this situation.

Naturally, the origin of each of the littoral's peoples is not attended by the same degree of certainty. This is true of other aspects as well; as we shall see, for some of them their presence on the seaboard, or even their very existence, is questionable. One should keep in mind that some of the peoples are mentioned only in poetic works and mostly in mythical contexts and their real

[26] A characteristic example is provided by the *Etymologicum Gudianum* (*s.v.* Ἅλυς), where we read that the Halys was a river in Paphlagonia separating the Chalybes from the Leucosyroi.

[27] *Cf.* Doonan 2019.

existence is not confirmed by any preserved historical or geographical work. Such is the case of the Philyres and perhaps also the Enetoi. Therefore, it is probable that they were just products of poetic imagination. On the other hand, there are peoples who may actually have existed in Anatolia, but either their alleged expansion as far as the Black Sea is doubtful, like the Bebryces, the Mares and the Choi, or one cannot be certain which part of the littoral they actually inhabited, which is the case of the Caucones. Nonetheless, given the fragmentary preservation of the ancient sources, we cannot be absolutely certain about any presumption. Consequently, such peoles are properly examined in their (supposed or real) geographical location according to the texts, but on the relevant maps (Fig. 10.1–3) their names are followed by a question mark.

Any information about the relations of indigenous peoples with others, whether neighbours or Greeks, is taken into consideration, as are any traditions about the migration of some of them to the west, to Greece or the Adriatic Sea. Gathering all the elements presented above, an attempt is made to shape an impression of the indigenous peoples that is as close to reality as possible – an effort that, considering the afore-mentioned difficulties and limitations, all too often results in a very fragmentary and subjective image – sometimes, I would say, even fictitious. In my opinion, among the most valuable aspects of this image are the anthropological data, which, although always emanating from Greek or Latin sources, tend to distinguish one people from another and particularise their identity. These data also prove that after the Greek colonisation these peoples 'retained some of their own social, economic, cultural and political institutions', as required by the definition of indigenous peoples cited in the beginning. Similar is the contribution of accounts of social or economic life, which unfortunately are meagre. What one can also observe through the relevant records is how the Greeks evaluated different familiar or unfamiliar behaviours, by seeing which foreigners are described as 'valiant', 'barbarous', 'warlike', 'rough', 'fair', 'weird', etc.

For the rest, political affairs and military events along the littoral appear almost exclusively in sources concerning the history of the Persian Empire or other foreign powers, and thus will be examined in the relevant sections. It is noteworthy that, with very few exceptions, among them the Bithynians and the Paphlagonians, the indigenous peoples of the southern Black Sea littoral did not play an important political role in their area or leave their mark on the ancient history of Anatolia, and many of them would be totally unknown to us were it not for a reference to them in one or two surviving texts.

When and how the end of those peoples came about is also unclear. From the written sources it seems that almost half of them still existed in the Roman period, while others must have been assimilated earlier. By the 6th century

AD, however, many of them seem to have disappeared, as is implied by the use of the word πρώην (former) by the anonymous author of the *Periplus* (32–38) composed in that century in the five cases of local peoples that he mentions (Chalybes, Tibarenoi, Mossynoikoi, Macrones/Macrocephaloi, Becheires).[28] This happens to be the period by which it has been suggested that 'the Greek language had triumphed over the various indigenous tongues of western and central Anatolia'.[29]

2. The Bithynians

Bithynia was the first country that one would encounter when entering the southern Black Sea littoral from the west, after clearing the Bosporus (Apollonius *Argonautica* 2. 345–348, 619–620; Strabo 12. 3. 2; Dionysius Periegetes *Orbis descriptio* 793–797; Procopius *De bellis* 8. 2). Unlike almost all the other peoples that inhabited the littoral, the Bithynians have the privilege of being shadowed by very little doubt concerning either the coastal boundaries of their territory or their origin.

2.1. The Territory (Fig. 10.1)

The western (natural) border of the Bithynians is given, and there is little discrepancy in the ancient sources concerning their eastern boundary, after which dwelt the Mariandynoi: it was formed either by the Sangarius (Strabo 12. 3. 7; *contra* Arrian *Alexexandri Anabasis* 1. 29; Arrian, in Eustathius *Comm. in Dion. Per.* 793 and *Comm. ad Hom. Il.* 2. 754, who included the river in Bithynia) or by the Hypius (Pseudo-Scylax 91), which are not far from each other.[30]

The only issue here relates to the possible presence of the Bebryces on the littoral. As we shall see below, the Bebryces have sometimes been mentioned as occupying part of the littoral to the west of the Mariandynoi, with whom they had several conflicts over an 'iron-bearing land' (Apollonius *Argonautica* 2. 137–153), which is assumed to have been the area between the Sangarius and the Hypius.[31] But these references occur only in mythical contexts, which

[28] For the exception of the Paphlagonians, see Section 2.6.4B.

[29] Vryonis 1971, 48. *Cf.* Janse 2002, 356–57.

[30] In any case, Calpe was in Bithynia (Theopompus in Aelius Herodianus, *s.v.* Κάλπη). The statement that the Mariandynoi border the Bithynians after the Rhebas (Dionysius Periegetes 793; *Paraphrases in Dion. Per.* 793; Nicephorus Blemmydes *Conspectus geographiae* 35–37) does not help much, since this river lies well to the west.

[31] See in detail in Sections 2.3 and 2.4.

probably belonged to local traditions created in Heraclea, while in any case the Bebryces may have disappeared from the region by the 8th century BC.

The proximity of the two rivers permits a fairly good perception of the extent of the Bithynian territory along the southern Black Sea coast, while the small discrepancy could even be due to occasional shifts of the border after specific conflicts between the two neighbouring peoples.[32] The territory of Bithynia included one of the Black Sea's very few islands, Thynias (Nymphis in Schol. in Apol. Arg. 2. 672; Apollonius *Argonautica* 2. 350, 672–673; Pseudo-Scylax 92; Menippus *Periplus* 8. 10; Strabo 12. 3. 7; Pomponius Mela 2. 98; Pliny *NH* 6. 13; Aelius Herodianus *De prosodia catholica*, *s.v.* Θυνιάς; Anonymous *Periplus* 6; Schol. in Apol. Arg. 2. 177), which, according to Ptolemy (5. 1. 15) was also called Daphnousia.[33] It also included Calpe (Stephanus of Byzantium, *s.v.* Κάλπη), which until the 4th century BC seems to have been the only Greek settlement on the entire coastline between Byzantium and Heraclea (Xenophon *Anabasis* 6. 4. 2).[34]

As for Bithynia's southern borders, these must have included the peninsula created between the Black Sea, the Bosporus and the Olbian gulf[35] (e.g. Stephanus of Byzantium, *s.v.* Ἀστακός; Νικομήδειον; Χρυσόπολις), as well as the next peninsula to the south, covered by Mt Arganthoneion. The border proceeded eastward to Lake Ascania, Nicaea (Menecrates in Plutarch *Theseus* 26; Pliny *NH* 5. 43), the land to the north of Mt Olympus (Strabo 12. 4. 10, 8. 1, 8. 8, 8. 10; Constantine Porphyrogenitus *De thematibus Asia* 4. 13; Nicephorus Blemmydes *Conspectus geographiae* 465. 1. 2–4) and the southern course of the Sangarius, which ran parallel to the sea (Strabo 12. 5. 2; Pliny *NH* 5. 43).

This means that the city of Chalcedon lay within Bithynian territory, which could explain Ptolemy's unique designation of the Bithynians as Chalcedonioi (5. 1. 11); Ptolemy never refers to them as Bithynians, although, probably influenced by the names of the administrative provinces, he calls the whole land, from the Bosporus to Cytoron in the east, Bithynia. Otherwise, references to Chalcedonioi in the sources mean the inhabitants of the city or its vicinity, including Byzantium (e.g. the *Suda*, *s.v.* Χαλκηδόνιοι; Nicephorus Blemmydes *Conspectus geographiae* 464. 2. 44–45).

[32] After all, Pseudo-Scylax, who mentioned the Hypius as the border, is much earlier than authors of the Roman period.

[33] Thynias Island was sometimes mentioned as being situated opposite or slightly after the land of the Mariandynoi (Apollonius *Argonautica* 2. 350–352; Pomponius Mela 2. 98; Schol. in Apol. Arg. 155).

[34] For Calpe, see in Section 4.4.1.

[35] The Gulf of Nicomedia: Pseudo-Scylax 92.

According to Strabo (12. 1. 1), the eastern neighbours of the Bithynians, the Mariandynoi, lived to the west of Paphlagonia (12. 3. 2; *cf.* Arrian *FGrH* 2b 156, F. 76a, who places the Mariandynoi to the east of the Sangarius). Elsewhere (12. 3. 9, 42 and 12. 4. 1), however, Strabo says that the Bithynians border upon the Paphlagonians to the west together with the Mariandynoi, and in any case he calls the land to the west of Paphlagonia Bithynia. This means that the whole area to the west of Paphlagonia had ended up being called Bithynia (*cf.* Arrian *Periplus* 13), initially probably as an administrative area, a coastal north-eastern part of which was inhabited by the Mariandynoi. Hence the sea to the west of Heraclea was frequently called 'the sea of Bithynia' (e.g. Nymphis in Schol. in Apol. Arg. 2. 729). This might be another reason for the appearance of the Sangarius in later authors (see above) not as the eastern border of Bithynia but merely as a navigable river, the biggest in that country (Eustathius *Comm. in Dion. Per.* 793), while, they add, the Rhebas, in the west, had the fairest water to sweep over the land (Dionysius Periegetes 796; *Paraphrases in Dion. Per.* 793–798). It could also explain the fact that Bithynium, a city whose name denotes that it belonged to Bithynia (Pliny *NH* 5. 43), was mentioned by Constantine Porphyrogenitus (as Claudioupolis: *De thematibus Asia* 6. 15), as a 'metropolis of the Mariandynoi.[36] That Mariandynia was the earlier name of Bithynia, which was founded by Phoenix (Georg. Syncellus *Ecloga Chronographia* 185), is probably an error.

As far as the natural environment of the Bithynian territory is concerned, Xenophon provides a fairly detailed description of the area around Calpe,[37] perhaps with the idea of promoting its appropriateness as a site for a new colony (see below). Xenophon stressed its ability to accommodate 10,000 people, as well as its good harbour and the abundantly flowing spring of fresh water close to the seashore. Nor did he fail to praise the richness of the area in timber of various sorts, barley, wheat, beans of all kinds, millet and sesame, figs, an abundance of grapes that yielded a good sweet wine, 'in fact everything except olives' (*Anabasis* 6. 4. 3–7, 6. 6. 1). Much later, Athenaeus (*Deipnosophistae* 2. 50E–F) collected references from several authors, who focused in their works on specific types of fruit produced in Bithynia, including bush-cherries, mulberries, myrtle-berries and strawberries. Other later authors also praised the fertility of Bithynia, which appears to have been a blessed region, able to provide its inhabitants with plenty of opportunity to develop a variety of economic activities.[38] Thus it stands to reason that the Bithynian lands

[36] See also in Section 2.4.1. *Cf.* Section 2.6.1.
[37] The passage is cited in its entirety in Section 1.3.1.
[38] See more in Section 1.3.1.

contained many inhabited villages (*Anabasis* 6. 4. 6). Judging from Xenophon's narrative (*Anabasis* 6. 3. 1), the first Bithynian villages were situated about 30 stadia (5.5 km) from the sea, at least in the vicinity of Calpe. This means in a terrain not very high, with an altitude of less than 100 m, but still not coastal. However, there must have been a coastal road from Calpe at least to the border with Mariandynia (see above), since this was the route chosen by the Spartan general Cheirisophus on his way from Heraclea to Calpe (*Anabasis* 6. 2. 18, 6. 3. 10).[39] Besides, later authors mention navigation as one of the occupations of the Bithynians (Eustathius *Comm. in Dion. Per.* 793).

The coastal part of Bithynia, being to a large extent uninhabited contrary to the close hinterland, provided enough space for the founding of a new city, which apparently occurred to some Greek soldiers, but once again the plan did not get off the ground (*Anabasis* 6. 4. 7).[40] Before it was abandoned, however, it seems to have winged its way afar, for lo and behold we read further along in the narrative that very soon

> market products came in from the Greek cities on all sides, and people coasting past were glad to put in, since they heard that a city was being founded and that there was a harbour. Even the hostile peoples who dwelt nearby began now to send envoys to Xenophon, for they heard that he was the man who was making a city of the place, to ask what they must do in order to be his friends (6. 6. 3–4).

2.2. On the Origin of the Bithynians

Let us now see who the people were that possessed this well-endowed country. The many and unanimous references of the ancient sources to the origin of the Bithynians leave no doubt that they were a Thracian nation, which spread from the Balkans into north-west Anatolia by crossing the Hellespont. Already Herodotus calls them 'Bithynian Thracians' (1. 28), or 'Thracians of Asia' (3. 90), or just 'Thracians' (7. 75). So do Thucydides (4. 75), Pherecydes (Schol. in Apol. Arg. 2. 181), Xenophon (*Hellenica* 1. 3. 2; *Anabasis* 6. 4. 2) and, directly or indirectly, all the authors who have dealt with the Bithynians,[41] and their territory was thus frequently called 'Bithynian Thrace' (e.g. Xenophon *Hellenica* 3. 2. 2). Herodotus further explains that the Bithynians took that name after crossing into Asia, whereas before they had been called Strymonians,

[39] For the Bithynian settlements, see Section 5.3. For the roads in Bithynia, see Section 5.5.

[40] Similar thoughts also occurred elsewhere to some Greeks (e.g. *Anabasis* 5. 6. 15–33). For this event, see Section 4.4.3.

[41] Pseudo-Scylax 92; *Oracula Sibyllina* 11. 176; Diodorus 14. 38. 3; Strabo 7. 3. 2, 12. 3. 2, 3; Arrian *Anabasis* 1. 29. 5; *Periplus* 13. 6; Galen *De purgantium medicamentorum facultate* 11. 336. 15; Porphyrius *Quaestionum Homericarum ad. Il.* 13. 3; Schol. in Apol. Arg. 97, 141; Schol. in Dion. per. 323; Schol. in Thucyd. 4. 75.

for they dwelt upon the Strymon.[42] From there, according to their own account, they had been driven out by Mysians and Teucrians (Herodotus 7. 75), who, before the Trojan War, crossed the Bosporus into Europe and conquered all the Thracian peoples (7. 20).

Nonetheless, judging from Strabo's words (12. 4. 8), it seems that before Herodotus, Scylax of Caryanda had affirmed, and in fact was the first to do so, that Bithynia was first inhabited by Mysians. If this is the case, then the Mysians who campaigned in Europe were most probably those living in the later Bithynia, where the people they had driven from their land (Strymon) ended up. Strabo further complicated things by noting that, 'according to most writers, the Bithynians, though formerly Mysians, received this new name from the Thracians (the Thracian Bithynians and Thynians) who settled the country in question' (12. 3. 3), and twice clarifying that the Mysians themselves must have been Thracians (12. 3. 3, 12. 4. 8).

Much simpler is the version narrated by Appian (*Mithridatica* 1), according to whom:

> the Greeks think that the Thracians who marched to the Trojan War with Rhesus fled to the outlet of the Euxine sea at the place where the crossing to Thrace is shortest. Some say that as they found no ships they remained there and possessed themselves of the country called Bebrycia. Others say that they crossed over to the country beyond Byzantium called Thracian Bithynia and settled along the River Bithyas,[43] but were forced by hunger to return to Bebrycia, to which they gave the name of Bithynia from the river where they had previously dwelt; or perhaps the name was changed by them insensibly with the lapse of time, as there is not much difference between Bithynia and Bebrycia. So some think. Others say that their first ruler was Bithys, the son of Zeus and Thrace, and that the two countries received their names from them.

Thus, Appian seems to ignore the Mysian (and Teucrian) campaign and connect the Thracian crossing to Asia with just the Trojan War. He also identifies Bithynia with Bebrycia geographically, in fact in a way that presents the Bithynians and the Bebryces as kindred.[44]

The connection, even identification, between the two peoples is also encountered in several later sources (Nicephorus Blemmydes *Phil. et Theol. Conspect. geogr*. 465. 1. 1–2), where we also read that Bithynia was formerly

[42] A piece of information which was adopted by Hesychius (*Lexicon*, *s.v.* Στρυμόνιοι), Stephanus of Byzantium (*s.v.* Στρυμών) and partially also Pliny (*NH* 5. 40). The latter, however, elsewhere (*NH* 4. 18) places the Thynians more to the east, in the area between Rhodope and the Hebrus.

[43] Perhaps the Bathynias river (Strabo 7. 1. 56; Pliny *NH* 4. 18; Ptolemy 3. 11. 6), close to Byzantium?

[44] On this matter, see more in Section 2.3.

called Bebrycia (Sallust *Histories*; Servius *Aeneidos commentarius* 5. 373; Georg. Syncellus *Ecloga chronographica* 212; Scholia in Oppianum Haleutica 1. 618). Moreover, the mother of Amycus, king of the Bebryces, was a Bithynian nymph (Pseudo-Apollodorus 1. 9. 20; Hyginus *Fabulae* 17).

Whatever the exact reasons for the movement of the Thracians to Asia Minor and their settlement in Bithynia may have been, the Thracian origin of the Bithynians appears to be certain. In fact, if Strabo's comment (12. 3. 3) that 'the Bithynians were actually Mysians that received their new name from the Thracian Bithynians and Thynians who settled the country in question' is correct, then his implication is that there were already Bithynians in European Thrace, just as there were Thynians. After all, the Bithynians' relation with the Thynians appears to be a major issue in their history. The Thynians were also a Thracian people, who – both they and their land (Thynia or Thynias) – are encountered together with the Bithynians in ways that leave it unclear whether they were the same nation or not.

Herodotus mentions the Thynian and Bithynian Thracians (1. 28) as if they were two different groups of Thracians, as does Strabo (7. 3. 2, 12. 3. 3),[45] while Callisthenes (Schol. in Apol. Arg. 2. 672) seems to have pointed out that the land and the island (see above) were called Thynias by the Greeks and Bithynia by the barbarians, which, however, is not confirmed by any other source.[46] Arrian has reportedly (Eustathius *Comm. in Dion. Per.* 793) attempted to shed some light on the issue: from the straits of Bosporus to the Rhebas (so in a rather narrow region) dwelt the Bithynians, while the mountainous land beyond (to the east of) the Rhebas was occupied by the Thynians, up to the River Cales, to the west of Heraclea.[47] However, the difference in size between the two 'neighbouring' territories is huge, unless the reference to the 'mountainous land' meant that a coastal strip was still Bithynian, which is highly unlikely.[48] In any case, Arrian's narrative clearly differentiates between

[45] And also Porphyrius (*Quaestionum Homericarum ad. Il.* 13. 3). *Cf.* Stephanus Diaconus *Vita Stephani* 27. 3.

[46] However, in another script of the specific fragment *Bi* is omitted from the word *Bithynia*, and the name used by the barbarians is *Thynia*. If the first version is correct (*Bithynia*), then we have in the commentary an equivalent to Xenophon's account, where the Bithynians are mentioned in Asia and the Thynians in Europe (see below, Section 2.2.3). The discrepancy may just be due to the fact that later authors tended to present such combinatory theories, precisely because they were aware of many different versions.

[47] *Cf.* also Pseudo-Scymnus 977, who mentions the Sangarius as crossing Thynias and flowing to the sea through it.

[48] The perception of the Rhebas as a border might have derived from Apollonius' words in *Argonautica* 2. 788–789 (see below). Domitius Callistratus is reported (Stephanus of Byzantium, *s.v.* Ψίλιον) to have mentioned a river called Psillion as the natural border between Thynia and

Thynians and Bithynians, adding that they were named after the prominent brothers Thynos and Bithynos, sons of Phineus (hence their Thracian origin).[49]

We should keep in mind that Thynias was also the name of a coast and a promontory in the European part of Thrace, to the south of Apollonia Pontica (Hecataeus *FGrH* 1a 1 F 166; Apollonius *Argonautica* 2. 177(?), 460; Schol. in Apol. Arg. 2. 178, probably quoting Hellanicus; Pseudo-Scymnus 728; Strabo 7. 6. 1, 42. 9; 12. 3. 3; Arrian *Periplus* 25; Ptolemy 3. 11. 4; Anonymous *Periplus* 87–88); its territory seems to have extended to the south as far as the Propontis and the Hellespont (Xenophon *Anabasis* 7. 2. 22, 7. 4. 2–5, 14, 18, 22; *cf.* Apollonius *Argonautica* 2. 529).[50]

Our area of interest was usually called Bithynia, in order to differentiate it from this European Thynias, for example in the *Argonautica* (2. 347, 619); however, some authors used the name Thynis to denote Bithynia (e.g. Pseudo-Scymnus 977; Anonymous *Periplus* 7). Memnon called Bithynia or part of it 'the Thynian territory' (*Hist. fragm.* 9. 4), probably meaning the geographic area as opposed to Bithynia as the name of the kingdom, while he specified that Zipoetes, ruler of Bithynia, ruled over Thynia in Thrace (9. 5). Conversely, the European Thynia was sometimes called Bithynia (Asclepiades fragm. 4. 17).

Much later, in the 12th century, Anna Comnena spoke about the 'country round Bithynia and Thynia' referring to the area of Propontis and Nicomedia (*Alexiad* 3. 11. 1, 4), and the description of her contemporary Eustathius (*De capta Thessalonica* 30. 26) is similar. It is obvious that the two place and ethnic names were never considered as denoting two clearly unrelated peoples,

Bithynia, which however was not the better known Psillis river in Bithynia that flows into the Black Sea.

[49] Nevertheless, right afterwards (*Comm. in Dion. Per.* 809) Eustathius cites another passage from Arrian, according to which Thynos (who founded Thynia) and Mysos were sons of the Nymph Arganthone (after whom the mountain in Mysia was named). Apollonius (*Argonautica* 2. 485) also referred to a Nymph Thynias, who however is always mentioned as the wife of Rhesus, not Phineus (Aelius Herodianus *De prosodia catholica* and Stephanus of Byzantium, both *s.v.* Ἀργανθών and Ἀργανθώνη; Parthenius *Narrationes amatoriae* 36; Asclepiades fragm. 1b. 18–20). This is just another example of genealogical myths of eponymous heroes with several variants that have appeared and predominated in various periods for various reasons. According to one such, Phineus had another son, Mariandynus (Schol. in Apol. Arg. 2. 140), which has led even to the spotting of a resemblance between the Mariandynoi and the Bithynians (Strabo 12. 3. 4).

[50] The Ladepsoi and the Tranipsoi, who are mentioned by Theopompus as nations of the Thynians (Aelius Herodianus *De prosodia catholica* and Stephanus of Byzantium, both *s.v.* Λαδεψοί καὶ Τρανιψοί) must have belonged to these European Thynians, judging by Xenophon's reference to them in the *Anabasis* (7. 2. 32). Dionysius Periegetes (*Orbis descriptio* 730) also mentions a people called Thynians in the area of the Caspian Sea, who are otherwise unknown.

even by those who mentioned them separately. Therefore, the Scholiast of Apollonius is not absurdly confused by the poet's words in *Argonautica* 2. 177, where the Argonauts, having left the land of the Bebryces, arrive at 'the opposite lying Thynian land'. He claims that it is unclear whether the poet here means the European or the Thracian Thynia, 'for both are opposite the land of the Bebryces',[51] but concludes that Apollonius must rather have meant European Thynia, since Phineus, whom the Argonauts meet there, lived in that place, close to Salmydessus (Schol. in Apol. Arg. 2. 177). Indeed, the Argonauts have just made their way through the Bosporus to arrive in Thynia and meet Phineus (2. 168), but since in order to reach the southern Black Sea coast (2. 317–318) they are going to have to pass between the Cyanean Rocks, which are in the northern part of the Bosporus, this necessarily places Thynias in 2. 177 in the European part (western Black Sea).[52]

What remains undisputed is that the Bithynians (and the Thynians) were Thracians.[53] It is, therefore, somewhat surprising that Pliny and Pausanias present several other theories regarding their descent: that they were Arcadians from Mantineia (Pausanias 8. 9. 7), for example, or that their country was 'at first called by the name of Cronia, after that, Thessalis and then Malianda and Strymonis, whose people were by Homer called Halizones, from the fact that it was a nation begirt by the sea' (Pliny *NH* 5. 40). While it is not unusual for Pliny to give several names for an area, this association with Homer's Halizones and the interpretation of their name agree with the account given by Arrian (in Eustathius *Comm. ad Hom. Il.* 2. 857)[54] of a people that constitute a very ambiguous case.[55]

But when did the Bithynians come to Asia Minor? As already mentioned, Herodotus relates that the Bithynians came to Asia from the area of the Strymon, where they were known as Strymonians. From that region they had been driven out by the Mysians and Teucrians (Herodotus 7. 75). A little earlier in the same book (7. 20), the historian mentions the expedition of the Mysians and Teucrians before the Trojan War, in which the two nations crossed the Bosporus into Europe and, after conquering all Thracian peoples, pressed

[51] See Section 2.3 (Mysia is meant here).

[52] One must not, of course, forget that geography in verse should not be taken too literally. See, for example, Manoledakis 2015b.

[53] According to the Christian chroniclers, the Bithynians belonged to the genealogy of Ham (Hippolytus *Chronicon* 132. 18; Epiphanius *Ancoratus* 113. 3; *Chronicon Paschale* 58. 12; Georg. Syncellus *Ecloga chronographica* 52. 14).

[54] Which was adopted by several late lexicographers and commentators: *Etymologicum Genuinum*, *Etymologicum Gudianum*, *Etymologicum Magnum*, *Etymologicum Symeonis*, Pseudo-Zonaras *Lexicon*, all *s.v.* Ἁλιζῶνες; Schol. in Homer. Il. 2. 856.

[55] See in Section 2.9.5.

forward till they came to the Ionian Sea, while sweeping southward as far as the River Peneus (7. 20). This places the Bithynian migration before the Trojan War, but with no greater precision than that, for Herodotus does not say how long before the war the Mysian and Teucrian expedition to Thrace took place. Appian, on the other hand, as we have seen, cites the Trojan War itself as the occasion of the Thracian migration to Bithynia.

During the speech of welcome that Lycus addresses to the Argonauts when the latter find themselves in the land of the Mariandynoi, to the east of Bithynia, he recounts the deeds of the brave Heracles, among them his subjugation of 'the tribes of the Bithynians and their land, as far as the mouth of the Rhebas' (Apollonius *Argonautica* 2. 788–789).[56] This places us in a time period before the 8th century BC – indeed, before the 12th – even if the Asiatic part as far as the Rhebas is very small, but the narrative still indicates a passing of the Bosporus. However, this is just a mythical tale, at least part of which must have been based on traditions created in Heraclea,[57] and one must not seek real historical sequence in myths.

Conversely, Arrian wrote in his Bithynian history (*Bithyniaca*) that the Thracians, under the leadership of a certain Pataros, passed from Europe to Asia when the Cimmerians were invading Asia, driving them out and taking possession of their land, Bithynia (Eustathius *Comm. in Dion. Per.* 322); Arrian's words, therefore, point to a dating in the 8th century BC. The Nicomedian historian is a reliable, if not the most reliable available, source regarding Bithynia.[58] Herodotus, on the other hand, admits that he is repeating what the Bithynians themselves say about their early history, and they would have good reason to place their coming to Asia as early as possible. An intermediate report presents the Bithynians as occupying Bebrycia, thereafter called Bithynia,[59] having crossed over from the Strymon in 972 BC (Eusebius = Jerome *Chronicon* 123).

None of this information permits a safe conclusion regarding the time of the Bithynians' emergence on the western side of the southern Black Sea littoral. E. Meyer considered that the Thynians inhabited the Black Sea coast on both sides of the Bosporus,[60] their migration to the Asian side occurring progressively rather than as a major population movement. The same holds true for the Bithynians, who gradually expelled the Bebryces. While Meyer's theory

[56] See also in Section 2.3.

[57] See also in Sections 2.3 and 2.4.1.

[58] *Cf.* Dana and Dana 2014, 27–32; Dana 2016, 219–23, with the discussion of the dating of the work and a modern bibliography on it in n. 282 there.

[59] *Cf.* Section 2.3.

[60] Meyer, *RE* III.1 (1897), *s.v.* Bithynia, 513–14.

seems at first sight to argue for a distinction between Thynians and Bithynians as two different peoples, he hastened to clarify that for him the assumption that the two peoples were actually one and the same was perfectly possible, in which case the two names would merely indicate a geographic differentiation. Whatever the case may be, it is in any case true that, as Meyer points out, the name of the Bithynians prevailed, at least as far as the political history of the two peoples is concerned, irrespective of whether they were actually the same people or not. Meyer's view seems entirely plausible, and the only specification one might make is that the Thynians and the Bithynians may have been two kindred branches of East Balkan Thracians who moved into northwest Asia Minor in the 1st millennium BC, probably in or around the 8th century.[61] The Bithynians may have completely abandoned Europe while the Thynians did not, the part of the latter group that passed into Asia becoming identified with or even incorporated into the land of the Bithynians.

2.3. The Bithynians in the Eyes of the Greeks

Bithynia stands out among the districts of the southern Black Sea littoral as the one whose history from the Hellenistic period on, and especially in the Roman era, is the best known and most studied, primarily through the written and epigraphic (less so through archaeological) sources. Personal names, descriptions of events and institutions, and diplomatic and military relations with Macedonians and Romans provide a fairly representative overview of the history of the Bithynians, as opposed to the rest of the indigenous peoples that inhabited the littoral in the 1st millennium BC, including the Paphlagonians, who in that respect come second and, one might say, last.[62]

Apart from the surviving ancient sources containing multiple references to Bithynia and the Bithynians, most of which are cited in this section, it happens that we are aware of several ancient authors who focused specifically on this area, sometimes, apparently, in laboriously painstaking works which are quoted by later authors, although the works themselves have been lost. Thus, we know of the poet Demosthenes, who composed a *Bithyniaca* comprising at least ten books, which is considered a source for Alexander Polyhistor (2nd–1st centuries BC), who also wrote about Bithynia; and of others who were themselves Bithynians, like the grammarian Asclepiades, who wrote a *Bithyniaca*, also in at least ten books, Nicander of Chalcedon, who wrote about Prusias and another work relating adventures of (probably but not certainly) other

[61] Meyer 1913, 689; *RE* III.1 (1897), *s.v.* Bithynia, 510–11; Prêteux 2005, 249.
[62] See Section 2.6.

Bithynian kings, and of course Arrian of Nicomedia, whose *Bithyniaca*, in eight books, is the source of some of the information presented here.

Given the consequent adequacy of the examination of Bithynian history in modern literature,[63] we shall limit ourselves here to an adumbration of some anthropological aspects of the ancient Bithynians, especially as they seem to have appeared in the eyes of the Greeks, who, after all, produced the majority of the texts by which we are introduced to them.

The first historical data relating to the Bithynians after their establishment in Asia Minor concern their subjugation to the Lydians under Croesus in the mid-6th century BC (Herodotus 1. 28). We meet them next in the 5th century BC, now incorporated into the Persian kingdom and fighting in the Persian Wars; their equipment is described by Herodotus (7. 75):

> They wore the skins of foxes upon their heads, and about their bodies tunics, over which was thrown a long cloak of many colours. Their legs and feet were clad in buskins made from the skins of fawns; and they had for arms javelins, with light targes, and short dirks.

The description matches the image of barbarian warriors like the Thracians, to whom the Bithynians were, after all, kindred. This perception of the Bithynians would be reinforced some decades later, this time through the account of an eyewitness: Xenophon records that between Byzantium and Heraclea there was no other city, either friendly or Greek, except for the harbour called Calpe at the mid-point of the voyage between the two cities (Xenophon *Anabasis* 6. 4. 2–3); there were only Bithynian Thracians. Indeed, this image did not change until the Roman period, although there are references to Greeks from Heraclea dwelling in the island of Thynias (Pseudo-Scylax 92; Pseudo-Scymnus 1026;[64] Anonymous *Periplus* 6). Xenophon seems to provide, even if unintentionally, an explanation for this absence of Greek settlements: 'The Bithynians are said to abuse outrageously any Greeks they may find shipwrecked or may capture in any other way' (Xenophon *Anabasis* 6. 4. 2).

In fact, not much after the entry of the Ten Thousand into their territory, in three divisions, the Bithynians made their intentions clear by attacking them, aided by a force of horsemen sent by Pharnabazus, the Persian satrap of Hellespontine Phrygia (*Anabasis* 7. 8. 25), in order to prevent the Greeks from entering their province (*Anabasis* 6. 4. 24).[65] During the battles the Bithynians proved themselves to be both sturdy warriors and tough negotiators, refusing

[63] More recently and indicatively: Marek 2003; Bekker-Nielsen 2008; Winter and Zimmermann 2013.

[64] Only in Diller (1952, 176); not in Müller (1855, 237).

[65] See also in Section 3.3.

to give the hostages demanded by the Greeks (Xenophon *Anabasis* 6. 3. 1–9, 6. 4. 23–5. 32. *Cf. Hellenica* 3. 2. 3–5). Later, too, when the Greeks had crossed into European Thrace, Xenophon did not fail to mention the rumours following the local Thynians, who 'were said to be the most warlike of all men, especially by night' (*Anabasis* 7. 2. 22).

Whether or not they were aware of Xenophon's accounts of the cruel Bithynians and Thynians, several other ancient authors missed no opportunity to highlight this savagery (e.g. Diodorus 12. 82); even in the Hellenistic period descriptions of battles where either the Bithynians killed their enemies (e.g. Memnon 9. 2; 20) or their king (Prusias I) subjected a city (Heraclea) to a severe siege and killed many of the besieged (Memnon 19. 2), were not rare. In fact, the Bithynians were sometimes reported even to have turned against Pharnabazus (Xenophon *Hellenica* 3. 2. 2), forcing him to take steps against them (Diodorus 14. 38).

The rough and warlike nature of the Bithynians, at least until the 4th century BC, has sometimes been explained as reflecting the fact that, since there were no Greek cities along their northern coast, the people were not influenced by the Greeks (Greek culture?).[66] But there were multiple contacts between Bithynians and Greeks in other parts of Bithynia, such as on the coasts of the Bosporus and the Gulf of Nicomedia, not to mention the fact that other indigenous peoples further east along the coast who did have contact with the Greeks did not seem to have altered their manners because of it, as we shall see.

But were the Bithynians really so cruel, or is it rather that what we have here is a (typical for Classical Greece) image of the barbarous 'other' who, especially having fought the Greeks in the Persian Wars, can only be uncivilised?[67] The latter is very probable. In fact, if we consider what (according to Athenaeus *Deipnosophistae* 6. 271B) Phylarchus says about the Byzantines' exploitation of the Bithynians as helots, one might even suspect that on occasion it was actually the Greeks of the coastal cities who oppressed the local Bithynians.[68]

On the other hand, there are many indications of 'Greek influence' from the Hellenistic period onwards. This influence concerns primarily the Bithynian rulers from the early 3rd to the early 2nd century BC (Zipoetes I, Nicomedes I, Ziaelas, Prusias I) and is manifest in their habit of founding cities, usually

[66] Meyer, *RE* III.1 (1897), *s.v.* Bithynia, 514. *Cf.* Scholten 2007, 18–19.

[67] *Cf.* Section 2.6.4.

[68] *Cf.* Scholten 2007, 18–20, who explains the state-building activities of the Bithynian kings later on as precisely a response to this pressure.

through synoecism, some of which were named after them.[69] The populations of these cities would have included both Greek and Bithynian elements, and it has been argued that no population was dominant or subordinate, in the model known as 'Peer Polity Interaction'.[70] The merger of Greek and Thracian elements in Bithynia is conspicuous in the mythical genealogical traditions as well, for example the one describing the eponymous ruler Bithys as the son of Zeus and Thrace.[71] In addition, we know that the Bithynian calendar of months included Greek month names.[72]

Such factors might sometimes have led to the creation of a perception of the Bithynians and/or the Thynians that was diametrically opposed to Xenophon's account of them. For example, Nicholas of Damascus (fragm. t127) says that 'the Thynians welcome the shipwrecked people and make them their friends, while from the strangers, they honour those who come inadvertently, but punish those who come willingly'; something that is not corroborated, even indirectly, by any other source.

It would be frivolous, however, to ascribe these things to a policy of philhellenism on the part of the Bithynian rulers,[73] who seem never to have renounced their Thracian origins. Polybius' thumbnail sketch of Prusias II is revealing in this regard:

> King Prusias was an ill-favoured man, and though possessed of fair reasoning power, was but half a man as regards his appearance, and had no more military capacity than a woman; for not only was he a coward, but he was incapable of putting up with hardship, and, to put it shortly, he was effeminate in body and mind through his whole life, a defect that no one, and least of all Bithynians, like to see in a king. In addition to this he was most incontinent in satisfying his sensual appetites; he was entirely a stranger to literature, philosophy, and all such studies, and generally speaking had no notion whatever of what goodness and beauty are, but lived by day and night the barbarous life of a Sardanapalus. So that all his subjects, the moment they saw the least chance of success, became irrevocably resolved not only to throw off allegiance to the king, but to exact punishment from him' (Polybius 37. 7; *cf.* Diodorus 32. 19).

Apart from Prusias' wantonness and profligacy, Polybius' description reveals how detestable this way of life was to the average Bithynian and, at the same time, how barbarous to the Greeks, who compared it to the life of

[69] For the Hellenistic character of the Bithynian kingdom, see indicatively Hannestad 1996; Scholten 2007. *Cf.* also Dana and Dana 2014, 30; Dana 2016, 218.

[70] Scholten 2007, 17. For the Peer Polity Interaction model, see Renfrew and Cherry 1986.

[71] See above, Section 2.2.2. *Cf.* Dana 2016, 222.

[72] See more recently, for example, Russell 2016, 177.

[73] As Scholten (2007, 21–22) correctly pointed out, providing bibliography for the 'philhellenism interpretation'.

Sardanapalus. And if for Polybius, who, after all, was not the average Greek, the most important reasons for criticising the Bithynian king seem to have been his sensual appetites and his unfamiliarity with literature, philosophy and beauty, his words suggest that what was worst for the Bithynians was his cowardice and effeminacy.

The barbarous nature of the Bithynians – always according to the Greeks – was further reflected in their celebration of the 'trieterides', three days during which they drank continuously and engaged in any kind of sexual intercourse (the *Suda*, *s.v.* τριετηρίδες). We know what a shocking custom open intercourse was in the Greeks' eyes, thanks to Xenophon's words on the Mossynoikoi.[74] Moreover, polygamy appeared to have been accepted among Bithynians, as might be implied by Arrian's reference (in Eustathius *Comm. in Dion. Per.* 322) to Dolonkos, eponymous hero of a Thracian nation and half-brother of Bithynos (Stephanus of Byzantium, *s.v.* Δόλογκοι, Βιθυνία, Θράκη), who had many children from many wives.

Also according to Arrian (Eustathius *Comm ad. Hom. Il.* 3. 275), the Bithynians used to dispense justice facing the sun, as if the god Helios was presiding. He further related that the Bithynians celebrated the 'nekysia', the days of the dead, during which they sought to communicate with deceased relatives (Eustathius *Comm ad. Hom. Od.* 9. 65), and that they called Zeus Papas on the mountains (Eustathius *Comm ad. Hom. Il.* 2. 110).

To sum up, the Greeks' image of the Bithynians was not substantially different from their typical image of the barbarian stranger, at least until the Classical period. Later, however, this image would be gradually modified, but only partially, for reasons that are usually ascribed to long co-existence with the Greeks of the coastal cities. Such reasons include, for example, the penetration of Greek cults into Bithynian religious life, as reflected in the names of some months and in numismatic iconography.[75]

It could also be argued that another factor in the improvement of the Greeks' initial image of the Bithynians was the appearance of certain men of letters and science who lived and worked in Bithynia in the Hellenistic period; apart from the authors named above, it is worth mentioning the physician Asclepiades and the mathematician, geographer and astronomer Hipparchus. Even if they were of Greek origin, the epithet 'Bithynian' that accompanied their names might have played its role.[76]

[74] See Section 2.11.3.

[75] Meyer, *RE* III.1 (1897), *s.v.* Bithynia, 523–24; Russell 2016, 177.

[76] Indicative is Scholten's (2007, 24) suggestion that Demosthenes the Bithynian (Stephanus of Byzantium, *s.v.* Ἥραια), a man with a Greek name and a Thracian ethnic (Βιθυνός Δημοσθένης), betokens multiple or merged identities as a characteristic of Hellenistic Bithynia. Like the

3. The Bebryces

In the first stages of the history of Asia Minor, after the 12th century BC, one of the peoples that are encountered in the written sources as having inhabited its north-western part are the Bebryces. Although most of the information obtained from these sources tends to place them on the eastern coast of the Hellespont and the southern coast of the Propontis, some references, mainly in a mythical context, present them as occasionally having spread to the westernmost part of the southern Black Sea littoral, as far as the River Hypius (Fig. 10.1). It is for this reason that we will be dealing with the Bebryces in this section, laying more stress on the specific references.

The first to speak of the Bebryces was Charon, a 5th-century BC historiographer from Lampsacus. His work is lost, but his references to the Bebryces survive in the works of later authors. In the most extensive of these passages, which is preserved by Plutarch (*Moralia* 255), we read of Mandron, king of the Bebryces, and his daughter Lampsace: The Phocaeans under the command of Phobus marched to the assistance of Mandron, king of the Bebryces of Pityussa,[77] who had been attacked by the neighbouring barbarians. As a reward for their service, Mandron granted the Phocaeans a part of the country and the city, and invited them to settle there. By their courage and conduct they had obtained many victories and had enriched themselves with great spoils, which made them first an object of envy and later also an object of fear to the Bebryces. Thus when Mandron went away on a journey, the Bebryces formed a resolution to massacre the Phocaeans. But his daughter Lampsace got wind of the plot and, as she could not prevent it, she secretly revealed it to the Greeks. The Phocaeans prepared a magnificent sacrifice in the suburbs and invited the barbarians to partake of it. They then divided themselves into two groups, one of which secured the walls and the other slew the banqueters and made themselves masters of the city. They afterwards rewarded Lampsace with honour and named the city Lampsacus after her (Plutarch *Moralia* 255; Polyaenus *Strategemata* 8. 37).

*

* *

Paphlagonians, the Bithynians have also been accused of being evil in Late Antiquity: for example, they were criticised for their pagan beliefs (Basil *Adversus Eunomium* 29. 505; Athanasius *Contra gentes* 23. 18) and difficulty in accepting the Christian teachings (Theodoretus *Interpretatio in xiv epist. St. Pauli* 82. 505).

[77] The earlier name of the city of Lampsacus (Strabo 13. 1. 18).

Another reference of Charon to the Bebryces is cited by the Scholiast of Apollonius (Schol. in Apol. Arg. 2. 2), who attributes to him the information that the land of Lampsacus was formerly called Bebrycia because it was inhabited by that people, who disappeared 'because of the wars'. Thus, Charon's accounts provide us with both a good geographic placement of the Bebryces and an aetiological myth for the city of Lampsacus, together with the name of a Bebrycan king. The placement of the Bebryces in the area generally known as Mysia, which includes Parion and Lampsacus, is corroborated by all the other sources that mention them.[78] Even after the disappearance of the Bebryces their memory remained strong in the area, especially in Lampsacus, as late as the Roman period, mainly due to the cult of Priapus.[79]

Some works, however, hint at an expansion of Bebrycan territory to the Black Sea coast. One of these is Apollonius' *Argonautica*, where we encounter the most extensive reference to this people in ancient literature. Their land lies on the Argonauts' way to Colchis and clearly to the south of the Propontis. Another of their kings, Amycus, son of Poseidon and Melia (*cf.* Hyginus *Fabulae* 17), was – according to the epic – the most arrogant of all men, who had laid down an insulting ordinance that no stranger should depart from his land until he had confronted him in a boxing match; and he had slain many of his neighbours.[80] Amycus challenged the Argonauts, too, but Polydeuces killed him and the rest of the Bebryces who rushed to avenge their king were put to flight by other Argonauts (*Argonautica* 2. 1–136). When the Mariandynoi and their king Lycus learned of the death of Amycus, they began destroying the vineyards and villages of the Bebryces, their steadings and farms, and driving off their countless sheep, because there had always been strife between them over an 'iron-bearing land' (*Argonautica* 2. 137–153).

Here there is an implication of contiguity between the Bebryces and the Mariandynoi,[81] who undoubtedly dwelt on the Black Sea coast, and thus an indication that the former also occupied part of the seaboard. Indeed, later in the poem, when the Argonauts have steered through the Bosporus and, after

[78] E.g. Apollonius *Argonautica* 2. 1–136; Theocritus *Idylls* 22. 27–29; Pseudo-Apollodorus 1. 9. 20; Valerius Flaccus *Argonautica* 4; Oppianus *Halieutica* 1. 618; Lucian *Dial. deorum*. 25; Dionysius Periegetes *Orbis descriptio* 804–806 and the *Paraphrases*; *Orphica Argonautica* 657; Eustathius *Comm. ad Hom. Il.* 1. 45; *Comm. in Dion. Per.* 805.

[79] See the detailed study of Prêteux (2005).

[80] Several authors of the 4th century BC (Scamo of Mitylene, Theophrastus of Ephesus, Cydippus of Mantinea, Antiphanes, Aristodemus and Aristotle) are reported by Clemens of Alexandria (*Stromata* 1. 76) to have mentioned the tradition that Amycus was the first inventor of boxing gloves. *Cf.* Euphorion fragm. 77; Eusebius *Praep Evang*. 10. 6. 10.

[81] See Section 2.4. For different versions regarding the Bebryces' territory, see also Dana 2016, 217–18.

passing the Sangarius river, reached the land of the Mariandynoi, the latter and their king Lycus welcome them warmly because they had fought their enemies, the Bebryces, who, according to Lycus, had long been severing great tracts of his land until they set their bounds at the meadows of the Hypius (*Argonautica* 2. 752–810). Thus we now have a concrete location for the Bebryces' presence on the coast: to the west of the Hypius, and so on its westernmost stretch. As for the 'iron-bearing land', which was said to be the apple of discord between them and the Mariandynoi, L. Robert has placed it between the Sangarius and the Hypius, in the area of Mt Çam Daği, where there are mines thought to have been worked in antiquity.[82] Robert concluded that the Hypius could have been the natural border between the Bebryces and the Mariandynoi, in an area rich in minerals that was claimed by both peoples.

Nonetheless, it was not Apollonius who introduced this tradition about the conflict between the Bebryces and the Mariandynoi, since Herodorus of Heraclea seems to have been aware of it already *ca.* 400 BC, as was his fellow countryman Nymphis in the 3rd century BC (Schol. in Apol. Arg. 2. 752). But the most important evidence for the Bebryces' possession of a stretch of the Black Sea coast was furnished by Theopompus (4th century BC), who, according to Strabo (12. 3. 4), wrote that Mariandynus (eponymous hero of the Mariandynoi)[83] ruled over a part of Paphlagonia, which was under the sway of many potentates. Among other lands he invaded and took possession of the country of the Bebryces, and he left the country that he had abandoned called after him. Furthermore, according to Pseudo-Apollodorus (2. 5. 9), Heracles was hosted by Lycus and took part in a battle between the Bebryces and the Mariandynoi, where he killed Amycus' brother Mygdon. Then he took much land from the Bebryces and gave it to Lycus, who called it all Heraclea (*cf.* Schol. in Apol. Arg. 2. 786).

Consequently, the tradition that the Bebryces had at some time dwelt in the western part of the southern Black Sea littoral had been certainly created by the late 5th century BC.[84] Nonetheless, the Bebryces are not mentioned in the *Periplus* of Pseudo-Scylax or any other geographic text or in Xenophon's *Anabasis* (Book 6) as actually living there. This would mean that their supposed possession of part of the Black Sea coast must have come to an end before the late 5th century BC, or even that it was merely a mythical tradition.

That the Bebryces, whose occupation of the land to the south of the Propontis is undoubted, were in any case one of the most ancient nations of

82 Robert 1980, 5–10.
83 See Section 2.4.2.
84 And not in the Roman period, as Prêteux (2005, 248) seems to suggest.

north-west Asia Minor is something that the sources tend to agree on. Indeed, they are supposed to have been living there well before the Trojan War. Dionysius of Halicarnassus (*Antiquitatum Romanarum* 1. 54. 2) relates that Aeneas had preserved Ilium from utter destruction at the time of its capture and sent Trojan allies safely to Bebrycia. Although such an incident is not mentioned by any other author, Photius (*Bibliotheca* 186. 12) does say that 'Ilos, who founded Ilion, defeated in battle the king of the Bebryces, named Byzos,[85] and raised Ilion up greatly.'

These may be mythical accounts, but Strabo, trying to reconstruct the historical past, claims that many changes took place in north-western Asia Minor after the Trojan War; 'for the parts round Cyzicus as far as the Practius were colonised by Phrygians and those round Abydus by Thracians; and still before these two by Bebryces and Dryopes' (13. 1. 8),[86] and also that 'the Bebryces took up their abode in Mysia before the Bithynians' (12. 3. 3). One might reasonably ask why, then, the Bebryces are not mentioned in the *Iliad*, for example in the Trojan Catalogue, where we find the Mysians instead? It might be because, as Strabo argued (12. 3. 27), Homer could scarcely have listed all the peoples and places of Asia Minor, since that was not his object, nor was there any need to do so, since many of them were in any case well known. Or perhaps it was, as Apollodorus is said to have claimed (Strabo 14. 5. 23), because the Bebryces were included, like the Doliones, among the Phrygians,[87] who do appear in the *Iliad*.

*

* *

It should be stressed that, whether the Bebryces extended to the Black Sea coast or not, they seem to have disappeared relatively early. Eratosthenes is reported (Pliny *NH* 5. 127) to have written that by his time many peoples had disappeared from Asia Minor, including the Bebryces. But this must have happened much earlier, and it is usually the Bithynians who are blamed for that. Apart from the above-mentioned passage in Strabo (12. 3. 3), this is also implicit in later sources, where we read that Bithynia was formerly called Bebrycia (Appian *Mithridatica* 1. 1; Sallust *Histories*, Serv. Aen. 5. 373; Georg. Syncellus *Ecloga chronographica* 212; *Scholia in Oppianum Haleutica* 1. 618). Further-

[85] *Cf.* Aelius Herodianus (*De prosodia catholica*) and Stephanus of Byzantium, both *s.v.* Βυσναῖοι.

[86] The Dryopes should be emended to Doliones, according to Leaf (1923, 61). *Cf.* Strabo 14. 5. 23.

[87] Maybe due to 'confusion' between Mygdon the brother of Amycus and Mygdon of Phrygia (Homer *Iliad* 3. 186)?

more, the Hypius, which Robert held to be a line dividing the Bebryces from the Mariandynoi, is presented by Pseudo-Scylax (*Periplus* 91) as the natural border between the latter and the Bithynians. Apollonius himself places the Bebryces only in Mysia, and after the Bosporus it is the land of the Bithynians that the Argonauts encounter, followed by that of the Mariandynoi. The events in the 'iron-bearing land' are supposed to have taken place earlier than the narrative time.

W. Leaf, studying Strabo's passages, attributed the big change in the population map of north-west Asia Minor after the Trojan War to the incursions of Thracian tribes from Europe through the Hellespont,[88] and specifically the irruption of the Bithynians, which also caused a dislocation of the Phrygians from the area around Lake Ascania, where they had been placed by Homer.[89] F. Prêteux argues that the replacement of the Bebryces by the Bithynians must have taken place in the 8th century BC,[90] while according to Leaf it was the Treres who invaded the Troad in this period.[91]

Leaf was also led to the conclusion that the Bebryces were one of those Phrygian nations that had been driven out of Bithynia, and based this theory *inter alia* on the fact that 'their name is apparently closely connected with that of the Phrygians; *Βέβρυκες*: *Βρύγες* almost exactly, and we may compare the *Βερέκυντες*, another legendary people'.[92] This seems to contradict Strabo's statement that the Bebryces were Thracians (12. 3. 3), who lived in the area before the Phrygians (13. 1. 8).

Thus we come to the issue of the Bebryces' origin. Were they Thracians, hence (among) the first Thracians that came to Anatolia from Europe, or Phrygians? Both seem possible and neither can be proven.[93] Indeed, for Strabo, the

[88] We previously mentioned a reference by the Scholiast of Apollonius (Schol. in Apol. Arg. 2. 2) to Charon's information that the Bebryces had 'disappeared because of the wars'. It is pobably the Trojan War that is meant.

[89] Leaf 1923, 60–61. *Cf.* Prêteux 2005, 247–49.

[90] Prêteux 2005, 249. *Cf.* W. Ruge, Bebrykes 2, *RE* III.1 (1897), 180. But if Plutarch's and Polyaenus' accounts (see above) refer to the foundation of Lampsacus as a Phocaean colony (see more recently, Morel 2006, 360–61), which took place in the 7th century BC (see Roebuck 1959, 113, with earlier bibliography), then we have a tradition that contradicts this dating. Moreover, during the speech of welcome that Lycus addresses to the Argonauts when the latter find themselves in the land of the Mariandynoi (see in Section 2.4), he talks about the subjugation of 'the tribes of the Bithynians and their land, as far as the mouth of Rhebas' by Heracles, which happened before the conflict between the Bebryces and the Mariandynoi (Apollonius *Argonautica* 2. 788–795), so the Bithynians cannot have come after the Bebryces and replaced them. See also in Section 2.2.2.

[91] Leaf 1923, 61.

[92] Leaf 1923, 62. On the comparison between Bryges and Berecyntes, see also Haas 1966, 19–21.

[93] See more in Prêteux 2005, 247–48.

two alternatives are actually not that different, since he considered the Brygians, and thus the Phrygians, to be a Thracian nation, like the Mygdones, the Bebryces, the Maedobithynians, the Bithynians, the Thynians and the Mariandynoi (7. 3. 2). The until recently long-lived and erroneous notion of 'Thraco-Phrygians', based to a degree on ancient passages like this one, unavoidably on occasion, although not frequently, influenced the study of the origin of the Bebryces as well.[94]

Only A. Baschmakoff attempted to develop the theory that the Bebryces were not immigrants from Europe but a local people of Asia Minor since the 3rd millennium BC, and more specifically kindred of the Paphlagonians.[95] He even attempted to use linguistic evidence to support this theory, but apparently with little success.[96] Neither the textual nor the archaeological evidence is able to lead to a firm conclusion concerning the Bebryces' origin; it just appears more plausible, taking together all the written evidence, that they had been a Thracian people.

*

* *

To return to the central question here, the Bebryces' possible occupation of Black Sea coastal territory, it seems that this must be excluded for the period after the 8th century BC, by which time they seem to have generally disappeared. But what about earlier in the Iron Age? Did they indeed inhabit the littoral or was that just a mythical tradition? If we look more closely at the texts pointing to a Bebrycan presence in the Black Sea littoral, we see that they all present a mythical context closely related either to the Mariandynoi or to Heracles. The Scholiast of Apollonius (Schol. in Apol. Arg. 2. 780–783a), slightly modifying Apollonius' narrative, reports that Heracles, going to the Amazons to seize Hippolyta's girdle, passed through the land of the Mariandynoi, who were beset by the Bebryces, whom he helped them defeat. Elsewhere in the Scholia we read that the Mariandynoi have always been in conflict with the Bebryces and had many times been defeated by them (Schol. in Apol. Arg. 2. 758).

Thus the view that ascribes these narratives to a local mythical tradition created in Heraclea[97] is quite plausible. It is in the area where Heraclea would be established that the Argonauts are warmly welcomed by the Mariandynoi in

[94] See Sakellariou 1958, 434.
[95] Baschmakoff 1948, 32.
[96] Prêteux 2005, 248.
[97] Delage 1930, 141; Vian and Delage 1974, 156–63; Robert 1980, 10.

the epos of Apollonius. Their king Lycus tells them of his admiration for Heracles, the great benefactor of the Mariandynoi and patron of Heraclea, in 36 lines (*Argonautica* 2. 774–810), before the feast begins. The conquest of a part of the Mariandynian land by the Bebrycan king Amycus is related precisely in this speech of Lycus. In this part of his poem Apollonius is considered, correctly I think, to have been using the works of important Heraclean scholars like Herodorus and Nymphis.[98]

It is beyond any doubt that the Bebryces were remembered for many centuries after their extinction, as their frequent appearance in the literature as late as the Late Byzantine period indicates.[99] Of special interest is Ptolemy's reference (5. 1. 13) to the inland cities of the Bebryces, which actually occupy most of Bithynia's hinterland.[100] Their early disappearance, in an obscure period marked by important events and with few sources, combined with the fact that most references to them (all dated much later) had a mythical context, resulted first in ambiguity and gradually in total confusion concerning their history, from their geographic area to their very identity. A confusion that is obvious in the Late Antique and Mediaeval texts, where the Bebryces appear occasionally as having occupied several parts of western Asia Minor, Bithynia, Mysia, the western part of the Black Sea coast, even Lydia, around Ephesus and Magnesia (Schol. in Apol. Arg. 2. 124),[101] or as being identified with the Bithynians (Nic. Blemmydes *Phil. et Theol. Conspect. geogr.* 465. 1. 1–2),[102] the Phrygians (Apollodorus in Strabo 14. 5. 23), the Trojans (Pseudo-Zonaras *Lexicon*, *s.v.* Βέβρυκες; *Scholia in Lycophronem* 1305; *Scholia in Oppianum Haleutica* 1. 618) and even the Greeks (*Lexicon Artis Grammaticae*, *s.v.* Βέβρυξ).

[98] Delage 1930, 140, 154, 280; Vian and Delage 1974, 156–59.

[99] Like the works of Eustathius (*cf.* also *De capta Thess.* 30) and those of many commentators and lexicographers that have been cited above. See also Prêteux 2005, 250, for the appearance of the name Bebryx in Roman inscriptions.

[100] On the issue of whether Ptolemy's work contains only historical or also mythical data, see in Section 2.5.

[101] The scholiast adds 'in those years': probably the mythical era of the Argonautic expedition? Kretschmer's view (1896, 206; *cf.* Prêteux 2005, 250) that, according to Polyaenus (*Strateg.* 8. 37), some Bebryces lived in Phocaea is erroneous, since this is not what the ancient author says (see above). Some authors mention the Bebryces with no indication of their exact location (e.g. Lycophron *Alexander* 512–515; 1304–1305; 1474).

[102] According to Appian, 'perhaps the name was changed from Bebrycia to Bithynia by the Thracians insensibly with the lapse of time, as there is not much difference between Bithynia and Bebrycia. So some think. Others say that their first ruler was Bithys, the son of Zeus and Thrace, and that the two countries received their names from them' (Appian *Mithr.* 1. 1, who also cites other theories relating to the Bebryces that existed in his time). Moreover, the mother of Amycus was a Bithynian nymph (Pseudo-Apollodorus 1. 9. 20; Hyginus *Fabulae* 17).

In anthropological terms, Amycus' mythical personality has led to a perception of the Bebryces in general as being of gigantic size (Arrian, according to Eustathius *Comm. in Dion. Per.* 805), arrogant (Apollonius *Argonautica* 2. 129, 758; *Orphica Argonautica* 657) and thick-haired (Theocritus *Idylls* 22. 76).[103]

We should bear in mind that we also meet a people called Bebryces in the Iberian Peninsula.[104] Their relation to the Asian Bebryces and indeed their very existence are obscure,[105] but it is worth noting that Pseudo-Scymnus (196) mentions only the Iberian Bebryces and not the Asian ones (*cf.* Silius Italicus *Punica* 3. 420; Cassius Dio *Hist. Rom.* 189; Joan. Zonaras *Epit. histor.* 2. 234).

In this misty landscape one cannot claim to be certain whether the Bebryces really occupied the western part of the southern Black Sea littoral at one time or not. But it would be reasonable to wonder whether the tradition of such an occupation – created most probably by the Heracleans – contains a degree of historicity, reflecting for example attempts to invade or temporarily inhabit a part of the coast.[106]

4. The Mariandynoi

Among the peoples that dwelt on the southern Black Sea littoral in the 1st millennium BC, the Mariandynoi undoubtedly occupy a significant place. Nevertheless, it should be pointed out that their importance, as reflected in their frequent appearance in the Greek and Latin written sources, is due less to factors relating to the Mariandynoi themselves than to the fact that they happened to live in an area where one of the most important Greek colonies in the whole Black Sea region was later founded and flourished: Heraclea. A colony that developed into a powerful commercial city as well as mother-city of her own colonies, including the equally significant Chersonesus in the Crimean peninsula. A colony which, moreover, was the home of several important intellectuals,

[103] The Bebryces appear several times in Greek versions of the Alexander Romance (*Historia Alexandri Magni, Recensio* α 1. 42. 6; 3. 19. 5–6, 9; 3. 20. 2; 3. 21. 7; 3. 22. 14), but in a context which is otherwise totally unknown, including the 'Pierian city of Bebrycia'.

[104] Aelius Herodianus (*De prosodia catholica*, *s.v.* Βέβρυξ); Stephanus of Byzantium, *s.v.* Βεβρύκων: 'two nations called that; the one in Europe in Iberia, and the other in Asia, on which several theories appear; that they come from Bebryx, or Bebryce, or others'.

[105] On the Iberian Bebryces, as well as their supposed relation to ours, see in detail Perez 2010, with earlier bibliography. Perez connects them with a historic people of the 3rd century BC.

[106] We also read about a city on the Asiatic side of the Bosporus called Amycus, where the homonymous king of the Bebryces is said to have been killed and where there was a heroon on his supposed tomb (Pliny *NH* 16. 239; Schol Apol. Arg. 137; *Etymologicum Genuinum* and *Etymologicum Magnum*, *s.v.* Ἄμυκος).

who recorded her history, figures like Herodorus, Promathidas, Nymphis and Memnon. Heraclea's powerful political presence, already from the classical period, combined with an equally strong literary presence, resulted in the creation of potent traditions about her early history, traditions which inevitably included the Mariandynoi and coloured intellectual and literary works such as Apollonius' *Argonautica*, one of the most important and influential works describing the southern coast of the Black Sea, with all that that implies.

4.1. The Territory (Fig. 10.1)

Let us begin by attempting to define the geographical territory of the Mariandynoi. This task is not as difficult as with other peoples on the Black Sea coast, especially its eastern stretch. The discrepancies among the written sources concerning the boundaries of their country, especially its western and eastern borders, are few and minor. The most important problems seem to be the definition of their southern limits and the ambiguous cases of the Bebryces and the Caucones[107] and their supposed geographical overlapping with the Mariandynoi on, respectively, the west and the east (see below).

Generally, the Mariandynoi are known to have possessed the land between the Bithynians and the Paphlagonians.[108] From the former they were separated by either the Sangarius or the Hypius, which are in any case not far from one another;[109] the same is true of the Callichorus and the Billaeus, which appear as their natural border with the Paphlagonians.[110] Several important sites were located in the land of the Mariandynoi, first and foremost Heraclea.[111] Other

[107] See in the relevant Sections 2.3 and 2.5.

[108] Pseudo-Scylax 90–92; Apollonius *Argonautica* 2. 350–352, 722–723; Strabo 12. 3. 2, 3. 9, 4. 1; Dionysius Periegetes 788; Pomponius Mela 1. 97; Pliny *NH* 6. 4; Arrian *FGrH* 2b 176 F 75 (Eustathius *Comm. in Dion. Per.* 791); Schol. in Apol. Arg. 157, 183.

[109] Sometimes, Thynias Island is mentioned as being situated opposite or slightly after the land of the Mariandynoi (Apollonius *Argonautica* 2. 350–352; Pomponius Mela 2. 98; Schol. in Apol. Arg. 155). The statement that the Mariandynoi border the Bithynians after the Rhebas (Dionysius Periegetes 793; *Paraphrases. in Dion. Per.* 793; Niceph. Blemmydes *Consp. Geogr.* 35–37) does not help much, since this river lies well to the west. Unique is Ptolemy's information (5. 1. 11) that the Chalcedonioi (see in Section 2.2.1) possess the coast from the Bosporus to the Hypius, and the Mariandynoi the coast from Heraclea to Cytoron.

[110] In the *Orphica Argonautica* (742) the Mariandynoi are erroneously placed much farther east, bordering the Macrones. They have always been numbered among the peoples on 'this side of the Halys' (Herodotus 1. 28, 3. 90, 7. 72; Ephorus in Strabo 14. 5. 23; Pseudo-Scymnus 943; Strabo 12. 3. 9; Anonymous *Periplus* 27). *Cf.* Burstein 1976a, 28; Counillon 2004, 130–33.

[111] Xenophon *Anabasis* 6. 2. 1; Pseudo-Scylax 91; Strabo 12. 3. 4; Ephorion *FGrH* 2a 70 F 44b; Nymphis fragm. 2. 4; Arrian *Bith. Fragm.* 43; Pausanias 5. 26. 7; Athenaeus *Deipnosophistae* 6. 84, 6. 87, 14. 11; Pomponius Mela 1. 103; Eustathius *Comm. ad. Dion. Per.* 791; Schol. in Apol. Arg. 155, 157, 185.

cities ascribed to their territory are Stephanis (Hecataeus *FGrH* 1a 1 F 198) and Aconae, a town in a rocky area (Theophrastus *Hist. Plant.* 9. 16. 4), where aconite grew in abundance from the bile that Cerberus vomited when he was brought up from the Underworld by Heracles, according to a mythical tradition created by the Heracleots. For this reason, the area has also been called 'the sacred plain of the Mariandynoi' (Dionysius Periegetes 792; *Paraphrases in Dion Per.* 792). Much later, Constantine Porphyrogenitus (*De thematibus Asia* 6. 15) refers to Claudioupolis (Bithynium) as a 'metropolis of the Mariandynoi'.

The land of the Mariandynoi, which was generally characterised by a fertile soil (Apollonius *Argonautica* 2. 723), also included other important places of (mythological) interest. Aconae, for example, was situated at the place where, according to Heraclean tradition, the gates of Acheron were located (Xenophon *Anabasis* 6. 2. 2; Apollonius *Argonautica* 2. 351–356, 728–751; Schol. in Apol. Arg. 123, 155, 157).[112] Here, too, were the Lycus and Lake Anthemoeisis (Apollonius *Argonautica* 2. 724; Aelius Herodianus *De prosodia catholica* 3. 1. 102; Stephanus of Byzantium, *s.v.* Ἀνθεμοῦς), later called also Anthemous (*Etym. Gen.* and *Etymologicum Magnum*) by confusion with the Macedonian place-name. Mariandynia, as the land was sometimes called,[113] was also the place where Idmon, one of the soothsayers accompanying the Argonauts, was killed (Pherecydes in Schol. in Apol. Arg. 1. 139; Apollonius *Argonautica* 2. 815–850; Pseudo-Apollodorus 1. 126; Schol. in Apol. Arg. 190). Finally, it was in the land of the Mariandynoi that Hierax lived, the 'man of justice and distinction' whom Poseidon turned into a falcon because, although he erected temples to Demeter, he neglected the sacrifices to the sea-god (Antoninus *Metamorphoseon synagoge* 3. 1).

There remains the question of the Caucones, the obscure people who are reported to have inhabited the area between the Mariandynoi and the Paphlagonians, around the River Parthenius.[114] On their location and very existence depend both the eastern limits of the Mariandynoi and the important matter of whether the significant Greek colony of Tieion belonged to the land of the Caucones (e.g. Strabo 12. 3. 5) or, if the Billaeus was their eastern border, the Mariandynoi (Pliny *NH* 6. 4[115]). Tieion has even been considered as belonging to the

[112] See in Part 1.

[113] First by Eupolis (fragm. 279), later by Aelius Herodianus *De prosodia catholica*, Stephanus of Byzantium (*s.v.* Μαριανδυνία), *Etymologicum Genuinum* and *Etymologicum Magnum*, *s.v.* Ἀνθεμοῦς. However, usually the whole land to the west of Paphlagonia was called Bithynia (see Section 2.2.1).

[114] See in detail in Section 2.5.

[115] In an ambiguous passage which mentions the River Coralius, a name otherwise totally unknown.

land of the Paphlagonians, if the Callichorus was a Paphlagonian river (Pseudo-Scylax 90; Schol. in Apol. Arg. 2. 904), unless the Caucones were an ancient people that had been replaced by the Mariandynoi at least in the specific area, which has sporadically come under the control of the Paphlagonians.

A similar difficulty attaches to the Bebryces and whether they had indeed possessed part of the southern Black Sea coast or not.[116] As we have seen, the natural western border of Mariandynia is sometimes reported to have been the Sangarius and sometimes the Hypius, and it is precisely the region between these two rivers which has been described by Apollonius (*Argonautica* 2. 137–153) as the 'iron-bearing land' that had been the apple of discord between the Bebryces and the Mariandynoi, most probably according to a tradition created by the Heracleots,[117] who were most likely also the source of the tradition that Mariandynus, the eponymous hero of the Mariandynoi, ruled over a part of Paphlagonia and invaded and annexed the country of the Bebryces (Theopompus in Strabo 12. 3. 4).

Finally, as far as the southern borders of the Mariandynoi's territory are concerned, the only hint is provided much later, by Constantine Porphyrogenitus (*De thematibus Asia* 6. 15), who mentions Claudioupolis (Bithynium) as a 'metropolis of the Mariandynoi'; his further observation, that the Mariandynoi have also been called Galatians, could be compared with the tradition that Heracles had once conquered the Phrygians for the Mariandynoi (Nymphis in Schol. in Apol. Arg. 2. 786 – see also below). Otherwise, nothing can be concluded from the written sources.

4.2. Who were the Mariandynoi?

Very little is actually known about the Mariandynoi and their origin, and, as is the case with several other peoples of the southern Black Sea littoral, it is mostly from mythology that one might attempt to draw some conclusions. But even this proves to be a difficult task.

Mariandynus was the eponymous hero of the Mariandynoi, who gave his name to the country (see above). Several versions arose regarding his father and his origin, and consequently the origin of his eponymous people. Most of these were collected and presented by the Scholiasts of Apollonius and Aeschylus. Thus, his father is often reported to have been Titias, who had also one or two other sons: Lycus and Priolaus or Bormus (Schol. in Aesch. Pers. 940–941 invoking Callistratus; Schol. in Apol. Arg. 1. 1126, 2. 140, 723, 780

[116] Which has been discussed in Section 2.3.
[117] See in Section 2.3.

invoking Promathidas). Titias appears as a son of Zeus, as does Mariandynus himself (Schol. in Apol. Arg. 1. 1126). Other versions present Mariandynus as the son of Cimmerius or Phrixus or Phineus (Schol. in Apol. Arg., *op. cit.*, who attributes the former version to Domitius Callistratus and the latter to Hesiod).

Each version could lead, and indeed has led, to a different theory of the origin of the Mariandynoi, especially since, as Strabo (12. 3. 4) complained, the ancient authors failed to say where these people came from. Cimmerius points to a relation with the Cimmerians,[118] who however tend to be connected with the formation of tribes on the coast eastward of the Leucosyroi, as we shall see. On the other hand, the presence of Phineus, and the resemblance to the Bithynians reported by Strabo (12. 3. 4) albeit with no clarification, leads to the possibility that the Mariandynoi were Thracians (Strabo 7. 7. 3, 12. 3. 4).[119] The Scholiast of Aeschylus' *Persians* (937) called them a Paphlagonian tribe, while of no less interest is the version describing Mariandynus as an Aeolian (Eustathius *Comm. in Dion. Per.* 787).

The obscurity surrounding the origin of the Mariandynoi has left the way open for the formulation of other theories in modern scholarship, such as the one that includes them among the local Asia Minor peoples.[120] The existence of the Mariandynoi in the area from very ancient times could also be inferred from Nymphis' assertion (according to the Schol. in Apol. Arg. 2. 786) that Heracles had once conquered the Phrygians for the Mariandynoi, although his account is likely to reflect Heraclean tradition in the matter of the Mariandynoi, and not necessarily the historical truth. What can be assumed is that the historians of Heraclea[121] tended to believe that the Mariandynoi were native in Anatolia well before the Trojan War, as indicated by, among other things, their inclusion in the Argonautic myth.

Arguing a more combinatorial theory, S.M. Burstein, to whom we owe one of the most detailed studies of the Mariandynoi, claimed that 'the Mariandynian population of historical times was not a unity but the result of the fusion of several originally distinct ethnic groups some time during the first

[118] Lehmann-Haupt *RE* XI, *s.v.* Kimmerier, 413. *Cf.* Burstein 1976a, 7.

[119] Strabo could have proven to be more helpful where he writes that 'Homer nowhere names the Pamphylians, Bithynians, Mariandynoi, Pisidians, Chalybes, Milyans, or Cappadocians – some because they had not yet settled in this region, and others because they were included among other tribes, as, for example, the Hidrieis and the Termilae among the Carians, and the Doliones and Bebryces among the Phrygians' (14. 5. 23), but he does not specify to which of the two categories the Mariandynoi belong.

[120] Meyer 1913, 702 §476; Lehmann-Haupt *RE* XI, *s.v.* Kimmerier, 413.

[121] See Dana 2011, 243–46.

half of the 1st millennium'.[122] This theory is probably not far from reality, but even so the most plausible scenario still seems to be that the basic element of the Mariandynian population was Thracian.[123]

The tradition of a problematic relationship between the Mariandynoi and the Phrygians is in any case not encountered elsewhere, unlike the former's well-known (and already mentioned[124]) conflicts with the Bebryces, especially over the 'iron-bearing land', where again Heracles played a decisive role. Before his intervention, the Bebryces had several times defeated the Mariandynoi, and during one of these battles Amycus had killed Mariandynus' brother Priolaus or Bormus (see above). Mariandynus composed a lament for his brother to be accompanied by the flute, and for centuries the Mariandynoi were reported to have mourned their hero to its strains each year at harvest-time. He also taught this song to Hyagis, the father of the famous musician Marsyas (Schol. in Aesch. Pers. 935–937, 940; Schol. in Apol. Arg. 2. 650, 758 citing Nymphis; Eustathius *Comm. in Dion. Per.* 791; *Carmina Popularia* 32; *Comica Adespota* 415. *Cf.* Hesychius *Lexicon*, *s.v.* Μαριανδυνός θρῆνος; Photius *Lexicon*, *s.v.* Μαριανδυνόν θρηνητήν), and the *borimos* thus became known as a song of the Mariandynian farmers (Pollux *Onomasticon* 4. 54. 55). It seems to have been remembered as a plaintive wail, as Aeschylus (*Persae* 935–940) implies, and so apt to be referenced derisively (Hesychius *Lexicon*, *s.v.* μαριανδυνίζεις).

The most extensive historical information on the Mariandynoi that we possess is connected with the foundation of Heraclea in their territory (see the next section). In addition, they are mentioned by Herodotus as having been subjugated by Croesus (Herodotus 1. 28) and campaigning with the Persians against Greece (7. 72). In that war they were armed in a manner similar to the Paphlagonians and led by Gobryes, the son of Darius and Artystone, who also commanded the Ligyes and the Syroi. The Mariandynoi belonged to the 3rd Persian satrapy, together with the Hellespontians (of the right-hand shore as

[122] Burstein 1976a, 6–11, with an analytic presentation of all theories.

[123] *Cf.* Lendle 1995, 372. Saprykin (1991, 243–46, with earlier bibliography) also leaned towards a Thracian origin for the Mariandynoi, since he sees a relation between them and the Paphlagonians, whom he also considered to be Thracians, which, however, as we shall see (in Section 2.6), is a rather problematic assumption. He later (Saprykin 1997, 27) presented the Mariandynoi as belonging to a native population of mixed Thraco-Anatolian origin, which is indeed plausible. An Indo-European origin is in any case indicated by their assignment to the generation of Japheth (Hippolytus *Chronicon* 80. 12; Epiphanius *Ancoratus* 113. 5; *Chronicon Paschale* 47). That Mariandynia was the earlier name of Bithynia, which was established by Phoenix (Georg. Syncellus *Ecloga Chronographia* 185), must be a mistake. Prêteux's assumption (2005, 248–49) that the Mariandynoi drove the Paphlagonians from the part of the coast they occupied when the Bithynians drove the Bebryces from their area, namely around the 8th century BC, is unsubstantiated.

[124] In Section 2.3.

one enters the straits), the Phrygians, the Asiatic Thracians, the Paphlagonians and the Syroi, and paid a tribute of 360 talents (Herodotus 3. 90). Although not clearly stated in the sources, the Mariandynoi might have been ruled by chieftains at the time when Heraclea was founded.[125]

Some further scattered information on specific customs or individuals is preserved by Athenaeus, according to whom Pherecrates (5th century BC) wrote that the Mariandynoi called black dried figs pots (*Deipnosophistae* 14. 67) and Clearchus (4th–3rd century BC) reported that Sagaris, a king of the Mariandynoi, used, from luxury, to eat out of his nurse's mouth until he arrived at old age, so that he might not have the trouble of chewing his own food; and that he never put his hand lower than his navel (*Deipnosophistae* 12. 40).

4.3. Mariandynoi and Greeks

It seems that our knowledge of the Mariandynoi would have been much more meagre were it not for the foundation of Heraclea in their territory. Of specific importance is the fact that this event, unlike the cases of other southern Black Sea colonies, attracted scholarly interest not only with regard to the identity of the founders, over which there is some debate caused by the ancient texts themselves, but also for the way in which the Mariandynoi and the Greek colonists, whoever they may have been, encountered each other and the relations that developed between them.

The question of the identity of the founders of Heraclea does not concern us here. Strabo's ascription of its establishment to the Milesians (12. 3. 4) stands out from the rest of the sources, which assert that the city was founded by Megarians, Boeotians, or both.[126] Although there is nothing of an archaeological, epigraphical or numismatic nature to support the Milesian theory, some scholars have discussed the possibility of a Milesian presence in the area before the founding of Heraclea by Megarians and Boeotians in the mid 6th century BC, as well as Milesian encounters with the local Mariandynoi.[127]

Whatever the case may have been,[128] the aspect of this colonisation that seems to have attracted the greatest interest on the part of several ancient authors is the relation between the colonists and the local people and how the

[125] Burstein 1976a, 10.

[126] Or even with the participation of Thessalians (Asheri 1972, 14). See also Saprykin 1997, 21–27; Avram *et al.* 2004, 956.

[127] E.g. Asheri 1972, 12–13, 19–28, with the relevant arguments; Saprykin 1997, 21–27.

[128] I also consider Strabo's information as probably erroneous, however surprising that may be given that Strabo is one of the most reliable sources we have on the history of the southern Black Sea littoral.

latter were treated by the incomers. From all the sources referring to this matter it appears that the status of the Mariandynoi under the Greek colonists of Heraclea was a kind of slavery, and it is remarkable that no other such case is mentioned anywhere else on the southern Black Sea coast. But let us first see exactly what the written sources say, and what may be deduced from this information.

The available relevant references start from the 4th century BC, and most of them have survived in citation by authors of the Roman period, especially Athenaeus. They all mention the case of the Mariandynoi, comparing them with similar cases of slavery in the Greek world, in a wider context of discussion about forms of slavery. Plato appears as the first to have compared the slavery of the Mariandynoi, who had been subjugated by the Heracleots, with the status of the Helots in Lacedaemon and the Penestae in Thessaly (*Leges* 6. 776c–d). In the part of his *Deipnosophistae* dealing with slaves and slavery, Athenaeus collected many relevant references from earlier authors, Plato's included, and it is in this work that we read the words of Poseidonius which add a new dimension to the explanation of the Mariandynoi's slavery, as well as the epithet δωροφόροι (gift-bearers) that sometimes accompanied them, at least since the 3rd century BC, and Euphorion:

> Poseidonius (2nd–1st century BC) says, in the eleventh book of his *Histories*: 'Many persons being unable to manage themselves on account of the weakness of their intellect, give themselves voluntarily to the service of more intelligent men, in order that they may secure from them provision for their daily needs, and in turn may themselves render to their patrons, through their own labours, whatever they are capable of in the way of service. And so in this manner the Mariandynoi put themselves in subjection to the Heracleots, promising to serve them continually so long as the Heracleots provided for their needs, though they stipulated in addition that there should be no selling of any of them beyond the Heraclean territory, but that they should stay right in their own territory.' Perhaps, therefore, it is for that reason that the epic poet Euphorion calls the Mariandynoi gift-bearers: 'Gift-bearers shall they be called, secretly dreading their masters.' Callistratus [2nd–1st century BC], too, the disciple of Aristophanes, says that they called the Mariandynoi gift-bearers to take away the sting in the term slave, as the Spartans did in the case of the Helots, the Thessalians in the case of the Penestae, the Cretans in the case of the Clarotae (Athenaeus *Deipnosophistae* 6. 84; *cf.* Hesychius *Lexicon*, *s.v.* δωροφόρους; Eustathius *Comm. ad Hom. Il.* 3. 944).

Strabo, who as we have seen considered the Milesians as the founders of Heraclea, pointed out that it was they who forced the Mariandynoi to serve as Helots, so that they sold them, but not beyond the boundaries of their country (for the two peoples came to an agreement on this), just as the Mnoan class, as it is called, were serfs of the Cretans and the Penestae of the Thessalians (Strabo 12. 3. 4). Although not certain, it is possible that Strabo drew this information from Theopompus (4th century BC), given that in the same passage the immediately preceding content, concerning Mariandynus, is quoted from him.

Later, Pausanias calls the Clarotae 'metoikoi, like the Mariandynoi in Heraclea Pontica, and the Helots in Lacedaemon, and the Penestae in Thessaly, and the Kallikyrioi in Syracuse' (*Attike onomaton synagoge*, *s.v.* Κλαρῶται; *cf.* Photius *Lexicon*, *s.v.* Κλαρώται; the *Suda*, *s.v.* Κλαρόται; Eustathius *Comm. ad Hom. Il.* 1. 456–457). Pausanias also mentions offerings that 'were brought to Olympia by the people of Heracleia when they had overrun the land of the Mariandynoi, their foreign neighbours' (Pausanias 5. 26. 7), probably referring to the time of the latter's subjugation by the Heracleots, although their characterisation as 'foreign neighbours' is obscure, as it would seem to imply an earlier status of a friendlier relation and co-existence between the two.

Finally, Pollux (2nd–3rd century) claimed that 'between free and slaves are from the Lacedaemonians the Helots, from the Thessalians the Penestae, from the Cretans the Klarotae and the Mnoans, from the Mariandynoi the dorophoroi, from the Argeans gymnetae, and from the Sikyonians the korynephoroi' (*Onomasticon* 3. 83).

What can be drawn from all these passages? First, and indisputably, that the newcomers from Greece who founded Heraclea managed to subjugate the Mariandynoi in a way that recalled other cases of slavery in the Greek world, like those of the Helots and the Penestae. Although Strabo clearly states that the Greeks forced the Mariandynoi to serve them as Helots, another reason for this slavery appears, according to Poseidonius, to be the inability of the Mariandynoi to manage for themselves. Being aware of this disability, the Mariandynoi themselves are supposed to have asked the Heracleots to provide for their needs and in return they would serve them, the only condition being that they should not be sold outside their territory. Maybe this was also a reason for calling them gift-bearers rather than the more demeaning slaves. The fact that the two peoples seem to have come to an agreement on this status, as Strabo comments, has led to the view that the Mariandynoi were something between free and slaves.

The controversy as to who asked for or demanded this status, and the frequency of reference to that status by the ancient authors, have attracted the interest of several modern scholars, roughly since the 1970s, who have attempted to shed light on this obscure, and for the southern Black Sea area unprecedented, situation.

To start with the possibility that the Mariandynoi did indeed themselves seek such an arrangement, it cannot pass unnoticed that Apollonius, in his narrative of the mythical Argonautic expedition, devotes a lengthy part of the second book to the words and actions of Lycus, the grateful king of the Mariandynoi, towards the Argonauts. On the one hand, he reminds them how much Heracles had helped him and his people against their various enemies, such as the Bebryces, and he promises that whatever requital he might now be able to make, he will do so gladly, adding that this is the rule for weaker men when the stronger have helped them (*Argonautica* 2. 799–801), and on the other he gives them many gifts (δῶρα φέρειν) before their departure (2. 813–814). The similarities with Poseidonius' words about the weaker people's dependence on the stronger and with the established epithet δωροφόροι are striking, as is the phrase 'they (the Mariandynoi) made a league with them (the Argonauts), because the latter had killed Amycus (king of the Bebryces, enemies of the Mariandynoi)' (*Argonautica* 2. 755).

This certainly tends to point to an agreement between the two parties: the Greek colonists, represented in the *Argonautica* by the Argonauts, and the indigenous Mariandynoi. An agreement which benefitted both sides: the Mariandynoi, because they would now have a strong ally and protector against their enemies, hence the statement of Poseidonius; and the Greeks, because the Mariandynoi would not only let them develop their colony unhindered but also help by serving them in various ways, such as working the land they had already conquered (and would thenceforth exploit) or serving as sailors in Heraclea's fleet, as Aristotle implies (*Politics* 1327b). This status would explain characterisations such as 'between free and slaves', 'neither slaves nor free' and 'voluntary subjection'.[129]

But this is just one point of view. For there is nothing to assure us that Apollonius' account of the enthusiastic welcome given the Greeks by the Mariandynian king and his people's subsequent voluntary subjugation to them corresponds to the actual course of events. One can hardly accept that there has really been a people in history that asked for its own subjugation, unless it had a reason to do so. And the Heracleots had a very good reason for this, since they had decided to associate their city with the Mariandynian past.[130] If two people(s) desire a relationship that could lead to mutual benefit, then we are talking about an agreement and not slavery – the two things are mutually exclusive.[131]

[129] Asheri 1972, 21–22; Kidd 1988, 297; Saprykin 1997, 28; Tuplin 2007b, 64; Thompson 2011, 195–96.

[130] Something that, according to Asheri (1972, 18), might have started at the time of the tyranny of Clearchus (4th century BC), for reasons of political propaganda. One should not, however, exclude an earlier date.

[131] See more discussion in Burstein 1976a, 28–30.

Strabo's words (12. 3. 4), even with the dubious naming of the Milesians as the colonists of Heraclea,[132] seem to describe the situation better and more clearly than any other: the Greeks *forced* the Mariandynoi to become their slaves; the latter did not agree to this and tried to avoid it, but the only thing they achieved was to persuade the Greeks not to sell them outside the borders of the country, which means that they *could* sell them within the country, which is slavery, pure and simple.[133] This was the only actual agreement between the two sides. The comment of Poseidonius, who lived centuries after the events in question, is nothing more than a justification of slavery as a natural institution, according to the perception of several ancient intellectuals, including Aristotle and the Stoics.[134]

Two questions remain to be answered: Did the enslavement of the Mariandynoi take place[135] immediately after the foundation of Heraclea or later? And did it apply to all the Mariandynoi? More or less plausible speculations can and have been made for both questions, but conjecture is all that can be offered.

Concerning the first question, scholars seem to agree that the enslavement of the Mariandynoi was not an immediate consequence of the colonisation. There might have been an initial acceptance of the colonists by the local people, perhaps indeed for reasons having to do with the latter's perception of the incomers as of possible assistance in dealing with their neighbours,[136] but the continuation could have been much worse, probably due to the rapid economic and political growth of Heraclea. It has been even argued that the enslavement of the Mariandynoi could actually have been the result of a conflict between the two sides.[137]

A clue to the time of this outcome could lie in Herodotus' information that the Mariandynoi campaigned with the Persians against Greece (7. 72), which would mean that the enslavement took place sometime after 480 BC, which coincides with the beginning of the westward expansion of the Heraclean territory.[138] It also seems to coincide with political changes in Heraclea,

[132] This has been used as an argument for the view that the tradition of the enslavement of the Mariandynoi was not invented by the historians of Heraclea but borrowed by them from the Milesians: Asheri 1972, 22, criticised by Saprykin (1997, 35).

[133] So Lotze 1959, 56–57. See more literature in Burtsein 1976a, 29.

[134] See Garnsey 1996, 147–50, with more bibliography. The case of Mariandynian slavery could also be treated as an example of social closure, since for the citizens of a Greek colony the indigenous peoples were equated to their slaves: McInerney 2014, 51. *Cf.* also Burstein 1976a, 91.

[135] Burstein (1976a, 28–29) claimed that the subjugation of the Mariandynoi to the Heracleots (whenever it took place) did not also mean the end of their culture, just that the land they cultivated now belonged to the oligarchs of Heraclea to whom they paid tribute.

[136] Relevant bibliography in Manoledakis 2015a, 85.

[137] Burstein 1976a, 23–26; Saprykin 1997, 28–56; Thalmann 2011, 106.

[138] Burstein 1976a, 23, 28.

where democracy was replaced by oligarchy (Aristototle *Politics* 1304b), which is supposed to have played a role in the subjection of the Mariandynoi.[139] But these are mere hints and cannot lead to safe conclusions, as has sometimes been attempted.[140] Even for the expansion of Heraclea to the west, some of the settlements that have been cited, such as Elaion, Cales, Chelae and Dia,[141] are first mentioned in texts of the Roman period and would thus appear to have been established in that era,[142] and therefore much later than the presumed expansion.

It has been emphasised that the 'arrangement' between the Greeks and the Mariandynoi seems to have been unknown, or at least of little interest, to both Herodotus and Xenophon.[143] But this does not really mean much, because for neither of those two authors is there any indication in their specific passages on the Mariandynoi that such a thing was of any concern to them. Herodotus' focus was on listing the peoples that had been subjugated by Croesus (1. 28), mentioning the division of the Persian Empire into administrative provinces (3. 90), and enumerating the peoples that had campaigned with the Persians against Greece (7. 72), while Xenophon mentions the Mariandynoi (*Anabasis* 6. 2. 1) solely as the people in whose land Heraclea was established and deals no further with them, contrary to his treatment of the Heracleots and the Bithynians. So there was no reason for either of them to have examined their status, and behind their silence we should not seek for any argument to place the Mariandynian enslavement after 400 BC.

Conversely, a dating before – not to say well before – 480 BC is possible, if we consider that the enslavement did not concern all the Mariandynoi, but only those who lived in the coastal area that had been seized by the Heracleots. If the territory of the Mariandynoi extended south as far as Bithynium,[144] there remains a large territory in the hinterland that was not necessarily under Heraclea's control, and the Mariandynoi of this territory might not have been

[139] Burstein 1976a, 23–26; Avram *et al.* 2004, 956.

[140] For example, Saprykin (1997, 28–56) offers a thorough and very interesting socio-political analysis of the events in question; however, he fails (naturally) to prove that 'at the time of the Greeks' appearance in Northern Asia Minor the Mariandynoi were in the stage of forming social groups and tribal nobility', as well as that it was 'native aristocracy that was deeply interested in close alliance with the Greeks' (1997, 30). Moreover, I fail to see how Strabo's (12. 3. 4) and Apollonius' (*Argonautica* 2. 780) words indicate that 'when the Greeks appeared in their country, the Mariandynoi were suffering a decay of their primitive community system' (Saprykin 1997, 24).

[141] Saprykin 1997, 33–34.

[142] See also in Section 4.4.1.

[143] Kidd 1988, 296.

[144] See above, in Section 2.4.1.

enslaved by the Greeks.[145] These would be the ones that campaigned with the Persians. In other words, the enslavement of the Mariandynoi (of the coast, around Heraclea) does not need to be either connected with the end of the Persian Wars or attributed to any political changes in Heraclea. Any delay in subjugating the local Mariandynoi after the founding of the city was due rather to the time it required to become sufficiently powerful than to the change of regime.

Finally, both Burstein and Saprykin, especially the latter in his extended socio-political analysis of the matter, gave much weight to trying to define what exactly the περίοικοι and γεωργοῦντες were in the Mariandynian community, for they were mentioned by Aristotle (*Politics* 1327b), as they thought, as 'the dependent population of Heraclea, who served as sailors' (see also above).[146] This, however, is not a safe conclusion. What Aristotle actually says, in his effort to argue for the advantages of cities that, having a substantial naval force, are able also to defend some of their neighbours, is that:

> it is not necessary for states to include the teeming population that grows up in connection with common sailors, as there is no need for these to be citizens; for the marines are free men and are a part of the infantry, and it is they who have command and control the crew; and if there exists a mass of villagers (περίοικοι) and tillers of the soil (γεωργοῦντες), there is bound to be no lack of sailors too. In fact we see this state of thing existing even now in some places, for instance in the city of Heraclea; the Heracleots man a large fleet of triremes, although they possess a city of but moderate size as compared with others.[147]

As one can see, Aristotle is speaking generally about the importance of the existence of περίοικοι and γεωργοῦντες in or round a city,[148] and finishes by citing Heraclea as an example. This does not necessarily mean that this was precisely the case with Heraclea, namely that there were both περίοικοι and γεωργοῦντες among the Mariandynoi serving in the city's fleet. Therefore, the whole attempt to identify both categories of people as undoubtedly Mariandynoi, and to define the status of each category under the Heracleots and the degree of their subjection to the latter, thus drawing conclusions concerning the social history of both the Mariandynoi and Heraclea, might be considered audacious. Unfortunately, the ancient sources do not provide us with any significant information concerning the political or social organisation of the

[145] As also Asheri (1972, 18–19, 23) has supported.

[146] Burstein 1976a, 29–30; Saprykin 1997, 30 and 36–39.

[147] Translation by H. Rackham, Cambridge, MA, Harvard University Press; London, William Heinemann Ltd, 1944.

[148] For the *perioikoi*, see also Tuplin 2007b, 66, n. 28.

Mariandynoi, or their economic activities, customs and manners, as is the case with other peoples on the littoral who were much more 'barbarous' and culturally 'weird' and alienated from the Greeks, such as the Mossynoikoi.[149]

Equally dubious is the interpretation of Domitius Callistratus' reference to the city Arciroessa as subject to Heraclea (Stephanus of Byzantium, *s.v.* *Ἀρκιρόεσσα*) as 'direct evidence on a local community which was paying tribute to Heraclea'.[150] The city is not mentioned elsewhere in the ancient literature. One cannot even say whether it was a coastal town (it is absent from all the *periploi*), and all these elements also characterise the village of Hodioupolis, apart from the fact that it is cited not as subject to Heraclea but as belonging to it (Stephanus of Byzantium, *s.v.* Ὁδιούπολις). The Greek names do not point to Mariandynian settlements, though, and they might just be settlements that were within Heraclea's sphere of influence, first mentioned (and probably established) in the Roman period.

5. The Caucones

One of the peoples reported by Homer to have fought in the Trojan War on the Trojan side is the Caucones. But although they are mentioned twice elsewhere in the *Iliad* (10. 429; 20. 329) they do not appear in the Trojan Catalogue;[151] and while neither passage specifies their geographic location, it is obvious that they cannot be identified with the Caucones of the Peloponnese.[152] It was only about three and a half centuries later, in the 4th century BC, that Callisthenes is mentioned by Strabo (12. 3. 5) as having been aware of two additional lines in the Trojan Catalogue, after line 2. 855 (Κρῶμνάν τ'Αἰγιαλόν τε καὶ ὑψηλοὺς Ἐρυθίνους):

> Καύκωνας δ'αὖτ'ἦγε Πολυκλέος υἱὸς ἀμύμων,
> οἵ περὶ Παρθένιον ποταμὸν κλυτὰ δώματ' ἔναιον[153]

> *And the excellent son of Polycles led the Caucones,*
> *who had their famed dwellings around the river Parthenius*

[149] See in detail in Section 2.11.3.

[150] Saprykin 1997, 37.

[151] As was observed in the Scholia on Homer (Schol. in Il. 10. 429).

[152] Who again are first mentioned by Homer (*Odyssey* 3. 366), and then probably by Hecataeus (*FGrH* 1a 1 F 119), Herodotus (1. 147) and several others, more than those mentioning the Asian Caucones. See below for more on them.

[153] Eustathius (*Comm. ad Hom. Il.* 4. 414) also knew of some slightly different variants:
Καύκωνας αὖτ'ἦγε Πολυκλέος υἱὸς Ἄμειβος,
οἵ περὶ Παρθένιον ποταμὸν κατὰ δώματα ναῖον.

This seems to have been the first known clear situating of the Caucones on the southern Black Sea littoral: around the River Parthenius, and so neighbouring the Paphlagonians to their east (Fig. 10.1). Another three and a half centuries after Callisthenes, Strabo became the first and only known ancient author to attempt to investigate the Caucones of the southern Black Sea, perhaps – among other things – because of his admiration for Homer.[154] Strabo (12. 3. 5) attributed to Callisthenes more information on the Caucones, namely that they dwelt from Heraclea and the Mariandynoi to the Leucosyroi ('whom we call Cappadocians'),[155] and that the tribe of the Caucones round Tieion extended to the Parthenius River, whereas the Enetoi, who held Cytoron, were situated next to them after the Parthenius River;[156] and 'still to-day certain Cauconitae [Cauconiatae in 8. 3. 17] live in the neighbourhood of the Parthenius River', added Strabo.

It is indeed noteworthy that this people was never mentioned by any other known ancient author, whether historian or geographer or even poet or mythographer, for example Apollonius, who speaks of several peoples on the coast. This silence regarding the Caucones, combined with Strabo's contradictions concerning their existence in his time (see below) and his statement that 'all people do not give the same account of the Mariandynoi and the Caucones' (12. 3. 4), provokes some questions: Who exactly were the Caucones of the southern Black Sea? Did they really exist – or, to put it another way, did the lines cited by Callisthenes (as reported by Strabo) correspond to reality? And why were they not mentioned by Homer himself in the Trojan Catalogue? And lastly, if the Caucones did exist, what was their relation with the Caucones of the Peloponnese?

According to Eustathius (*Comm. ad Hom. Il.* 1. 571), clearly referring to Strabo's words in 14. 5. 23, Strabo considered the Caucones, like others mentioned in the *Iliad*, as unknown peoples. I think, however, that a closer reading of the passage shows that Strabo is saying that it was Apollodorus who considered the Caucones and the others to be unknown peoples (Strabo 14. 5. 23; *cf.* 14. 5. 28).[157] Indeed, Strabo does not appear to have been unaware of them, for he discusses them in other parts of his work (in Books 7, 8, 12, 14 – see

[154] It was only Eustathius and some scholiasts who, much later, treated Homer's and Strabo's passages on the Caucones (see below). Strabo also dealt with the Caucones of the Peloponnese (8. 3. 3, 10–11, 16–17, 29–30; 8. 7. 5).

[155] For the Leucosyroi, who in Strabo's time were usually called Cappadocians, see Section 2.8.

[156] See Section 2.7.

[157] Apollodorus was, therefore, aware of the interpolation mentioned by Strabo (12. 3. 5) – see at the end of this section.

below). Taking together all he says in all these passages, it is evident that he encountered great difficulty in trying to draw conclusions about a people that for him could not have been fictional; knowing Strabo's admiration for Homer and his repeated and extensive replies to the view of Eratosthenes and Apollodorus that Homer could not have been acquainted with the seacoast and the rest of the territory of Pontus (7. 3. 6–8; 12. 3. 26–27; 14. 5. 27–29), we suspect that he would have found it unthinkable for the poet to have mentioned a non-existing nation. On the whole, Strabo seems to have been trying to balance the Homeric references, the variant used by Callisthenes and the situation in his own era. But his attempt, which appears and reappears on several occasions – in the descriptions of Epirus (Book 7), the Peloponnese (Book 8), the southern Black Sea littoral (Book 12) and Cilicia (Book 14) – did not produce a concrete result.

On the contrary, while Strabo speaks twice of the existence in his time of certain Cauconitae (12. 3. 5) or Cauconiatae (8. 3. 17) in the neighbourhood of the River Parthenius and bordering the Mariandynoi, elsewhere (7. 7. 2; 12. 3. 9) he falls into contradiction, stating that 'the race of the Caucones has been destroyed' and 'they are nowhere to be found'. This inconsistency is in itself especially indicative. Equally indicative is the inconsistency in Callisthenes' geographical siting of the Caucones, as repeated by Strabo (12. 3. 5 – see above), since at the same time they appear on the one hand to have dwelt from Heraclea as far to the east as the Leucosyroi, which brings us to the River Halys, and on the other to have had the River Parthenius as their eastern border, the latter separating them from the Enetoi, who held Cytoron.[158]

The obscurity characterising the Greeks' perception of the Caucones might find parallels in the cases of the Leleges and the Pelasgians, who often appear together with the former, in Homer as well as in Strabo, who cites Hesiod's words about the Leleges, concluding that 'from earliest times they were a collection of mixed peoples and that this was why the tribe disappeared. And the same might be said of the Caucones' (Strabo 7. 7. 2). The Leleges are an equally nebulous people, who are sporadically encountered in various parts of Asia Minor and Greece, especially – once again – the Peloponnese.[159] Coincidentally, these are precisely the two peoples who are listed as allies

[158] Compare the contradiction in Herodotus, who initially named the Halys as the western border of the Leucosyroi (1. 6; 1. 72) but later said that they lived about the rivers Thermodon and Parthenius, neighbouring the Macrones (2. 104). See in 2.8.1.

[159] Eratosthenes included the Leleges among the peoples that were extinct in his day (Pliny *NH* 5. 127).

of the Trojans in *Iliad* 10. 427–434 but are missing from the Trojan Catalogue.[160]

In an attempt to explain the absence of the Caucones from the Trojan Catalogue, Porphyrius (3rd–4th century AD) claimed that they had not been engaged in the war from the outset but joined in later; this, he says is implied by the second passage from the *Iliad* (20. 329), which describes them as being arrayed in order to enter the war (Porphyrius *Quaestionum Homericarum ad Iliadem* 20. 329). The weakness in this argument, however, is that in focusing on the second passage relating to the Caucones (20. 329) it ignores the earlier one (10. 429), where the Caucones are mentioned together with another nine peoples who fought on the Trojan side, eight of which do appear in the Catalogue.[161] Of those ten, only the Leleges are, like the Caucones, missing from the Catalogue. Could this mean that they, too, delayed entering the war?[162]

*

* *

Aware of the obscurity shrouding the Caucones (and the Leleges), Strabo tried to provide an explanation by highlighting

> the confusion of the tribes this side the Halys River, and the fertility of the country there, particularly that of the seaboard, on account of which attacks were made against it from numerous places and continually by peoples from the opposite mainland, or else the people nearby would attack one another. Now it was particularly in the time of the Trojan War and after that time that invasions and migrations took place, since at the same time both the barbarians and the Greeks felt an impulse to acquire possession of the countries of others; but this was also the case before the Trojan War, for the tribe of the Pelasgians was then in existence, as also that of the Caucones and Leleges. And, as I have said before, they wandered in ancient times over many regions of Europe' (Strabo 12. 8. 4; *cf.* Eustathius *Comm. ad Hom. Il.* 1. 561, 4. 414–415, 465; *Comm. ad Hom. Od.* 1. 133).

And he continues:

> But the obscurity has arisen, not on account of the changes only, but also on account of the disagreements of the historians, who do not say the same things about the same subjects, calling the Trojans Phrygians, as do the tragic poets, and the Lycians Carians; and so in the case of other peoples' (12. 8. 7).

[160] The others are the Carians, the Paeonians, the Pelasgians, the Lycians, the Mysians, the Phrygians, the Maeonians and the Thracians (*Iliad* 10. 427–434).

[161] See the previous note.

[162] Conversely, from the peoples of the Catalogue (2. 840–877) only four are missing in 10. 427–434: the Cicones, the Paphlagonians, the Enetoi and the Halizones. But this does not necessarily mean much, since these four could have been arrayed in another place, and not 'towards the sea', like those of 10. 427–434.

An example of such disagreements relating to the Caucones is found in 12. 3. 2, where Strabo presents the disagreement on whether the land between the Bithynians and the Paphlagonians belonged to the Mariandynoi or the Caucones. With regard to the origin of the latter, 'some people say that they were Scythians, others that they were a certain people of the Macedonians, and others that they were a certain people of the Pelasgians' (12. 3. 5). Unfortunately, Strabo does not specify who these people were or when they lived. The attribution of Scythian origin to the Caucones was also known to the Scholiasts of Homer (Schol. in Hom. Il. 10. 429), while in the modern literature some scholars ascribe to them an ethnic kinship with the Phrygians,[163] although with no justification. The assertion of a Scholiast of Homer (*Iliad* 10. 429) that the Caucones were called by some Cyanioi remains unique and unsubstantiated.

Let me now move to the possible connection with the Caucones of the Peloponnese. Strabo says in 8. 3. 17 that:

> there are several accounts of the Caucones; for it is said that, like the Pelasgians, they were an Arcadian tribe, and, again like the Pelasgians, that they were a wandering tribe. At any rate, the poet tells us that they came to Troy as allies of the Trojans. But he does not say whence they come, though they seem to have come from Paphlagonia; for in Paphlagonia there is a people called Cauconiatae whose territory borders on that of the Mariandynoi, who are themselves Paphlagonians [another option for their origin].

It is noteworthy that the Caucones of the Peloponnese are mentioned in the *Odyssey* (3. 366) as great-hearted, while the leader of the Paphlagonians in the Trojan Catalogue (*Iliad* 2. 851) was 'stout-hearted'. The Caucones were not the only people of Asia Minor, and specifically of the southern Black Sea littoral, who have been linked to Greece or Europe. Traditions about similar connections have developed with regard to the Enetoi and the Paphlagonians as well, and even the Bebryces.[164] Strabo himself noted (7. 7. 1) that

> Hecataeus of Miletus had said of the Peloponnesus that before the time of the Greeks it was inhabited by barbarians. Yet one might say that in the ancient times the whole of Greece was a settlement of barbarians, if one reasons from the traditions themselves: Pelops brought over peoples from Phrygia to the Peloponnesus that received its name from him; and Danaus from Egypt; whereas the Dryopes, the Caucones, the Pelasgians, the Leleges, and other such peoples, apportioned among themselves the parts that are inside the isthmus – and also the parts outside.

Over time, as usually happens, these traditions mutated in several ways, sometimes even becoming reversed. Thus, according to a Scholiast of the *Iliad*

[163] Kretschmer 1896, 207.

[164] See in Sections 2.7, 2.6 and 2.3 respectively.

(10. 429; 20. 329), from the Peloponnesian Caucones came those who settled on the Black Sea, around Paphlagonia, while he also distinguishes yet another group of Caucones, 'a Trojan people from Caria' (20. 329). In other commentaries on the same passage we meet also a homonymous 'city in Lycia and Caria', which elsewhere appears as 'Caucanos, a coastal city of Lycia and Caria', namely in south-west Anatolia.[165] Moreover, Aelius Herodianus and Stephanus of Byzantium (both *s.v.* Ἀράβυζα) mentioned Arabyza as a city in Cauconis, the latter adding that the city 'is now called Pessinus', so Cauconis was again placed in another area. Ptolemy's reference to the Caucones (5. 1. 11) is significant, for he places them not on the coast but inland, to the south of Mt Orminion (probably Mt Köroğlu), which, however, is mentioned by no one else in the entire body of ancient literature.[166] Finally, in the *Scholia vetera in Apol. Arg.* (156–157) we read that 'the Enetoi were called Caucones before the Trojan War', despite the fact that Homer mentions them separately.

A similar ambiguity also characterises the geographic location of the Peloponnesian Caucones, who are sometimes placed in the south-west Peloponnese, in Triphylia (e.g. Homer *Odyssey* 3. 366; Callimachus *Hymni* 1. 39; Strabo 8. 3. 17), sometimes in the north-west Peloponnese, in Dyme (e.g. Antimachus fragm. 27; Strabo 8. 3. 10–11) and sometimes as occupying the whole area between Messenia and Dyme (Strabo 8. 3. 17).

*

* *

The paucity of information about the Caucones makes it impossible to draw a firm conclusion about their existence. Judging by the hesitancy of some scholars to accept the existence of other peoples mentioned in the Trojan Catalogue, such as the Halizones, one would expect this reservation to include the Caucones as well, which however is not the always the case.[167] All the obscurities and disagreements surrounding them do indeed make us doubtful of their precise geographic location and even of their very existence.[168] But this does not actually mean that they did not exist. Ptolemy's reference to the Caucones (5. 1. 11) is of particular interest, for he speaks of them as if they still existed in his time, contrary to half of Strabo's statements (7. 7. 2; 12. 3. 9) on the

[165] Stephanus of Byzantium (*s.v.* Καυκώνεια) lists a Cauconia, land of the Caucones, which he says was named for either a king or a river, but without further specification.

[166] See more about Ptolemy's Caucones below.

[167] See more recently Dan 2012–13 on the Halizones, and especially 38 on the Caucones. See also here, Section 2.9.5.

[168] Claudius Aelianus (fragm. 282) speaks of 'the Homeric Caucones and Leleges', as if Homer had invented them, but then he adds that they 'plot against people'.

matter. We shall bear in mind that Ptolemy emphasised that he was presenting a map of the *Oecumene* as it was in his day, not creating an historical atlas. This is made clear in his introduction to the *Geography*, one of the most important examples of an exposition of scientific methodology in all of ancient literature. There Ptolemy points out that he takes care to collect the latest information from many different reliable sources, such as people who have visited the places, and to crosscheck that information in order to create a world map as close to reality as possible, at the same time being well aware that the danger of mistakes is ever present, since 'many parts of the world are different now from what they were before, because features have ceased to exist or have changed' (Ptolemy 1. 5; 1. 19).[169]

Ptolemy's reference to the Caucones[170] is indeed one of the best arguments in favour of their historicity, especially if it is combined with the other half of Strabo's statements (8. 3. 17; 12. 3. 5). Like the Leleges and the Pelasgians,[171] it seems that the Caucones might have belonged to those very ancient peoples for which things were not perfectly clear after the Archaic period. The several theories about them and the geographic areas to which they were native might have been encouraged by some sporadic reminiscence or remaining traces of them, or perhaps even claims by later inhabitants here and there. Some of these theories may have contained elements of fiction, but the Caucones themselves need not necessarily have been fictitious.

*

* *

The obscurity surrounding the Caucones is reflected in the creation of the variant of the *Iliad* that includes them in the Trojan Catalogue, which was presented at the beginning of this section as having been known to Callisthenes (Strabo 12. 3. 5). First of all, we are not dealing with an insertion made by Callisthenes himself, as we read in most translations of Strabo,[172] which seem to have influenced several modern scholars.[173] The original text says Καλλισθένης δὲ καὶ ἔγραφε τὰ ἔπη ταῦτα εἰς τὸν διάκοσμον, μετὰ τὸ..., which means that he found (and not invented) these two lines in a manuscript of the *Iliad*, which remains unknown to us. According to Bolling, it could have been

[169] Also interesting is the fact that in some manuscripts of Ptolemy's *Geography* the Caucones appear as Cycones or Cyllones (Stückelberger and Grasshoff 2006, 5. 1. 11 and n. 6).

[170] See also in Section 2.3, on his reference to the Bebryces.

[171] See also Sakellariou 1958, 414–26.

[172] As in the one by Jones used here (Loeb), or in that by W. Falconer (London 1903), etc.

[173] Including myself: e.g. Burstein 1976b, 340–41; Ivantchik 1998, 319; Manoledakis 2013a, 33.

a text that used 'an old Argonaut epos',[174] though this is just speculation – probable but unproven. In any case, these two lines have been placed after 2. 855 and thus called 2. 855a,b.

Whatever Callisthenes' source may have been, this 'tendency to make the Trojan Catalogue complete'[175] from an early period of time, which this variant reflects, probably indicates incomplete Greek knowledge of the population map of the southern Black Sea littoral before the composition of the *Iliad* in the early 7th century BC – which is considered not the era but just the *terminus ante quem* to which the epos refers.[176] After all, the variant with lines 2. 855a,b involves a geographic contradiction to the dominant variant without them,[177] since in the latter (with 2. 853–855) the Parthenius belonged to the territory of the Paphlagonians (or the Enetoi), while in the former it was the Caucones whose land it traversed. This might be one reason, among others, why the insertion of 2. 855a,b was generally disregarded in antiquity, as is apparent from the work of Apollonius, for example, who totally ignored it.[178]

Both lines 2. 853, 855 and lines 2. 855a,b were interpolations to the Homeric Catalogue,[179] which displays an almost standard pattern in the presentation of the contingents, naming nations, leaders and region, but not cities (as in 2. 853 and 855). Who in antiquity accepted which of these lines as genuine and who not is another matter. For example, Apollonius and Strabo seem to have accepted lines 2. 853–855, but not lines 2. 855a,b; while Apollodorus seems to have rejected lines 2. 853–855, or to have been unaware of them, which has led to the view that they were a 2nd- or 1st-century BC interpolation.[180]

However, if Callisthenes knew of lines 2. 855a,b, then there appears to be a problem with the dating of the interpolation of lines 2. 853–855 in the 3rd century BC, and before Apollonius (as I have proposed[181]), who mentions

[174] Bolling 1925, 77–79. See more in Rengakos 1993, 129–130, n. 5.

[175] As Bolling (1925, 78) has put it.

[176] See the bibliography on the relevant issue in Manoledakis 2013a, 30, n. 71.

[177] Παφλαγόνων δ' ἡγεῖτο Πυλαιμένεος λάσιον κῆρ (851)
ἐξ Ἐνετῶν, ὅθεν ἡμιόνων γένος ἀγροτεράων,
οἵ ῥα Κύτωρον ἔχον καὶ Σήσαμον ἀμφενέμοντο (853)
ἀμφί τε Παρθένιον ποταμὸν κλυτὰ δώματ' ἔναιον
Κρῶμνάν τ' Αἰγιαλόν τε καὶ ὑψηλοὺς Ἐρυθίνους (855).

[178] For the interpolation, see also Kirk 1985, 259 who calls it implausible; Rengakos 1993, 129, 129–30, n. 5, with additional bibliography. Strabo also seems not to accept them.

[179] As Bolling (1925, 78) correctly noted.

[180] Mainly Allen 1921, 156–59, who, however, later revised this view, pointing to the pre-Alexandrian period (1924, 348–50). See more in Rengakos 1993, 129, n. 2; Manoledakis 2013a, 31, n. 82.

[181] Manoledakis 2013a, 31–33 in detail, with earlier bibliography. The first to attribute lines 853–855 to an interpolation (and specifically from the *Cypria*) was Allen (1921, 157–59). He was followed by several scholars, but not by Burstein (1976b), who was not persuaded by his

(*Argonautica* 2. 941–942) exactly the same settlements, since Strabo's formulation seems to indicate that 2. 855a,b were added to the specific lines.[182] But I think that the problem is not necessarily substantive, for, as we have pointed out, Strabo did not say that Callisthenes himself inserted lines 2. 855a,b after line 2. 855 (as Strabo knew it) but that he (Callisthenes) knew of a source which included those lines in that position. Strabo wrote in the late 1st century BC and early 1st century AD, thus had access to all the versions of the Trojan Catalogue with all their interpolations, but without knowing (or at least without saying) when each of them had appeared or by whom it was inserted.[183] The version predominant in his time (as also in Apollonius' day)[184] is the one with lines 2. 853–855 but without 2. 855a,b; and so he treated the latter (irrespective of when they had appeared: according to him, by the 4th century BC), as inauthentic to the text *that he knew and accepted* (in the 1st century), which does not necessarily require that lines 2. 853 and 855 pre-existed lines 2. 855a,b.[185]

6. The Paphlagonians

The Paphlagonians are without doubt an exceptional case in the southern Black Sea littoral. This particularity has to do mainly with the plethora of written references to them, uninterruptedly from the Classical age until Late Antiquity, which among other things provide the names of several Paphlagonian individuals who played crucial roles in the politics of their times and therefore attracted the criticism of others, often representing their whole nation. One should, of course, keep in mind that almost everything we know about the Paphlagonians is derived from Greek written sources (and a very few Latin

arguments and considered the lines original. However, I have proposed a new argument for a 3rd-century BC interpolation.

[182] Which would force us date the interpolation of 2. 853–855 before Callisthenes. See, for example, Ivantchik 1998, 318–20. One might also suspect that Strabo's information was simply erroneously transferred or was even erroneous itself (see more in Burstein 1976b, 339–40), but this is unlikely, since Strabo is generally considered reliable concerning his information on the Homeric texts.

[183] As he also knew of the mentioning of the dubious Crobialos (12. 3. 10; *cf.* Apollonius *Argonautica* 2. 942), which he attributes to 'some people'.

[184] Even in the modern bibliography we find the opinion that lines 2. 853–855 were original: Drews 1976, 20–22.

[185] In this case, it was not only the unknown interpolator used by Callisthenes who altered line 2. 854 and replaced it after line 2. 855a, which he inserted (therefore as 2. 855b), but it was also the interpolator of the lines mentioning the cities (2. 853 and 855) who put 2. 854 between his new lines. Thus 2. 854 could even have been original, since, unlike cities, rivers, lakes, mountains, etc., are not rare in the Trojan Catalogue (e.g. 2. 824, 829, 839, 845, 849, 863, 865–866, 868, 869, 877).

ones, which add nothing crucial to this knowledge[186]) and nothing from the Paphlagonians themselves. Thus, the image we acquire of them is actually the Greeks' image, with all that this implies. But let me first attempt to determine the territory of the Paphlagonians, a task that could itself be suggestive.

6.1. The Territory called Paphlagonia and the Cultural Impact of the Paphlagonians (Fig. 10.1–2)

The area bearing the name Paphlagonia in antiquity is impossible to determine accurately, since even contemporary sources describe different borders, even for its coastal region. In some sources the western border of Paphlagonia is the River Parthenius (Arrian *Periplus* 13–14), in others the River Billaeus (Constantine Porphyrogenitus *De thematibus Asia* 7)[187] or even the Callichorus (Pseudo-Scylax 90; Schol. in Apol. Arg. 2. 904), although in all cases it borders on Bithynia.[188] To the east it is separated from Cappadocia by the River Halys (Herodotus 1. 6. 1; 1. 72. 2. *Cf.* Strabo 12. 1. 3; 12. 3. 9, 12; Con. Porph. *De thematibus Asia* 7), while in other accounts it is the River Euarchus, east of Sinope, that forms the border (e.g. Menippus *Periplus* 9; Stephanus of Byzantium 356; Anonymous *Periplus* 23. 18–20).[189] Hecatonymus' description of Paphlagonia in Xenophon's *Anabasis* (5. 6. 9; *cf.* 6. 2. 1) implies that all four rivers, the Thermodon, the Iris,[190] the Halys (*cf.* Apollonius *Argonautica* 4. 245) and the Parthenius, lay within Paphlagonian territory, which according to Xenophon's account seems to have extended as far as the region of Cotyora, east of the Iris (*Anabasis* 5. 5. 6). The largest territory emerging from the texts thus extends from the Callichorus to Cotyora, while the smallest lies between the rivers Parthenius and Euarchus.

[186] And usually speak about events that took place in Paphlagonia and not about the Paphlagonians.

[187] On the disagreement, see Menippus *Periplus* 8; Anonymous *Periplus* 13. 9–12. In the Trojan catalogue (Homer *Iliad* 2. 851–854) the Parthenius is mentioned as part of Paphlagonia.

[188] According to Strabo (12. 1. 1), to the east of Paphlagonia lies Cappadocia and to the west live the Mariandynoi (12. 3. 2. *Cf.* Arrian *FGrH* 2b 156, F. 76a, who places the latter to the east of the Sangarius). But in 12. 3. 9, 42 and 12. 4. 1 he says that it is the Bithynians who neighbour the Paphlagonians to the west, together with the Mariandynoi. This could mean that the Mariandynoi inhabited part of coastal north-east Bithynia. See also in Section 2.2.1.

[189] For the different borders of Paphlagonia in different periods, see also C. Marek, Paphlagonia, *Der Neue Pauly* 10, 477. *Cf.* Lendle 1995, 335.

[190] It is worth mentioning here Eustathius' misconception (*Comm. in Dion. Per.* 976), who, copying the relevant passage of the *Anabasis* (5. 6. 9) erroneously transcribes the River Ἶριν as Τίγριν! One might think that this was a copyist's error, were it not for a whole chapter (976) in which Eustathius talks about the Tigris of Mesopotamia and clearly adds that there is a river of the same name in Paphlagonia, citing the above-mentioned passage of Xenophon.

Even greater vagueness attaches to Paphlagonia's southern border, where we know only of some place-names, many of them unidentified.[191] What is certain is that Paphlagonia covered an extensive territory, comprising both broad plains and lofty mountains, and was traversed by important rivers. Apart from those named above, some authors also mention the rivers Aeginetes (Aelius Herodianus *De prosodia catholica* 3. 1. 74; Stephanus of Byzantium 43. 7), Hypios (Eustathius *Comm. ad Hom. Il.* 1. 119; 3. 916) and Callichorus or Oxinas (Pseudo-Scylax 90; Dionysius Scytobrachion *FGrH* 1a 32, F 13; Schol. in Apol. Arg. 193. 17).

There could be various reasons for this uncertainty about the borders of Paphlagonia. For example, several scholars agree that 'ethnic groups are not static or monolithic, but dynamic and fluid. Their boundaries are permeable to a degree, and they may be subject to processes of assimilation and differentiation.'[192] If this is the case, then the changing borders of what is called Paphlagonia could be due to assimilation or other kinds of connections between the Paphlagonians and neighbouring peoples.

There are several interesting references in the ancient Greek sources that lead us to examine the relation of the Paphlagonians to their neighbours on both east and west and their relationships with those nations. For example, it is interesting that, while Xenophon states that Cotyora was in the territory of the Tibarenoi (*Anabasis* 5. 5. 3; *cf.* 6. 1. 1), he goes on to say that when his army was there they got their provisions partly from Paphlagonia and partly from the estates of the Cotyorites (*Anabasis* 5. 5. 6). This could mean two things: either that the lands of the Tibarenoi and the Paphlagonians were adjacent and Cotyora was close to the border or, more probably, that the whole region, including the part where the Tibarenoi lived, had come to be called Paphlagonia.[193]

[191] According to a theory, the southern borders of Paphlagonia (with Phrygia), at least in the last centuries of the Phrygian state's existence, could have been around the area of Gangra (modern Çankırı) (Matthews 2009b, 154). Sometimes in modern bibliography Paphlagonia is treated as an area lying entirely inland, for example as 'a mountainous region between the Pontic Alps in the North and the Anatolian plateau in the South' (Summerer and von Kienlin 2010, 195). But this could only refer to Paphlagonia as the name of an administrative province during a specific part of the Roman period, when the kingdom of Pontus became the united province of Pontus and Bithynia, which comprised the maritime parts of Paphlagonia and Cappadocia, 'leaving these territories only as interior provinces, given to vassals' (see Diller 1952, 158–59, 161). From Strabo (e.g. 7. 3. 6, 12. 3. 41, etc.), it is obvious that Paphlagonia as a geographic area was both coastal and inland. For the cities of Paphlagonia, see the entries in Aelius Herodianus, Stephanus of Byzantium, Hesychius, and of course the relevant data given by Strabo and Ptolemy (among other sources, see also Athenaeus *Deipnosophistae* 3. 23).

[192] Hall 1997, 29, 33, with previous bibliography.

[193] See also in Section 2.10.1.

But the Tibarenoi are not the only nation to the east of the Paphlagonians with which there appears to be a geographic overlap. The same holds true also for the Leucosyroi. According to Xenophon (*Anabasis* 6. 1. 15; *cf.* Nicephorus *Geographia, GGM* II 464b, 30–32), Sinope was in Paphlagonia, although Pseudo-Scylax (89–90) says that both the Halys and Sinope were in the land of the Assyrians and places Paphlagonia farther west (*cf.* Herodotus 1. 72; Dionysius Periegetes *Orbis descriptio* 787). We could get around this by saying that what is meant here by Paphlagonia is simply a geographic region and citing the text of *Etymologicum Gudianum* (*s.v.* Ἅλυς), which states that the Halys was the border between the Chalybes and the Leucosyroi in Paphlagonia. However, we also have the impressive evidence of the Scholiast in Dionysius Periegetes (*Orbis descriptio* 65–66) that 'ἡ Σινώπη τῶν Ἀσσυρίων ἐστίν ἤτοι Παφλαγόνων', which seems to present the two peoples as kindred.[194] Even the enigmatic Halizones, who are mentioned in the Trojan Catalogue clearly as a different people from the Paphlagonians, and whose identification has generated numerous views,[195] are also sometimes mentioned as a nation of Paphlagonia (Hesychius *Lexicon*, *s.v.* Ἁλιζῶνες) or even of Paphlagonians (*Etymologicum parvum*, *s.v.* Ἁλιζῶνες).[196]

But there is much more. The Paphlagonians are closely associated with the Enetoi,[197] concerning whom things are far from clear. Some texts refer to the cultural influence of the Paphlagonians on their neighbours as manifest through similarities in their armour. According to Herodotus (7. 72), the Ligyes, the Matienoi, the Mariandynoi and the Syroi were equipped like the Paphlagonians (*cf.* Stephanus of Byzantium 653). Similar information is provided by Hecataeus (1a, 1, F 287. 1), one of Herodotus' sources, namely that the Matienoi wore clothes like those of the Paphlagonians, while according to Herodotus (7. 72) both had the same commander, Dotus son of Megasidrus, which would indicate that their regions lay in close proximity. The Matienoi are mentioned by Hecataeus as neighbours of the Gordians (F 287) and the Moschoi (F 288). Herodotus places them in the 18th satrapy of the Persian Empire (3. 94).[198] Finally, Xenophon says that both the Drilae and the Mossynoikoi, two peoples of the eastern part of the coast, wore helmets that were or

[194] This question is analysed in Section 2.8.3C.

[195] For the Halizones, see Section 2.9.5.

[196] *Cf.* Saprykin 1991, 250.

[197] See Section 2.7.

[198] The description of the course of the Halys in Herodotus 1. 72 indicates that the Matienoi did not live on the coast, but inland, to the south of the Leucosyroi, who bordered the Paphlagonians to their west, and thus to the south-east of the latter, the Halys flowing between them. Herodotus also mentions a homonymous nation between Armenia and Kissia (5. 49; 5. 52).

resembled Paphlagonian ones (*Anabasis* 5. 2. 22; 5. 4. 13; *cf.* the *Suda sigma* 772).

It is noteworthy that in all these accounts it is always other peoples that had armour resembling the Paphlagonian and never that the Paphlagonians had armour resembling that of another people or that all these peoples wore similar armour. There are, as well, peoples to the west of the Paphlagonians that are connected with them, such as, for example, the Caucones.[199] Finally, we also have the Mariandynoi. Theopompus, who lived in the 4th century BC, said (according to Strabo 12. 3. 4) that king Mariandynus governed a part of Paphlagonia[200] which was subject to many masters, while another version presents the Paphlagonians as having been subdued by Heracles, for the sake of the father of Lycus, king of the Mariandynoi (Apollonius *Argonautica* 2. 790; Asclepiades fragm. 2 = Dinias fragm. 1).[201]

All the above could indicate a cultural influence, if not a cultural dominance, of the Paphlagonians over their neighbours, some of whom may have even had an ethnic kinship with them.[202] This dominance, which has even led to expression of the view that all the local peoples of the central part of the southern Black Sea littoral were probably Paphlagonians,[203] may probably have established the name Paphlagonia in a large area extending from the Billaeus to east of the Iris river. The name is purely cultural and geographic, devoid of any administrative character, and the territory is not to be confused with the homonymous Roman province, which in some periods even lacked access to the sea. In passing, it is worth noting that the Paphlagonians are among the very few southern Black Sea peoples that managed to give their name to the territory they inhabited, and perhaps an even larger one.[204]

[199] For the Caucones, see Section 2.5.

[200] For the relation between the Mariandynoi and the Paphlagonians, see also Saprykin 1991, 245.

[201] For the presence of Heracles in Paphlagonia, especially in works of art, see Summerer 2009.

[202] Impressive is the phrase of the *Etymologicum magnum* (340. 23–25): ''Ενετοί γάρ οἱ Παφλαγόνες οἱ Λυδοί'. But would this imply a relationship of the Paphlagonians even with the Lydians?

[203] See most recently Tsetskhladze 2007, 193; 2012b, 236, invoking Saprykin (1991).

[204] Thus, it is striking that Apollonius refers very rarely to the Paphlagonians and never to Paphlagonia. In the first of the four relevant passages, the 'hills of the Paphlagonians' are vaguely placed in the huge area between the Billaeus and the Halys or the Thermodon (*Argonautica* 2. 357–358; *cf.* 2. 790–791; 4. 245, 300), although elsewhere Sinope, to the west of the Halys, is said to belong to the land of Assyria (2. 946).

6.2. On the Origin of the Paphlagonians

But who were these Paphlagonians who managed to dominate such a large part of the southern Black Sea littoral and influence so many peoples around them? There are several testimonies concerning the (mythical) origin of the Paphlagonians in the ancient written sources. These testimonies are anything but unanimous and tend rather to create confusion regarding the origins of this people. One cause of this confusion may be that their roots stretch far back in history. A comprehensive study of the written sources leaves no doubt that the Paphlagonians have always been considered a most ancient nation, as even their harsh critic Constantine Porphyrogenitus (*De thematibus Asia* 7) acknowledges, as did Justinian (*Novellae* 218) before him.

We are taken deepest back in time by the historian Flavius Josephus, writing in the 1st century AD. In his work *Antiquities of the Jews* (1. 122–126. *Cf.* Eustathius *Comm. in hex.* 756), he presents the Paphlagonians as descended from Riphath, grandson of Noah's son Japheth (*cf. Genesis* 10:1–3). This account, and the fact that the Paphlagonians are the only southern Black Sea nation for which Josephus makes such a statement,[205] are surely suggestive. But there is more. Herodotus, in a well-known story (2. 2), tells of Psammetichus' experiment showing that the most ancient language in the world was Phrygian (and thus that the Phrygians were the most ancient nation), because the first word uttered by the two babies used in the experiment was βέκος (bread in Phrygian). However, in the texts of some later Scholiasts, the Paphlagonians replace the Phrygians in the narrative and the word βέκος is ascribed to their language (Schol. in Aristoph., Schol. in nub. 398c, where Psammetichus is also replaced by Sesonchosis). Elsewhere, we meet the Phrygians quarrelling with the Paphlagonians about who is more ancient, and in this case it is the Phrygians who emerge as the winners (Schol. in Ael. Aristid. Rhet. 3. 7). The same scholiast of Aristophanes (Schol. in nub. 398d) cites another version,[206] according to which three nations were quarrelling, the Arcadians (who in 398c are also mentioned as προσέληνοι), the Persians and

[205] He also mentions the Cappadocians in a way that connects them with the Moschoi, who may be connected to the Mushki. However that may be, the Moschoi did not inhabit the coastal area of the southern Black Sea.

[206] These versions are presented by the Scholiast on the occasion of the word βεκκεσέληνε met in Aristophones *Nubes* 398, which is interpreted in various ways, indicating old age as well as folly: Schol. in nub. 398b, d: βεκκεσέληνε = 'you ancient and fool'; 398c: 'the word means archaism'; 398f: 'both the bread and the moon are most ancient'; 398g: βεκκεσέληνε = 'you moonstruck and trivial'; 398h: 'you unfeeling'.

the Phrygians, when the intervention of Psammetichus proved that the Phrygians were the most ancient of all.[207]

The sources may have agreed on the antiquity of the Paphlagonians, but they did not agree on their origin. Josephus (see above) refers to the Greek habit of inventing several kinds of relationships with other peoples in their mythology, which could even go so far as to attribute a Greek origin to a foreign nation, even their traditional enemies, the Persians. As Josephus colourfully puts it:

> some of the nations do still retain the denominations which were given them by their first founders; but some have lost them also, and some have only admitted certain changes in them, that they might be the more intelligible to the inhabitants. And they were the Greeks who became the authors of such mutations. For when in after-ages they grew potent, they claimed to themselves the glory of antiquity; giving names to the nations that sounded well (in Greek) that they might be better understood among themselves; and setting agreeable forms of government over them, as if they were a people derived from themselves.[208]

Thus in our case, always according to Josephus, from Riphath came the Riphathaioi, 'who are now called by the Greeks Paphlagonians'.[209] And, of course, the link with Japheth as ancestor of the Paphlagonians is consonant with the identification of Japheth as ancestor of the Indo-Europeans, to whom the Paphlagonians belong.

However, there is also an element that connects the Paphlagonians with the Greeks. In one version of the origin of Pelops (Diodorus 4. 74. 1), the hero who founded the Olympic Games and gave his name to the Peloponnese was a Paphlagonian,[210] and indeed, according to Apollonius (*Argonautica* 2. 358–359, 790. *Cf.* Schol. in Apol. Arg. 157; *Etymologicum Magnum* 340. 23–25), was king of the Paphlagonians from Enete, just like Pylaemenes in the *Iliad* (2. 851–852), the most ancient Greek source mentioning the Paphlagonians. What we have here is the reverse of what Josephus says. The eponymous hero of one of the most ancient Greek regions is presented as having an Asian origin. Thucydides (1. 9. 2), citing persons familiar with Peloponnesian traditions, explains the power Pelops acquired in the Peloponnese by the great wealth he

[207] On these versions, see also *Scholia graeca in Aristophanem. Cum prolegomenis grammaticorum, varietate lectionis optimorum codicum integra, ceterorum selecta, annotatione criticorum item selecta, cui sua quaedam inseruit Frg. Dübner*, Paris 1842, 431, *ad nub*. 397.

[208] *Antiquities of the Jews* 1. 121. *Cf.* Mitchell 2007, 180–84; Zacharia 2008, 30; Gruen 2011, 223, and especially on the Persians 224–27.

[209] Indeed, Hornblower (2008, 39; *contra* Xydopoulos 2010, 19) points out that: 'Greeks who settled new areas were noticeably and sometimes ingeniously prone to recategorize as Greeks the people whom they found there.'

[210] Other versions also mention a Phrygian or Lydian origin. All of them in Gruen 2011, 227–29.

brought with him from Asia. Possible reasons for the desire of the Greeks to establish such a connection to the Paphlagonians could be detected in the general perception that they had of them, as this is traced through the Greek texts.[211]

Other mythical traditions, however, connect the Paphlagonians with the Thracians, since the eponymous hero Paphlagon is reported to be the son of Phineus, king of Thrace, who once lived in Paphlagonia, in the town called Sesamos (Arrian *FGrH* 2b 156, F. 77a; Aelius Herodianus *De prosodia catholica* 3. 1. 297, 382; Stephanus of Byzantium 513. 5, 562. 22–23; Constantine Porphyrogenitus *De thematibus Asia* 7). Although most authors say that it was just the region of Paphlagonia that took its name from this hero, there is a passage of the Scholiast of Apollonius (Schol. in Apol. Arg. 2. 181) where the Paphlagonians are presented as Thracians from Asia.[212] Such a passage and an attempt – not actually a successful one – to compare Paphlagonian names with Thracian ones, had led to the theory that the Paphlagonians were Thracians who came to Asia Minor from the Hellespont.[213]

The confusion that prevailed over the years concerning the real origin of the Paphlagonians is further indicated by the fact that some later authors (e.g. Schol. in Apol. Arg.) mention both the Phineus and the Pelops versions, while others (e.g. Constantine Porphyrogenitus *De thematibus Asia* 7), although tracing their origin to Phineus, call the Paphlagonians an Egyptian nation! This last may be due to the version of the myth that presents Phineus as son of Belus, king of Egypt (Apollodorus *Bibliotheca* 2. 1. 4, citing Euripides).

Other references to the Paphlagonians in ancient Greek literature, such as the passage of Herodotus (7. 73), in which we read that the dress of the Phrygians closely resembled that of the Paphlagonians, combined with the presence of the rock-cut tombs (*Felsgräber*) that are often encountered in the Phrygian as well as the Paphlagonian landscapes, and the presence of Phrygian pottery (which he attributes to the 'North-East Phrygians') in many areas of Paphlagonia, led to another theory, namely that the Paphlagonians had an ethnic kinship with the Phrygians and spoke the same or similar languages.[214] Apparently,

[211] See more in Section 2.6.4.

[212] The passage is as follows: "Ὅτι δὲ ἦρχεν ὁ Φινεὺς μέχρι τοῦ Βοσπόρου Θρᾳκῶν πάντων τῶν ἐν τῇ Ἀσίᾳ (εἰσὶ δὲ οὗτοι Βιθυνοί τε καὶ Παφλαγόνες), Φερεκύδης ἐν τῇ ϛ′ φησί'. Since the crucial words are in parenthesis, it seems that the Thracian origin of the Paphlagonians is supported by the Scholiast and not Pherecydes himself. *Cf.* Ruge 1949, 2518.

[213] Leonhard 1915, 298–308; Saprykin 1991, 249–50. In an extension of this theory, it was even argued (Leonhard 1915, 304–08) that there was also an Illyrian invasion in Paphlagonia. Ruge (1949, 2518–19) finds Leonhard's arguments weak.

[214] Haas 1966, 9–18, 57, 61, 173, 179–84, 231, 235, 241–42, 249. *Cf.* Akurgal 1955a, 93, 126; 1956, 57. See also in Section 3.4. On the evidence about relations between Paphlagonians and Phrygians, including a criticism of Haas's arguments, see Manoledakis 2021d.

this theory also connected the Paphlagonians with population movements from the Balkans to Anatolia in the beginning of the Early Iron Age. The Phrygians were indeed considered another very ancient nation in antiquity, as the abovementioned myth of the Psammetichus experiment indicates. And there are undeniably several other similarities between the two peoples, who also frequently appear together in the literature from the Imperial period on. But none of these is enough to lead to a conclusion of a common origin for these two peoples, and it might have been an aspiration of the Paphlagonians to be connected to the Phrygians in an attempt to establish a weightier national identity.[215]

All the theories mentioned so far have been based on sources dating from the Roman through the Byzantine period, which present only the Greek and Roman traditions relating to the antiquity and the origins of the Paphlagonians, and possibly in some cases Paphlagonian traditions, though of a very late period. One can easily understand that these traditions were shaped according to the perceptions of their creators and might well be totally unrelated to reality, which should be sought centuries before their creation. The problem is that, as is the case with almost every other nation of the southern Black Sea region, there are no solid archaeological data to tell us about the Paphlagonians.[216] Consequently, the emergence of these people in the specific area in the centuries that preceded the arrival and establishment of the Greeks there remains obscure.

According to one widely accepted theory, the Paphlagonians should be associated with those Bronze Age people who in Hittite texts are said to have occupied the land called Pala. Geographically, this land roughly coincides with the territory of the later Paphlagonia as defined above. The name of the Paphlagonians would thus have evolved from the word *pala*. The language spoken by the people of Pala, the so-called Palaic language, was a 2nd-millennium BC Anatolian tongue related to Hittite and Luwian of which practically nothing is known. Relations between the people of Pala and the Hittites are thought to have been interrupted until the Kaska people invaded and occupied the southern Black Sea littoral, where the Palaic people lived, when the latter were displaced by the Kaska and actually disappear from the historical sources.[217]

[215] Manoledakis 2021d, in detail.

[216] Marek 1993, 14–15. *Cf.* also Summerer and von Kienlin 2010. On some types of funerary monuments that have been attributed to the Paphlagonians, but also to others, see Manoledakis, 2021c; 2021d, with earlier bibliography.

[217] Forrer 1921, 24; Ruge 1949, 2511, 2521; Maksimova 1956, 19; Mellaart 1978, 53; Melchert 1994, 10; Finkelberg 1997, 5; Luraghi 1998, 169–72; Popko 2008, 60–62; Taracha 2009, 58–59; Bryce 2006, 140; 2009, 552, who claims that the land of Pala was subject to the

The plausibility of this theory could be heightened if our knowledge of the Paphlagonian language were not restricted to a few personal names recorded by Strabo and in inscriptions,[218] which of course are not enough to give us much of an idea of this language, which most probably lacked a script. It is, however, a theory far more plausible than all the rest. The clear cultural dominance of the Paphlagonians over the whole central part of the southern Black Sea littoral could be an argument for it, and besides, we should not regard the peoples we are dealing with in this book as coming into an empty space. It is possible that the people who became known as Paphlagonians in the 1st millennium BC derived from Palaic populations that survived the Kaska occupation and were mixed with other local elements of northern Anatolia.[219]

The matter could, of course, be better investigated if we had archaeological remains safely attributed to the Paphlagonians. But unfortunately this is impossible, as is the case with any other nation of the southern Black Sea. In the in any case difficult issue of the archaeological identification of an ethnicity,[220] there are no archaeological data to tell us about the Paphlagonians.[221]

6.3. Ancient Literature and the Paphlagonians

In the most ancient testimony concerning the Paphlagonians, which occurs in the earliest (first half of the 7th century BC[222]) Greek literary work preserved, the *Iliad* (2. 851–855), we read that the Paphlagonians took part in the Trojan War as allies of the Trojans. They were led by 'stout-hearted' Pylaemenes, who came 'from the lands of the Enetoi', known, according to this passage, as the place from which the wild mules come. The Paphlagonians held Cytoron, Cromna, Aegialos and 'lofty' Erythinoi and lived in the country around Sesamos, in places bathed by the River Parthenius. Later in the poem we are told that Pylaemenes was killed in the tenth year of the war, by Menelaus (Homer *Iliad* 5. 576–579. *Cf.* Hyginus *Fabulae* 113; Tzetzes *Allegoriae Iliadis* 5.

Hittites; *contra* Ruge 1949, 2515. See also Section 2.1. For the Palaic language and its relation to Hittite and Luvian, see also Steadman 2011; Beckman 2011.

[218] Strabo 12. 3. 5: Bagas, Biasas, Aeniates, Rhatotes, Zardoces, Tibius (*cf.* 7. 3. 12), Gasys, Oligasys and Manes. We also know of a woman called Boa, who was a flute–player and a courtesan, a Paphlagonian by birth (Athenaeus *Deipnosophistae* 13. 38; Caryst. *FGrH* 12). For Paphlagonian personal names, see Avram 2018, with the previous bibliography.

[219] *Cf.* Mellaart 1978, 79.

[220] For the problems and risks associated with trying to trace ethnicity through archaeological data, see also Antonaccio 2001, 124–33. *Cf.* Jones 1997, 124; Tsetskhladze 2012a; 2012b; 2014, 316. For the general problem of the ethnic identification of the Paphlagonians, see also Maksimova 1956, 19.

[221] Marek 1993, 14–15. *Cf.* also Summerer and von Kienlin 2010.

[222] See the discussion in Manoledakis 2013a, 29–30.

81–82), while other sources say he was slain by Achilles (Dictys 3. 5) or Patroclus (Nepos *Datames* 2. 2). According to Pliny (*NH* 6. 5), the region of Paphlagonia was called Pylaemenia, after Pylaemenes.[223]

The next source of information about the Paphlagonians is the *Histories* of Herodotus, who records that the Paphlagonians took the field with Xerxes against Greece. They wore woven helmets and sandals covering the lower leg and carried lances, daggers, small shields and spears (Herodotus 7. 72). The Paphlagonians, who like most peoples west of the River Halys had earlier been conquered by Croesus (Herodotus 1. 28; *cf.* Xenophon *Cyropaedia* 6. 2. 10), passed under the dominion of Cyrus the Great (Xenophon *Cyropaedia* 1. 1. 4) after the fall of the Lydian kingdom; they belonged to the 3rd province (satrapy) that – according to Herodotus – Darius established and paid 360 talents in tribute (Herodotus 3. 90; *cf.* Xenophon *Cyropaedia* 8. 6. 8). About a century later, *ca.* 361–360 BC, the Paphlagonians were still on good terms with the Persians, since Artabazus is reported to have tried to draw supplies from the 'friendly lands of Upper Phrygia, Lydia and Paphlagonia' (Demosthenes *Against Aristocrates* 155) for his war against Charidemus in Hellespontine Phrygia.[224] This does not mean that there were no exceptions: at the end of the 5th century BC some Paphlagonians refused to present themselves when the Persian king summoned them, as their ruler was too proud to obey (Xenophon *Anabasis* 5. 6. 8).

Returning from Cunaxa in 400 BC, Xenophon and the rest of the Ten Thousand made their way along the southern Black Sea coast. Inevitably, they encountered the Paphlagonians, since they passed through their territory. Xenophon (*Anabasis* 1. 8. 5) calls them barbarians, of course, but takes care to remind his readers that they had fought alongside the Greek mercenaries for Cyrus against his brother Artaxerxes (*cf.* Diodorus 14. 22). During the Greeks' passage through Cotyora, a colony of Sinope, tensions arise between the Sinopeans and Xenophon's army. The Sinopean envoy Hecatonymus threatens to join forces with the Paphlagonians of Corylas to drive them out of his city's colony (*Anabasis* 5. 5. 12). Xenophon, however, replies that not only is he not afraid to fight, but also that he too could ally himself with Corylas against the Sinopeans, since he knows that Corylas has expansionist designs on them (5. 5. 22–23); later on, however, Xenophon asserts that Hecatonymus was a friend of Corylas and his official representative at Sinope (5. 6. 11). There appears to be some confusion here: if everything Xenophon says is true, Hecatonymus

[223] More about Pylaemenes and his name in Manoledakis 2013a, 21.

[224] For the dating and location of these events, see Heskel 1997, 119. More about this period with regard to the Paphlagonians in Xenophon *Hellenica* 4. 1.

appears to be at once an envoy of Corylas' Paphlagonians in Sinope and an envoy of the Sinopeans to Xenophon's army (see more in the next section).

Another source of hostility between Greeks and Paphlagonians during the Ten Thousand's stay in the Cotyora district was the mutual plundering of food supplies (5. 5. 6; 6. 1. 1. *Cf.* Diodorus 14. 31). The problem was eventually overcome, and both peoples made common sacrifices with oxen that they had earlier scrounged from the area (probably from each other!) and ended up celebrating together with a feast followed by a great deal of dancing (6. 1. 4–13), before the Greeks departed for Sinope with the assistance of the Paphlagonians (6. 1. 14). During the several dances performed by Greeks from various regions[225] the Paphlagonians cheered, but also thought it most strange that all the dancers were armed, and asked whether women also fought alongside the men.

Finally, there are only very few direct references to the religious life of the Paphlagonians. According to Strabo (12. 3. 40), Mt Olgassys in Paphlagonia was full of temples erected by the Paphlagonians; he does not, however, name the deities to which these temples were dedicated. In the same era, Plutarch (*De Iside et Osiride* 69) states that Dionysus was worshipped by the Phrygians, the Paphlagonians and the Greeks as a god of the year-cycle. This is the sum of our knowledge about the religious life of the Paphlagonians.

6.4. The Greeks' Perception of the Paphlagonians

A. The Positive Side

One of the most interesting things about the Paphlagonians, which distinguishes them from the rest of the southern Black Sea peoples, is the profusion of references to them in ancient Greek literature. Whatever the reasons for this abundance may have been, it is important because it enables us to form an indicative image of the Greeks' perception of this people until very late in antiquity. Therefore, and since the way in which the Greeks saw and thought of the peoples of our area of interest is one of our concerns in this section of the book, we shall dedicate some pages to it.[226]

Having fought against the Greeks in both the great wars – mythic and historic – that marked forever both Greek history and the Greeks' relationship with the barbarians of the East, one would expect a very negative impression to attach to the Paphlagonians in Greek eyes. A widely accepted theory holds

[225] For this feast, see Lane Fox 2004a, 191.

[226] A thorough examination of this issue is attempted in Manoledakis 2021c, on which the present section is based.

that the Persian Wars played the most crucial role in the Greeks' conception of their ethnic identity.[227] As a result, in almost all the works of Tragedy and Old Comedy, that is, in the period immediately after the Persian Wars, the opposition between Greeks and barbarians plays a primary role.[228] This is evident in many of the works of Aristophanes. In one of them, especially (*Knights*, 424 BC), the main protagonist, the politician Cleon, whom the poet detested and wished to insult, appears under the name Paphlagon. Although it is generally assumed that Aristophanes chose this name because of the sense of the verb παφλάζω, meaning to splutter,[229] a tendency that Cleon was noted for, we cannot exclude the possibility that the Paphlagonians were generally looked down on by a segment of 5th-century Athenian society, as was the case with other regions that provided the city with slaves.[230]

Rhetoric against peoples of the East, and specifically the Persians, of which Isocrates was a famous practitioner, was dominant until at least the late 4th century BC. However, only a few years after Aristophanes presented his *Knights*, another Athenian, Xenophon, devoted hundreds of pages, especially in the *Cyropaedia*, not only to describing Persia and the Persians, as well as many peoples of northern Anatolia, but also to praising Cyrus the Great, the most important king of the Persian Empire.[231] This shows that anti-Persian sentiment was not universal after the Persian Wars.[232] Xenophon's *Anabasis*,

[227] See, for example, bibliography in Hall 2002, 175. Also, Hornblower 1991, 11. For a different approach, see Mitchell 2007, 77–112; Hornblower 2008, 38–39; Vlassopoulos 2013, 35–36.

[228] See Hall 1989, 3–13; 2002, 172–89, both with more bibliography. Hall (1989, 133) also adds that the tragic poets were influenced by the descriptions of foreign lands and peoples provided by historians, especially Hecataeus and Herodotus (who, unfortunately, is not as helpful in our case as with other peoples). For the barbarians, especially in Greek Comedy, see Long 1986. On Greeks and barbarians of the Black Sea, see also Dana 2011, 339–55.

[229] As several ancient scholiasts and lexicographers noted (e.g. Schol. in Aristoph., *sch. in eq.* 2a. 4, 2e. 1, 2f. 1; *sch. in vesp.* 34; Eustathius *Comm. ad Hom. Il.* 1. 566; 3. 532; the *Suda omicron* 444), long before the modern scholars who expressed the same view, often without citing the former (Long 1986, 114; Bäbler 1998, 94; Lewis 2011, 100). For the portrait of the Paphlagonian in the *Knights*, see Lauriola 2006, with previous bibliography.

[230] Avram 2007; Tsetskhladze 2008a; Lewis 2011, all with previous bibliography. As these authors – and others before them – correctly suggest, it is not enough to rely on personal slave names to conclude the real ethnic origin of the slaves. An ethnic slave name may indicate the place where the slave was sold and not his real place of origin, not to mention that a personal name could be popular in more than one area, as for example the name Manes, which was common in Phrygia, Paphlagonia and other parts of Asia Minor (Özlem-Aytaçlar 2010, 525–26; Lewis 2011, 97–98). However, especially as far as Paphlagonia is concerned, we cannot fail to mention Lauffer's view that most of the mining slaves from Asia Minor came from areas with mines, such as Paphlagonia (Lauffer 1979, 140).

[231] See indicatively Hirsch 1985, especially 61–100.

[232] *Cf.* Gruen 2011, 53.

narrating events of 401–400 BC, is the first to provide important information about the Paphlagonians and some aspects of their relationship with the Greeks. He may call the Paphlagonians barbarians (*Anabasis* 1. 8. 5), but in the very same passage he records that they had fought side by side with the Greek mercenaries for Cyrus against his brother Artaxerxes. So there was already contact between the two peoples as allies. A contact that continued during the passage of the Ten Thousand through Paphlagonia on their way back to Greece.

The *Anabasis* contains several interesting references to that. We have already remarked on the unclear relation between the Sinopean envoy Hecatonymus and the Paphlagonian ruler Corylas (Xenophon *Anabasis* 5. 5. 22–23; 5. 6. 11). The relevant passage in Xenophon points to political/diplomatic relations between the Greeks of Sinope and the local Paphlagonians, which, however, are disturbed by the expansionist views Corylas entertains over the territory of Sinope. What exactly was the role of Hecatonymus, and what was his relationship with Corylas, who, still according to Xenophon (5. 5. 23), had designs on the former's city? Apparently we have here a delicate aspect of the complex issue of the relationship between Greeks and locals in the colonial world.[233] In any case, this relationship was multi-dimensional and often close, judging by the Sinopeans' excellent knowledge of Paphlagonian territory, stressed twice in the text (5. 6. 2, 6).

But the most important narrative is that recounting the feast shared by Greeks and Paphlagonians, which ended with common sacrifices, a rich dinner and a lot of dancing (6. 1. 4–13). This account is important, not only because it is the first one recording a meeting of a cultural nature between the two peoples, but also because it seems to indicate an equality or mutuality concerning the actions of both sides, from the mutual depredations to the common feast and the interest of one in the customs of the other. This narrative gives no sense of a distinction between civilised and barbarian people, according to the spirit of the age mentioned above, since both peoples do the same things and end by celebrating together in two sacred aspects of Greek culture: hospitality (ἐπί ξένια ἐδέχοντο) and banqueting. Indeed, it is the Paphlagonians who are shocked by the warlike nature of the Greek dances.

In the first part of our examination of the Paphlagonians we saw that the information of the Greek written sources often suggests a cultural dominance of the Paphlagonians over their neighbours. If, then, the Greeks recognised a cultural superiority of this sort, could one perhaps speak of 'more and less barbarian' peoples? Indeed, studying the Greek texts we can assume that for

[233] For this issue, see recently Manoledakis 2015; 2017b.

the Greeks there must have been various degrees of barbarism, that some peoples were considered more barbarian than others, as indicated by Diogenianus' adage, 'Lydians are bad; Egyptians are worse; but Carians are worst of all.'[234]

A very significant element in this graduation was surely religion, since, as we have seen, Herodotus considered it an important criterion for determining ethnicity. Herodotus himself mentioned the Budini, who had sanctuaries where they worshipped Greek gods (4. 108). According to F. Hartog, 'temples, statues, and altars are signs of Greekness and may function as a criterion of Greekness'.[235] Therefore, having seen that Mt Olgassys in Paphlagonia was full of temples erected by the Paphlagonians and that Dionysus was worshipped by the Paphlagonians in the same way as by the Greeks (see above), it is tempting to think that in Greek eyes the Paphlagonians did not belong to the higher levels of barbarism; on the contrary, they should have been considered among the least barbarous of the non-Greek world. About no other nation of the southern Black Sea is it said that they worshipped Greek deities. This is the case only with the inland-dwelling Phrygians (e.g. Plutarch *De Iside et Osiride* 69), about whom it also emerges from the sources that they were considered by the Greeks 'much less barbarian' than many others, and this not without reason.[236]

Another important element in assigning gradations of barbarism is, of course, a people's customs and way of life generally. Comparing Xenophon's descriptions of the several southern Black Sea peoples that he met, we note that there is nothing reprehensible about the Paphlagonians, as there is, for example, with the weird Tibarenoi, the tough Chalybes, the shameless Mossynoikoi, or the dreadful and warlike Drilae.[237] The way Xenophon treats the Paphlagonians, in contrast to their barbarian neighbours, shows that for someone who had himself met foreign peoples, the classification as 'more or less barbarian' had nothing to do with Greek tradition or history (for example, if they had fought against the Greeks), but with his personal experience of them.

But it was not always necessary for a Greek to travel to Paphlagonia and meet the Paphlagonians there in order to form the opinion that they were not the ultimate in uncivilised barbarians. The opposite could also be the case. A 4th-century BC epitaph from Laurion is indicative.[238] It commemorates a man called Atotas, from Paphlagonia, who was a miner. Although a manual

[234] Long 1986, 129–31; Hall 2002, 180–81; Xydopoulos 2010, 19–21, with more bibliography and the addition of the view that 'barbarians had their own barbarians'; Bonfante 2011, 8.

[235] Hartog 1988, 176.

[236] See Manoledakis 2016 in detail.

[237] See in the relevant sections.

[238] *IG* II^2 10051.

worker – with all that that would imply, especially in classical Athens (lack of education, low social class, etc.) – and possibly even a slave, at least for some period of his life, Atotas seems to have been perfectly familiar with the *Iliad*: he knows the reference to Pylaemenes, the leader of the Paphlagonians in 2. 851, and indeed likens himself to him,[239] and names Achilles rather than Menelaus as the slayer of Pylaemenes (*Iliad* 5. 576–578), which has been interpreted as an effort to increase the importance of his mythical ancestor.[240] The Achilles version is found in Dictys (3. 5), a much later source (4th century AD), but given the epigram we may suspect that it occurred in another source as well, unknown to us but known to Atotas, which for the reason suggested he preferred.

What can this signify about the education of this Paphlagonian workman in classical Athens, the centre of the civilised world, and the formation of the distinction between Greeks (especially Athenians) and barbarians in the classical period? And how would Atotas' Athenian contemporaries have seen him? As a barbarian? I would rather suspect that, at least after his death, Atotas would have been a person highly regarded (in terms of *hellenicity*) by the Athenians, who – or at least some of whom – would not immediately think of either the Persian Wars or Aristophanes' Paphlagon when they read his epitaph.

All these things might have influenced the Greeks' perception of the Paphlagonians so greatly that they gradually led to the creation of traditions that brought the two peoples very close together. The tradition of the Paphlagonian origin of Pelops,[241] which must have pre-existed its earliest currently known source,[242] is in my opinion an indication that the polarisation between the Greeks and the Anatolian 'barbarians' was not as strong as may sometimes have been thought, despite the traumatic experience of the Persian Wars. In particular, the connection of Pelops with Pylaemenes is impressive as well as suggestive. Moreover, the Caucones, who lived in Paphlagonia and were even considered to be Paphlagonians,[243] were, according to one version, settlers of the homonymous nation in the Peloponnese (Schol. in Hom. Il. 10. 429).

[239] Pylaemenes is called stout-hearted (λάσιον κῆρ), while Atotas calls himself great-hearted (μεγάθυμος).

[240] Lauffer 1979, 199–203; Bäbler 1998, 94–97; Vlassopoulos 2013, 130–31.

[241] See Section 2.6.2.

[242] As a typical Hellenistic mythographer, Apollonius worked with traditions that existed well before his era (e.g. the Argonautic myth). Therefore, the myth of Pelops should with relative (but not absolute) certainty also be considered much earlier than the 3rd century BC.

[243] See Section 2.5.

Furthermore, one of the criteria determining membership in an ethnic group, according to J. Hall,[244] is 'a putative subscription to a myth of common descent and kinship'. If we accept this view, then the Paphlagonians, according to this version, are not disgusting Anatolian barbarians but on the contrary have what amounts to an 'ethnic kinship' with the Greeks (of the Peloponnese) – a close relationship that was introduced into their tradition by the Greeks themselves, not the Paphlagonians, who are unlikely even to have been aware of it.[245] As L. Mitchell characteristically says, 'having been rooted in Asia, some of the Greek genealogies subsequently re-acquire Greek credentials, apparently marking a shift in the Greeks' confidence in their relationship with Asia and the Asian civilizations'. Thus, the Greeks 'also found the confidence to locate Asia on the edges of their world, and to bring it under their intellectual control'.[246] So, despite the fact that the Paphlagonians had fought against the Greeks in both the Trojan and the Persian Wars, they were not culturally disparaged by the latter in the way one would expect, compared to other peoples.

B. The Negative Side

This Greek image of the Paphlagonians, probably not universal but certainly not negative for a 'barbaric' people, did not last throughout the whole of antiquity. For in the Roman period a totally different perception of them began to take shape, which would last until the Late Byzantine period, as becomes visible through an almost obsessive attitude towards them in the majority of the written sources, usually mocking, belittling or even cursing them.

In the 1st century AD, Curtius Rufus (*Alexander* 6. 11. 4) recounted the accusation levelled by the Macedonian soldier Bolon against the general Philotas, that he ridiculed men from the country, calling them Phrygians and Paphlagonians, although he, Macedonian-born, was not embarrassed to use an interpreter to listen to men who spoke his own language. In his famous account of the 2nd-century fraud Alexander of Abonou Teichos, Lucian may seem to be focusing on the false prophet but he is actually deriding the credulous Paphlagonians, calling them fat-heads and simpletons, superstitious, wretched, mad and persons with neither brains nor individuality, who bore no resemblance to men that live by bread and had only their outward shape to distinguish them from sheep (Lucian *Alexander* 9, 11, 15, 45).

[244] Hall 2002, 9.

[245] The same could be claimed for the Phrygians or the Lydians as well. See the relevant traditions about the origin of Pelops mentioned above.

[246] Mitchell 2007, 182, 184.

Only the 5th-century church historian Socrates of Constantinople, known as Socrates Scholasticus, praised the Paphlagonians, saying that,

> while the inhabitants of the East are addicted to sensual pleasures, the Paphlagonians and Phrygians are prone to neither of these vices [...] and it is well known that there is no race of men on the face of the earth who more rigidly govern their passions in this respect than the Phrygians and Paphlagonians (*Historia Ecclesiastica* 2. 38. *Cf.* 4. 28).

But his was an isolated voice.

Returning to negative views, in the 10th century Constantine Porphyrogenitus (*De thematibus Asia* 7), describing the province of Paphlagonia, uses horrible words for the Paphlagonians, calling them blameworthy, shameless and rude, and concluding that they are the derision, the shame and the contempt of the human race. Indeed, he goes so far as to claim that Homer himself implied their wickedness and brutality by saying that the race of mules originated from their area (Homer *Iliad* 2. 851–855), which of course was not Homer's intention at all. It is characteristic that the Scholiast of Aristophanes (*sch. eq.* 47) thinks that the poet deliberately chose to present Cleon in the *Knights* as a Paphlagonian tanner, to mock him for both his origin and his craft. This means that being Paphlagonian was considered derogatory in the Scholiast's time.

Moreover, Constantine the Rhodian (10th century) wrote 42 lines (*In Theod. Paphlag.*) vilifying the eunuch Theodore of Paphlagonia and through him the Paphlagonians, as pork-butchers, bacon-curers and so on, suggesting that there must have been a political motivation behind this detestation of the Paphlagonians. Similar expressions were used by the Scholiast of Lucian, who claimed that the Paphlagonians were not quiet people, but rather capable of inventing any kind of evil and using it in shameless ways (42. 9) and that they could turn good into rubbish (56. 27). He also called them pig-assed (80. 7. 3), fat, uneducated, insensible and stupid (42. 22–35). Later, Georgius Cedrenus (11th–12th century), drawing on Constantine the Rhodian, called the Paphlagonians detestable (*Compendium historiarum* 1. 566) and averred that during a famine they emasculated their children and sold them (1. 590). Finally, in the 12th century Eustathius of Thessalonica (*De capta Thessalonica* 32–34) also used disparaging words for the Paphlagonians, calling them a presumptuous and barbarian nation.

What were the reasons for such a sudden assault on and hatred of the Paphlagonians? It was P. Magdalino[247] who convincingly suggested that this attitude had political origins, mainly having to do with their highly visible role in Byzantine society and their great influence on politics in the 10th and

[247] Magdalino 1998.

11th centuries.[248] Indeed, we know of many important political personalities of Paphlagonian origin, including the emperor Michael the Paphlagonian, with whom the empress Zoe fell in love, the emperor Constantine Doukas, and other individuals from the 9th to the 11th centuries, among them members of the houses of the Macedonians, the Doukai and the Komnenoi. Already from the 9th century Paphlagonia was well known as the homeland of the famous empress Theodora, wife of Theophilus, remembered especially for the restoration of the veneration of icons. But among them all, Magdalino lays stress on seven Paphlagonian eunuchs of the imperial household, who had a great influence on political life and are, according to the author, to be blamed for the bad reputation of the Paphlagonians in this period.[249]

However the case may be, this detestation of the Paphlagonians cannot have been universal, since we also find positive words not only about Paphlagonian political figures like the empress Theodora (Ephraem *Chronicon* 2327–2332), the emperor Michael (e.g. Ephraem *Chronicon* 3003–3014; John Skylitzes *Synopsis*, Mich. 4 and other Byzantine sources) and John Doukas, brother of the emperor Constantine Doukas (Mich. Psellos *Chronicon* 7.Mich.VI26, MichVII16–17), but also about some of the Paphlagonian eunuchs, like Constantine Phagitzes (John Skylitzes *Synopsis*, Mich. 4) and John Orphanotrophus (Mich. Psellos *Chronicon* 4. 12–13, who includes the negative aspects of his character).

The Paphlagonian influence in politics might suffice to explain the negativity of the Byzantine texts. But what about the disparaging words of Curtius Rufus and Lucian in the 1st–2nd centuries? Actually, these are not of the same nature as the Byzantine ones. Lucian just mocks the Paphlagonians in a rather humorous text, while Curtius recounts the disparagement of the Phrygians and the Paphlagonians by a particular Macedonian general, which could have had several, more or less important, causes. There seems to be no animosity behind these words, as is the case in the Byzantine texts.

Apparently, in Late Antiquity some people detested the Paphlagonians while some continued to admire them. The only certain thing is that from early on the Paphlagonians obtained an at least culturally dominant position in the southern Black Sea littoral, which they maintained for many centuries. Their continued appearance in the literature over such a long period of time may be due to Paphlagonia's very strong and influential culture, as was pointed out in the beginning. But it seems that a strong element of this culture must, in turn, have been politics, or, more specifically, the ability of the Paphlagonians, or at

[248] See also Manoledakis 2019.
[249] Magdalino 1998, 143–50.

least of those who ruled them, to adapt to changing political circumstances and to manoeuvre among the powers that influenced their destiny, striving to maintain the best possible situation for themselves.[250] One might therefore allege that deft political manoeuvring and effective diplomacy may have characterised the Paphlagonians more than any other people of the southern Black Sea littoral, not only in the 1st millennium BC, but also in the first half of the next.

7. The Enetoi

The Enetoi of Paphlagonia are one of the peoples who are first mentioned by Homer, in the *Iliad*'s Trojan Catalogue (2. 851–852), and all later references to them seem to have been based on these lines:

> Παφλαγόνων δ' ἡγεῖτο Πυλαιμένεος λάσιον κῆρ
> ἐξ Ἐνετῶν, ὅθεν ἡμιόνων γένος ἀγροτεράων.
>
> *And the Paphlagonians did Pylaemenes of the stout heart lead*
> *from the land of the Enetoi, whence is the race of wild she-mules.*

In this passage it is made quite clear that the Enetoi belonged to the Paphlagonians, but nothing more is specified, apart from the fact that their territory was known as the breeding place of the wild and unruly half-ass, a cross between the donkey and the horse (*cf.* Schol. in Hom. Il. 2. 852). From the preserved texts it was only Strabo who attempted to find out more about the Enetoi, they having by then totally disappeared from Paphlagonia (12. 3. 8), along with their devotion to the breeding of horses (5. 1. 4).

Thus, according to the geographer (and loyal admirer of Homer), the Enetoi were the most noble tribe of the Paphlagonians (12. 3. 8). He refers to them at several points in his work in addition to the description of Paphlagonia in Book 12 and in these passages he also preserves for us the words of earlier authors, which have otherwise been lost. Strabo's main assertion is that, after the death of their leader Pylaemenes[251] and the fall of Troy, the Enetoi crossed over to Thrace, whence their wanderings took them to 'today's' land of the Enetoi, namely the Veneto (12. 3. 8. *Cf.* 1. 3. 2; 1. 3. 21; 3. 2. 13; 5. 1. 4; 13. 1. 53). He goes on to say that 'according to some writers, Antenor and his children took part in this expedition and settled at the inner recess of the gulf

[250] For all this, see Manoledakis 2019 in detail.

[251] Who was killed in the tenth year of the war by Menelaus (Homer *Iliad* 5. 576–579. *Cf.* Hyginus *Fabulae* 113; Tzetzes *Allegoriae Il.* 5. 81–82) or, according to other sources, by Achilles (Dictys 3. 5) or Patroclus (Cornelius Nepos *Datames* 2. 2), and is said to have been buried in the Troad.

of Adria', and closes with the conjecture that this is the reason why the Enetoi have disappeared and are no longer to be seen in Paphlagonia (12. 3. 8).[252]

The tradition of the migration of the Enetoi from Paphlagonia to the Adriatic was not invented by Strabo. The earliest preserved record of it is found in Pseudo-Scymnus' citation of earlier authors – among them Theopompus – references to the Enetoi 'who came from the land of the Paphlagonians to inhabit the shores of the Adriatic'.[253] It had, therefore, appeared by the 4th century BC.

Strabo also records other theories regarding the origin of the Adriatic Enetoi, such as one presenting the Henetoi as being colonists of those homonymous Celts who lived on the ocean-coast, somewhere in Transalpine Gaul (5. 1. 4). However, no source is provided for this theory, nor do we meet it elsewhere in ancient literature, and Strabo himself seems sceptical of it.[254]

Strabo also observes that Zenodotus, who in the 3rd century BC produced the first critical edition of Homer, emended ἐξ Ἐνετῶν to ἐξ Ἐνετῆς, identifying it as a city and specifically as Amisos (12. 3. 8, 25); Enete was considered to be Amisos by Hecataeus as well, at least according to Apollodorus, as reported by Strabo (12. 3. 25). This identification of Homer's supposed Enete, however, would conflict with Amisos' geographic location in the land of the Leucosyroi, thus outside Paphlagonia.[255] Stephanus of Byzantium (*s.v.* Ἐνετοί) and Eustathius (*Comm. ad Hom. Il.* 1. 567. 18) mention the city Enetos, which we read as Enetoi in Hesychius (*Lexicon, s.v.* ἐξ Ἐνετῶν).

Speaking of geography, if the Enetoi were indeed a Paphlagonian tribe, a question arises concerning their geographical placement. Did they generally live all over Paphlagonia or in a specific part of it? The place-names mentioned in the continuation of the Trojan Catalogue (*Iliad* 2. 853–855), namely Cytoron, Sesamos, the Parthenius, Cromna, Aegialos and Erythinoi, strongly indicate a location in the western part of Paphlagonia (Fig. 10.1). Even if the specific lines are considered a 3rd-century BC interpolation,[256] there is another argument for this placing of the Enetoi: Strabo points out that, according to

[252] Compare also Cornelius Nepos' theory, according to Pliny (*NH* 6. 5).

[253] Pseudo-Scymnus 369–390 = Theopompus *FGrH* 2b 115 F 130. The same author is also said to have reported that the Enetoi who lived on the Adriatic (and only they, not the Paphlagonians) had the custom of sending gifts to the jackdaws, so that they would not destroy their crops (*FGrH* 2b 115 F 274a; *cf.* Aristotle *De mirabilibus* 841b; Antigonus *Mirabilia* 173; Aelian *De natura animalium* 17. 16).

[254] Contrary to what Thallon (1924, 54) saw. Strabo referred only to the Italian Enetoi in 5. 1. 1, 3, 8–10.

[255] See also Schol. in Eurip. Hippol. 231d, 1132b. Apollonius' reference to Πέλοψ Ἐνετήιος (*Argonautica* 2. 358–359; *cf.* also *Etymologicum Magnum* 340. 20; Schol. in Apol. Arg. 156) is ambiguous, but adds to our mythical knowledge of the Enetoi the reign of Pelops, whose Paphlagonian (*inter alia*) origin is discussed in Section 2.6.2.

[256] Manoledakis 2013a, 31–33.

Callisthenes, the Enetoi lived to the east of the Parthenius, neighbouring the Caucones who inhabited the area to the west of the river and held the city of Cytoron, which means they possessed the most western part of Paphlagonia (12. 3. 5, 8). Moreover, Strabo also mentions (12. 3. 8) that some people consider Henete to be a coastal city to the west of Amastris, which again brings us to the western Paphlagonian border region, while according to Pliny (*NH* 6. 5) Cornelius Nepos also placed them in much the same area. This information seems to have influenced the post-Ptolemaic cartographers, who always placed the Enetoi in the western part of the Paphlagonian territory.[257]

But there appears to have been a contrary theory as well. For Strabo does not omit to record the opinion of Maeandrius that the Enetoi first set forth from the country of the Leucosyroi and allied themselves with the Trojans and that they sailed away from Troy with the Thracians and took up their abode around the inner recess of the Adriatic, but that the Enetoi who did not have a part in the expedition had become Cappadocians (12. 3. 25).[258] Elsewhere (12. 3. 8), however, he invokes other, unspecified, sources saying that it was after an expedition with the Cimmerians that the Enetoi were driven out to the Adriatic; those sources mention the Enetoi as neighbours of the Cappadocians.

Irrespective of which of the last two traditions is the more reliable, the fact is that both present the Enetoi as living in the eastern, and not the western, part of Paphlagonia. Indeed, the first actually implies a kinship between the Enetoi and the Leucosyroi/Cappadocians,[259] even if in a rather confusing way. Confusion seems to have featured largely in the case of the Enetoi, as in cases of other peoples mentioned in the Trojan Catalogue,[260] since it characterises not only their geographic location[261] but also, judging from the above-mentioned theories, their possible alliance with Trojans, Thracians and Cimmerians – which last case must, however, refer to a later period – and even their connection with the Italian Enetoi, the tradition for which begins appearing in the preserved literature from the 4th century BC. As Strabo himself notes (1. 3. 2; 3. 2. 13), it belongs to the general tradition about vanquished heroes, like Aeneas and Antenor, who had survived the Trojan War and then migrated to the West, where they established several cities – a tradition that was surely

[257] Just indicatively: from the maps of the pioneer A. Ortelius, *Pontus Euxinus* (1590) to the work of the well-educated H.T. Menke and K. Spruner, *Pontus Euxinus et quae ei adjacent* (1865).

[258] See also in Section 2.8.3C.

[259] On the general issue concerning possible connections between the Paphlagonians and the Leucosyroi, see in Sections 2.6.1 and 2.8.3C.

[260] See for example the Halizones in Section 2.9.5.

[261] In the *Scholia vetera in Euripidem* (1132) and the *Scholia in Hippolytum* (1131) we even read of a city in Epirus called Enete, known for its good half-asses.

created after Homer,[262] but developed and was chiefly promoted during the Roman period, when several variations appeared.[263] About a millennium later, Eustathius of Thessalonica attempted to compress most of them into his Commentary on Dionysius Periegetes (*Orbis descriptio* 378), where he includes a passage from Arrian (which, however, is not preserved and is otherwise unknown) affirming that the Enetoi went to Europe and settled by the River Po after having fought against the Assyrioi (Leucosyroi). He also treats Maeandrius' version in much the same way, speculating on it or making assumptions that are not clearly justified by Strabo's account, for example about a war between the Enetoi and the Leucosyroi.[264]

All this uncertainty, combined with the fact that we have practically no further information about the Enetoi,[265] such as historical events, customs, etc., makes us hesitant to draw any firm conclusions of a historical nature regarding this people, although such attempts have been made in the past.[266]

*

* *

Apparently, apart from the question of their possible migration to Italy, it was only the wild mules of the Enetoi, which Stephanus of Byzantium called Ἐνετίδες ἵπποι (*s.v.* Ἐνετοί), that attracted any interest, and this happened in the Byzantine period. We have already presented the particularly deprecatory comments which the majority of the Byzantine authors directed towards the Paphlagonians.[267] Euripides' reference to 'Enetian horses' twice in his *Hippolytus* (231, 1131) also caused some discussion. Eustathius (*Comm. ad Hom. Il.* 1. 567–568)

[262] Probably around the 6th century BC, for which there is evidence that the Greeks were active in the Adriatic: Beaumont 1936. However, it is worth noting that Herodotus (1. 196; 5. 9) mentions only the Illyrian tribe of Enetoi, with no reference to connections with Paphlagonia, as does Pseudo-Scylax (19).

[263] E.g. Curtius Rufus *Alexander* 3. 1. 22. According to Livy (*Hist. ab urb. cond.* 1. 1), Antenor sailed into the furthest part of the Adriatic, accompanied by a number of Enetoi who had been driven from Paphlagonia by a revolution, and after losing their king Pylaemenes at Troy, were looking for a settlement and a leader. Dio Chrysostom (*Orationes* 11. 138) specifies that Antenor acquired dominion over the Enetoi in the Adriatic.

[264] About a century later, George Pachymeres (*Hist.* 4. 27. 23), called the Enetoi great-hearted, influenced by the stout-hearted Pylaemenes of Homer.

[265] On the Enetoi as Paphlagonians, see also Aelius Herodianus *partit.* 33. 4–5; the *Suda, s.v.* Ἐνετοί; Schol. in Eur. Hippol. 231a, 1131–1132a; Schol. in Hom. Il. 2. 851.

[266] For example, Thallon (1924), based on risky and unfounded assumptions regarding the Enetoi's language and 'racial characteristics', tried to establish historical dimensions for the above-mentioned traditions, presenting among other arguments the frivolous but in previous decades not uncommon view of a Thracian origin for both the Paphlagonian and the Italian Enetoi, since the Thracians were supposed to have migrated 'from Asia Minor to Italy'.

[267] See in Section 2.6.4.

provides more information on these 'Enetian horses', giving a complete account of everything that had been said about them up to his day. He tells us that the she-asses of this species were ideal for farm work, but thinks, as does the Scholiast on Euripides' *Hippolytus* (231a), that the reference to them in Euripides is an anachronism, arguing that these animals were unknown to the Greeks of the poet's day, having been introduced into Greece after the 85th Olympiad when, according to Polemon, Leon of Lacedaemonia won the race with a team of these animals.[268] As for Homer's description of them as ἀγροτεράων, Eustathius interprets this not as meaning that they were wild, but as a reference to their habit of living in large nomadic herds, while adding that some interpret the term as meaning 'hard to domesticate' while others translate it as 'useful in the fields'.

The difficulty with Eustathius' remarks about the supposed anachronism in Euripides is that that Leon won the chariot race in the 89th games (424 BC), not the 85th (440 BC). Both took place during Euripides' lifetime (480–406 BC), and even if there was another Leon who participated in the 85th games the *Hippolytus* (428 BC) was in any event written after them, in which case Eustathius was incorrect in identifying an anachronism. Finally, it must be stressed that Leon is considered to have won using horses from the land of the Italian Enetoi and not the Paphlagonian,[269] so the question is 'did Euripides really mean horses from the Veneto or rather from Paphlagonia?'[270]

8. The Leucosyroi

One of the most peculiar cases in the geography of the peoples of the southern Black Sea littoral is that of the Leucosyroi. Mentioned in the ancient sources also as Assyrioi, or Syroi, or Cappadocians, these people purportedly inhabited one of the largest, most central and in several aspects most important parts of the littoral, held within their borders two of the most significant Greek colonies and were connected to one of the most important ethnic groups and homonymous empires of the world's prehistory and ancient history, as their very name indicates. Remarkably, despite these impressive attributes the ancient authors who mention the Leucosyroi – many more than for most of the southern Black Sea peoples – do not provide any further information about them, as is for example the case with the Paphlagonians or the Mossynoikoi.

[268] Müller 1878, 122.
[269] Devereux 1964; Hodkinson 2009, 313, 329, n. 12.
[270] Devereux 1964.

There is nothing about their customs,[271] nothing about their political or social structure, no names of historical persons. There is not even a clear explanation of their connection with the Assyrians of the Near East, which seems to have been the only matter of concern to these writers. As a result, the modern bibliography on the Leucosyroi has also concentrated almost exclusively on this connection, omitting other aspects of the history of these people which could be of some interest.

As is the case with many other peoples of the coast, most of what we know about the Leucosyroi stems from the works of the 'usual suspects' – the geographers, Herodotus and the later lexicographers and commentators. Our first task will be to seek to define the geographical area they inhabited.

8.1. The Territory (Fig. 10.2)

The land of the Leucosyroi can be situated only generally, without precise demarcation, which is, after all, the usual case on the southern Black Sea littoral. According to both Herodotus (1. 6; 1. 72) and Strabo (12. 1. 1; 12. 3. 2, 9, 12), the River Halys, flowing north to the Black Sea, was the border between 'Syria' and Paphlagonia[272] (*cf.* Anonymous *Periplus* 25; Constantine Porphyrogenitus *De thematibus Asia* 7), the former thus lying to the east of the latter.[273] Herodotus seems to fall into contradiction in his second book, however, where he writes that the Syroi lived about the rivers Thermodon and Parthenius (2. 104), since the latter stream is, of course, much farther west, and according to some sources (Arrian *Periplus* 13–14; *cf.* Menippus *Periplus* 8; Anonymous *Periplus* 13. 9–12) actually forms the western border of the Paphlagonian territory! How could Herodotus have named both the Halys and the Parthenius as western borders of the Syroi?[274] This could perhaps be taken together with the statement of the 4th-century BC Callisthenes, who, according to Strabo (12. 3. 5), wrote that the Caucones[275] extended from Heraclea and the Mariandynoi to the Leucosyroi, seemingly ignoring the Paphlagonians.

Assuming that we can overcome this problem, we are in fact little farther ahead, for the Halys is not unanimously reported as the western border of the

[271] The only exception being Herodotus' account (2. 104) that the Leucosyroi practised circumcision, a custom 'lately' learned from the Colchians. See also Sections 2.13.3 and 2.15.2.

[272] And not, of course, between Syria and Cappadocia, as Alexander (*De figuris* 39. 24–25) erroneously writes, citing Herodotus (1. 6).

[273] According to one theory, a tribe of Paphlagonians called Enetoi neighboured the Cappadocians (Strabo 12. 3. 8). See more on them in Section 2.7.

[274] Actually, the Parthenius is not even mentioned as a border of the Syroi, only as belonging to them.

[275] For them, see in Section 2.5.

Leucosyroi, since several authors mention not only Amisos (commonly considered a city in Leucosyrian territory) but also Sinope as lying in their land (Pseudo-Scylax 89; Apollonius *Argonautica* 2. 946; Apollodorus fragm. 176. 19–20; Pseudo-Scymnus 917, 941–943; Valerius Flaccus *Argonautica* 5. 109; Anonymous *Periplus* 26–27. Tzetzes,[276] writing in the 12th century, said that everyone called Sinope Assyria) and Sinope of course lies to the west of the Halys. The chronologically closest to Herodotus of these authors, Pseudo-Scylax (89), places the border between Paphlagonia and Assyria somewhere between Harmene (in Assyria) and Stephane[277] (in Paphlagonia). Pomponius Mela's (1. 105) placing of Amisos and Sinope in the territory of the Chalybes might be a result of the author's rather perfunctory work.[278] But the discrepancies do not end here, since Apollonius, who had placed Sinope in the land of the Assyrioi (*Argonautica* 2. 946), later in the same work placed the Halys estuary, which is located to the east of Sinope, on the Paphlagonian coast (4. 245), while others, finally, mention the River Euarchus, slightly east of Sinope, as the western border of the Leucosyroi (Mennipus 9; Stephanus of Byzantium 356; Anonymous *Periplus* 23. 18–20).

As for their eastern neighbours, things are even more complicated. In the 4th century BC Ephorus (*FGrH* 2a 70 F 43) stated that the Leucosyroi and the Chalybes bordered the Tibarenoi, without clarifying their relative positions. He seems, however, to have placed the Chalybes in the hinterland (Strabo 14. 5. 24), which could lead to the conclusion that, at least in his time, they occupied a small inland area with the Leucosyroi to their north, west and south and the Tiberenoi to their east, who in this way had both the Leucosyroi and the Chalybes on their western border.[279]

But the problem here is that Xenophon does not even mention the Leucosyroi,[280] while Pseudo-Scylax, writing some decades after the events narrated in the *Anabasis* and at the same time as Ephorus, clearly lists in his *Periplus* (87–90) as coastal peoples (from west to east) the Paphlagonians, the Assyrioi, the Chalybes and the Tibarenoi, curiously without attributing any coastal settlement to the last of these. More confusion is created by Herodotus' statement (2. 104) that the (eastern) neighbours of the Syroi were the Macrones, who in the rest of the sources appear to have dwelt in an area more to the

[276] *Chiliades* 12. 56. *Cf.* also *Etymologicum Magnum*, *s.v.* Ἀσσυρία; Eustathius *Comm. in Dion. Per.* 772, 775.

[277] On these and all the other cities mentioned here, see in Part 5.

[278] See in the Introduction Section 2.2.

[279] See in Section 2.9.1.

[280] See Section 2.8.3C.

east.[281] Strabo does not particularly help when he says (12. 1. 1) that to the east of Cappadocia are Armenia, Colchis and the intervening peoples who speak different languages (ἑτερόγλωττοι), who might include the Tibarenoi, Chalybes, Macrones, Appaetae and Mossynoikoi, mentioned by him later, in 12. 3. 18–19.[282] It is only much later that Eustathius (*Comm. in Dion. Per.* 767, 772) clearly mentions the Chalybes as bordering the Assyrioi.

The territory of the Syroi seems to have extended south as far as Phrygia (south-west) and Cilicia (Herodotus 5. 49, referring to the tablet of Aristagoras; Strabo 12. 1. 1 *cf.* Nepos *Datames* 1. 2). When Strabo (12. 2. 10) mentions a mountain parallel to the Taurus to its north as a boundary of Pontus and Cappadocia, and thus seems to fall into contradiction with his words on the Cappadocians in Pontus (12. 3. 2), this is due to the nature of his work, which is to provide both political/administrative[283] and geographical/ethnographic information about the areas he deals with.[284]

Hecataeus (*FGrH* 1a 1 F 200, 201) accounts to the Leucosyroi the cities of Chadisia and Teiria.[285] Herodotus also mentions the still-disputed Pteria as a city of the western part of Syria, immediately to the east of the Halys and not far from Sinope (1. 76), while Pseudo-Scylax (89) assigns to Assyrian territory the towns of Themiscyra, Lykastos, Caroussa, Sinope, Kerasous, Harmene and Tetrakis as well as the River Thermodon (*cf.* Strabo 12. 3. 9, 25; Pliny *NH* 6. 3; Appian *Mithridatica* 292–293). In the 2nd century AD, in his famous *Geography* (5. 6. 2, 9), Ptolemy mentions the Leucosyroi as a nation inhabiting the coastal part of Cappadocia in the area of Ancon and the Iris estuary.[286]

Then there is the case of Curtius Rufus, a Latin author not known for geographical accuracy: describing the area towards Hyrcania, which Alexander the Great reached in 330 BC (in his narrative of the latter's campaign), Curtius says that to the south of the Caspian Sea there is a continuous valley, 'to which arms, so to speak, of that land jut forth', and, bending in the middle, form a curve like a half-moon. The Kerketae, the Mossynoikoi and the Chalybes

[281] See in the homonymous section.

[282] Janse's (2002, 347) attribution of the ἑτερόγλωττοι to Cappadocia itself, which thus 'was a multi-lingual region', is rather erroneous.

[283] For Strabo's statement (12. 1. 4) that Cappadocia was divided by the Persians into two satrapies, of which one was named 'Cappadocia Proper' or 'Cappadocia near Taurus' or even 'Greater Cappadocia', and the other 'Pontus' or 'Cappadocia Pontica', see also in Section 3.3.

[284] In Section LC.3 we will see that the (As)syroi were sometimes thought to have inhabited generally a vast area from the Black Sea to Mesopotamia. Here, we are talking only about the possible borders of the Leucosyroi. On the issue, see Section 2.8.3C.

[285] *Cf.* Summerer 2005, 130.

[286] Nöldeke's (1871, 447) harsh criticism of Ptolemy's placement of the Leucosyroi is totally unjust and indicative of the wrong expectations people may have when reading sources of a specific period and scope.

live on the left (western) part, and on the other (eastern) side are the fields of the Leucosyroi and the Amazons (Curtius 6. 4. 16–17).

Obviously, Curtius is in a geographical muddle: first of all, a little later in his text (6. 5. 24) he makes it clear that he is speaking of the Amazons of Hyrcania and the episode with their queen Thalestris, about which other authors are dubious.[287] He says, however, that in Hyrcania dwelt the Amazons that inhabited Themiscyra, near the Thermodon, whose queen lived and reigned in the area between the Caucasus and the River Phasis. This is impossible, since Themiscyra lay on the southern Black Sea coast, far from the Caucasus and the Phasis as well as from Hyrcania. Curtius has confused two different groups of Amazons, those of Themiscyra and those of the Caspian, and in his effort to consider them as one he has brought together two areas that are about 1000 km apart. When Strabo objects that 'of the numerous historians, those who care most for the truth do not make the assertion, nor do those who are most trustworthy mention any such thing' (11. 5. 4), he implies that others before Curtius also had a problem with this matter. Moreover, in the first part of the geographic account cited above, Curtius also says that the Kerketae, the Mossynoikoi and the Chalybes live to the west of the Leucosyroi and the Amazons (of Themiscyra), while in reality the opposite is true, and in any case none of these peoples lived anywhere near the Caspian and Hyrcania.

8.2. One People, Four Names

The first known author to have mentioned the people under consideration is Hecataeus, in the 6th century BC, who called them Leucosyroi (*FGrH* 1a. 1 F 200, 201). A century later, Herodotus became the first not only to mention them as Syroi and Cappadocians, but also to give an explanation for the combination of names; the relevant passages are: 'The Cappadocians are known to the Greeks by the name of Syroi' (1. 72); 'the Cappadocians, whom we Greeks know by the name of Syroi' (5. 49) and 'the Syroi, or Cappadocians, as they are called by the Persians' (7. 72). From these passages it becomes clear that in Herodotus' time the same people were usually called Syroi by the Greeks and Cappadocians by the Persians (to whom they belonged administratively).[288]

Strabo cited the work of Herodotus and noted that 'by Syroi Herodotus means the Cappadocians, and in fact they are still today called Leucosyroi,

[287] Thalestris went there to meet Alexander and bear children by him (Diodorus 17. 77; Curtius 6. 5. 24–32; Justin 12. 3. 5–7. *Contra* Plutarch *Alexander* 46. 1–2; Strabo 11. 5. 4, who doubt that this account is factual).

[288] 'Katpatuka' was the Persian name that in Greek became 'Cappadocia'. See, for example, Herzfeld 1968, 101.

while those outside the Taurus are called Syroi. As compared with those this side the Taurus, those outside have a tanned complexion, while those this side do not, and for this reason received this appellation' (12. 3. 9). This passage of Strabo has always been considered important, since he gives an explanation for the name Leucosyroi – seemingly an anthropological one. But this is not the only noteworthy element of his account. First of all, although quoting the 'Syroi' of Herodotus, he afterwards ignores this name and continues using the appellation Leucosyroi, which he says was still used in his day, about four centuries after Herodotus, while 'Syroi' was used only for the people outside the Taurus. Moreover, he distinguishes the two peoples as living on either side (north and south) of Mt Taurus, thus giving the impression of a geographic proximity, if not neighbourhood. Such closeness could raise a question about the difference in complexion, but we will return to this later.

Another observation should be made here: although Strabo prefers the name Leucosyroi (12. 3. 9, 12, 24), at some point in his work he refers to Callisthenes' additions to the Trojan Catalogue (*Iliad* 2. 855), among them 'that the Caucones extended from Heraclea and the Mariandynoi to the Leucosyroi, whom we call Cappadocians' (12. 3. 5). Given the way in which Strabo presents the words of Callisthenes, we might conclude that the latter used the name 'Leucosyroi',[289] while Strabo is admitting that although 'Cappadocians' was the prevailing name in his day, he preferred not to use it.[290] It seems that he was wavering between a name that had become established and the name he favoured as an intellectual who had made a thorough study of Greek literature.

Not long after Herodotus, things apparently changed: Xenophon used only the name 'Cappadocia' for the area of the homonymous Persian satrapy, and by 'Syria' or 'Assyria' he always meant the area of Mesopotamia. Similarly, the people of that satrapy he always called Cappadocians (e.g. *Cyropaedia* 1. 5. 2–3; 2. 1. 5), never Leucosyroi, Syroi or Assyrioi.

According to Clement of Alexandria (*Stromata* 1. 16. 77), several authors of the 4th–3rd centuries BC, such as Scamo of Mitylene, Theophrastus of Ephesus,

[289] Strabo's text is preserved as follows: 'Callisthenes in his treatise on *The Marshalling of the Ships* was for inserting after the words "Cromna, Aegialos and lofty Erythinoi" the words "the Caucones were led by the excellent son of Polycles – they who had their famed dwellings around the Parthenius River," for, he adds, the Caucones extended from Heraclea and the Mariandynoi to the Leucosyroi, whom we call Cappadocians.' As we see, the part that concerns us is not in quotation marks, like the previous words of Callisthenes, but is rather reported by Strabo, together with his final clarification after the comma.

[290] However, he used it sometimes (e.g. in 12. 3. 2), and he used 'Cappadocia' to refer to the geographic region (12. 1. 1).

Cydippus of Mantinea, Antiphanes, Aristodemus, Aristotle, Philostephanus and Strato the Peripatetic, reported that 'the Cappadocians first invented the instrument called the nabla and the Assyrians the dichord'. Although it is not specified whether these words were used by all or just some of these authors, it is noteworthy that all of them belong to the age of Xenophon, who always called the people in question Cappadocians and never anything else, distinguishing them from the Assyrians (of Mesopotamia), just like the authors mentioned by Clement. This is a non negligible indication that in the 4th century BC the 'Persian' name for the Leucosyroi was increasingly being used by Greek authors, possibly due to the long establishment of Persian political power in Anatolia. Especially in the case of Xenophon, a man familiar with Persian culture and administration, we may easily conceive that he purposely always used the appellation 'Cappadocians', which also allowed him to be clear when he wanted to refer to either of the two 'Syroi' of Asia. Similar distinctions were made by Plutarch (*Lucullus* 26. 1), Lucian (*Macrobii* 13; *de Syria dea* 10), Ptolemy (5. 13. 4–5; 8. 19. 2), Philostratus (*Vitae sophistarum* 1. 518), Hippolytus (*Chronicon* 199–200), Epiphanius (*Ancoratus* 113. 2) and other later authors.

What can also be observed, looking at the rest of the sources, is that from the Hellenistic period on those who write about the Cappadocians, meaning the Leucosyroi, are usually referring to the people of the homonymous satrapy of the Persian or Alexandrian empire, within the context of political or military events, with mention of their rulers. Herodotus himself wrote about the 'Cappadocian Syroi' (1. 72), meaning the Syroi who live in Cappadocia. Here, he clearly distinguishes between the Persian administrative province (satrapy), for which he uses the name given by those who established it, and the people inhabiting it, as they were called by his fellow Greeks.

As regards the terms 'Syroi' and 'Assyrioi', the use of 'Syroi' by Herodotus as the Greek name for the specific people led F. Jacoby to claim that this was also the case with Hecataeus, who is considered to have influenced Herodotus. According to F. Jacoby, 'Leucosyroi' was the name used by Stephanus of Byzantium and Aelius Herodianus, who preserve the fragments from Hecataeus' work, and not Hecataeus himself, who 'called them certainly Syrians, just like Herodotus'.[291] That the name Syroi was used in the 5th century BC is inferred also from the fact that Strabo quoted Pindar's words that the Amazons 'swayed a Syrian army that reached afar with their spears', interpreting them as a clear indication that the Amazons lived around Themiscyra (12. 3. 9). Besides, the name is also mentioned by Nicholas of Damascus (fragm. 48), probably with

[291] Jacoby 1957, 356. *Cf.* Dan 2012, 77.

reference to the 5th-century Xanthus of Lydia and, according to Stephanus of Byzantium,[292] by Sophocles. But after that, this appellation fell into disuse for the Leucosyroi. We meet it again, with reference to the Syrians, the early inhabitants of Sinope, in Pseudo-Scymnus (941), an author who, as we saw, also used the name Leucosyroi (917), and after that only in few Late Antique writers, who cited all the names that had been used for that people. Indeed, Strabo says that in his day only those 'outside the Taurus' were called Syroi (12. 3. 9).

Moreover, the name Leucosyroi never stopped being used or even being popular. In the 4th century BC we have the words attributed to Callisthenes, Ephorus and Andron,[293] and in the 3rd, words attributed to Maeandrius (Strabo 12. 3. 25); in around 100 BC, Pseudo-Scymnus used both 'Leucosyroi' (917) and 'Syroi' (941),[294] while from the Roman period on the name continued to be widely used (Nepos, Strabo, Curtius Rufus, Ptolemy, Appian, etc.), even if simultaneously with the name 'Cappadocians' (e.g. Pliny, Plutarch). In the Byzantine age, several authors reminded their readers that the Cappadocians were called Leucosyroi in antiquity. Consequently, I would not be as certain as Jacoby about Hecataeus' use of the word 'Syroi'.[295]

According to E. Herzfeld, the name Katpatuka (the Greek Cappadocia) was chosen by the Persians in the early sixth century BC for their new satrapy, which did not extend beyond the Halys in the south; this was the later Cappadocia ad Pontum. But after 500 BC, once the Achaemenids had added Cilicia (above the Taurus, with its capital Mazaka) to this Katpatuka, the name spread southwards and this southern part, as politically more important, was called 'Great Cappadocia'; this gradually became 'Cappadocia', leaving for the northern part, 'Cappadocia ad Pontum', only the name 'Pontus'. Thus, always according to Herzfeld, 'the name Cappadocia survived in a region that before the Achaemenid period did not belong to Cappadocia'.[296]

With this view taken into consideration, one can comprehend first of all Xenophon's exclusive use of the name Cappadocians; secondly, how, from the Roman period on, divisions of several 'Syrias' like those mentioned by Dionysius and his Commentators appeared, with Cappadocia being located in the

[292] *S.v.* Χαλδαῖοι.

[293] For Callisthenes (Strabo 12. 3. 5) and Ephorus (*FGrH* 2a 70 F 43), see above. According to the Scholiast of Apollonius Rhodius (2. 946), Andron of Teos reportedly called the Assyrioi Leucosyroi in his *Periplus*.

[294] And also 'Cappadocians' (939), but only designating those living in the Anatolian hinterland, 'near Pisidia', like Dionysius Periegets (*Orbis descriptio* 974).

[295] Thus, Langella's view (1997, 99) that the name Leucosyroi first appeared in the Hellenistic period is also erroneous.

[296] Herzfeld 1968, 101–02.

specific inland area instead of north of the Halys;[297] and thirdly, that the Leucosyroi, who had always been clearly placed on the Black Sea littoral above the Halys before the Roman period,[298] were none other than the inhabitants of the 'original' Cappadocia and thus the Cappadocians, as they were called by the Persians. What remains incomprehensible is the total nonappearance of the Cappadocians in Xenophon's *Anabasis*, but this we shall return to later.[299]

As for the name 'Assyrioi' applied to the Leucosyroi, we first encounter it in Pseudo-Scylax (89) in the form 'Assyria, a nation', and afterwards in Apollonius (*Argonautica* 2. 946), Valerius Flaccus (*Argonautica* 5. 109) and Dionysius Periegetes (*Orbis descriptio* 772, 975). It is noteworthy that what we have here are a geographer, or rather the compiler of a geography based on Scylax's original,[300] a mythographer clearly influenced by geographic texts, among which Pseudo-Scylax's work would be of the utmost importance[301] since it dealt with the same area as Apollonius, a Latin poet who retold Apollonius' story, and lastly, a traveller writing a geographic text obviously influenced by works like Pseudo-Scylax's *Periplus*.[302] Consequently, here again we cannot speak of a name established in literature for the Leucosyroi,[303] although it was quite often used to denote generally the people living in the vast area from the Black Sea to Mesopotamia.[304] Meanwhile, Herodotus' statement regarding the Syroi of Palestine, 'this people were called by the Greeks Syroi and by the barbarians Assyrioi' (7. 63), is unconfirmed, while the report attributed to Arrian (*FGrH* 156 F 74) that the Assyrioi who were also called Cappadocians changed their name after Cappadox, son of Ninyas (*cf.* Diodorus 2. 7, 20–21) is found nowhere else.

The issue of multiple names also attaches to the name of a river that rises in the Taurus and flows into the Mediterranean. Aelius Herodianus (3. 1. 200) and Stephanus of Byzantium (*s.v.* Πύραμος) mention that the River Pyramos of Cilicia (modern Ceyhan) was previously called Leucosyros, which, however, is not confirmed by Strabo in his account of the river (12. 2. 4). The same

[297] See in Section 2.8.3A.

[298] *Cf.* Nöldeke 1871, 468.

[299] See Section 2.8.3C.

[300] On the use of the name by Pseudo-Scylax, see Counillon 2004, 113, and generally Debord 1999, 85, n. 11.

[301] The influence is rather obvious in the specific passages of both authors mentioning Sinope in the land of Assyria.

[302] We also meet the specific appellation in works by authors of Late Antiquity, in the same context as the one mentioned for the name 'Syroi'. See also in Section 2.8.3A.

[303] Dan (2011, 83–84) offers three possible reasons for the choice of the specific name by Pseudo-Scylax and Apollonius, none of which, however, as Dan herself admits, can be proven.

[304] On this issue, see Section 2.8.3.

river is called Cappadox by Pliny (*NH* 6. 9), who adds that 'the Cappadocians, who were formerly called Leucosyroi, took their name from it'. Later, Procopius (*De bello* 1. 17) ascribes the name Leucosyroi to the people that in his time were called Lesser Armenians, their first city being Melitene, thus far away from the Black Sea littoral.

Any conclusions drawn can thus be of a general character only. Apparently, the authors, especially from the Hellenistic period on, were themselves confused by the different uses of all four names by their predecessors, and this confusion increased geometrically as time went by, adding to the authors' need to show that they were familiar with the issue of the identification of peoples and names.

Strabo, for example, in his long and rather impassioned response to Apollodorus, who is reputed to have claimed that Homer's knowledge of the southern Black Sea coast was minimal, says that Homer 'mentions the Amazons, but not the Leucosyroi, or the Syroi, or the Cappadocians, or the Lycaones, though he repeatedly mentions the Phoenicians and the Egyptians and the Ethiopians' (12. 3. 27). This use of all three names together in the specific context and alongside the neighbouring (to the south-west) Lycaones is rather astonishing, and one might wonder whether we have here an implication of slightly different 'Syrian' groups or just an exaggeration resulting from Strabo's bad temper while criticising Apollodorus, combined with his wish to show that he is aware of the issue of the many names.

Similarly strange is the fact that, much earlier, Xenophon had referred to the Mesopotamian area using both the names 'Syria' and 'Assyria' together, without clarification. For example, he says in the *Cyropaedia* (1. 1. 4) that 'Cyrus conquered Syria, Assyria, Arabia, Cappadocia, etc.' (*cf.* also 4. 2. 31; 6. 2. 10); the case at the end of the *Anabasis* (7. 8. 25) is similar, even if the specific passage is considered a later interpolation.[305] Obviously, the linguistic issue of the origin and interrelation of the names Syria and Assyria, which has been a matter of debate from as early as the 17th century,[306] does not concern us here, although we cannot help but notice that some of the scholars who have attempted to pinpoint differences in the use of the names Assyrios and Syros, or Assyria and Syria, laid much stress on the texts by authors of Late

[305] Tuplin 2004, 176–77.

[306] See, indicatively, Selden 1617; Nöldeke 1871; Tvedtnes 1981; Frye 1992; Rollinger 2006, with bibliography. Rollinger argues that an 8th-century BC inscription found in 1997 in Çineköy (Cilicia) 'settles the problem once and for all', and proves that 'Syria' and 'Assyria' were used as interchangeable toponyms.

Antiquity, who, however, as noted above, were victims of the confusion created by their predecessors rather than reliable aids to overcoming it.[307]

8.3. On the Origin of the Leucosyroi

A. Some Issues Arising from the Ancient Texts

We have noted that Strabo was the first to give an explanation for the name Leucosyroi (12. 3. 9), which was thought to have been used in order to distinguish the Syroi of the Black Sea littoral from the darker-skinned Syroi beyond the Taurus.[308] He also pointed out that

> apparently the name of the Syroi extended not only from Babylonia to the Gulf of Issus, but also in ancient times from this gulf to the Euxine. At any rate, both tribes of the Cappadocians, both those near the Taurus and those near the Pontus, have to the present time been called 'Leucosyroi', as though some Syroi were black, these being the Syroi who live outside the Taurus (16. 1. 2).

Thus he seems to imply that there was an uninterrupted Syrian occupation of the land from the Black Sea to Coele Syria. Elsewhere, he says of the Cataones in the southern part of Cappadocia towards Mt Taurus that although they were considered by 'the ancients' as a different people from the Cappadocians, in his time they had same language and customs as the latter, adding that it was surprising how utterly all signs of their being a different nation had disappeared (12. 1. 2).

In the 2nd century AD Appian (*Proemium* 1. 2) wrote that:

> the Cilicians come next to the Syroi, and their neighbours are the Cappadocians, and that part of the Armenian country called Lesser Armenia. Along the Euxine are other nations called by the common name Pontic, subject to the Roman rule. The Syroi and Cilicians border on the Mediterranean, the Armenians and Cappadocians extend to the Pontic nations and to the interior as far as Greater Armenia.

Several authors after Strabo repeated, in one way or another, his assertion that Syria extended from the Mediterranean to the Black Sea. Dionysius Periegetes (*Orbis descriptio* 970–975) reports that

[307] As an indicative example we could mention Photius' definition of 'Leucosyroi' in the homonymous entry of his *Lexicon* that 'they were the Cappadocians, whom the Ionians called Syroi'. So who called them Leucosyroi? Dan (2011, 78) reads 'Milesians' instead of 'Ionians', but this is not the appropriate place to discuss the frequent equation of Milesians, Ionians and Greeks, an equation that is often considered natural but which I would characterise as frivolous and destructive, even if the historical context seems occasionally to favour it.

[308] An explanation repeated later by Menippus *Periplus* 9; Anonymous *Periplus* 23; Photius *Lexicon*, *s.v.* Λευκόσυροι; Eustathius *Commentary on Dionysius Periegetes* 772–775; Constantine Porphyrogenitus *De thematibus* 2. 33–35.

> beyond Libanus towards the rays of the sun there lies stretched the extensive land of the other Syria, reaching as far as sea-washed Sinope. In the middle of this deep land there dwell the Cappadocians, experts in horsemanship, and the Assyrioi near the sea, by the mouth of the Thermodon.

The *Scholia in Dionysium periegetam* (970) add that the north-eastern part of this huge land was Armenia, inhabited by the Armenians, and in the *Paraphrases* (970–976) the Assyrioi of Pontus are also called Leucosyroi. In the *Scholia* on another passage of Dionysius' work (897) a distinction is made between three Syrias: Coele Syria, beyond Antiochia and Apamea; coastal Syria, which is the former Cilicia; and the Syria beyond Cappadocia. The scholiast also points out that both the inhabitants of this last Syria and those living round Babylon were called Assyrioi.

Thus, by Late Antiquity the name 'Syria' had seemingly come to apply to the whole land from the Black Sea to the Mediterranean and the Euphrates, inhabited by people who might be divided into three apparently kindred nations, of which the two bordering the seas were called Assyrians – those on the Black Sea also Leucosyroi – and the third, occupying the interior between them, Cappadocians (*cf.* Pseudo-Scymnus 935–940).[309]

In the 12th century Tzetzes (*Chiliades* 12. 56) wrote that 'all people of Mesopotamia are called Assyrians; but Syrians are only those living between Casius and Lebanon'. And he continues:

> But everybody calls Sinope Assyria. And according to some etymologists it has the same boundaries as Syria. And the land of Assyrians they call Assyria together with Syria but not the hollow part of it.

Rather astonishing is the entry on Assyria in the contemporary *Etymologicum Magnum* (*s.v.* 'Ασσυρία), where we read, first, that Assyria is the land around Sinope as far as Apamea and southward from Pontus to Persia and India, and then, with references to the 3rd-century BC Callimachus, that Syria was called so because it was swept away by the deluge; that Assyria was the part that was close to the swept-away land (and this was Sinope) from not being swept away (so the first letter of 'Assyria' is an *alpha* privative?); and that the nation in Pontus was called Assyrioi because they were no longer Syrians.[310]

[309] Nöldeke (1871, 454) offered the not very enlightening explanation that when people (the Greeks?) became more familiar with Asia after the fall of the Assyrian Empire, they realised that the name Assyria covered a very wide geographic area, and so they started using the names Syria and Assyria to denote different parts of this area.

[310] Nöldeke (1871, 442, n. 1) evaluated this definition as 'sehr unglücklich'. Many of those narratives are presented by Eustathius (*Comm. in Dion. Per.* 772–775).

B. Modern Theories

Following and parallel to the confusion created by the use of the four names, further confusion apparently arose, with respect to the origin of the Leucosyroi, engendering such questions as: How many Syrians actually existed between the Black Sea and Mesopotamia? What was their connection with each other? And how, if Strabo's theory about the name of the Leucosyroi (12. 3. 9) is correct (although in 16. 1. 3 he wonders about the existence of two different Syrian complexions), did the Syrians of Pontus become more light-skinned? Such questions have led to the formulation of several theories about the origins and ethnic identity of the Leucosyroi, most of which first appeared in the literature more than a century ago.

According to one such theory, expressed by several scholars with slight variations, the existence of the Leucosyroi on the southern Black Sea littoral indicates a northward expansion of the Assyrians. The variations pertain to the time, the place and the nature of this expansion, namely whether it took place in the 2nd millennium BC or in the 1st, i.e. during the Old or the Neo-Assyrian Empire; whether it reached the Pontic coast or just the Halys; and whether or not it could be considered a 'colonisation'. Over more than a century and a half almost all the above have been propounded. The establishment of Assyrian settlements in Cappadocia,[311] at least as far as the Halys, is unanimously accepted. As A. Goetze wrote, expansion towards the West and the Mediterranean was one of the characteristic aims in the history of Mesopotamia.[312] Whether this expansion corresponded to a kind of colonisation,[313] and whether it had the character of a political occupation[314] or was just the outcome of trading activity,[315] is another matter. Whatever these Assyrian settlements were, it has been assumed that 'the descendants of these settlers and merchants were in sufficient number to cause the Greeks to identify them as Syrians', because 'they spoke the same language as the inhabitants of Syria and Mesopotamia'.[316]

In a step further, it was argued that the Assyrian expansion to Cappadocia had gone beyond the Halys and reached the Black Sea coast as early as 1600 or even the 19th century BC.[317] Although there is no archaeological or other

[311] As cuneiform records from the 2nd millennium BC found in Kültepe indicate (see bibliography in Frye 1992, 281).

[312] Goetze 1957, 64.

[313] Garelli 1963, 370–77, who tries to compare it with other colonial movements in world history.

[314] Garelli (1963, 363–70) did not avoid the temptation to characterise the Assyrian expansion 'imperialistic'.

[315] See, for example, Goetze 1957, 69–81, presenting the theory of the 'Handelskolonien'.

[316] Frye 1992, 281 and Janse 2002, 347, respectively.

[317] Nöldeke 1871, 444; Meyer 1913, § 435, 474. *Cf.* Maksimowa 1956, 38–39.

confirmation of this view,[318] scholars in favour of it often invoked the mythical tradition on the relation between Sinope and Syrus: Plutarch (*Lucullus* 23. 6) related that the Syrians who were in possession of Sinope were descended from Syrus, son of Apollo, and Sinope, daughter of Asopus.[319] This tradition would imply the existence of a very early Assyrian colony in Sinope, which would account for the use of the name 'Syroi' in the littoral, regardless of any Assyrian occupation of Cappadocia.[320] Such an interpretation of this myth has been criticised,[321] while it has also been suggested that the creation of the myth was reinforced by the influence, already in antiquity, of Herodotus, who used the name 'Syroi', and that the image of a 'Syrian' Sinope reached its peak in the Hellenistic period and became a good example of the transfer of geography into mythical genealogy.[322]

Not long after the establishment of the 'Assyrian origin' theory critical responses appeared, sometimes affirming certainty about the Hittite origin of the Leucosyroi.[323] According to one of these, the Syrians came from the same Hittite tribe as the non-Indo-European populations of early Anatolia and, since the Greeks knew the Syrians, they attributed their name to the whole of Anatolia.[324] In order to circumvent the problem created by the myth of Sinope, the above-mentioned interpretation of which was still popular, it was claimed that the Assyrians did not constantly occupy the southern Black Sea littoral but invaded it sporadically and that the result of one such invasion was a transient settlement in the area of Sinope. The explanation proposed – albeit with some reservations – for the name of the Leucosyroi in that case was that these Cappadocians of Hittite origin had probably influenced the Syrians so strongly, especially those that had spread northwards, that they had come to resemble each other physically ('Körperähnlichkeit'). When the Greeks settled on the

[318] See, for example, Garelli 1963, 31–79; Dan 2012, 88–89.

[319] Who had thus evidently lost the virginity divinely guaranteed according to Apollonius (*Argonautica* 2. 946–954), where we read that the Argonauts 'landed on the Assyrian shore where Zeus himself gave a home to Sinope, daughter of Asopus, and granted her virginity, beguiled by his own promises. For he longed for her love, and he promised to grant her whatever her heart's desire might be. And she in her craftiness asked of him virginity. And in like manner she deceived Apollo too who longed to wed her, and besides them the River Halys, and no man ever subdued her in love's embrace' (*cf.* also Valerius Flaccus *Argonautica* 5. 110–113; Dionysius Periegetes 773–779).

[320] H. Philipp, review of Leonhardt 1915, in the *Berliner philologische Wochenschrift* of 1917, 709–10.

[321] E.g. Ivantchik 1998, 304–05, n. 20.

[322] Dan 2012, 86.

[323] Leonhard 1915, 291; Bilabel 1920, 34–40; Neumann 1961, 28–33.

[324] Leonhard 1915, 291.

southern Black Sea coast they met people who looked like those they had seen during their travels in the South (in Syria), just a bit more light-skinned, and so they called them Syroi or Leucosyroi.[325]

More recently, it was asked why one should continue to talk about Pontic Syrians after the fall of the Assyrian Empire, assuming an Assyrian expansion as far as the Black Sea had indeed taken place. And why would only Cappadocia be called Syrian, when the Syrian language was also spoken elsewhere in Asia Minor and the Middle East? One proposed possible answer is that 'Syroi' and 'Amazons' could be the Greek names, respectively historical and mythical, for those whom we call Hittites; the problem then would be that there is no confirmed presence of the Hittites in Pontus whom the Greeks could have encountered there.[326]

Linguistics inevitably entered the debate. Thus, a Hittite origin for the Cappadocians in general has been deduced from examination of the remains of their language, thought to be of Hittite-Luwian origin.[327] Conversely, it has also been suggested that the first part of the word Leucosyroi does not derive from the Greek word for white (designating the light-skinned Syroi, as opposed to the dark-skinned in the South, as we read in Strabo), but is rather the Anatolian stem *Lukki-* or *Lyko-*, which via Greek folk etymology became *Leuko-*.[328] In an effort to keep the emphasis on the word 'Leucosyroi' itself while retaining the connection with mythology, it was argued that the Leucosyroi were an Iranian nation who worshipped the sun god Syros, identified with Lycus, the predecessor of Apollo at Delphi, the name Leucosyroi thus deriving from the combination of these two mythical figures.[329] Another possibility would be that the root word *Lukk* was connected to the hydronym Lykastos, the name of the river in the land of the Leucosyroi after which, according to Apollonius (*Argonautica* 2. 999), the Amazons were called.[330]

The different theories on the origin of the Leucosyroi reflect the difficulty in even outlining the situation in eastern Anatolia in the specific periods. Not long ago, in response to the theory of the Assyrian settlements in north Anatolia, including Sinope, it was alleged that that these settlements ended with the establishment of the Hittite Empire; in addition, any historical justification of

[325] Bilabel 1920, 34–40.
[326] Dan 2012, 89–91.
[327] Neumann 1961, 28–33.
[328] Winckler 1903, 23. *Contra* Bilabel 1920, 39.
[329] Krappe 1946. *Contra* Janse 2002, 355.
[330] Dan 2012, 91.

a connection between north Anatolia and Mesopotamia was excluded and the possibility that the Leucosyroi were an Anatolian nation was presented.[331]

Recently, the whole issue was treated from a different viewpoint:[332] the presence of Syroi/Assyrioi/Leucosyroi in both north-eastern Anatolia and the Mediterranean and Mesopotamia was regarded as the effect of 'a narrowed perception of the *Oecumene*'. In such a perception of a world surrounded by Ocean, where the distances from North to South and from West to East are considerably decreased, the distances between Sinope and Mesopotamia are minimal. Such geographic motifs, derived from Homeric geography, are to be met in works of the Classical tragic poets (e.g. *Prometheus Bound*). Results of this 5th-century BC perception of the 'smaller world' could also be seen in the case of the Arabs and the Chaldaeans and Chalybes from the south-east Pontic region, as well as in the relation between the Colchians and the Egyptians narrated by Herodotus (2. 104–105),[333] whose influence is indisputable. It was concluded that in both cases, that of the Leucosyroi/Assyrioi/Syroi and that of the Chalybes/Chaldaeoi/Chaldoi, who interestingly enough were neighbours, an initial ethno-linguistic identification spread over a large geographic but actually heterogeneous zone, according to a Greek ecumenical perception of the whole area from the Taurus to the Black Sea.[334]

This theory is attractive and seems able to offer neat and easy answers to questions like those asked in the beginning of this section. However, it is at the same time highly speculative as well as problematic in one of the elements of its argumentation: the Homeric perception of the world surrounded by Ocean had been overthrown by the Classical period, when the name 'Syroi' appeared for the Leucosyroi,[335] and Herodotus' work itself, among others, offers good proof of this. It would, therefore, be a mistake to consider that in the Classical period such a perception still obtained in Greece. And of course there is no need to seek the geographic perception of the 5th century BC in tragic poetry when we have other kinds of literature. Besides, the tragic poets used myths to serve their – mostly political – purposes (Aeschylus' *Prometheus Bound* is a prime example of this),[336] careless of the geographic or other scientific validity of what they portrayed.

[331] Summerer 2005a, 132–35, 151–52; 2008, 63–64, with more bibliography on the issue; Dönmez 2006, 39.

[332] Dan 2012.

[333] See Section 2.15.2. *Cf.* Braund 1994, 17–18.

[334] Dan 2012, 93–94.

[335] See Section 2.8.2.

[336] See for example Manoledakis 2005, 104–06, 111–12, 139–40.

C. Who, then, after all were the Leucosyroi?

Apparently it is the texts of three ancient authors that have substantially influenced modern scholarship on the question of the origin of the Leucosyroi: Herodotus' account of the identity of the Syroi and the Cappadocians; Pseudo-Scymnus' information that the early inhabitants of the region of Sinope before the coming of the Greeks[337] were Syroi; and Strabo's musings on the name Leucosyroi and the expansion of the Syroi in Anatolia; the myth of Sinope, daughter of Asopus, was also sometimes considered in connection with the second text.

There are, however, two important remarks that must be made here: first, that the Greek myths about the origin of heroes, nymphs or other figures that gave their names to peoples or cities were created in order to explain or justify Greek connections to or activities in those peoples' lands or cities. They are purely Greek creations that were born *after* the encounters of the Greeks with the peoples described in each myth and have nothing to do either with the local peoples' perception of themselves or with any events supposed to have taken place well before these encounters; at best, they carry some traces of apposite narratives, whose plausibility, however, cannot be assessed.[338] The second observation has to do with the notorious difference of complexion. It is remarkable, especially in the ancient authors, that nowhere in the literature on the light-skinned Leucosyroi is the reasonable question asked: 'How did the Syroi who moved to the north became lighter-skinned?' Presumably there could be no explanation other than intermarriage. The dark-skinned Syroi settled in an area that was inhabited by light-skinned people and thus gradually their complexion changed. North Anatolia would naturally evoke an image of paler complexions than Mesopotamia, and one cannot help remembering the children of the wealthy Mossynoikoi,[339] not far to the east of the Leucosyroi on the coast, who according to Xenophon's description (*Anabasis* 5. 4. 32) were white to an extraordinary degree.[340] But then another reasonable question arises: who inhabited this area before the coming of the Syroi?

Before addressing this question, let us recall our previous comment that the total non-appearance of Cappadocians in Xenophon's *Anabasis* is remarkable; even, we would say, incomprehensible.[341] We should not forget that Xenophon passed through the southern Black Sea littoral in 400 BC and met its peoples.

[337] On the issue of the inhabitants of the region of Sinope and the founders of its city, see Section 4.3.2.

[338] *Cf.* Dan 2012, 86–87.

[339] See in the homonymous section.

[340] As Dan (2011, 85–86) has insightfully done.

[341] See in Section 2.8.2.

He spent five days in Sinope (*Anabasis* 6. 1. 15–33), which some years later is mentioned (Pseudo-Scylax 89) as belonging to the Assyrioi, but he never speaks about them there in the *Anabasis*, not even under the name Cappadocians, which one would expect: he merely states that the Sinopeans lived in Paphlagonia (6. 1. 15; *cf.* Nicephorus *Geographia*, *GGM* II 464b, 30–32), which extends to the east at least as far as Cotyora (5. 5. 6–6. 1. 13). In this important work, a personal record of his own experiences, neither Assyrioi nor Leucosyroi or even Cappadocians (frequently encountered in other works of his) are ever mentioned, unlike several other minor peoples in north-eastern Anatolia. This is the most remarkable absence of an important ancient nation or settlement of the southern Black Sea littoral from the works of the ancient historians and geographers that are considered reliable, the second being the absence of Amisos from the *Periplus* of Pseudo-Scylax. And if the latter could be explained, even if not totally satisfactorily, by a gap in the manuscript of the work,[342] Xenophon's silence regarding the Leucosyroi/Cappadocians in the *Anabasis* is indeed a mystery.

The truth is that throughout Xenophon's report of the Greeks' stay in Sinope (6. 1. 15–33) the narrative is limited to the soldiers' discussion about the continuation of their journey home to Greece and there is nothing regarding any meetings with local people, as had often been the case before. But this might not be so crucial here, since there had been many references to meetings between Greeks and Paphlagonians earlier in Cotyora, which is situated to the east of Sinope, rendering it obvious that Sinope would also belong to the land of the Paphlagonians. If this is the case, then where would the Leucosyroi have been living in Xenophon's time, bearing in mind that in the slightly later Pseudo-Scylax they occupied the coast at least from Themiscyra to Harmene? Could it be that by Xenophon's time the Leucosyroi had been subjugated by the Paphlagonians and thus become just a part of coastal Paphlagonia, as had been the case with the Chalybes, of whom Xenophon records (*Anabasis* 5. 5. 1), in the same period, that they were few in number and subject to the Mossynoikoi? Or was there a more substantive and closer relation between Leucosyroi and Paphlagonians?

However frivolous such speculation might initially appear, reports, direct or indirect, connecting the Paphlagonians with the Leucosyroi are not negligible in the written sources. Xenophon's attribution of the whole land where the Leucosyroi lived to Paphlagonia is a first strong element. Furthermore, Strabo tells us (12. 3. 25), citing the 3rd-century BC Maeandrius, that the Enetoi, whom he considered the most noble tribe of the Paphlagonians (12. 3. 8), first

[342] Counillon 2004, 116–17; Shipley 2011, 159.

set forth from the country of the Leucosyroi and allied themselves with the Trojans and that they sailed away from Troy with the Thracians and took up their abode around the inner recess of the Adriatic, 'but those Enetoi who did not take part in the expedition had become Cappadocians'.[343]

Strabo continues by saying that this account seems to agree with the fact that in the whole western part of Cappadocia near the Halys river, which extends along Paphlagonia, two languages are spoken, which abound in Paphlagonian names (12. 3. 25). One may easily imagine that the two languages would be Cappadocian and Paphlagonian, although what we actually know about the Cappadocian language is more or less the same as what we know about the Paphlagonian: virtually nothing;[344] and the names mentioned by Strabo have some parallels in Phrygian.[345]

However, there is also the impressive formulation of the Scholiast in Dionysius Periegetes (*Orbis descriptio* 65–66) that ἡ Σινώπη τῶν Ἀσσυρίων ἐστίν ἤτοι Παφλαγόνων, which seems to present the two peoples as kindred. Could all these wisps of information not be worked into another, even if far-fetched, theory, which would present the Leucosyroi simply as... Paphlagonians? Actually, this possibility has already been weighed and considered strong, but has never been backed by reference to the specific passages or other substantiation.[346]

Consequently, the answer to the question of who the Leucosyroi encountered when they arrived in the central part of the southern Black Sea coast could simply be: those whom the Paphlagonians found there, if not the Paphlagonians themselves.[347] In the 6th century BC, when we first read about the Leucosyroi in Greek literature, the populations that succeeded the Kaska had long since been settled on the coast – probably in the early 1st millennium BC – and the Assyrian provenance of the Leucosyroi (which is accepted here)

[343] Elsewhere (12. 3. 8), however, he invokes other, unspecified, sources saying that it was after an expedition with the Cimmerians that the Enetoi were driven out to the Adriatic Sea. See more in Section 2.7.

[344] *Cf.* Janse 2002, 352 and 354 with an attempt to clarify whether Cappadocian was an Indo-European language.

[345] And thus have been used in the extended argumentation of Haas (1966) about the kinship between Phrygians and Paphlagonians. On the names mentioned by Strabo, see Brixhe 2013; Avram 2018. On the relation between Phrygians and Paphlagonians, including a criticism on Haas' theory, see Manoledakis 2021d.

[346] See most recently Tsetskhladze 2007, 193; 2012b, 236, invoking Saprykin (1991). On the possible relations of the Paphlagonians with other southern Black Sea peoples, see Manoledakis 2021c. Leonhard (1915, 290–95) considered that the name Leucosyroi was falsely given to the non-Indo-European population of Paphlagonia. Ruge (1949, 2517) rejected this view.

[347] On this issue, see in Section 2.6.2.

was known to the Greek authors, some of whom invented the new name for them to distinguish them from the southern Assyrians.

The antiquity of the Leucosyroi on the coast is best documented by Pseudo-Scymnus (941–950), who, speaking of the timeline for the population sequence of Sinope, reports that the area was first settled by the Syroi, then by Thessalian Greeks; then came the first Milesians, who were killed by the Cimmerians, after whose descent on Asia came a new Milesian wave. We note that the Syroi are mentioned as having settled the area three phases (i.e. three generations or more) before the Cimmerians, who came to Anatolia roughly in the early 7th century BC, which takes us back to the 8th century and earlier. From all the references of the ancient literature to the early history of the local peoples of the southern Black Sea, this is the one that provides us with the earliest time-frame. We have, of course, the reference to the Paphlagonians, the Enetoi and the Halizones in the *Iliad* (2. 851–857), but that is a text of purely mythical context and with numerous issues concerning its creation, dating, etc. It is also worth pointing out that Pseudo-Scymnus calls the Syroi εὐγενεῖς (noble), but since this nobility is nowhere else implied in the sources, C. Müller was probably right to suggest that the correct epithet in the manuscript would be ἐγγενεῖς (indigenous),[348] which would precisely describe the status of the Leucosyrians in the period in question.

On the other hand, from several elements that were presented in the relevant section[349] it appears that the Paphlagonians had become a dominant element in the area situated roughly between the Parthenius and the Iris, which had thus come by many to be called Paphlagonia. At the same time, the name Leucosyria, which would denote the land of the Leucosyroi, was not used by any ancient author; and only once, in Pseudo-Scylax (89), do we have 'Assyria'.[350]

If we accept that the Paphlagonians originated from the Palaic populations that survived the Kaska occupation and became mixed with Hittite/Anatolian elements,[351] then the Assyrioi who came from Mesopotamia encountered them

[348] Müller 1855 I, 236.

[349] See in Section 2.6.

[350] It is only in the *Scholia in Apollonii Rhodii Argonautica (scholia vetera)*, which date from the Roman period and later (Dickey 2007, 62–63), that we read of Leucosyria or Leucosyriake (196, 198, 200, 201). In one of these passages (196) it is written that 'some of the ancients called Assyria Leucosyria', but it is not specified who these were, and since our preserved texts do not include such a word, I would be hesitant about taking this sentence literally – it could simply result from the popularity of the ethnic name (Leucosyroi) in the ancient sources. The attribution of the words of commentary nr. 201 to Hecataeus by the *Thesaurus Linguae Graecae* is rather risky, since the words clearly attributed to Hecataeus by the Scholiast in 201 refer to the Chadisiae Amazons in the very next line, and not to Lycastos, 'the village of Leucosyria', which is apparently called that only by the Scholiast himself.

[351] See in Section 2.6.2.

in the eastern parts of their territory, which would come to be called Paphlagonia, and this meeting resulted in the gradual emergence of the Leucosyroi. We cannot, of course, fail to stress that this is just another hypothesis, but, analysing the available written evidence and absent any archaeological data, I would say that it is the most plausible one.

*

* *

Closing this section, let me turn to another people that have been linked with the Leucosyroi: the Scythians. Any connection between them must have been a product of their encounters during the Scythians' sojourn in Asia Minor,[352] after the establishment of the Leucosyroi in the area. As is usual in such cases, mythical traditions appeared that brought the two peoples together. Diodorus (2. 43. 6) recounts that by some (mythical) Scythian kings:

> many of the conquered peoples were removed to other homes, and two of these became very great colonies: the one was composed of Assyrians and was removed to the land between Paphlagonia and Pontus, and the other was drawn from Media and planted along the Tanaïs, its people receiving the name Sauromatae.

While Diodorus' narrative may sound preposterous and probably reflects a Scythian tradition, a (Leuco)Syrian influence on the Scythians has nonetheless been proposed in the area of religion. Herodotus (4. 59. 2) tells us that in the Scythian language Apollo was called (G)oetosyrus, and it has been suggested[353]

> that the name is a loan word, which came with the god from the region south of the Black Sea, and that it is a compound adjective, the first part meaning 'white' (οἶο, γοιτο) and the second 'Syrian'. This title would then have the same significance as Λευκόσυρος.

In this case, Apollo must have come to Scythia from the land of the Leucosyroi, which would mean that the god was important there. Elderkin further justified this view by invoking Strabo's reference (12. 2. 6) to the sanctuary of the Cataonian Apollo, 'which is held in honour throughout the whole of Cappadocia, the Cappadocians having made it the model of temples of their own', as well as the mythical tradition that Apollo was the father of the Amazon Sinope (who gave her name to the city in the land of the Leucosyroi). What better explanation could there be for his epithet 'Leucosyrian'?[354]

[352] See Section 3.2.

[353] Elderkin 1935, 342.

[354] Elderkin 1935, 342–43, who points out that when Apollo appeared among the Greeks he was blond.

If this interpretation is correct, then (still according to Elderkin) the cult of Apollo may have come to Scythia from Leucosyria, either via the Black Sea, just as the Amazons found themselves in Scythia, according to Herodotus (4. 110), or around the eastern Black Sea coast to Lake Maeotis.[355] In any case, however, one cannot claim an ethnic kinship between the Leucosyroi and the Scythians.

9. The Chalybes

9.1. The Territory (Fig. 10.2–3)

The location of the territory that belonged to the people known as Chalybes might be the best example of uncertainty or discrepancy in the sources concerning the exact district inhabited by a nation on the southern Black Sea coast. It is even questioned whether they were always a coastal people. Hecataeus (*FGrH* 1a F 202–203), the first known author to have mentioned the Chalybes, as well as many other peoples, calls them a nation in Pontus, at the Thermodon, as does Eudoxus (fragm. 282), and ascribes to their territory the coastal town of Stamene (*cf.* Stephanus of Byzantium, *s.v.* Χάλυβες), adding that to the south they neighboured the Armenians. Pseudo-Scylax (88), who apart from Stamene also assigns to the territory of the Chalybes the harbour of Genetes and Iasonia,[356] places them between the Leucosyroi (to the west) and the Tibarenoi (to the east). So, too, does Dionysius Periegetes (767–772),[357] who, however, places the Thermodon within the territory of the Leucosyroi, thus restricting the Chalybes to a smaller district farther east, around Cape Iasonion. Similarly, Apollonius places the Chalybes on the coast to the east of the Thermodon and the Tibarenoi right after them, beyond Cape Genetes (*Argonautica* 2. 370–378, 1000–1010. *Cf.* 1. 1323; 4. 1475; *Orphica Argonautica* 741; Valerius Flaccus *Argonautica* 5. 140–142; 4. 601–610; Pliny *NH* 6. 11; *Scholia in Aeschylum Theb.* 941; 942). Conversely, in the much later *Etymologicum Gudianum* (*s.v.* Ἅλυς) we read that the Halys was the border between the Chalybes and the Leucosyroi.

Herodotus' inclusion of the Chalybes among the peoples dwelling west of the Halys (1. 28) is in total disagreement with the above as well with the evidence of most sources; but he is referring to the era of Croesus and his incursion, which could mean that the territory of the Chalybes had considerably

355 Elderkin 1935, 343–44.
356 On these settlements, see in Part 5.
357 *Cf. Paraphrases in Dion. Per.* 762–771.

decreased between the 6th and the 4th centuries BC. Unique, and of course erroneous, is the statement of Pomponius Mela (1. 105) that the Chalybes occupied the renowned cities of Sinope and Amisos. Equally erroneous is Curtius Rufus' placing of the Chalybes to the west of the Leucosyroi (Curtius 6. 4. 16–17).

In 400 BC, after a march of eight stages through the land of the Mossynoikoi, Xenophon's army reached the land of the Chalybes (*Anabasis* 5. 5. 1). Nothing more is said about that country except that next came the land of the Tibarenoi, which was much more level and had coastal settlements (*Anabasis* 5. 5. 2). However, Xenophon does provide two significant details: first of all, that the Chalybes were few in number and subject to the Mossynoikoi, and secondly, that after proceeding through a friendly country for two days the Greek army arrived at Cotyora, in the land of the Tibarenoi (5. 5. 3). This places the Chalybes farther east than our previous sources, and between the Tibarenoi and the Mossynoikoi, not between the Leucosyroi and the Tibarenoi as in the above-mentioned texts.

However, if we consider the distances given by Xenophon, we will see that they cannot be accurate unless the army had followed an inland route. Eight stages seems an excessive distance along the coast to be considered as the way between even the eastern borders of the land of the Mossynoikoi and the land of the Chalybes where Xenophon places it. And it becomes even less likely when we add another two days of marching (more or less another two stages)[358] in the land of the Tibarenoi to Cotyora. In order to account for it one would have to assume either that the unit of measurement given by Xenophon was wrong or that the Greek army turned aside from the coast and followed a long inland curve before reaching Cotyora. If the latter is the case, then there is no reason to assume that the land of the Chalybes necessarily had access to the sea, at least during the whole of the 4th century BC, and when it had, that this coastal part was small. The possibility, examined by O. Lendle, that the eight stages marching reported in 5. 5. 1 included the two days mentioned in 5. 5. 3,[359] is very small, because Xenophon clearly states that between the two marches the Greeks had entered the country of the Tibarenoi and been offered

[358] According to many scholars, Xenophon always used stages to denote a day's march. See recently Rood 2010, 52 and n. 6, where he cites several passages that support this view. One of the examples he cites is the specific case of the eight stages mentioned in the *Anabasis* 5. 5. 1 between the Mossynoikoi and the Chalybes, which are mentioned as eight days by Diodorus (14. 30. 7). However, there are two differences between the two passages: first of all, Diodorus mentions the Tibarenoi and not the Chalybes (whom he ignores) after eight days march from the land of the Mossynoikoi, and secondly, he says that the Greeks passed through the land of the Tibarenoi in three days, whereas Xenophon says two stages (*Anabasis* 5. 5. 3).

[359] Lendle 1995, 333–34.

gifts, and had only set out on the two days' march after deciding to accept the gifts and not attack the Tibarenoi.

The Chalybes we are speaking of here, of course, are those who lived on the Black Sea littoral or at least close to the sea; there were also Chalybes who clearly lived in the hinterland, and these are examined next. We should not close this section without a mention of by far the most peculiar geographic siting of the Chalybes, which belongs to Aeschylus: in *Prometheus Bound* (710–735) he places the Chalybes somewhere to the north of the Caucasus, in a geographic description of the whole eastern part of the Black Sea area which is, of course, entirely fictional and lacks any intention of geographical accuracy. Equally peculiar is the single reference of Pseudo-Aristotle (*De mirabilibus* 832a. 26) to an island above Cyprus where people called Chalybes lived.[360]

9.2. Ethnography

The Chalybes have always been a people famous for their metal-rich country[361] and their metalworking skills, as indicated by the Greeks' use of their name to denote steel.[362] Already in the works of the 5th-century tragedians they appear as ironworkers (Aeschylus *Prometheus Bound* 714–715; Euripides *Alcibiades* 980–981), and they continue to be famous for that as well as for exchanging their wages for daily sustenance (Xenophon *Anabasis* 5. 5. 1; Eudoxus fragm. 282; Apollonius *Argonautica* 2. 75–76, 1001–1006; Dionysius Periegetes 768–770; Virgil *Georgics* 1. 58; Ammianus Marcellanus 22. 8. 21; Libanius *Orations* 59. 68; the *Suda*, *s.v.* Χαλύβοις; *Etymologicum Magnum*, *s.v.* Χαλκός). Indeed, in several texts they are supposed to have been the inventors of ironworking (Callimachus *Aetia* 110. 48–50; Arrian *Bithynicorum fragm*enta 52; Arrian *FGrH* 2b 156 F 73; Eustathius *Comm. in Dion. Per.* 767; Schol. in Callim. Aet. 110. 48; Schol. in Aesch. Prom. 132–133, 301–302; Schol. in Aesch. Theb. 942; Schol. in Eur. Alc. 980; Schol. in Hesiod. Opera 142; Schol. in Lycophr. 1109).

Their occupation as ironworkers soon created the image of the Chalybes as 'the most wretched of men, who live in a barren country' and who 'never know a day of ease, but labour ceaselessly amid sooty flames and smoke' (Apollonius *Argonautica* 2. 374–376; 1001–1008; Dionysius Periegetes 768–770;

[360] Another detail is recorded only by Nymphodorus (Schol. in Apol. Arg. 4. 1470), namely that the Argonaut Polyphemus, the mythical founder of Cios (*cf.* Apollodorus 1. 9. 19), fell fighting against the Chalybes.

[361] A description of which has been made in the previous part of the book.

[362] *Cf.* Bryer and Winfield 1985, 3. For the origin of the name (probably Hittite or 'Anatolian'), see Chantraine 1999, *s.v.* Χάλυβες.

Schol Apol. Arg. 136, 159; Schol. in Dion. Per. 768). This picture of a hard, miserable life may have been responsible for another image, that of the savage people, who hate foreigners (Aeschylus *Prometheus Bound* 716; Schol. in Aesch. Prom. 716), and the tradition that they were named after a son of Ares called Chalyps (Schol Apol. Arg. 159) might have been inspired by this supposed cruelty of theirs.

What we have here is not a case of strangers who are considered by the Greeks as 'barbarians' because of their unusual or shocking habits, like the Mossynoikoi, for example.[363] It is rather an impression of people who work hard on tough metals in a barren country, which in a rather deterministic way creates the perception, so vividly conveyed by Aeschylus (*Seven Against Thebes* 727–741; *Prometheus Bound* 714–716), of people who *must* be savages since they live like that, not who *are* barbarians because the Greeks have met them and experienced their customs.

Strabo (12. 3. 19) makes it clear that the mines were in the mountainous part of the country and he is the only author who informs us that in earlier times there were also silver mines in this land.[364] He also mentions the occupations of the people living on its seaboard, which although very narrow had the natural advantage of wonderful pelamydes fisheries, and actually it was there that this fish (a kind of tunny) was first caught (*cf.* 7. 6. 2; 12. 3. 11). They also fished for dolphins and in fact, always according to Strabo, were the only people who cut up the dolphins, which are caught with bait, and used their abundance of fat for all purposes.

9.3. Chalybes: One, Two or Three?

As mentioned above, there were also Chalybes inhabiting the hinterland of the southern Black Sea littoral. These are mentioned by Xenophon, whose army encountered them on their way from Armenia to the Black Sea. The first reference to Chalybes in the *Anabasis* is in 4. 4. 18, when the Greeks had reached the Teleboas river in western Armenia. There, a Persian soldier told them that the Armenian ruler Tiribazus had Chalybian mercenaries. Later (4. 5. 34), having crossed a mountain and marched at least 154 km (28 parasangs in total are mentioned), they were informed by the headman of a village that they were still in Armenia and that the neighbouring land belonged to the Chalybes. After some days, they reached the 'Phasis' (4. 6. 4), which must be the

363 See in Section 2.11.3.

364 Actually, more to the east, near Tripolis, Arrian (*Periplus* 16; *cf.* Anonymous *Periplus* 36) mentions a settlement called Argyria. See more in Part 1.

Araxes,[365] where they met Chalybian soldiers, and proceeded to the country of the Taochoi (4. 7. 1). After this they reached the country of the Chalybes (4. 7. 15), then crossed the Harpasus river and the land of the Scythenoi (4. 7. 18), where they came to Gymnias (4. 7. 19). Finally, five days marching from this city brought them to Mt Theches, from where they could see the sea (4. 7. 20–21).

The two different passages mentioning Chalybes (4. 5. 34 and 4. 7. 15) have led to the view that there might have been two different inland nations with the same name, the coastal group being a third one.[366] The marching distances given by Xenophon, combined with the reported sequence of the countries, create doubts about whether the Greeks followed a continuous northwards route and thus the Scythenoi were to the north of the Chalybes, who were to the north of the Taochoi,[367] or whether they made a rather long curve to the east before again turning north and west, in which case, the Chalybes and the Scythenoi would be almost east-west neighbours, both having the Taochoi to their south.[368] Furthermore, Xenophon nowhere mentions where the Greeks entered or left Chalybian territory, and it is only in 4. 7. 15 that he clearly states that the Greeks *are* in the land of the Chalybes. These considerations could lead to the assumption that the Chalybes of 4. 5. 34 and 4. 7. 15 were not far enough apart to be considered different peoples.

What is obvious from Xenophon's narrative is that the coastal Chalybes (5. 5. 1) lived at some distance from the inland ones. The question is whether or not the two peoples always lived far from each other, separated by others. The truth is that there are different testimonies for different periods. Xenophon's account may attest to a distance between the two Chalybes, but not necessarily a very great one, since most of the intervening lands and cities mentioned (Colchians, Drilae, Trapezous, Kerasous, Mossynoikoi) were situated on the coast. This means that the Greeks had already turned west after the aforementioned curve, hence might have been advancing parallel to the inland Chalybes, who may not have been very far to the south.

Strabo's references to the Chalybes, on the other hand, create some confusion, for he mentions the Chalybes (who, he notes, were called Chaldaeoi in his day[369]) as living above Trapezous and Pharnakia as far as Lesser Armenia and close to the Tibarenoi (12. 5. 18–19, 28–29), and his reference to the Armenian kings who enlarged their territory (11. 14. 5) by cutting off Carenitis

365 See, for example, Rood 2005, 212, n. 95.
366 Lendle 1995, 246, 266.
367 As Lendle (1995, 251 map 4, *cf.* 266) seems to think.
368 Which seems to be preferred by Rood (2005, map on p. xliv).
369 See below in this section.

and Xerxene, which border on Lesser Armenia, from the land of the Chalybes and the Mossynoikoi does nothing to clarify the matter. Hecataeus (*FGrH* 1a F 203), one must remember, mentioned the Chalybes as a nation both on the coast and bordering the Armenians to the south.

Furthermore, Strabo (14. 5. 23–24) cites Ephorus' statement that the Tibarenoi bordered the Leucosyroi and the Chalybes (*FGrH* 2a 70 F 43), with no further specification,[370] which Strabo seems to interpret by saying that Ephorus placed the Chalybes inland and not on the coast. Indeed, he wonders why Ephorus should do so, adding that 'it would have been better to divide them and set one part of them on the sea and the other in the interior'. Finally, he states that 'the meridian through the land of the Chalybes would be drawn through Lesser Armenia and the Euphrates, cutting off on this side of it the whole of Cappadocia, Commagenê, Mt Amanus and the Issic Gulf', which again does not help.

Equally vague is Pseudo-Scymnus' outline of the Anatolian peninsula, where the Chalybes are mentioned as being inland and near them the Cappadocians who dwell in Pisidia and the Lydians and next to them the Mysians and the Phrygians (938–940. *Cf.* Anonymous *Periplus* 27, but the contrary in 31) and one wonders whether it is the Chalybes of the southern Black Sea littoral that are meant or Pliny's Armenochalybes (*NH* 6. 29, different from the coastal Chalybes of 6. 11), who were certainly an inland-dwelling people.[371]

The issue created by Xenophon's account of different groups of Chalybes has attracted some attention in modern literature. According to one theory, the Greeks used the name Chalybes for several nations that inhabited different parts of Anatolia and worked with iron.[372] M.I. Maksimova's claim[373] that Xenophon's inland Chalybes lived in north-east Asia Minor and were in fact those mentioned by Hecataeus, thus justifying the name Chalyboi instead of the Chalybes used by the latter, is unconvincing. She then observes, however, that the coastal and the inland Chalybes seemed to have little in common, for while we are told that the former practised metallurgy, our only information about the latter is of a military nature, concerning for example their fortresses; their economy would probably have been based on activities followed by other mountain people, such as cattle breeding.

[370] See also Section 2.8.1.

[371] On the issue, see also Kavtaradze 2002, 75–77.

[372] Müller 1855, I 65. See also Maksimova 1956, 27 for the acceptance of this theory by other scholars and her own criticism.

[373] Maksimova 1956, 29.

Still according to Maksimova, it was probably the Mossynoikoi, arriving from the South, who subdued the Chalybes (as reported by Xenophon *Anabasis* 5. 5. 1), who were there before them as an ancient local nation, and exploited their metallurgy, selling its products to the Greek cities. The inland Chalybes were initially one nation with the coastal ones, but changed their economic activities after it was broken up by the Mossynoikoi.[374] While the kinship of the two Chalybes is not implausible, Maskimova's theory suffers at the point where she takes for granted the southern origin of the Mossynoikoi and the local prehistory of the Chalybes: neither is substantiated by the sources; on the contrary, as we shall see, there is evidence that could connect both peoples to Scythia.[375]

While Maksimova ascribed the division of the Chalybes to the Mossynoikoi, another theory, which appeared in 1995, blames the Colchians. M. Lipka was not wrong in noting that in 400 BC, as clearly indicated by Xenophon's narrative and confirmed by Arrian (*Periplus* 11), the Colchians inhabited part of the eastern littoral of the southern Black Sea.[376] According to him, by Xenophon's time the Colchians had expanded their territory to the west as far as Kerasous.[377] This intrusion into the land of the Chalybes divided the latter into two parts; the smaller, clinging to the coast, was soon subjugated by the Mossynoikoi (*Anabasis* 5. 5. 1) who exploited their iron products, while the larger group remained in the hinterland.[378] Thus, Lipka agrees with Maksimova on the economic exploitation of the coastal Chalybes by the Mossynoikoi, but not on who it was whose expansion divided the Chalybes and their country into two parts.

The problem with this theory is that if the cause was indeed a Colchian expansion, then what happened (and when) to all the nations that by most sources, beginning with Hecataeus, are reported to have inhabited the area between Kerasous and Colchis, namely the Mares, Philyres, Macrones, Becheiroi, Sapeires and Dizeres?[379] Lipka anticipates this question, suggesting that these nations were displaced to the south or west; he also leaves open the possibility of an eastward displacement, while acknowledging that there is no evidence of such a shift in the written sources.[380] But neither is there of a displacement to the south or west, the only exception being the fact that Xenophon

374 Maksimova 1956, 30–31.
375 See Sections 2.9.4 and 2.11.4.
376 Lipka 1995, 65–67, but see more below.
377 *Cf.* Lendle 1995, 283. On the identification of this Kerasous, see Sections 2.11.1; 5.1.
378 Lipka 1995, 67.
379 See in Sections 2.13, 2.14 and 2.16.
380 Lipka 1995, 66 and n. 9.

encounters the Macrones inland (*Anabasis* 4. 8. 1–8) whereas Hecataeus and other authors place them on the coast.[381]

Finally, in the same year (1995) O. Lendle published a commentary on Xenophon's *Anabasis* where, like the previous scholars, he pointed out that the coastal Chalybes were part of the big inland homonymous nation(s), from which they had been detached.[382] As to who was responsible for dividing the Chalybes and separating the coastal ones from the others, he names the Armenians and posits an Armenian invasion.[383] His commentary being broadly based on Lehmann-Haupt's work on Armenia,[384] he cites the latter's view that the Tibarenoi had invaded from the east the territory of the most western (coastal) Chalybes, occupied it, and confined the Chalybes to its eastern extremity. The Mossynoikoi took advantage of this to subdue the already weak Chalybes (Xenophon *Anabasis* 5. 5. 1) and exploit their economy.[385]

Consequently, while there seems to be agreement on the fact that both coastal and inland Chalybes belonged to the same people, there are different opinions as to who was responsible for their partition. We have seen Mossynoikoi, Colchians, Armenians and Tibarenoi proposed for the role. Concerning these last, I really cannot imagine how – or why – someone invading a foreign people from the east would occupy the western part of their territory and not the eastern part, first encountered.[386] I consider this explanation untenable; besides, there is absolutely no substantiation for it in the ancient literature. Nothing was ever said or even implied about an invasion of the Tibarenoi from the east. In fact, that people was always described as rather friendly, especially compared with some of their eastern neighbours (e.g. Xenophon *Anabasis* 5. 5. 2–3).[387]

The Mossynoikoi, who have always been mentioned in the ancient literature as a coastal nation, we have already discussed. The remaining cases of the Colchians and the Armenians deserve further attention. The former are mentioned as having spread westward at least as far as Kerasous in Xenophon's time (*Anabasis* 5. 3. 2) and it seems that some memory of this persisted even in Strabo's day, since we read, for example, about 'the Tibarenoi and Chaldaeoi,

[381] See in Section 2.13.
[382] Lendle 1995, 331.
[383] Lendle 1995, 266.
[384] Lehmann-Haupt 1931, especially 687–833.
[385] Lehmann-Haupt 1931, 803–04.
[386] It should also be noted here that the Tibarenoi appear in some sources as living to the west of the Mossynoikoi (e.g. Xenophon) and in some as living to their east (see Sections 2.9.1 and 2.10.1). Lehmann-Haupt obviously refers to Xenophon's geographic order, but in any case this is not important here, since the description of such a proceeding is in itself implausible.
[387] See in Section 2.10.2.

who extend as far as Colchis' (12. 3. 29). The Armenians, too, are mentioned by several authors, including Hecataeus, Xenophon and Strabo, as neighbouring the Chalybes (to the south). This fact, combined with the strength of these two peoples and the size of their territory, makes it more probable that if there was indeed an invasion that split the Chalybes into parts, it would have been made by one of them.

As far as the Armenians are concerned, both Xenophon and Strabo imply that they bordered the more southern Chalybes, that is, those living in the interior, and thus there is nothing to suggest that they entered a northern part of Chalybian territory from the east. The Colchians, on the other hand, already inhabited a coastal territory and it would be more feasible for them to move west. The fact that Hecataeus mentions the Chalybes both on the coast and bordering the Armenians to the south may indicate that their partition took place sometime between the 6th century BC and Xenophon's day.[388] As to the problem that is created by this assumption, namely the question of the existence and fate of all the other coastal nations mentioned there instead of the Colchians by the rest of the sources (e.g. Hecataeus, Pseudo-Scylax, Apollonius, Pseudo-Scymnus, Strabo, Flaccus, Mela, Pliny, Dionysius), we could posit a temporary Colchian occupation of the coast in Xenophon's day. After all, he is the only author to place the Colchians on the coast, at least in the specific area. Arrian (*Periplus* 11) may seem to agree with Xenophon's information (see above), but if we read his words carefully, we notice that there is a difference. Arrian says that 'we sailed by the Colchians, who, as Xenophon observes, border on the Trapezuntians', while Xenophon (*Anabasis* 4. 8. 22) had written that Trapezous was built *in* the Colchian territory. Similar slight differences are also encountered immediately after this in Arrian's text.

Consequently, one might hypothesise that the Colchians, pressing westwards, divided the land of the Chalybes into two parts, coastal and inland; occasionally, they also occupied part of the southern Black Sea coast, but apparently not for long. The Chalybes, on the contrary, might never have been reunited. As for their initial territory, if they had spread as far south of Gymnias as Xenophon's description indicates, then they would have possessed a territory stretching from the Black Sea almost as far as the Lake Van district to the south. After their partition, recollections of their name Chaldoi were preserved in several areas of their former territory, as for example those recorded by Xenophon.[389]

[388] *Cf.* Lipka 1995, 66.
[389] Lendle 1995, 266.

The difficulty in identifying a standard territory associated with the (coastal) Chalybes in ancient literature, which was mentioned in the beginning, might well be a result of such events, and it should not lead to conclusions such as that the Chalybes were a legendary community, a conclusion encouraged by the fact that there were several areas (and of course there would have been) where iron could be mined.[390]

*

* *

A related issue that appears in the written sources is that the Chalybes were sometimes called Chaldaeoi, for example by Strabo (12. 3. 19), who mentions that in his day this was the name of the people called Chalybes in ancient times – and it is clear that he is referring to the coastal Chalybes. Xenophon on the other hand uses the name Chaldaeoi (*Anabasis* 5. 5. 17) for the Chalybes of the hinterland, whom he had mentioned in 4. 7. 15, while in the last passage on them (7. 8. 25), which however has been considered a later interpolation,[391] he mentions both the Chalybes and the Chaldaeoi, separately. Stephanus of Byzantium (*s.v.* Χαλδαῖοι) mentions the Chaldaeoi as a nation close to Colchis, their territory being called Chaldaea, while Eustathius (*Comm. in Dion. Per.* 767) accounts for Strabo's words by pointing out that the name Chalybes became Chaldaeoi, who were actually the Chaldoi, that is, those who lived in Chaldia (while the Chaldaeoi had been named after Chaldaeus).

Would this synonymy denote an ethnic kinship? There is nothing to support such a possibility. In the modern literature, from early on it has been claimed that the Chaldaeoi of Xenophon (5. 5. 17) and Strabo (12. 3. 19) were the Chaldoi, as Eustathius had correctly clarified – namely the Chalybes. The fact that the Chaldoi were called Chaldaeoi in the classical literature is due to a confusion that had taken place,[392] the causes of which remain obscure.[393] We have already commented[394] on the view that explains the seeming assimilation of different peoples through a common name as a result of the classical Greek perception of the narrowed *Oecumene* surrounded by Ocean, a view that has been expressed with regard to the Leucosyroi/Syroi and has also been applied

[390] De Planhol 1963, 303–04, followed by Dan 2012–13, 52–54. See the correct criticism by Counillon (2004, 105).

[391] Tuplin 2004, 176–77.

[392] Lehmann-Haupt 1931, 709, with references to earlier works of his, followed by Lendle 1995, 266.

[393] See also A. Baumstark, Chaldaioi 2, *RE* III.2 (1899), 2061–62.

[394] In Section 2.8.3B.

to the case of the Chalybes/Chaldaeoi.[395] It should just be noted that the only classical author, at least according to the works that are preserved, who seems to have confused the Chaldaeoi with the Chaldoi/Chalybes was Xenophon, and only in *Anabasis* 5. 5. 17. After him, it is only Strabo who states that in his era the Chalybes were called Chaldaeoi, while the slightly earlier Diodorus seems to make a point of saying 'the Chaldaeoi, *as they are called*'. Could this have been a momentary confusion on Xenophon's part, which had an impact in later periods? We will never know.

9.4. Chalybes and Scythians

We turn now to the question of the origin of the Chalybes. Their purported connection to the inland Chalybes and the similarity with the corresponding issue of the Leucosyroi[396] could lead to the assumption that the Chalybes were another Anatolian nation that evolved from an earlier local (Hittite? Kaskan?) population. However, in Hesychius (*s.v.* Χάλυβοι), in the *Suda* (*s.v.* Χαλύβοις), in the *Etymologicum Magnum* (*s.v.* Χαλκός), as well as in several commentaries of Scholiasts, we clearly read that the Chalybes were a Scythian nation (Schol. in Aesch. Prom. 298a, 301a; Schol. in Theb. 727, 817; Schol. in Apol. Arg. 119, 159; Schol. in Callim. 110. 48). Some of these Scholiasts add that there is a lot of bronze in Scythia (Schol. in Aesch. Prom. 298a), although this would not necessarily mean that there was also a lot of iron. Elsewhere, they mention the Scythian city of Chalybos (Schol. in Aesch. Theb. 728) or refer to the 'Chalybian land of Scythia' as 'iron-mother' (Schol. in Aesch. Prom. 301).

The information provided by these late sources would be less noteworthy were it not based on the references to the Scythians in two tragedies of Aeschylus, which thus indicate a perception of the Chalybes as Scythians from as early as the 5th century BC. In *Seven Against Thebes* (728) the person called Chalybos is said to come from Scythia, while in *Prometheus Bound* (709–716) the Chalybes are described as neighbouring Scythia, a peculiar geographic placement as noted above.[397] Apparently the Scholiast(s) of Aeschylus deduced from these passages that the Chalybes were clearly a Scythian nation, bearing in mind as well the famous metallurgy of the Scythians, which would encourage such a conclusion. What is even more interesting is that the Scholiasts of other authors adopted the information and repeated it, although nothing of the

[395] Dan 2012, 93–94.
[396] See Section 2.8.3.
[397] See Section 2.9.1.

sort had been said by the authors whose works they were commenting:[398] for example, neither Apollonius nor Callimachus mentioned anything about a Scythian origin of the Chalybes; this could be a useful element for students of ancient Greek scholarship. It is interesting that Eustathius (*Comm. in Dion. Per.* 767) says that according to some people the Thermodon, which flows close to the land of the Chalybes, was considered a 'Scythian river'.[399]

What could be the reason why the Chalybes were thought to be Scythians? This idea seems to have emerged very early, unlike the case of other coastal peoples that have been mentioned as Scythians but only in late sources.[400] Was it the common element of working with metals that made the Chalybes appear akin to the Scythians, or were they indeed kindred? For want of relevant historical information, our only option, given the available texts, is to examine any anthropological characteristics of the Chalybes mentioned – an exercise that might actually be of some assistance.

Indeed, Eudoxus, in the 4th century BC, purportedly mentioned a nation living in the land of the Chalybes, who were called Chabarenoi. These were accustomed to take foreign women, eat their breasts raw and devour children (fragm. 283; *cf.* Aelius Herodianus and Stephanus of Byzantium, *s.v.* Χαβαρηνοί). These shocking customs remind us of similar habits reported for Scythian peoples or peoples living in the vicinity of the Scythians: the Androphagoi, for example, used to eat men, as their very name indicates (Herodotus 4. 106), and although Herodotus states that the Androphagoi were not Scythians, Pseudo-Scymnus (810) says that they were.[401] Moreover, we know, of the Massagetae, that 'when a man is very old all his family meet together and kill him, with beasts of the flock besides, then boil the flesh and feast on it. This is held to be the happiest death' (Herodotus 1. 216. 2–3); and of the Issedones (related to the Scythians), that 'whenever a man's father dies, all the nearest of kin bring beasts of the flock and, having killed these and cut up the flesh, they also cut up the dead father of their host and set out all the flesh mixed together for a feast' (Herodotus 4. 26. 1). It seems that eating human flesh was a habit encountered among peoples in Scythia, and since, as already stated, the customs of a people had been considered since antiquity to constitute a significant

398 For the dating of the Scholia to Aeschylus, see Dickey 2007, 35–38; see also 62–63 for the Scholia to Apollonius, and 66 for the Scholia to Callimachus.

399 It is only in the *Scholia in Lycophronem* (1109) that the Chalybes are called a Paphlagonian nation.

400 See Section 2.11.4.

401 About this and other divergences between the two authors concerning peoples of this specific region, as well as on whether Herodotus went so far north or learned about this area from merchants or slaves, see Corcella 2007, 548–63, and 592 and 656 on the Androphagoi.

indication of ethnicity, we might suspect that working with metals was not the only element that connected Chalybes with Scythians.

But there could be more. If the coastal Chalybes did indeed belong to the same nation as the inland ones, then it is worth looking at the latter's customs too, since Xenophon does record something of them, unlike those of the coast. And indeed, among other things we read that the inland Chalybes 'slaughtered whomever they might be able to vanquish and then they would cut off their heads and carry them along on their march and they would sing and dance whenever they were likely to be seen by the enemy' (4. 7. 16). The similarity to the habits of both the Scythians and the neighbouring Tauroi is striking: Herodotus (4. 64–66) relates that the Scythian soldiers cut off the heads of their enemies, carry them to the king and then hang the scalps from their bridle-reins. And for the Tauroi that they used to kill all shipwrecked men or any Greeks they captured in sea-raids, decapitate them and place every head on a pole (Herodotus 4. 103). Additionally, it is significant that just before reaching the coastal Chalybes, Xenophon himself mentions that during a battle between the Greeks and the Mossynoikoi, the latter cut off the heads of the dead men and displayed them to the Greeks, dancing at the same time to a kind of strain that they sang (*Anabasis* 5. 4. 17). Finally, both the Chalybes (Strabo 12. 3. 19) and the Mossynoikoi (Xenophon *Anabasis* 5. 4. 28) made extensive use of dolphin blubber. And for the Mossynoikoi there are also references in ancient literature to their being a Scythian nation.[402]

9.5. The Question of the Homeric Halizones

We cannot close our examination of the Chalybes without mentioning a question that has beset research not only in the modern period but since antiquity: the question of the identification of the people called Halizones in the Trojan Catalogue (*Iliad* 2. 856–857). The whole to-do seems to have been caused by the fact that these people are encountered in Homer's celebrated epic. All the other writers who mention them are clearly referring to the people named by Homer and attempting to place them geographically, albeit with considerable differences of opinion, since Homer himself is of no help in the matter.

We thus have various versions of where the Halizones lived: in Bithynia (Arrian, in Eustathius *Comm. ad Hom. Il.* 2. 857),[403] on the coast between Mysia, Caria and Lydia (Ephorus *FGrH* 2a 70 F 114) or even above the

[402] See Section 2.11.4.
[403] See in Section 2.2.2.

Borysthenes (Hellanicus, Herodotus and Eudoxus, in *FGrH* 1a 4 F 186 and Strabo 12. 3. 21).

Strabo, who identified the Halizones with the Chalybes (12. 3. 19–25, *cf.* also 14. 5. 24), presents a more reasoned account of his opinion. Feeling the need to respond to the views of a number of writers, such as Apollodorus and Demetrius of Scepsis who argued, respectively, that the Halizones could not have lived east of the Halys and that they could not have had any connection with the Chalybes, Strabo gives an analytical argument for placing the Halizones beyond the Halys and identifying them with the Chalybes. He does not think that the aural similarity between the country called 'Alybe where the Halizones lived', according to the Trojan Catalogue (*Iliad* 2. 856–857), and the people called Chalybes is accidental, and argues that either the phrase τηλόθεν ἐξ Ἀλύβης (*Iliad* 2. 857) was a misreading of the original τηλόθεν ἐκ Χαλύβης, or that in Homer's time the Chalybes were called Alybes, offering several examples of names that had undergone alteration. For Strabo, the fact that the Chalybes were known for their mines accords perfectly with Homer's statement that Alybe was the 'native place of silver' (*Iliad* 2. 857).[404]

This view, however, was not shared by everyone in antiquity. According to Arrian (in Eustathius *Comm. ad Hom. Il.* 2. 857), Alybe has nothing to do with mines or with the Chalybes. Hellanicus (in Stephanus of Byzantium, *s.v.* Ἀλύβη) calls it a Pontic lake, Apollonius (Lexicon Homericum *s.v.* Ἀλυβή, accented on the final syllable) and Hesychius (*Lexicon*, *s.v.* Ἀλύβη) a city of the Troad, while the *Etymologicum Magnum* (*s.v.* Ἀλύβας) describes it as a mountain, citing also two alternative views, that it was a lake or a city. According to the Scholiast of the Iliad 2. 856, it was a village in Bithynia, the native place of silver.[405] The range of opinion is as diverse as that concerning the Halizones themselves. Eustathius says (*Comm. ad Hom. Il.* 1. 571) that for Alybe some wrote Chalybe, some – like Menecrates – Alobe and some Alope: here, in other words, we have two of the principal opinions concerning the location of the Halizones, in Bithynia and in the land of the Chalybes.

Unfortunately, the sources tell us nothing more about the Halizones (or Alybe). It is obvious that the ancient writers themselves were confused about who they were, as is clear from Eustathius (*Comm. ad Hom. Il.* 1. 571–572),

[404] For all the views and more argumentation, see Dan 2012–13; Manoledakis 2013a. For the view that the name Alybe is the Greek form of a Hittite name, see Allen 1921, 159–61; Thomas and Stubbings 1963, 304; Simpson and Lazenby 1970, 177. For the definition of the name Halizones as denoting people who live surrounded by sea, see Schol. in Hom. Il. 1. 570.

[405] *Cf.* Thomas and Stubbings 1963, 305, who hold that both the Paphlagonians and the Halizones were thought to have lived north of Troy.

who tried to investigate every reference he knew of. Curiously, the *Etymologicum Guidanum* has an entry for Ἀλίζωνες, who are described as Bithynians, followed immediately by an entry for Ἀλιζώνων (genitive case), who are described as Paphlagonians. Hesychius mentions the 'Alizoni' (with the smooth breathing) as a Paphlagonian tribe (*Lexicon, s.v.* Ἀλιζῶνες). Others call them a tribe of Thracians (Schol. in Hom. Il. 5. 39), and even a Pontic tribe from Boeotia (Schol. in Hom. Il. 2. 856). According to Eustathius (*op. cit.*), Strabo himself had trouble identifying the Halizones, saying that Homer speaks of some 'unknown peoples' (Strabo 14. 5. 23).[406]

The thorny question of the Halizones, which has plagued commentators since antiquity, has led to the view that in the end the Halizonian contingent was simply a figment of the poet's imagination.[407] In a step further, it has even been argued that the Halizones were a product of 'intentional history', a fictional people invented and used by scholars 'who acted out of patriotism, in a quest for prestige for their fatherlands', Strabo, judging by his impact, being the most successful of these.[408]

On the other hand, several attempts to find a geographic order in the list of the Trojans' allies have not led to persuasive conclusions. It has been observed that the Trojan Catalogue consists of two parts; one lists the contingents from Troy and the Troad (some consider this as two parts), and the other follows four radii, the last place on each being described as 'far off': across the Hellespont, from Thrace westwards; towards the southern Black Sea; towards south-east of the Troad; and towards south-western Anatolia.[409] Even if this observation appears correct, which would reinforce the possibility that the Halizones are the Chalybes, the order of these areas does not correspond to any reasonable geographic, historical, or other order. But what is more interesting is that scholars who invoke this theory[410] present a different geographic order, in which all the Asian contingents appear as one part, and thus place the Halizones between the Paphlagonians and the Mysians (who follow them in the Catalogue), namely in north-western Asia Minor.[411] But this would shape an even more confusing and senseless route. I am afraid that trying to find a logical geographic order in the Trojan Catalogue is more or less as vain as

[406] An overview of the sources is offered by Olshausen (2012).

[407] Kirk 1985, 259.

[408] Dan 2012–13. The first to speak of the 'patriotic' motives of Strabo was Allen (1921, 160). However, Dan seems not to be equally sceptical about the existence of the Caucones (2012–13, 38), although there is actually not much more evidence for their existence than for that of the Halizones. See Section 2.5.

[409] Kirk 1985, 250; Bryce 2006, 127–28.

[410] Dan 2012–13, 36 with n. 10.

[411] Dan 2012–13, 37–40.

attempting to trace a real geographical course in the imaginary voyage of Odysseus.[412]

Furthermore, I find the view on Strabo's patriotic intentions rather exaggerated and formulated too absolutely. In the first place, I am not persuaded that Strabo would have felt such a strong need to invent a whole story and insert it in a very long passage in his work, which in other respects does not belong to fiction – not that I would totally exclude it. But even if that really was the case with Strabo and all the other ancient authors who wrote on the Halizones, if every one of them placed these people according to his 'patriotic' wish, why should this preclude the fact that Homer was indeed referring to a real people, as in all the other cases of his catalogues? For we cannot help but notice that none of the other peoples mentioned in the Trojan Catalogue are fictional.[413] What intention, in the early 7th century BC, could have lain behind the invention of a people who lived in a distant place (τηλόθεν is the word used, as in other cases) called Alybe and had indeed fought *against* the Greeks in the Trojan War? It is difficult to find a logical answer, unless the specific passage is an interpolation, for which there is no evidence.

This does not mean that we should necessarily support one or another of the theories that have been expressed to date about who the Halizones were. It is indisputable that the peoples called Halizones in the *Iliad* cannot be safely identified with any peoples that we know of, and most probably they will never be. But this is not so strange and should not cause irritation; they are not the only people mentioned in ancient literature who remain unidentified. The possibility that they did not really exist remains, but there is no justification for turning it into a certainty. Nevertheless, for reasons that I have already analysed elsewhere,[414] the most likely connection of the land of Alybe seems to me to be the one with the land of the Chalybes.[415] In that case, the Halizones could have been, not necessarily the Chalybes themselves, but perhaps a people who lived there before them, or a Chalybian tribe (like the Chabarenoi, who are mentioned only by Eudoxus – see above), and this is why they have been given a place in this section.[416] But this is still merely conjecture.

[412] Manoledakis 2015b.

[413] Even if the identification of some of them is ambiguous, e.g. the Pelasgians (e.g. Bryce 2006, 138), which however does not mean that they did not exist. On the view that the Trojan as well as the Achaean Catalogues belong to the original core of the *Iliad*, see Kullmann 2011, 92–94, with earlier bibliography.

[414] Manoledakis 2013a, 25–27.

[415] *Cf.* Page 1959, 141–42.

[416] *Cf.* also their examination in the respective Section by Counillon (2004, 104).

10. The Tibarenoi

10.1. The Territory (Fig. 10.3)

One of the significant peoples one would have met after crossing the Halys were the Tibarenoi. Their precise territory is ambiguous, as are their eastern and western neighbours, reflecting the problem created in the area between the Leucosyroi and the Mossynoikoi by the uncertainty attending the exact location of the Chalybes and whether they had always possessed part of the seaboard.[417] Documenting the westward progress of his troops, Xenophon says that after proceeding eight stages through the country of the Mossynoikoi they reached the land of the Chalybes, who were few and subject to the Mossynoikoi, and right after that the land of the Tibarenoi, where the city of Cotyora was situated (*Anabasis* 5. 5. 1–3).

Hecataeus, by contrast, says (*FGrH* 1a F 204) that the eastern neighbours of the Tibarenoi were the Mossynoikoi, which puts the Chalybes to their west. He does not name any city in their land, but since he places the Chalybes at the Thermodon (*FGrH* 1a 1 F 203) and places the town of Choirades in the land of the Mossynoikoi (*FGrH* 1a 1 F 204), the country of the Tibarenoi must therefore have lain around Capes Iasonion and Genetes and included Cotyora, which tallies with Xenophon's account. The same order (*viz.* from west to east: Chalybes – Tibarenoi – Mossynoikoi) also appears in the work of Pseudo-Scylax (86–88), slightly later than the *Anabasis*, where again, however, no city is mentioned in the territory of the Tibarenoi; Cotyora is absent from the *Periplus*. This order is further asserted by Apollonius (*Argonautica* 2. 374–380, 1000–1018), who puts the Tibarenoi immediately to the east of Cape Genetes, Pliny (*NH* 6. 11), Dionysius Periegetes (*Orbis descriptio* 766–768) and Anonymous (*Periplus* 33), while Pseudo-Scymnus 'solves' the problem by placing the Chalybes inland (900–915) and Pomponius Mela (1. 105–106) is ambiguous on the matter. Herodotus, although not providing a geographic order, mentions the Tibarenoi as belonging to the 19th Persian province along with the Moschoi, the Macrones, the Mossynoikoi and the Mares, all dwelling to the east of the Tibarenoi, the Mossynoikoi being their nearest neighbours (Herodotus 3. 94); and in fact they had the same leader as the Moschoi in the Persian expedition (Herodotus 7. 78).[418] Aelius Herodianus (*De prosodia catholica, s.v.* Τιβαρηνία) mentions the Tibarenoi as neighbours of the Chalybes and

[417] See in Section 2.9.1 in detail.

[418] *Cf.* Menippus' reference (fragm. 2) to the land of the Tibareboi together with those of Chaldaeoi and Sannoi.

the Mossynoikoi, but this is not necessarily of much help, as is the case, too, with Plutarch's passages (*Lucullus* 14. 3; 14. 6; 19. 1).

It seems, then, that Xenophon is alone in placing the Tibarenoi to the west of the Chalybes and only Strabo seems to agree with him (12. 3. 18, 28–29),[419] although this is not clearly stated; on the contrary, elsewhere he implies that Trapezous was in the neighbourhood of Tibarania and Colchis (7. 4. 3; *cf.* 11. 14. 1; 12. 1. 3; 12. 3. 1).[420] However, Xenophon makes it clear that Cotyora lay two day's march to the west from the eastern borders of the Tibarenoi's land (and within their territory), but since he is vague in his overall account of the country of the Chalybes and their spatial relation to the land of the Mossynoikoi, we cannot place these borders precisely. As for the western Tibarenian border, there is no indication of its position, since from Cotyora they took ship for Sinope and in so doing deprive us of an account of the littoral between the two cities.

The fact that Cotyora appears to belong to the land of the Tibarenoi not only in Xenophon but also in Hecataeus suggests that the discrepancy in the relative positions of the Tibarenoi and the Chalybes among the authors[421] is not after all so significant, especially if we take into account the possibility that in Xenophon's time the Chalybes may have occupied a very small part of the coast. But the discrepancy remains. It has been argued that Xenophon must be considered one of the most reliable sources that we have for the littoral, and one might be willing to accept his geographic order were it not for what appears to be one of the most enigmatic cases in the ancient sources concerning the geographic distribution of peoples in the southern Black Sea littoral. The problem is that while he clearly places Cotyora in the territory of the Tibarenoi (*Anabasis* 5. 5. 3), he immediately afterwards says that during the 45 days that he and his soldiers remained in that city they got provisions partly from Paphlagonia and partly from the villages of the Cotyorites (5. 5. 6; 6. 1. 1), while in his account of events in Cotyora, which occupies the rest of Book 5 (after 5. 5. 6) and up to 6. 1. 14, it is only the Paphlagonians who appear, in a narrative that recounts mutual depredations, negotiations and a final peace sealed by a feast between them and the Greeks,[422] with never a word about any Tibarenoi.

[419] Ctesias (*FGrH* 3c 688 F 56) also mentioned the Tibarenoi as neighbours of the Mossynoikoi; however, we have no reference of his to the Chalybes.

[420] Ephorus (*FGrH* 2a 70 F 43) states that the Leucosyroi and the Chalybes bordered the Tibarenoi, but is unclear as to their precise respective positions.

[421] See also Peretti 1961, 28–29.

[422] See in Section 2.6.

One not unnaturally wonders how is it possible that an army could remain for 45 days in a city situated 'in the land of the Tibarenoi' and never have a noteworthy meeting with any Tibarenian, while having so much to relate about the Paphlagonians there. Upon their arrival in Cotyora, ambassadors from Sinope, having heard that the land of their colony was suffering from depredations by Xenophon's soldiers, came to them and told Xenophon, among other things, that they had taken this land away from the barbarians and established Cotyora (*Anabasis* 5. 5. 10). The next question, therefore, is which barbarians the Sinopeans had taken the land from – the Paphlagonians or the Tibarenoi? Moreover, during the negotiations between the Sinopean ambassadors and Xenophon, both sides threaten to form an alliance with the Paphlagonians (5. 5. 12–23). Meanwhile, we should also keep in mind that Xenophon never mentions the Leucosyroi.

An easy explanation to this puzzle could be that Xenophon simply made a mistake when he wrote that Cotyora was in the land of the Tibarenoi, given that the latter had last been mentioned two days' march previously. Or perhaps Cotyora was situated at the western edge of the Tibarenoi's land, near the border with the Paphlagonians,[423] and the nearest villages (for provisioning the Greek camp) were in Paphlagonian territory. This might be reinforced by the fact that the Greeks consider the assistance of the Sinopeans valuable, 'by virtue of their acquaintance with Paphlagonia' (5. 6. 1), suggesting that it was Paphlagonian territory that they had to march through after Cotyora; indeed, Xenophon says that the Greeks sailed from Cotyora to Sinope having Paphlagonia on the left (6. 1. 14). But we could also consider the possibility that by Xenophon's time the whole wider region, including the part where the Tibarenoi lived, had come to be called Paphlagonia (*cf.* 6. 1. 1–13), given the general dominion of the Paphlagonians in that part of the coast.[424] We have a similar case with the Drilae,[425] who appear to live in the wider area belonging to the Colchians (5. 2. 1–3), or we might even think of the Chalybes who were few and subject to the Mossynoikoi (5. 5. 1) and speculate that there was a similar relation between the Tibarenoi and the Paphlagonians.

However that may be, it seems fair to say that the territory of the Tibarenoi cannot have been large. If the western borders of the Mossynoikoi were in the area between Ischopolis and Kerasous (Giresun),[426] and if the Chalybes occupied a small area between the Mossynoikoi and the Tibarenoi (5. 5. 1–2) and

423 Apollonius, as we saw, placed the Tibarenoi directly to the east of Cape Genetes in his poem, which does not crucially disagree with this.

424 See Section 2.6.1; also, Manoledakis 2021c.

425 See Section 2.12.

426 See also Section 2.11.1.

Cotyora indeed lay on the latter's western border, then the stretch of coast that remains for the land of the Tibarenoi is very modest and could well be traversed in two days march (5. 5. 3). Xenophon actually reports that this country was much more level (than that of the Mossynoikoi) and had villages upon the seacoast that were less well fortified (5. 5. 2), and this image fits the terrain to the east of Cotyora.[427] Nor does the country of the Tibarenoi seem to have been densely populated, since it was here that Xenophon had thought of establishing a new colony, an idea that he later dropped (5. 6. 15–33).[428] This renders all the more curious the contrasting impression created by Strabo in the passages cited above (Strabo 7. 4. 3; 11. 14. 1; 12. 1. 3; 12. 3. 1), of an important coastal area occupied by the Tibarenoi covering an unspecified part of the seaboard between Sinope and Colchis.[429]

Worth mentioning, too, is Cicero's report (*Ad Familiares* 15. 4. 10) of a people called 'Tebarani' in the area round Mt Amanus, between Cilicia and Syria, with reference to events that took place in 51–50 BC.[430]

10.2. Idiosyncrasy and Customs

The fact that, despite their small territory and presumably modest numbers, the Tibarenoi achieved a lasting place in the ancient literature might be justified by their strange customs, which provided authors, especially poets like Apollonius, with exciting material, but also by an idiosyncrasy which was in sharp contrast to that of their neighbours, especially those to the east.

First of all, the Tibarenoi[431] were the only people on the coast who were not at all hostile towards Xenophon's army. All along their march the Ten Thousand had trouble with the local peoples in whichever new land they entered, and fighting, or in the best case hard negotiating, was needed in order for them to pass through. This had been and would be the case with the Mossynoikoi, the Paphlagonians, the Bithynians and even the Greek ambassadors of Sinope

[427] There is nothing in Hecataeus, Pseudo-Scylax or Apollonius that disagrees with this general placing of the Tibarenoi's land; only Anonymous (*Periplus* 33) shifted the country of the Tibarenoi a bit farther west, between Cotyora and Polemonion, thus practically around the promontory of Capes Iasonion and Genetes.

[428] See more on this matter in Section 4.4.3.

[429] Strabo further adds that the Tibarenoi were governed by various potentates, who at one time were friends of the Armenians (12. 3. 28), while in his day their ruler was Pythodoris, the wife of Archelaus, the last king of Cappadocia (12. 3. 29). More about the land of the Tibarenoi in the previous part of the book.

[430] See more on them later, in Section 2.10.3.

[431] Or Tibaranoi: Aelius Herodianus *Peri orthographias* 3. 2. 483; *Peri paronymon* 3. 2. 878.

in Cotyora.[432] But with the Tibarenoi things were utterly different. Not only had they no thought of attacking the Greek army when it entered their land, but on the contrary they sent gifts of hospitality. This must have caused consternation among the Greeks, judging by Xenophon's account, because the generals had been ready to attack the local settlements in order to get supplies for the army. So they did not accept the gifts, but directed the Tibarenoi to wait until they should take counsel (*Anabasis* 5. 5. 2). This was not the most sensible course of action, given this rare good fortune of encountering a friendly people after all the adventures they had had, but perhaps they were carried away by their own momentum. So they proceeded to offer sacrifices, and it was only after many offerings had been made and the seers had declared their opinion that the gods would in no case permit a war, that the generals accepted the gifts and proceeded 'as through a friendly country' for two days, before reaching Cotyora (*Anabasis* 5. 5. 3).

Xenophon says nothing more of the Tibarenoi, but their unexpectedly peaceful attitude seems to have intrigued others and inspired fuller accounts. Ephorus (*FGrH* 2a 70 F 43), for example, reported that the Tibarenoi attached great importance to games and laughter, considering these elements of well-being, while Nymphodorus added they were extremely just and fair and did not engage in hostilities without first announcing the date, the place and the hour fixed for the battle (Schol. in Apol. Arg 2. 1010) – a most unusual attitude for that coast, and beyond. Ephorus' information was repeated by Pseudo-Scymnus (914–916) and, thanks to him, by Anonymous (*Periplus* 33). Xenophon's account of the Tibarenoi has led to the suggestion that the latter were skilful at negotiating with outsiders, which 'may reflect a long tradition of relations between mobile fishing communities from around the Black Sea interacting with local indigenous communities'.[433]

But the most curious custom of the Tibarenoi was this: 'when wives brought forth children to their husbands, the men lay in bed and groaned with their heads close bound; but the women tended them with food and prepared childbirth baths for them' (Nymphodorus, according to Schol. in Apol. Arg. 2. 1010; Apollonius *Argonautica* 2. 1011–1014; Valerius Flaccus *Argonautica* 5. 147–148; Plutarch *Prov. Alex.* 10. 12; Sch. Apol. Arg. 202; Zenobius *Epit. Luc. Tar.* 5. 25). The same custom was reported by Diodorus (5. 14. 2) for the inhabitants of Cyrnus (Corsica) and Strabo (3. 4. 17) for the Spanish Basques,

[432] See in the relevant sections on the Mossynoikoi, Paphlagonians and Bithynians.
[433] Doonan 2019, 97.

while it has also been encountered in later periods among several communities around the world.[434]

We first acquire the information attributing this custom to the Tibarenoi from Nymphodorus and Apollonius, whose sources, apart from Nymphodorus, also included Ephorus, who however is not reported to have mentioned it; neither did Apollonius mention the pleasant temperament of the laughing Tibarenoi noted by Ephorus. It is noteworthy, but not unprecedented, that this characteristic of the Tibarenoi was not reported by Xenophon, the only one of all those authors who actually met them, and who has been considered as a source of Ephorus, which makes one wonder how Ephorus (and Nymphodorus) knew about it and whether they had actually heard such a thing or were just embroidering on the friendly disposition reported by Xenophon.[435]

Conversely, some later sources (Porphyrius *De abstin*entia 4. 21; Eusebius *Praeperatio evan*gelica 1. 4. 7; Theodoretus *Graecarum affectionum curatio* 9. 36), presumably copying from one another, strike a different note on the behaviour of the Tibarenoi, with their assertion that they threw their very old men over cliffs while they were still alive – something for which there is no evidence in any ancient text. Such a reversal in behaviour is indeed astonishing, and one might reasonably ask whether this information actually corresponds to reality or was an exaggeration of authors of the Early Christian period seeking to revive the contrast between the civilised Greeks and the 'barbarians' of Asia Minor and misjudging the latter, an attitude encountered in the case of the Paphlagonians.[436]

10.3. On the Origin of the Tibarenoi

It may by now have become clear that several and long-lasting attempts have been made to trace the origins of the peoples that inhabited the southern Black Sea littoral in the 1st millennium BC. The case of the Tibarenoi could not be absent from these ventures. Thus, several theories emerged concerning the identity and origin of this people, too; and, inevitably, this quest entered the obscure world of the first stages of Anatolia's post-Hittite history, where

[434] See W.P. Thayer, 'Couvade' in *1911 Encyclopaedia Britannica online*.

[435] For the use of Ephorus and Nymphodorus by Apollonius and Pseudo-Scymnus, the use of Pseudo-Scymnus by Anonymous as one of his sources, as well as the issue of Ephorus' use of Xenophon, see in Section 2.11.2. See also Vian and Delage 1974, 282; Morrison 2020, 156. Hoefer's claim that whatever Xenophon reported of the Mossynoikoi was reported by Ephoros and Pseudo-Scymnus for the Tibarenoi (1904, 545) must be an error.

[436] See Section 2.6.4.

many theories jostle in the darkness created by the deafening silence of written and archaeological data.

From early on, the Tibarenoi have been often identified with Tabal, a Neo-Hittite kingdom of South Anatolia in the wider area between Cataonia and Cilicia, which appeared during the Early Iron Age. Similarly, the Moschoi have been identified with the Mushki. Both Mushki and Tabal are mentioned in Assyrian annals from the 12th until as late as the 8th century BC, and are associated with Japheth's sons Meshech and Tubal respectively, mentioned together in the Bible (*Genesis* 10:2; *cf.* also *Ezekiel* 38:2; 39:1). The earliest Assyrian annals seem to refer to Mushki and Tabal as Phrygian peoples and lands, who might have allied themselves with the Kaska against the Hittites and contributed to the overthrow of the Hittite Empire.[437] Otherwise, the Phrygian Empire, which was of course established after the collapse of the Hittite one, something that indicates one of the difficulties in such theories, was 'a federation or coalition of several tribes or elements, in which the Phrygians seemed to have represented the western element; the eastern was formed by Mushki and Tabal'.[438]

If the Tibarenoi and the Moschoi of northern Anatolia (who, nevertheless, are distinguished from the Phrygians), were indeed no other than the Tabal and the Mushki respectively,[439] then it has been assumed that they had 'swarmed southwards over the central plateau into the Halys bend, joining hands with the Phrygians advancing from the north-west'.[440] This would explain the existence of Cicero's Tebaranoi in Cilicia.[441] However, the opposite route has also been proposed, namely a migration of (parts of) both peoples from Cilicia north to the Black Sea.[442] In another theory, the Tibarenoi are considered a Kartvelian people who migrated to Pontic Georgia.[443]

G.A. Wainwright argued that Cicero's Tebarani were probably relatives of the Tibarenoi, and supported his view by comparing the proximity of the Tibarenoi and the Moschoi with the proximity of Tabal and Mushki.[444] This

[437] Barnett 1975, 417, 420; Miller 1995, 288.

[438] Barnett 1975, 421.

[439] See also Meyer 1913, 698. Flavius Josephus (*Antiquities of the Jews* 1. 124–125) identified the Mosochenoi founded by Mosoch with the Cappadocians, while he mentions Tubal as Thobel, who founded the Thobelites, who are identified with the Iberes.

[440] Barnett 1975, 420.

[441] *Cf.* Meyer 1913, 698.

[442] Wainwright 1936.

[443] See indicatively Rapp and Crego 2012, nn. 4a and 6, with earlier bibliography. Barnett (1975, 420) examined this possibility as well, but considered the connection with the Moschoi and Tiberenoi more plausible (see above).

[444] Wainwright 1936.

similarity is indeed noteworthy, and so might be the likeness with the previously demonstrated case of the Leucosyroi, since they and the Tibarenoi were neighbours. However, these do not necessarily mean much. The resemblances between the words Tabal and Tibarenoi and between the words Mushki and Moschoi, especially the former, are not striking enough to exclude the possibility that they are just accidental. Therefore, there can be no real comparison with the case of the Leucosyroi, where there is a clear association with the Syrians of Mesopotamia, both in the name itself as well as in several written sources. What cannot pass unremarked is the resemblance between Tibarenoi and Cicero's Tebarani (see above), but the problem here is that nowhere else in his earlier writings or contemporary literature are the latter mentioned in the specific area, nor are they mentioned by the slightly later Strabo, who offers such a detailed description of the area and its history.

Southern Asia Minor, however, is not the only place that has been connected with the origin of the Tibarenoi: in the *Scholia on Apollonius* (2. 378, 1010) we read that they were a Scythian nation. As we shall see, the *Scholia on Apollonius* attribute Scythian origin to almost all the peoples that inhabited the littoral to the east of the Iris. For some of them, like the Mossynoikoi and the Chalybes, such an ascription was adopted by several other sources (almost exclusively later ones, like this), but in the case of the Tibarenoi the Scholiast of Apollonius stands on his own among the ancient authors, including Apollonius himself, in calling the Tibarenoi Scythians. There has never been any evidence for such an attribution, unless it were the fact that they were surrounded by Scythian nations – the Chalybes, whom most ancient sources, including Apollonius, placed to their west were widely considered Scythians,[445] as were many peoples to their east – that led the Scholiast to assume a Scythian origin for the Tibarenoi as well.

Be that as it may, the Scholiast's attribution has led to a far-fetched – indeed, I would say barely conceivable – explanation of their origin that incorporates the Tabal theory as well. By this account, the Tibarenoi must have come to Asia Minor from Scythia or Thrace, and in fact only from Thrace. How is this possible? Because the *Book of Jubilees*, probably based on or accompanied by a world map, mentions the region of Japheth's son Tubal (7. 19; 9. 11) as probably situated in Thrace, and 'the only ethnic name that fits them is that of the Triballoi, to the south of the Danube'. Therefore, the Tabal must have been a people that migrated from Thrace before the time in which they first appear in Asia Minor,[446] that is, 'before the era of the Assyrian king Shalmaneser I

[445] See in Section 2.9.4.

[446] Herrmann 1931, 81–82; *RE* VI A.1 (1936), 764 (*s.v.* Tibarenoi).

(13th century BC), in the annals of which the Tabal first appear'.[447] And indeed there is a parallel that would reinforce such a theory, namely the similar migration to Asia Minor of the Treres, who according to Strabo (13. 1. 8) came from Thrace and were thus neighbours of the Tubal people.[448]

The biggest challenge of dealing with such a theory is where to start one's criticism. First of all, the proposed geographic placement of the Tubal in the *Book of Jubilees* is highly disputable. As is the linguistic connection of the Tubal with the Triballoi; I doubt that one would have considered it were it not for the need to justify such a theory. Moreover, the Tabal and their kings do not first appear in the annals of king Shalmaneser I, but in those of Shalmaneser III (9th century BC),[449] if this is of any importance. And the fact that the Treres (who, by the way, are elsewhere called Cimmerians, not Thracians, by Strabo himself, namely in 1. 3. 21 and 14. 1. 40), had also migrated from Thrace to Anatolia cannot, of course, be cited as evidence for a similar activity on the part of any neighbouring people. Finally, one might well wonder how the reference of the Scholiast of Apollonius to the Tibarenoi as Scythians 'leads to the conclusion' that they came from Thrace and hence the need to invent all this theory.[450]

It is worth noting that Strabo (2. 5. 31) speaks of 'Tibaranic nations', as does the *Orphica Argonautica* (741), a description that suggests an ethnic group rather than a separate people. This statement is made, however, in the general description of the world in the second book of his *Geography*, and is not repeated in the twelfth book, where he focuses on the particular area, or in the other passages of his work where he mentions the Tibarenoi.[451] In all these passages he clearly places the Tibarenoi on the coast, even if not with the desired accuracy, and never says anything about their origin or any subdivision into several tribes. The reference to 'Tibaranic nations' does not necessarily indicate a group of different nations, as for example is the case with the Scythians, but could derive from the general vagueness concerning the size of the Tibarenian territory in Strabo, between Sinope and Colchis, as we have already seen.

In any case, even speculating about the origin of the Tibarenoi seems rather frivolous. The encounter with the savage custom mentioned by Porphyrius among several peoples in the Scythian neighbourhood, as the author himself mentions, is not in itself enough to warrant a firm conclusion.

447 Herrmann, *RE* VI A.1 (1936), 764 (*s.v.* Tibarenoi).
448 Herrmann 1931, 82.
449 E.g. Barnett 1975, 422–23.
450 Herrmann, *RE* VI A.1 (1936), 764 (*s.v.* Tibarenoi).
451 See above, Section 2.10.1.

11. The Mossynoikoi

Of all the peoples of the eastern part of the southern Black Sea littoral it is for the Mossynoikoi that we have the most information, albeit largely derived from a single author (Xenophon). It includes material of a social, urban and even anthropological nature; with regard to the last, in fact, it seems that the customs of the Mossynoikoi had irritated the Greeks, since we are told in the fifth book of the *Anabasis* that the Mossynoikoi 'were set down by the Greeks who served through the expedition as the most uncivilised people (βαρβαρώτατοι) whose country they traversed' (*Anabasis* 5. 4. 34). The use of the superlative degree for the adjective βάρβαρος, which was scarcely ever employed in ancient Greek literature, and indeed by a person who generally approached the term βάρβαρος in a different way from his predecessors, demands an explanation. In the following paragraphs we will be examining all the details we have for this people, starting, as usual, from the general location of their land.

11.1. The Territory and its Topography (Fig. 10.3)

According to most sources, the Mossynoikoi lived to the east of the Tibarenoi, although in Xenophon (*Anabasis* 5. 5. 1–2) they appear to have been separated from the latter by the Chalybes.[452] Hecataeus says that the town of Choirades was founded in the country of the Mossynoikoi (Aelius Herodianus 3. 1. 58, *cf.* also 3. 1. 151, 177, 298, 397; Stephanusof Byzantium, *s.v.* Χοιράδες = *FGrH* 1a 1 F 204), as does Pseudo-Scylax (86), who also adds the Zephyrios Limen and the Island of Ares. To the east of the Mossynoikoi was Colchis and the land of the Drilae, according to Xenophon (*Anabasis* 5. 2. 1–5. 4. 1), but other authors placed different peoples there, including the Mares (Hecataeus *FGrH* 1a 1 F 205; Stephanus of Byzantium, *s.v.* Μᾶρες), the Macrocephaloi (Pseudo-Scylax 85; Mela 1. 107; Pliny *NH* 6. 11), the Philyres (Apollonius *Argonautica* 2. 393, 1231; Dionysius Periegetes 766–767) and the Macrones (Valerius Flaccus *Argonautica* 5. 151–153).[453] Anonymous (*Periplus* 35) says that the Mossynoikoi lived in the area between the cities of Kerasous and Cotyora, while Curtius Rufus erroneously places them to the west of the Leucosyroi (Curtius 6. 4. 16–17).[454] The ambiguous reference of the *Orphica*

[452] See more in Section 2.9.1.
[453] See the relevant issue discussed in Section 2.9.3.
[454] See in detail Section 2.8.1.

Argonautica (739–740) to 'neighbouring peoples who live mixed together in the region of the Mossynoi' is of no help.

The difficulty in defining the exact borders of the coastal part of the Mossynoikoi's land derives to a considerable degree from a combination of three factors: Xenophon's placing of the Colchians' territory as far to the west as Kerasous (*Anabasis* 5. 3. 2); the fact that there was more than one city by this name on the coast; and the problem of calculating distances from the stages and parasangs given by Xenophon in this part of his narrative. In Section 2.9.3 we referred to the first matter and the concomitant problem of the existence of the peoples mentioned by several other sources between the Colchians and the Mossynoikoi. As far as Kerasous is concerned, the one mentioned by Xenophon and Diodorus (14. 30. 5) as the colony of Sinope situated three days march west of Trapezous should, it seems, be placed in the area of Giresun, and not, as many scholars suggest, of Vakfikebir.[455]

Moreover, Xenophon says that after Kerasous they reached the land of the Mossynoikoi (5. 4. 1), but does not specify how far after Kerasous the border between the Colchians and the Mossynoikoi was. Similarly, he says that from the place where the events with the Mossynoikoi took place (for which see below) they needed eight days to reach the land of the Tibarenoi, but since he does not specify where those events took place, we have no starting-point for estimating the eight stages. He merely implies that Cotyora was situated within Tibarenian land, two days' march west of its eastern border. If one stage usually denotes a single day's march,[456] then we could place the land of the Mossynoikoi generally between the wider area from Ischopolis to the other Kerasous (Giresun) in the west and, roughly, Coralla in the east. The problem is complicated by the fact that Xenophon does not mention any known settlement in this land, as, judging by their appearance in Pseudo-Scylax's work, Choirades and Zephyrios Limen would have been. Nor are the rest of our sources any more helpful in building a firm conclusion about the borders of the Mossynoikoi, for while Pseudo-Scylax (86) puts the Island of Ares within their territory, Apollonius places it about a day's sailing to the east (*Argonautica* 2. 1001–1034). If, however, we take the geographical text to be more reliable in terms of geography than the epic, and considering also Apollonius' unique alteration of the area's geography in the specific passage,[457] we find a general agreement with Xenophon's account concerning the western borders

[455] See Section 5.1, comment 14 on Table 2. *Cf.* Manoledakis 2022b, with extensive argumentation and the whole bibliography.

[456] See, for example, Rood 2010, 52, and our comments here, in the third note of Section 2.9.1.

[457] For this alteration, see in Section 2.14.1 in detail.

of the Mossynoikoi, which we may thus place between Ischopolis and Kerasous (Giresun). The eastern border remains problematic.

The southern borders of the territory of the Mossynoikoi are wholly unclear, but it is obvious that these people lived mainly in the mountains (e.g. Pseudo-Scylax 86; Strabo 12. 3. 18), which suggests a considerable inland expansion.[458] Strabo tells us that the Mossynoikoi occupied the heights of the mountain Scydises, which brings us at least 70 km inland, while the fact that some of their land, specifically Carenitis and Xerxene, which bordered on or was part of Lesser Armenia, had been cut off and annexed to Armenia in the 2nd century BC as kings Artaxias and Zariadris set about enlarging their territories (Strabo 11. 14. 5), indicates an even larger initial expanse.

Herodotus tells us that the Mossynoikoi belonged to the 19th Persian province (νομός), together with the Moschoi, Tibarenoi, Macrones and Mares, and were ordered to pay 300 talents in tribute (Herodotus 3. 94. 2). They were equipped like the Moschoi, and the commander who marshalled them (together with the Macrones) was Artayctes son of Cherasmis, who was governor of Sestos on the Hellespont (Herodotus 7. 78. 1).

The settlements of the Mossynoikoi were not all of the same type. Xenophon's narrative makes clear that there were cities, villages, strongholds and a chief citadel called Metropolis, which occupied the highest ground in the territory, for control of which the Mossynoikoi fought among themselves. There was also the acropolis of the Metropolis, where the king lived in his wooden tower. The average distance between settlements was 80 stadia, or 14–15 km, but Xenophon noted that 'the inhabitants could hear one another shouting from one town to the next, such heights and valleys there were in the country' (*Anabasis* 5. 4. 31), thus giving a vivid impression of the topography of the region, whose settlements must have been built on steep mountain slopes. While some of these would have been more easily accessible and assailable than others, (5. 4. 14; 5. 4. 30), it seems reasonable that the Mossynoikoi were very confident of their strongholds and did not at first let the Greeks pass through their land (5. 4. 3), as well as that the Greeks had great difficulty in finally defeating them (see below).

The buildings in these settlements were almost entirely made of wood. Indeed, the very name of the Mossynoikoi (or Mossynoi in some sources) derives from their characteristic wooden towers (called *mossynes*), which, like

[458] According to Maksimova (1956, 30–31), they were indeed a tribe that had come from the south, but this is far from provable.

their houses, were built of logs.[459] According to Diodorus (14. 30. 6) these towers had up to seven storeys, suggesting a height for which no parallel is known anywhere on the southern Black Sea littoral. The number seven is also encountered in the later name of the Mossynoikoi, which according to Strabo (12. 3. 18) was 'Heptacomitae'. Strabo reports that 'Mossynoikoi' was a name used in earlier times, but the change must not have taken place much before Strabo, since both Pseudo-Scymnus and Diodorus mention only the old name. Besides, the name 'Heptacomitae' does not appear elsewhere in ancient Greek literature, apart from Stephanus' homonymous entry in his *Ethnica*, where he clearly cites Strabo, although with the addition that some of them were called Byzeres,[460] while in another part of his work (11. 14. 5) Strabo refers to our people only as Mossynoikoi. Consequently, the appearance of the 'new' name raises a number of questions. Did it appear after the Armenian annexation of Mossynoikan territory in the 2nd century BC? If yes, then the old name had not been abandoned, since it is used in sources that postdate this period. And was it in fact a name actually applied to the Mossynoikoi or was it a creation of Strabo's, product of confusion or invention, as perhaps indicated by a possible connection between the seven storeys and its 'seven villages'?

11.2. Social Hierarchy, Life and Economy

A significant aspect of the Mossynoikoi's life recorded by Xenophon as well as other ancient authors is their social hierarchy. The Mossynoikoi had a king and *archontes* and their society was characterised by economic inequality. It seems that the duties of the king were social rather than political in nature. Although Xenophon only says that the king lived in a wooden tower built upon the citadel, and that all the people jointly maintained and guarded him in his dwelling place there (*Anabasis* 5. 4. 26),[461] other authors specify that from this place he dispensed upright judgments to the multitude. 'But if he made a mistake in his decrees, for that day they kept him shut up in starvation' (Apollonius *Argonautica* 2. 1026–1029; Pseudo-Scymnus 904–910; Anonymous

[459] Hecataeus *FGrH* 1a 1 F 185. 3; Apollonius *Argonautica* 2. 379–381, 1016–1017; Callimachus *Aet.* 43. 68, and his scholiast; Lycophron *Alexander* 433, 1432, and his scholiast; Pseudo-Scymnus 900–903; Dionysius Periegetes 765; Strabo 12. 3. 18; Anonymous *Periplus* 35; Hesychaeus *Lexicon*, *s.v.* Μόσσυνες, μοσσυνικά; Sch. Apol. Arg. 124; 159. For the 19th-century parallels of these wooden houses in the same region, see Hamilton 1842, 317, and, more recently, Bryer and Winfield 1985, 1. Kavtaradze (2002, 74) saw in the name of the Mossynoikoi a Kartvelian element.

[460] See in Section 2.14.5.

[461] On whether the king was guarded by his people or himself guarded his land (φυλάττουσιν or φυλάττοντα – it is the former reading that prevails), see Hoefer 1904, 542–43.

Periplus 35; Diodorus 14. 30. 6–7); and indeed Diodorus states that tradition required the king to remain in the capital for his entire life.[462]

Conversely, all the negotiations that Xenophon and the Greeks had to engage in with the Mossynoikoi in order be allowed to pass through their land were made with the *archontes*, and it is these who initially deny and later accept the Greek request (*Anabasis* 5. 4. 1–10; 5. 4. 26). The negotiations were mediated by Timesitheus the Trapezuntian, who was the official representative of the Mossynoikoi at Trapezous. His existence shows that the Greeks of the colonies maintained relations with the Mossynoikoi, at least of an economic nature. Timesitheus talks with the *archontes* of the Mossynoikoi, which reinforces the view that it was they who took the political decisions, rather than the king, who seems to have been restricted to the above-mentioned duties and is almost ignored by Xenophon in his narrative. This has led Maksimova to argue that the king of the Mossynoikoi was the sort of 'king-priest' or 'god-king' found in several other cultures,[463] as described by J.G. Frazer in his famous work,[464] an institution that arose from the perception of a king with supernatural qualities. A similar approach was formulated by U. Hoefer, who pointed out that the king's obligations stemmed from his sacredness.[465]

But let me remain a little longer with the king. His justice-related duties and punishment are mentioned by several authors, but not Xenophon. The Scholiast of Apollonius (Sch. Apol. Arg. 2. 1029) attributes the information to Ephorus and Nymphodorus, who are considered basic sources used by Apollonius.[466] Ephorus was, of course, also the main source for Diodorus in the part of his work that refers to the period of the events narrated in the *Anabasis*. The same holds for Pseudo-Scymnus,[467] who in turn was clearly used by Anonymous.[468] Thus, the first to offer the specific information appears to have been Ephorus.

Since Xenophon refers to the king's wooden tower (*Anabasis* 5. 4. 26), it would be reasonable to ask why he failed to mention the custom pertaining to the king's judgments and punishment. And conversely, of course, why Ephorus did not mention all the other information recorded by the slightly earlier

[462] This is also mentioned by later authors: Nicolaus *Fragmenta* 126; Anonymous *Periplus* 35; Pomponius Mela 1. 106.

[463] Maksimova 1956, 140–41.

[464] Frazer 1894.

[465] Hoefer 1904, 553.

[466] See, for example, Hoefer 1904, 555–63; Stylianou 2004, 80. Hoefer, however, also underlined the similarities between the geographic order of the peoples mentioned by Apollonius and the relevant passage of Pseudo-Scylax's *Periplus*.

[467] Hoefer 1904, 545, 554; Schwartz 1907, 4.

[468] On the several sources used by Anonymous, see Diller 1952, 102–17.

Xenophon, especially since the latter is considered by some scholars[469] to have been his main source. Besides, how did Ephorus know about a custom that was not mentioned by Xenophon?

The answers to all three questions can only be speculative. As regards the first, it may be that Xenophon was simply unaware of the custom. He and his soldiers spent most of their time in the land of the Mossynoikoi either negotiating or fighting with them. But they did, once the fighting was over, have time to walk through the settlements of the Mossynoikoi, talk with the people and learn about their habits (*Anabasis* 5. 4. 27–34), and so had opportunity enough to learn about the specific custom. Another explanation could be that some decades after the march,[470] Xenophon did not remember or feel the need to focus on something irrelevant to the military events. On the other hand, he narrates details relating to many other customs of the Mossynoikoi and this one was not insignificant, as indicated by the impression it made on other Greek authors (see below).

As for the second question, the truth is that Ephorus' use of Xenophon is not proven; indeed, some scholars think that it should be excluded,[471] although I would hesitate to accept this so readily. Besides, Ephorus was never reputed to be a reliable narrator, especially of events that took place within military contexts.[472] Moreover, we must remember that the work of Ephorus has not survived in itself and what we have are parts of it copied mainly by Diodorus. We cannot, therefore, totally rule out that Ephorus said more about the Mossynoikoi than we know of today. If this is the case, then he might after all have used Xenophon, but not necessarily only him.[473] Apollonius not only mentions the information Ephorus gives about the Mossynoikoi, but also that provided by Xenophon, which could mean either that he used both of them and not only Ephorus, or that Ephorus had said more about the Mossynoikoi in his work than the later authors have handed down.

An interesting view has been expressed by M. Flower, who thinks that

> the differences between Xenophon's and Diodorus' accounts may be due to the fact that Ephorus would have rhetorically elaborated his sources in order to make his own version look original, and that Diodorus often introduces inaccuracies in the process of abridgement.[474]

[469] E.g. Hoefer 1904, 547, 549, 554; Stylianou 2004, 80. The latter's certainty is not shared either by Flower (2012, 71), or, even less, by Schwartz (1907, 11–12).

[470] On the dating of the *Anabasis*, most probably in the early 360s, see Cawkwell 2004, 47.

[471] Schwartz 1907, 11–12.

[472] See, for example, Polybius 25f. For criticism on Ephorus' work, see also Strabo 9. 2. 2, 4; 9. 3. 12; 13. 3. 6. *Cf.* Schwartz 1907, 7, 11.

[473] *Cf.* Hoefer 1904, 550.

[474] Flower 2012, 70.

If this is true, then we can think of a possible answer to our last question: Ephorus did not *know* about the custom of the king, he just *invented* it.[475] We cannot help noting that the custom of a king who is kept by the multitude confined in starvation in his tower to punish him for making a mistake in his decrees is not only very peculiar but is not recorded for any other people in antiquity (see below). This could also explain its 'omission' by Xenophon, the only known author to have met the Mossynoikoi before the 4th century BC, who thus cannot be so easily excluded as a source for Ephorus concerning these people.

Finally, there is the remarkable insertion of a new element concerning the king of the Mossynoikoi by Pomponius Mela (1. 106), namely that they chose their kings by vote. Since Mela is generally listed among the less trustworthy sources,[476] this information, which only he in the whole known body of ancient literature records, surely surprises. The rest of his account of the king's life resembles pretty much what we know from the rest of the sources, and unlike other cases of historical or ethnographical information provided by the author is detailed, indeed more detailed than in many earlier texts: 'The Mossyni keep their kings in chains and under the closest guard, and when the kings have earned blame for exercising some power wrongfully, the people punish them by depriving them of a whole day's food.'[477] Hence one may reasonably wonder what Mela's source for the elections of the king was. Was there indeed a text mentioning this information – the other preserved ones ignore it – or was it an arbitrary assumption of his?

In any case, the king and other rulers of the Mossynoikoi seem to have displayed a high degree of dignity and self-sacrifice, since we read that when the Greeks invaded the Metropolis, the king refused to come forth, as did also the commander of the stronghold, which had been captured earlier, preferring to perish in the flames of their towers (*Anabasis* 5. 4. 26).

[475] See also in Section 2.10.2. According to Westlake (1987), the immediate source for Diodorus as regards the expedition of Cyrus was indeed the work of Ephorus, who, in his turn, had used the *Hellenica Oxyrhynchia*. As for the source of the latter, Sophaenetus from Stymphalus has been proposed, who also participated himself in the expedition, like Xenophon, although it is only Aelius Herodianus and Stephanus of Byzantium that attributed a work on the expedition of Cyrus to Sophaenetus. Thus one could claim that Ephorus took the information from Sophaenetus, who just paid more attention to the custom than Xenophon. However, such a possibility should be excluded, since Sophaenetus had been embarked on the ships in Trapezous (*Anabasis* 5. 3. 2) and continued the voyage by sea up to Cotyora (5. 4. 1). Therefore, he had not been at all in the land of the Mossynoikoi.

[476] See in the Introduction Section 2.2.

[477] Translation by Rommer (see in Romer 2001).

Another interesting aspect of Mossynoikan society that emerges from Xenophon's account is that there were apparently varying degrees of wealth. After the end of the fighting between the Greeks and the Mossynoikoi, some of those who had allied with the Greeks exhibited to them fattened children of the wealthy local inhabitants, who had been nourished on boiled nuts and were soft and white to an extraordinary degree, and of pretty nearly equal length and breadth. Their backs were decorated with many colours and their foreparts with tattoos in flower patterns (*Anabasis* 5. 4. 32). Here we have an extremely rare (in Greek literature) instance of locals showing off compatriots of a higher economic status to the Greeks, and it is one of those cases where we wish that Xenophon had written his text immediately, because it would be intriguing to see whether there had been a trace of irony – on either side – in the demonstration of the rich people. The reference to the children's thickness as well as the unnatural equality of their measurements could hint at such an attitude. Apart from that, it seems that tattooing was something done only by the rich and thus could have been a class marker in Mossynoikan society.

Nor were there only social differences separating the Mossynoikoi from each other. From the very beginning of Xenophon's account we learn that they were divided into at least two different groups, which were fighting among themselves for control of what was apparently their most important urban centre, the Metropolis (*Anabasis* 5. 4. 3, 15). These opposing groups might denote different tribes of Mossynoikoi. It was, in fact, this civil war that Xenophon's Greeks took advantage of in order to pass through their country and proceed west: they gained the aid of one group by promising assistance against the other. Initially, the Greeks and their allies were unsuccessful; worse still, their opponents cut off the heads of the fallen and displayed them to the Greeks and their own enemies, while dancing to a kind of strain that they sang. This caused great dismay and anger among the Greeks, forcing Xenophon to make a speech to encourage them, and the next day the Greeks managed to attack successfully, besiege the Metropolis, hand it over to their Mossynoikan allies, and depart.

Xenophon seems to have been impressed by the Mossynoikan arms, for he describes them in detail:

> They carried wicker shields covered with white, shaggy ox-hide and like an ivy leaf in shape, and each man held in his right hand a lance about six cubits long, with a spearhead at one end and a round ball at the butt end of the shaft. They wore short tunics which did not reach their knees and were as thick as a linen bag for bedclothes, and upon their heads leathern helmets just such as the Paphlagonian helmets, with a tuft in the middle very like a tiara in shape; and they had also iron battle-axes (*Anabasis* 5. 4. 12–13).

Xenophon's narrative also reveals some aspects of the Mossynoikoi's economic activities. Carpentry was patently a major craft, and timber must have been the strong point of the land's natural resources and consequently of the economy of its inhabitants. From this timber they constructed not only their houses and fortifications, but also ships, since those of the Mossynoikoi who agreed to ally with the Greeks sent to them as many as 300 wooden ships, each made out of a single log (*Anabasis* 5. 4. 10–12). These ships contained three men each, two of whom disembarked and fell into line under arms, while the third remained in the canoe and sailed back with it. This also provides a clue as the size of the population. Nine hundred men are mentioned, 600 of whom would fight alongside the Greeks; and these belonged to the one group of the Mossynoikoi, the friendly ones. One might suppose that the fat white wealthy children did not belong to a group whose future would include such duties.

Wood was also used in the making of several kinds of weapons (*Anabasis* 5. 4. 12), and we have as well the testimony of Aristophanes for the Μοσσυνικά μαζονομεῖα (fragm. 417; Aristophanes *Holkades* fragm. 16),[478] namely wooden plates or paddles (Hesychius *Lexicon*, in the homonymous entry) on which they made or served bread.[479] Interestingly enough, the Mossynoikoi are the only people whose μαζονομεῖα are mentioned in ancient literature. Nor should we not forget that, at least in Xenophon's day, the Chalybes were subject to the Mossynoikoi, who might thus have been exploiting the formers' iron mines and metallurgical products,[480] at least for a period of time between the 5th and the 4th centuries BC, and we might suspect that the iron axes wielded by the Mossynoikoi in battle (*Anabasis* 5. 4. 13) were made of this iron.

M.I. Maksimova argued that the basis of the Mossynoikan economy must have been livestock-raising, since they lived in a mountainous area with little arable land, a lack that could also lie at the root of their civil wars.[481] This may make sense, but there are several references in Xenophon's narrative to other fields of economic activity, such as agriculture and fishing.[482] When the Greeks were plundering the strongholds of the Mossynoikoi, 'they found in the houses ancestral stores, as the Mossynoikoi described them, of heaped up loaves, while the new corn was laid away with the straw, the most of it being spelt.

[478] *Cf.* Julius Pollux *Onomasticon* 10. 84; Photius *Lexicon*, *s.v.* Σκαφίδας μίξας Μοσσυνικὰ μαζονομεῖα.

[479] *Cf.* Michail Psellos *Epistulae* 6. 436; Pseudo-Zonaras *Lexicon*, *Etymologicum Magnum* and the *Suda*, *s.v.* μαζονομεῖον.

[480] See more in Section 2.9.3.

[481] Maksimova 1956, 136–37.

[482] See in Section 1.3.3, in detail.

They also found slices of dolphin salted away in jars, and in other vessels dolphin blubber, which the Mossynoikoi used in the same way as the Greeks use olive oil; and on the upper floors of the houses there were large quantities of flat nuts, without any divisions. Out of these nuts, by boiling them and baking them with a lot of grain, they made the bread which they used most. The Greeks also found wine, which by reason of its harshness appeared to be sharp when taken unmixed, but mixed with water was fragrant and delicious' (*Anabasis* 5. 4. 27–29).[483]

So agriculture was not absent from their everyday life, neither was fishing, especially for dolphin, which seems to have been widely used by the Mossynoikoi, and in a variety of ways.[484] Actually, although living mainly in the mountains, the Mossynoikoi must have been perfectly familiar with the sea, since apart from fishing we know that they made and used large numbers of boats.[485] Finally, they also practised viniculture, and indeed with noteworthy results, at least for the Greek taste.

In any case, the picture of the Mossynoikoi painted by Xenophon is very different from that given by Strabo, who presents them even as tree-dwellers, adding that they lived on the flesh of wild animals and on nuts (Strabo 12. 3. 18), which almost implies a foraging economy. Perhaps they suffered a general crisis and decline after the loss, as noted above, of some their territory to the Armenians in the 2nd century BC (Strabo 11. 14. 5). The Mossynoikoi must in any case have disappeared by the 6th century, since Anonymous (*Periplus* 35) uses the word πρώην (former) for them.

11.3. Anthropological Features

The importance of all this social information notwithstanding, one might argue that the most interesting part of Xenophon's account of the Mossynoikoi is that which refers to their customs and habits. As mentioned at the start, only once in all of ancient Greek literature is a specific nation called βαρβαρώτατοι, and this is the Mossynoikoi. More specifically, Xenophon states that the Mossynoikoi 'were set down by the Greeks who served through the expedition as the most barbarous people whose country they traversed' (*Anabasis* 5. 4. 34). This same character is also ascribed to the Mossynoikoi by Pseudo-Scymnus (901), Diodorus (14. 30. 7) and the Anonymous author of the *Periplus*

[483] Translation by C.L. Brownson (Loeb ed.) (London 1922), with a slight variation – see below, Section 2.11.3.

[484] Athenaeus' report on the Mossynes who fed their oxen fish (*Deipnosophistae* 8. 345e) clearly refers to the Mossynes of Thrace and not to our people (Counillon 2004, 101).

[485] *Cf.* Valerius Flaccus *Argonautica* 5. 151–152.

(35). Inevitably, the question arises: what led Xenophon to use this word for the Mossynoikoi, a word that had never before and would never again be used by a Greek author, not even by Aristophanes or Isocrates, to describe a specific foreign nation? According to Herodotus' definition of Greek ethnicity (8. 144. 2), in the classical period at least the peoples who could – even tangentially – be considered barbarians were those that differed from the Greeks in terms of *blood*, *language*, *religion* and *customs and way of life*. Thus, we would expect that the *most barbarous* peoples would be those most at variance in those aspects. Looking at the description of the Mossynoikoi, we notice that the ancient literature is completely silent on the first three elements (*blood*, *language* and *religion*). All the particularities of the nation cited refer to their *customs and way of life*. Thus the answer to our question must be sought in the anthropological features of these people provided by the ancient literature.

Most of these are presented at the end of Xenophon's adventure in the land of the Mossynoikoi. The war between them and the Greeks is over, the Greeks and their Mossynoikan allies have defeated the hostile Mossynoikoi, and they are now being guided by their local friends in the villages. Xenophon recounts:

> And when the Greeks, as they proceeded, were among the friendly Mossynoikoi, they would exhibit to them fattened children of the wealthy inhabitants, who had been nourished on boiled nuts and were soft and white to an extraordinary degree, and pretty nearly equal in length and breadth, with their backs adorned with many colours and their fore parts all tattooed with flower patterns. These Mossynoikoi wanted also to have intercourse openly with the women who accompanied the Greeks, for that was their own fashion. And all of them were white, the men and the women alike. They were set down by the Greeks who served through the expedition as the most uncivilised people whose country they traversed, the furthest removed from Greek customs. For they habitually did in public the things that other people would do only in private, and when they were alone they would behave just as if they were in the company of others, talking to themselves, laughing at themselves, and dancing in whatever spot they chanced to be, as though they were giving an exhibition to others (*Anabasis* 5. 4. 32–34).[486]

Earlier in his narrative Xenophon noted other customs of the Mossynoikoi, and we have also mentioned the duties of their king and his punishment in case of erroneous judgment, added by other authors, as well as their habit of cutting off the heads of those killed in battle and displaying them to their enemies while dancing and singing. Strabo records (12. 3. 18) that the Mossynoikoi used to attack wayfarers, leaping down upon them from their (obviously wooden) scaffolds, and adds that during Pompey's invasion the Mossynoikoi (Heptacomitae in his day) mixed bowls of the crazing honey which is yielded

[486] Translation by C.L. Brownson (Loeb ed.) (London 1922).

by the tree-twigs and placed them in the roads, and then, when Pompey's soldiers drank the mixture and lost their senses, they attacked and easily disposed of them.

Apparently, then, there were several reasons for the Greeks to consider this nation barbarous. And it was not only Xenophon who called them *βαρβαρώτατοι*, as we have seen: Strabo wrote that the Mossynoikoi were the most savage of all the peoples who lived in the mountains of the wider area (12. 3. 18), Apollonius called them overweening (*Argonautica* 2. 1117) and Pomponius Mela (1. 016) described them as 'rough, crude and absolutely vicious (*asperi, inculti, pernoxii*) to those who put in to shore there'.

There is, however, one report on the Mossynoikoi that stands in sharp contrast to Strabo's and Pomponius Mela's statements that they were totally hostile towards strangers who approached them, respectively, by land or sea. Nicholas of Damascus (fragm. 126) says that the Mossynoikoi divided their food in equal portions and offered part of it to the arriving strangers. Clearly this has to do with hospitality, which was a typical element of the Greek culture. One cannot help wondering why Nicholas, writing in the 1st century BC, should have been the first to mention this hospitable custom,[487] how he knew about it, and why it is not confirmed by any other author, especially Xenophon. But even if there are reservations about the real intentions of the Mossynoikoi when sharing their meal with strangers, otherwise a 'gesture characteristic of the finest hospitality',[488] we should note, too, that there was at least one Greek city situated within their territory (Choirades: Hecataeus *FGrH* 1a 1 F 204; Pseudo-Scylax 86), as well as an official Greek representative of the Mossynoikoi at Trapezous (Timesitheus the Trapezuntian: Xenophon *Anabasis* 5. 4. 2). In any case, these factors do not seem to have influenced the widespread negative opinion on the Mossynoikoi among the Greeks.[489]

In attempting to determine the reasons for the pejorative image of the Mossynoikoi in Greek eyes,[490] it is worth examining in particular two aspects of their customs: first, which of these were highlighted by more authors, apart from Xenophon; and second, which of their manners and customs are also reported for other peoples in antiquity, especially in the vicinity of the Mossynoikoi, and how these were presented and assessed by the authors in those cases. Starting with the first question, most of the authors seem struck by the reported habit of the Mossynoikoi of doing in public things that other people

487 It is also repeated later by Joannes Stobaeus (*Anthology* 4. 2. 25).

488 Schmitt Pantel 2015, 226–27.

489 According to Brennan (2015, 106), this barbarism of the Mossynoikoi was the reason for the fact that they have not been claimed as ancestors by any of the later inhabitants of their land.

490 See a thorough analysis of this in Manoledakis 2021a.

do only in private, especially having intercourse.[491] Xenophon and Diodorus seem to regard this as the main reason for which they call the Mossynoikoi βαρβαρώτατοι. Ephorus' information about the king and the punishment imposed on him for a faulty judgment is equally mentioned.[492] Of the other anthropological information Xenophon provides about the Mossynoikoi, their habit of adorning much of their body with coloured tattoos in several patterns attracts some interest (Diodorus 14. 30. 7; Pomponius Mela 1. 106; Pliny *NH* 6. 11). The rest of the comments, about their eating or drinking habits, or even about their hostility to foreign people who approach their land, do not seem to have been considered so strange by the Greeks.

P. Cartledge[493] argued that the main reason for the Greeks' characterisation of the Mossynoikoi as the most barbarous people was 'not a matter of sexual usages, but of their food habits. For the Mossynoikoi did not normally eat bread.' To justify this view, Cartledge pointed to the importance of the earth-mother goddess of grain, Demeter, to all Greeks, regardless of origin or social status. The trouble with this view lies less in the fact that Cartledge chooses a food habit rather than a sexual one to explain the Greeks' cultural contempt for the Mossynoikoi than in the fact that Xenophon does *not* actually say that the Mossynoikoi did not normally eat bread. He just points out that their bread was made by baking grain with boiled nuts (*Anabasis* 5. 4. 29). So, the grain was there, and in fact it was dominant (πλείστῳ σίτῳ), and, after all, was found in abundance in their stores (5. 4. 27). Cartledge may have been influenced by a serious inaccuracy in a widely used English translation of Xenophon's *Anabasis*,[494] where the passage τούτων (καρύων) καὶ πλείστῳ σίτῳ ἐχρῶντο ἕψοντες καὶ ἄρτους ὀπτῶντες is rendered as 'out of these nuts, by boiling them and baking them into loaves, they made the bread which they used most', instead of 'out of these nuts, by boiling them and baking them with a lot of grain, they made the bread which they used most'. The error is important, because it suggests that there was no grain at all and the bread was made only with boiled nuts, which is hardly implied by the original text. The same erroneous interpretation was also used by Maksimova,[495] who wrote that

[491] Ephorus = Schol. in Apol. Rhod. II 1029; Apollonius *Argonautica* 2. 1018–1025; Pseudo-Scymnus 904–910; Diodorus 14. 30. 7; Artemidorus *Oneirocritica* 1. 8; Anonymous *Periplus* 35; Pomponius Mela 1. 106.

[492] Apollonius *Argonautica* 2. 1026–1029; Pseudo-Scymnus 904–910; Diodorus 14. 30. 7; Nicholas of Damascus fragm. 126; Stobaeus *Anthology* 4. 2. 25; Pomponius Mela 1. 106; Anonymous *Periplus* 35.

[493] Cartledge 1993, 44–45.

[494] By C.L. Brownson (Loeb ed.) (London 1922).

[495] Maksimova 1956, 138–39.

the Mossynoikoi baked bread with chestnuts,[496] which served as a substitute for cereals, thus she concludes that the latter were in short supply in the land of the Mossynoikoi.

Furthermore, it cannot pass unremarked that the Mossynoikoi drank wine and not beer, as the barbarians were considered to do,[497] and in fact a kind of wine that attracted the interest of Xenophon and his Greek soldiers, who described it as fragrant and delicious when mixed with water, although by reason of its astringency appearing sharp when taken unmixed (*Anabasis* 5. 4. 29). This evaluation of the Mossynoikan wine is particularly important, since we know that the Greeks laid stress on the flavours and other qualities of wine, and also on the therapeutic virtues of each kind of wine, as we read in several texts (e.g. Athenaeus *Deipnosophistae* 26a–27d; *cf.* Aristotle *De anima* 2. 10). So an astringent wine (like the unmixed Mossynoikan one) had the virtue of being refreshing and smoothed the stomach (Plutarch *Quaestiones convivales* 3. 5. 2), while a sweet one (like the mixed Mossynoikan one) was also good for the stomach (Athenaeus *Deipnosophistae* 27a, c).[498]

It can surely, therefore, not have been the dietary habits of the Mossynoikoi that led to their characterisation as the most barbarous people. In reality, according to D. Braund, Xenophon's account suggests 'admiration of the Mossynoikan diet more than alienation or an attitude of blunt hostility'.[499] In addition, we should note that no ancient author writing of the Mossynoikoi seemed to have been bothered or even surprised by their food habits, as for example is the case with their sexual customs.

As far as the second question is concerned, namely which of the customs of the Mossynoikoi are encountered in other 'barbarian' peoples and what impression do they seem to have made on the Greeks, the use of dolphin blubber was not something unheard of for the Greeks; on the contrary, in the northern Black Sea littoral, an area with numerous Greek colonies and an abundance of fish, sea creatures were the most substantial and readily available source of oil.[500] Neither was the decoration of the body with tattoos, which was a well-known custom among several peoples.[501] From Strabo's words, we might assume that tattooing the body was popular only among the wealthy Mossynoikoi, and

[496] The word used by Xenophon is κάρυα. It must be stressed that this word has always been used by the Greeks of Pontus to denote walnuts, and still is to this day. For the nuts of this part of the littoral, see Section 1.3.2.

[497] See indicatively Sulimani 2011, 243–46, with relevant sources and bibliography.

[498] See further Boulay 2015, 277–79.

[499] Braund 2015, 307, *contra* Cartledge 1993, 44–45.

[500] Braund 2015, 304, referring especially to Olbia.

[501] See, indicatively, Tsiafakis 2015, 93–96, with bibliography.

it may have been a characteristic of social status that differentiated them within their society. Attacking passing wayfarers or seafarers was also common among barbarians, thus something the Greeks expected. Xenophon himself ascribes this habit to the Bithynians, whom he met, without any sign of astonishment (*Anabasis* 6. 4. 2), while Diodorus (4. 40. 4) mentions that the natives on all the shores of the Black Sea were in the habit of slaying the strangers who landed on its shores – an exaggeration, of course, and referring to the period of the Argonautic expedition. Even the habit, encountered among some peoples, of cutting off the heads of their enemies was known to the Greeks at least as early as the time of Herodotus, who records that the Tauroi used to kill all shipwrecked men or any Greeks they captured in sea-raids, decapitate them and place every head on a pole (Herodotus 4. 103).

What is left to consider is the punishment of the king in the event of an incorrect judgment, and the tendency to do in public the things that other people do only in private. It was these two customs that astonished the Greeks most, judging by the frequency of their mention in the written sources cited above. While for the first I am not aware of any precise parallels,[502] as regards open intercourse there are references that reveal that it was common among other peoples as well. It is the Libyan Auseans, of whom Herodotus says that 'the intercourse of men and women there is promiscuous; they do not cohabit but have intercourse like cattle (*κτηνηδόν*)' (Herodotus 4. 180); and some tribes of the Caucasus, where 'men and women have intercourse openly, like beasts of the flock' (Herodotus 1. 203). Slightly different were the customs of the Libyan Nassamones and Gindanes, as well as the Massagetae north-east of the Caspian and the Agathyrsoi (Scythians or Thracians): the Nassamones' custom was that every man could have many wives:

> their intercourse with women is promiscuous, as among the Massagetae; a staff is placed before the dwelling, and then they have intercourse. When a man of the Nassamones weds, on the first night the bride must by custom lie with each of the whole

[502] There are some impressive similarities in the description of the life of the Egyptian kings, as well as the kings of the Arabian tribe of the Sabaeans, by Diodorus (1. 70–71 and 3. 47 respectively), but no mention of such a punishment them for a wrong judgment. Specifically as regards the Egyptian kings, it was said that their every action was regulated by prescriptions set forth in laws, not only their administrative acts, but also those that had to do with the way in which they spent their time from day to day, and with the food which they ate (Diodorus 1. 70. 1). The purpose of all this was to educate a king so that he would be afraid of the gods and live a life pleasing to them, and to be accustomed to a proper manner of conduct, not by sharp admonitions, but through praises that were agreeable and most conducive to virtue (1. 70. 8). As far as the kings of the Sabaeans are concerned, 'it was unlawful for them ever to leave the palace, and if they did so they were stoned to death, in accordance with a certain ancient oracle, by the common crowd' (Diodorus 3. 47. 4). For similar customs that have been observed in other parts of the world, such as Polynesia or Africa, see Hoefer 1904, 552–53.

> company in turn; and each man after intercourse gives her whatever gift he has brought from his house (Herodotus 4. 172. 4 and 1. 216. 1 on the Massagetae).

Furthermore, the Agathyrsoi have promiscuous intercourse with women, 'so that they may be consanguine with one another and, all being relations, not harbour jealousy or animosity toward one another' (Herodotus 4. 104). As for the Gindanes, 'Every woman there wears many leather anklets, because she puts on an anklet for every man with whom she has had intercourse; and she who wears the most is reputed to be the best, because she has been loved by the most men.'[503]

But none of these peoples lived in the areas traversed by those who called the Mossynoikoi βαρβαρώτατοι. Even if some of them, including Xenophon, may have read Herodotus' work, they had not met the peoples mentioned in it and encountered their customs. Of the Mossynoikoi, on the other hand, they had personal experience. One who had merely read or heard about the Auseans or the Caucasian tribes would not feel the same shock as the soldiers of Xenophon who themselves saw the Mossynoikoi ready to have intercourse openly with their women. So, apart from the negative impression created by this habit, personal experience was an additional factor that led the Greeks call the Mossynoikoi βαρβαρώτατοι. And neither of these two customs is reported for any other of the peoples whose territories Xenophon's soldiers traversed, although these, covering as they did almost all of Asia Minor and more, were fairly extensive.[504]

Even Apollonius, who had not personally experienced the customs of the Mossynoikoi, devoted several lines to the fact that

> their laws and customs are strange. Whatever it is right to do openly before the people or in the market place, they do it in their homes; but whatever acts we perform at home, these they perform out of doors without blame, in the middle of the streets. And they have no reverence for the marriage-bed, but, like swine that feed in herds, and without feeling ashamed in others' presence, they lie on the earth and have intercourse with women (*Argonautica* 2. 1018–1025).

Although at first sight Apollonius seems to repeat Xenophon's account, he actually adds here the very important element of the astonishing disregard of marital fidelity among the Mossynoikoi.

So, what does the impact on the Greeks of these two 'barbaric' customs of the Mossynoikoi indicate? First of all, as already stated, being 'barbarian' is more a matter of strange and astonishing customs than one of a different

[503] For the slight differences between these customs and thus the slight differences in levels of barbarism, see Cartledge 1993, 77–78.

[504] See, indicatively, *Anabasis* 7. 8. 25, irrespective of whether the passage is an interpolation or not.

origin, language or religion. Moreover, if Herodotus was the first to indicate who, for the Greeks, was a barbarian (Herodotus 8. 144. 2), Xenophon, by reason of his comments on the Mossynoikoi, could be considered the first to indicate which 'barbaric' qualities were thought to be more significant than others. Secondly, it indicates that for the ancient Greeks custom and legislation – customary and statute law – were equally important, as indeed can be inferred from several written sources.[505] It is not without significance that the characterisation *barbarian* is given based on the *ethos*, *dike* and *nomoi* of the several peoples, as it appears through the relevant passages/definitions in Herodotus, Xenophon and Pseudo-Scymnus, the last being the only one who adds *erga* (deeds) to the criteria.[506]

But there is also another conclusion that could be drawn: for the Greeks, at least in Xenophon's time, 'barbarian' could mean 'the furthest removed from the Greek customs' (*Anabasis* 5. 4. 34), but this 'barbarism' did not necessarily or primarily mean ἀγριότης (wildness, cruelty). There were other things, such as having intercourse in public, which seem to have been considered more remote from Greek customs than wildness and brute force. This is implied not only by Xenophon, but also by Herodotus,[507] while Artemidorus, in his *Oneirocritica* (1. 8), speaking of how common individual or ethnic customs may differ from place to place and from people to people, uses the habit of the Mossynoikoi (he calls them Mossynes) to have sexual intercourse in public and 'mingle with their wives just as dogs do', as an example of a behaviour that in the eyes of other men is considered shameful.

The superlative ἀγριώτατος (utterly wild) is again very rarely used in classical Greek literature, and usually to refer to animals or brutal human behaviours. But there is one case where it refers to the customs of a specific nation (ἀγριώτατα πάντων ἀνθρώπων ἔχουσι ἤθεα): this was the Androphagoi, who, as their name indicates, used to eat men, and in fact, according to Herodotus (4. 106) they were the only people in the hinterland to the north of

[505] *Cf.* Pindar's famous phrase (in Herodotus 3. 38. 4 and Plato *Gorgias* 484b) 'custom is lord of all'. Compare also Redfield 1985, especially 98–99, 116. Indeed, one could say that each of the two 'barbaric' customs we are dealing with here seems to refer to one of the two sources of law. Even if we might have many reasons to doubt whether the king's punishment in case of a wrong judgment (if indeed such a custom existed) was based on Mossynoikan legislation and not custom, for the Greeks of the classical period, who are here attributing barbarism to foreigners, the rules for dispensing justice were considered a matter of legislation and no longer of custom.

[506] Herodotus 4. 106: Ἀνδροφάγοι δὲ ἀγριώτατα πάντων ἀνθρώπων ἔχουσι ἤθεα, οὔτε δίκην νομίζοντες οὔτε νόμῳ οὐδενὶ χρεώμενοι; Xenophon *Anabasis* 5. 4. 34: τούτους ἔλεγον οἱ στρατευσάμενοι βαρβαρωτάτους διελθεῖν καὶ πλεῖστον τῶν Ἑλληνικῶν νόμων κεχωρισμένους; Pseudo-Scymnus 82–83: ἀξενώτατοι, ἔθεσι, τρόποις ἔργοις τε βαρβαρώτατοι.

[507] Cartledge 1993, 60, 76–80. See also Hirsch 1985, 89.

the Black Sea who did so.[508] But although this might be certainly appear to be the reason why they were called utterly wild, Herodotus immediately follows this characterisation with the statement that the Androphagoi knew no justice and obeyed no law (*ethos* and *nomos* again), only later adding that – by the way – they ate men,[509] thus reinforcing the impression that custom and legislation were of equal importance to the Greeks.

Neither was being extremely warlike the pre-eminent sign of barbarism, since Xenophon himself says (*Anabasis* 5. 2. 2) that the most warlike people of all that dwell upon the Pontus (πολεμικώτατοι – a superlative again) were the Drilae,[510] so not the βαρβαρώτατοι Mossynoikoi. What Xenophon and his soldiers did not experience is an encounter with people who ate men. This could be a reason for his never, in his entire work, having applied the word *ἄγριος* (in any degree) to peoples or human customs or deeds, but only to animals.[511]

It seems, therefore, that cruelty may be a manifestation of barbarism, but not the only one. We could say, though, that it is the worst manifestation of barbarism, thus the answer to the question 'which is worst, βαρβαρώτατος or ἀγριώτατος?', is 'the second'. Because βαρβαρώτατος primarily indicates difference, while ἀγριώτατος has a definitely negative sense. This is implied by Dionysius of Halicarnassus' (*Antiquitatum Romanarum* 1. 89. 4) description of the Achaeans who were settled near the Black Sea as the most savage of all barbarians (βαρβάρων συμπάντων ἀγριώτατοι).[512]

Remarkably, in the same period, namely living long after the classical era and thus detached from the issue of the use of the word *barbarian*, Strabo points out that the Greeks referred to all foreigners as barbarians (14. 2. 28), and cites calling the Mossynoikoi utterly savage (ἄγριοι τελέως) among the other *barbarians* of the region (12. 3. 18). In this way he both indicates that he

[508] Herodotus says that the Androphagoi were not Scythians (Herodotus 4. 106); Pseudo-Scymnus (810) says that they were; for this and other divergences between the two authors concerning peoples of this specific region, as well as on whether Herodotus actually went so far north or learned about this area from merchants or slaves, see Corcella 2007, 548–63, and 592 and 656 on the Androphagoi.

[509] Variations of the custom in Herodotus 1. 216. 2–3 on the Massagetae: 'when a man is very old all his family meet together and kill him, with beasts of the flock besides, then boil the flesh and feast on it. This is held to be the happiest death.' And Herodotus 4. 26. 1 on the Issedones (related to the Scythians): 'whenever a man's father dies, all the nearest of kin bring beasts of the flock and, having killed these and cut up the flesh, they also cut up the dead father of their host, and set out all the flesh mixed together for a feast.'

[510] See in Section 2.12.

[511] Only once does he mention the word *ἄγριος* with reference to humans, in *Cyropaedia* (1. 6. 34), talking about the upbringing of children.

[512] *Cf.* Diodorus 4. 40. 4: ὑπό ἐθνὼν βαρβάρων καί παντελῶς ἀγρίων.

was aware that the Mossynoikoi were more savage than the others in the area and confirms that ἀγριώτατος was considered worse than βαρβαρώτατος.

Nevertheless, we should not expect the term 'barbarian' to have had a standard meaning in ancient Greek literature. As already mentioned, personal experience is very important when it comes to characterising someone. I think that the sense of the word 'barbarian' (or its comparative and superlative) would always carry a degree of subjectivity from the person who used it, irrespective of the general perception of the period. To focus on the βαρβαρώτατοι Mossynoikoi, we should not forget that the Greeks experienced a very unfortunate incident in their land. As they and their Mossynoikan allies began to engage the hostile Mossynoikoi, the latter sallied forth and put the Greeks to flight, killing a considerable number of them and pursuing the rest until they saw other Greeks coming to the rescue; they then cut off the heads of the dead men and displayed them to the Greeks and to their own enemies, at the same time dancing and singing according to a custom they had. The Greeks were exceedingly angry, not only because the enemy had become bolder, but also because the Greeks involved in the clash had fled, although their numbers were very considerable – 'a thing which they had never done before in the course of the expedition'. This disappointing and unprecedented performance of the Greek soldiers and the anger that followed forced Xenophon to make a speech to encourage them (*Anabasis* 5. 4. 16–21). Let us remember Xenophon's precise words: that the Mossynoikoi were considered βαρβαρώτατοι 'by the Greek soldiers', namely those who had experienced the above-mentioned humiliation. A humiliation that a little later was followed by the unconcealed desire of the Mossynoikoi to have intercourse openly with the women of the Greek soldiers. Subjectivity is something that should always be taken into account in discussing any ancient Greek's use of the word *barbarian*.[513]

11.4. Mossynoikoi and Scythians

In closing our examination of the Chalybes we mentioned that several sources considered this people to be Scythian.[514] This idea seems to have been launched by Aeschylus and was repeated in some later sources, such as Hesychius (*s.v.* Χάλυβοι), the *Suda* (*s.v.* Χαλύβοις), the *Etymologicum Magnum* (*s.v.* Χαλκός) and several commentaries of Scholiasts to Aeschylus, Apollonius and Callimachus. Similar is the image created by some authors for the Mossynoikoi, although this time it was only Hesychius (*Lexicon mu* 1704, *s.v.* μόσσυνες)

[513] See Manoledakis 2021a, from which most of this sub-section was taken.
[514] See in Section 2.9.4.

and the Scholiast of Apollonius of Rhodes (159. 13) who called them a Scythian nation, again with no explanation; there is no pre-Christian source. Both authors also called the Chalybes a Scythian nation, while for the latter the same also applies to the Tibarenoi, Macrones, Becheires, Sapeires and Byzeres (or Dizeres) (Schol. in Apol. Rhod. 159. 13, 201. 20; 161. 7; 161. 17; 161. 19; 161. 21), that is, almost all the peoples who inhabited the eastern part of the southern Black Sea littoral east of the Iris river.

In an attempt to detect elements from accounts of the Chalybes that would justify a connection with the Scythians, we focused on some of their habits, such as the outrageous customs of the Chabarenoi who ate human flesh, metalworking, the use of dolphin oil, and the habit of the inland Chalybes to 'slaughter whomever they might be able to vanquish, and then cut off their heads and carry them along on their march, singing and dancing whenever they were likely to be seen by the enemy' (Xenophon *Anabasis* 4. 7. 16). The similarities of these last three with what has been reported for the Mossynoikoi is striking. And they can all be related to the Scythians.

Now, we have already mentioned the wide use of fish oils on the northern Black Sea littoral,[515] as well as metallurgy and the display of enemies' severed heads.[516] But concerning the Mossynoikoi especially, there are two more things worth noting. The one is the so-called 'mad honey', which is mentioned by Strabo (12. 3. 18) and also by Xenophon (*Anabasis* 4. 8. 20–21). It was examined in the previous part of the book. It seems that the honey of this general area and the symptoms it caused in those who ate it were well known in the region down the ages, as Pliny's report confirms (*NH* 21. 77),[517] and similar was the reputation of the honey of the Scythians among the Greeks of the Mediterranean.[518] Finally, the reaction of the king and the commander of the Scythian stronghold captured by the Greeks, who refused to come forth and were burned along with their towers (Xenophon *Anabasis* 5. 4. 26), resembles the resolve of the kings of the Cimmerians during the advance of the Scythians not to flee with the people but to fight with each other until all were killed by each other's hands, so that they would lie dead in their own country (Herodotus 4. 11. 3–4).

Could all this indicate a Scythian origin of the Mossynoikoi, or at least a kinship with the Scythians? Or is Maksimova's view that these people had come from the south more plausible?[519] Unfortunately, we have no further

[515] See Section 2.11.3.
[516] See Section 2.9.4.
[517] Pliny places it in the area of the Sannoi – see Section 2.13.
[518] Braund 2015, 298. Also, Polybius (4. 38. 4) reports that Pontus had plenty of honey.
[519] Maksimova 1956, 30–31.

information about the Mossynoikoi that could lead to a safe conclusion regarding their origin.[520] Nor do we have any material evidence of them, which could have been very helpful. According to a view expressed by S. Brennan,[521] the Mossynoikoi, despite their material wealth,

> did not choose to fashion durable artworks or construct memorial architecture or otherwise seek to preserve their own memory for posterity. [...] It is conceivable that at some level they believed their deeds would endure in the world of memory.

This, however, is the case with almost all the peoples of the southern Black Sea littoral,[522] and we should bear in mind that different peoples could have different philosophical perspectives concerning the memory of posterity itself. Besides, we cannot be sure that the wooden constructions of the Mossynoikoi, for example, were not elaborate or even 'memorial', especially their seven-storey towers. In any case, the origin of these people or their relation to others, such as the Scythians, remains obscure.

12. The Drilae

12.1. Geographical Location (Fig. 10.3)

The Drilae are a people who lived very near the coast, although not precisely on it, and so may be considered as inhabitants of the broader littoral. Virtually the only information we have about them comes from Xenophon (*Anabasis* 5. 2. 1–27). His army has already reached Trapezous, which he says is situated in the territory of Colchis (4. 8. 22). Although the Trapezuntians have supplied a market for the army, received the Greeks kindly, and given them oxen, barley-meal and wine as gifts of hospitality (4. 8. 23), these are not enough for the maintenance of the soldiers, and Xenophon advises them to obtain provisions by plundering the territory around Trapezous, which, however, is hostile and therefore dangerous (5. 1. 6–7).

We are at the beginning of Book 5 of the *Anabasis*. At the end of Book 4, Xenophon had recounted that his army had arrived at Trapezous, where they remained about 30 days in the villages of the Colchians, and from these as a base plundered Colchis (4. 8. 22). Apparently, their great number, which is

[520] Even in the *Chronicon* of Hippolytus, the general origin of the Mossynoikoi is not made clear, since they appear as descendants not only of Japheth (80. 15), which would be the most probable, but also of Ham (132. 23) and maybe also of Shem (169. 9). The uncertainty seems, however, to be removed by Epiphanius (*Ancoratus* 113. 5). *Cf. Chronicon Paschale* 47.

[521] Brennan 2015, 104–05.

[522] With some exceptions in the wider areas of Paphlagonia and Bithynia.

estimated at that point to have been about 8600 (4. 8. 15; 5. 3. 3), prevented the Greek soldiers from being accommodated in the friendly Greek city but was sufficient to dissuade the Colchian villagers from repelling them; on the contrary, they were driven out of their houses by the Greeks and are still considered hostile and dangerous. Furthermore, Xenophon advised his soldiers to continue their march to Greece by land rather than by sea, a proposal that enraged the majority of the soldiers, who were tired of marching and preferred the sea route. But since the roads there were difficult to travel, he recommends that they direct the cities that are situated along the sea to repair them; 'for they will obey, not only from fear, but also from the desire to be rid of us' (5. 1. 13).

Nevertheless, very soon it was no longer possible for them to obtain provisions and return to the camp on the same day, and therefore Xenophon took some Trapezuntians for guides and led forth half the army to the country of the Drilae, leaving the other half behind to guard the camp – because the Colchians, having been driven out of their houses, were now gathered together in one great body and had taken up a position on the heights above the camp (5. 2. 1). It is worth noting that the Trapezuntians did not lead the Greeks to Colchian places, from which provisions could be obtained easily, because they were friendly towards their inhabitants (unlike Xenophon's soldiers), but were eager to lead them into the territory of the Drilae, at whose hands they were continually suffering losses, though their country was mountainous and difficult to traverse and its inhabitants 'the most warlike (*πολεμικώτατοι*) of all that dwell upon the Euxine' (5. 2. 2).

We shall return to the warlike nature of the Drilae and their settlements later, but for now let us try to comprehend just where these people lived, in relation to both Trapezous and the Colchians. Xenophon's words (5. 2. 1–28) leave no room for any possibility that the Drilae lived anywhere but on the highlands and the mountains to the south of Sinope's colony. Their areas were inaccessible, with deep ravines, steep and narrow paths. It has already been said that attacking the Drilae was dictated by the Greeks' inability to obtain provisions and return to camp on the same day (5. 2. 1). Consequently, the Drilae lived further than half a day's march from the sea. And after the victorious battles against their settlement, the Greeks do not leave the latter immediately to return to their camp, but on the following day (5. 2. 28). This does not necessarily indicate a very great distance from the sea, nor from the Colchian settlements, for the Trapezuntian guides do not seem to traverse much distance from there to the Drilae (5. 2. 2); besides, the return from the Drilae's village to the camp (5. 2. 28–32) does not appear to have lasted more than a day.[523]

[523] Similar are the opinions of Lendle (1995, 298–99) and Lipka (1995, 67–68).

Apparently the Colchians inhabited only the coastal area around Trapezous, and the Drilae possessed the mountainous but not very distant hinterland. Thus, W. Tomaschek's suggestion that the Drilae lived in the Zigana mountains,[524] which rise about 50 km from the sea is unsubstantiated.[525] The terrain described by Xenophon starts some 15–20 km from the coast. On the face of it, some assistance may be afforded by the words of the much later Arrian, in his *Periplus of the Euxine Sea* (11), where he identifies the Drilae with the Sannoi:

> The Colchians, as Xenophon observes, border on the Trapezuntians; and he calls Drilae those people, whom he records to be of a warlike disposition and very hostile to the Trapezuntians; but they seem to me to be more properly called Sannoi. Because they are still nowadays very warlike and much hostile to the Trapezuntians. They dwell in strongly fortified places, and do not live under a monarchical government. They were formerly tributary to the Romans; but of late, being addicted to plunder, they do not pay the tribute regularly. However, now, by the Gods' assistance, we will either oblige them to be more punctual, or exterminate them. The Machelones and the Heniochoi border on these people, the latter of whom have a king called Anchialus. Next to these lie the Zydretae.[526]

Arrian's account, however, seems rather to confuse than to help us, when he identifies the Drilae with the Sannoi, since in most sources, starting from Strabo, the Sannoi are identified with the Macrones.[527] The latter, however, cannot be easily connected to the Drilae, especially in Xenophon, who clearly mentions the two as separate peoples encountered on different occasions.[528]

Arrian, who only in this part of his work (11) mentions peoples,[529] has clearly studied Xenophon, but apparently not Strabo, thus he appears to be unaware of the latter's information that 'Sannoi' is the new name of the Macrones. He does not mention the Macrones at all (albeit a student of Xenophon),[530] nor does he seem to be sure about the identification he makes ('they seem to me to be…'), basing it on the warlike nature and hostility towards the Trapezuntians of the people he calls Sannoi, as well as their

[524] W. Tomaschek, *RE* V.2 (1905), 1706, *s.v.* Drilai.

[525] See also Maksimova 1956, 125–26; Lendle 1995, 298.

[526] Translation by W. Falconer, London 1805.

[527] Strabo 12. 3. 18; Aelius Herodianus and Stephanus of Byzantium, both under *s.v.* Μάκρωνες and quoting Strabo; Eustathius *Comm. in Hom. Od.* 2. 73; *Comm. in Dion. Per.* 765, mentioning also the name 'Tzannoi'. This name was mostly used by Procopius (*De bellis* 8. 1. 8–9, 2. 30), who adds that they later became Christians. *Cf.* Danoff 1962, 1020. Probably they were the same as the Sanegai of Memnon (Phot. *Bibl.* 238a. 39).

[528] See in Section 2.13.

[529] *Cf.* Counillon 2004, 98, n. 312.

[530] According to Counillon (2004, 99), this indicates that he put the Macrones in the hinterland, but this is rather an *argumentum ex silentio*.

strongly fortified places, both of which coincide with what Xenophon had said about the Drilae five centuries earlier.

But actually, in Xenophon's narrative the Macrones (*Anabasis* 4. 8. 1–21) do not appear to live far from the Drilae.[531] They are clearly placed in the hinterland, far from the sea; the historian then mentions the Colchians, as possessing the coast around Trapezous, and then the Drilae, but in the meantime he has turned inland again: the Drilae and the Macrones might well, therefore, be neighbours. Xenophon's short account of the Drilae, coupled with the fact that there is no further mention of them after Xenophon (apart from Arrian and Stephanus of Byzantium,[532] both of whom are clearly citing the historian), may lead to the suspicion that the Drilae (who must anyway have been small in number) ceased to exist well before the Roman period or were assimilated by others, probably the Macrones.[533] This must have happened before Strabo at the latest, and probably even before Apollonius, since the third-century poet, well aware of Xenophon's work, mentions only the Macrones and not the Drilae. Some time before Strabo, the Macrones might have been renamed Sannoi,[534] and a *terminus post quem* for this name change might again be Apollonius' 3rd century BC, because the poet seems to be aware only of the old and not the new name. Moreover, we cannot fail to observe that Herodotus, tracing events back from the Persian Wars, mentions the Macrones several times (2. 104; 3. 94; 7. 78) but never the Drilae. Therefore, the latter might have been a rather small and historically insignificant tribe of the Macrones that happened to have found themselves in Xenophon's path and thus gained a place in history.

If this is the case, then for their exact geographic placement in the wider district around Trapezous we have only Xenophon's words to go on, as the only author who actually mentions them. This leads us, as outlined above, to the hinterland not far to the south of Trapezous.[535]

12.2. Settlements, Economy and Idiosyncrasy

Xenophon's account contains little information about the residential environment of the Drilae, or their economy and society. Invading their country, the

[531] See in Section 2.13.

[532] *S.v.* Δρίλαι: 'a district of Pontica, not far from Trapezous'.

[533] Lipka (1995, 67) argued that the Drilae might have been a subtribe of the Macrones or could even be identified with them. Lendle (1995, 298–99) is more reserved.

[534] But the old name was not forgotten, as indicated by its appearance in later sources. Or, more probably, Strabo's identification was erroneous. See in Section 2.13.2.

[535] As for the Sannoi, see in Section 2.13.2.

Greeks first encountered some villages, but were unable to secure anything except an occasional pig or ox or other animal, because the Drilae themselves set fire to those places which seemed to them easy to capture and fell back (*Anabasis* 5. 2. 3). Apparently, these villages were not fortified; but this is not the case with their main stronghold, which is mentioned as being their metropolis, to which they had all streamed. This place was difficult even to approach, since it was very well protected by both nature and its inhabitants. It was surrounded by an exceedingly deep ravine, while there was additionally a wide trench backed by a rampart, and upon the rampart palisades had been set and wooden towers were constructed at frequent intervals (5. 2. 5).

Xenophon's description further reveals the existence of a citadel, even more impregnable, approached by a road lined with wooden houses (5. 2. 17–27). Seizing the village was no easy matter, for it was stoutly defended, but the Greeks finally managed to achieve their goal, those wooden houses proving to be its weak point. The whole city was burned down – houses, towers, palisades and everything else except the citadel.[536] But the Greeks had acquired the provisions they wanted and left the place.

The wooden houses and towers inevitably remind us of the Mossynoikoi, who, after all, lived not far from the Drilae. The natural environment of both peoples' lands was more or less the same, and the similarities in the pattern of the settlements remarkable. Attempts have been made to compare this pattern with other areas of the southern Black Sea littoral[537] or the Caucasus region,[538] and while these may be reasonable they cannot necessarily lead to safe conclusions, especially concerning matters of origins or ethnic kinships.

As for the economy and the occupations of this people, they seem certain to have practised cattle breeding and carpentry; unlike the Mossynoikoi, however, there is no indication of any sort of familiarity with the sea and related activities.[539] Working with metal and leather could also be inferred from the description of their armour, which included 'wicker shields, spears, greaves, and Paphlagonian helmets' (*Anabasis* 5. 2. 22). The last of those would also be encountered later among the Mossynoikoi (5. 4. 13), and one might be reminded of the strong Paphlagonian influence over a broad swathe of the

[536] Noteworthy is Xenophon's attribution of the Greeks' victory to the gods: 'In the midst of their fighting and perplexity some god gave to the Greeks a means of salvation' (*Anabasis* 5. 2. 24). A very good examination of how the divine influenced Xenophon and his work is offered by Haywood (2016).

[537] Doonan 2019.

[538] Maksimova 1956, 128–36.

[539] See Section 2.11.2.

littoral that was mentioned earlier,[540] unless this is just a similarity noticed by Xenophon without further significance. For the rest, any attempt to draw conclusions about the social structure of the Drilae based only on the differentiation between the unprotected villages and the metropolis with the citadel[541] would be rather risky.

We shall close with the bellicosity of the Drilae. Actually this is the first element that Xenophon records, even before recounting the battles against them. He makes it clear that the Drilae were ill disposed towards the Trapezuntians, who suffered at their hands, and adds that they were the most warlike of all the tribes that dwelt upon the Pontus (*Anabasis* 5. 2. 2). This is borne out immediately afterwards in his account of the army's difficulty in fighting the Drilae and entering their land.

Examining the Mossynoikoi, we highlighted the fact that the Greeks who served through the expedition considered them the most uncivilised people whose country they traversed (*Anabasis* 5. 4. 34).[542] The very rare case of the superlative (βαρβαρώτατοι) used by Xenophon was justified in the Mossynoikoi's customs and manner, and especially their tendency to do in public the things that other people do only in private, notably open intercourse, and has nothing to do with belligerence. Here, we have another adjective in the superlative degree (πολεμικώτατοι), comparing the Drilae with all the other peoples of the southern Black Sea littoral. A reasonable question would be whether Xenophon used this adjective in the notes he kept during the expedition or in the text he composed long after his return. If the former is the case, then the perception of the Drilae as the most warlike of all the peoples of Pontus would reflect their reputation among the people he met then: obviously, the Trapezuntians, who suffered at their hands, for Xenophon and his army had just arrived in Pontus and all the other Pontic tribes were yet to be encountered, so no comparison among them could be made at that time.

If Xenophon used the characterisation when he wrote his book, then we should remember that before leaving Pontus, and much later than the encounters with the Drilae, Xenophon would meet the Bithynians, of whom he wrote that 'they are said to abuse outrageously any Greeks they may find shipwrecked or may capture in any other way' (*Anabasis* 6. 4. 2), like their kindred Thynians in European Thrace, who 'were said to be the most warlike of all men, especially by night' (*Anabasis* 7. 2. 22).[543] Thus, I would ascribe a greater

540 Section 2.6.1.
541 See, for example, Maksimova 1956, 136–37.
542 See Section 2.11.3.
543 See Section 2.2.3.

likelihood to the hypothesis that the characterisation of the Drilae as the most warlike people in Pontus was something that Xenophon heard when he arrived at Trapezous. After all, the historian rarely expresses his personal assessments of peoples just through an adjective, but usually attributes such evaluations to others: in the case of the Mossynoikoi (5. 4. 34) it was the Greek soldiers who called them most barbarous (ἔλεγον οἱ στρατευσάμενοι), while in the case of the Bithynians and the Thynians it was just generally said that they were wild and barbarous (λέγονται and λεγόμενοι εἶναι respectively). In any case, the Drilae went down in history as a very warlike and belligerent people who lived in the mountains, a perfect example of environmental determinism.

13. The Macrones and the Sannoi

13.1. The Territory (Fig. 10.3)

The Macrones are a people mentioned by several authors, but these various accounts are liable to create confusion with regard, first of all, to their precise territory over time, including whether they were (always) a coastal people or not, and secondly, to their possible identification with peoples referred to by different names. To start with the question of location, the first to speak of the Macrones was Hecataeus (*FGrH* 1a 1 F 206), as we learn from Aelius Herodianus and Stephanus of Byzantium (both *s.v.* Μάκρωνες), but he preserves no further information concerning their exact placement. The first particulars are furnished by Herodotus, who counts the Macrones among the peoples that belonged to the 19th satrapy, together with the Moschoi, Tibarenoi, Mossynoikoi and Mares (Herodotus 3. 94). In the Persian Wars they had the same leader as the Mossynoikoi, and were moreover equipped like them, as were the Tibarenoi and the Moschoi (7. 78). Elsewhere, he mentions the Macrones as neighbours of the Syroi, who live by the River Thermodon (2. 104; *cf.* Eustathius *Comm. in Dion. Per.* 772). While these descriptions are not without problems,[544] they clearly point to the eastern stretch of the littoral. A little later, Ctesias allegedly placed the Macrones in the area of Hieron Oros, between Kerasous and Trapezous (Schol. in Apol. Arg. 2. 1015).

The first to write more extensively on the Macrones after having visited them himself was Xenophon. It is noteworthy that in his account there is nothing to indicate that they lived on the coast. Moving north towards the Black Sea through the land of the Scythenoi, Xenophon's army reached a mountain

[544] For example, the reference to the Thermodon as belonging to the Syroi's territory (see more in Sections 2.8.1, 2.9.1, 2.10.1), thus their adjacency to the Macrones.

called Theches, from where they could finally see the sea. Going down this mountain they reached the country of the Macrones, which was separated from that of the Scythenians by a river.[545] After their encounters with the Macrones (for which see below), the Greeks proceeded north to the territory of the Colchians, whose southern border (with the Macrones) was a mountain and the northern the sea (*Anabasis* 4. 7. 19–8. 9, 8. 22 – briefly re-narrated also by Diodorus 14. 29. 4–5). This description, very detailed compared with most of our sources, suggests that the Macrones lived on the northern slopes of a mountain facing the sea, perhaps the mountain that separated them from the Colchians, but did not possess any coastal territory.[546] The only possibility for the Macrones to have been living on the seaboard is if they possessed a small stretch of coast to the east of the Colchians, but on the one hand nothing like that is implied by Xenophon's narrative and on the other, whenever the Macrones are mentioned as living on the coast, they are always placed to the west of the Colchians.[547]

According to Lipka, this contradiction of the sources regarding the location of the Macrones' land is a solid argument for the theory that the Colchians must have expanded their territory to the west between the days of Hecataeus and those of Xenophon.[548] But then we are faced with the problem of explaining what happened to the rest of the nations that most sources, starting with Hecataeus, report as inhabiting the area between Kerasous and Colchis, namely the Mares, Philyres, Becheiroi, Sapeires and Dizeres.[549]

Furthermore, some decades after Xenophon the Macrones would again be presented – by Apollonius (*Argonautica* 2. 393–394, 1231–1242) – as living on the coast, to the east of the Philyres and the west of the Becheires.[550] Similar is the geographical order presented by Valerius Flaccus (*Argonautica* 150–153) and Dionysius Periegetes (*Orbis descriptio* 765–766), with slight differences concerning their neighbours. In the late 2nd century BC, Pseudo-Scymnus (938) most probably placed them between the Mossynoikoi and the Colchians.[551] It is noteworthy that Strabo mentions the Macrones only once in his

[545] *Cf.* Aelius Herodianus, *s.v.* Σκυθηνός; Stephanus of Byzantium, *s.v.* Σκυθηνοί. This river cannot have been a major one, since, as Lendle (1995, 281–82) points out, all the big rivers of the littoral flow naturally northwards, into the Black Sea, while this one must have flowed parallel to that coast, if it separated two countries lying one north of the other.

[546] Lendle (1995, 277) places the area where the Greeks entered the territory of the Macrones around Hamsiköy.

[547] See, for example, above. Also, Lipka 1995, 66–67.

[548] Lipka 1995, 66–67.

[549] See also Section 2.9.3.

[550] See Sections 2.14.1 and 2.

[551] According to Müller (1855, 234), the Macrones were omitted by Pseudo-Scymnus, but in his newer study Diller (1952, 109, 173) concluded that the passage of Anonymous (*Periplus* 37)

work, and in a way that indicates no direct acquaintance with them: he just says that 'above Trapezous and Pharnakia are situated the Tibaranoi, the Chaldaeoi and the Sannoi, who in early times were called Macrones, as well as Lesser Armenia (12. 3. 18).[552] This is the first time since Xenophon that the Macrones are placed in the hinterland. Whether this could imply another movement of that people, or a confusion on the part of the ancient authors based on different sources, or even the existence in a given period of time of both a coastal and an inland population of Macrones, on the model of the Chalybes, remains an open question.

13.2. On the Origins of the Macrones and their Identification with Oher Peoples. The Sannoi

In any case, it seems that the ambiguity in the definition of the exact territory of the Macrones throughout antiquity sparked several efforts to connect them with other equally indistinct peoples of the eastern end of the southern Black Sea littoral. Inevitably, the origins of the Macrones are also veiled in obscurity. Already from Xenophon's era, Herodorus, and about a century later Philostephanus, are reported (Schol. in Apol. Arg. 1. 1024) to have written that the Macrones were a nation of the Becheires.[553] While this is not confirmed by any other testimony, it is worth noting that it is in the Commentaries on Apollonius' *Argonautica* that we learn about these words of Herodorus and Philostephanus, who are also supposed to have said that the Macrones were neighbours of the Doliones, a people known from the Argonautic myth as living on the southern coast of the Propontis, in the region of Cyzicus. The specific commentary refers to *Argonautica* 1. 1024, where Apollonius recounts how the Doliones were disturbed in the night and, seeing the Argonauts approaching their land, deemed that Macrieis were coming, a Pelasgian nation of warriors. It seems quite possible that the Macrones have been confused with the Macrieis, by reason of their similar names. An additional possible cause of this confusion could be the name of the rock around which the Argonauts threw their ship's hawsers upon arriving at the Doliones' land: the *Hiere petra*, the Sacred Rock (Apollonius *Argonautica* 1. 1019); this strongly recalls the *Hieron Oros*, which Ctesias mentioned as being situated in the land of our Macrones (see above), again according to the Scholiast of the *Argonautica* (2.

mentioning them and identifying them with the Macrocephaloi (see below) must derive from Pseudo-Scymnus' work.

552 On his identification of the Macrones with the Sannoi, see Section 2.12.1.

553 See in Section 2.14.2. Epiphanius (*Ancoratus* 113. 3) includes the Macrones in the descendants of Ham.

1015). Another cause for confusion could be the passage in Herodotus' influential work where the leader of the Macrones (and the Mossynoikoi) Artayctes, son of Cherasmis, was governor of Sestos upon the Hellespont (Herodotus 7. 78), not far from the land of the Doliones.[554]

The identification of the Macrones with the Macrieis connected them to Euboea, also called Macris, and indeed they had sometimes been considered as colonists from Euboea (e.g. Dionysius of Chalcis, according to Schol. in Apol. Arg. 1. 1024; *cf.* also 161. 7) – hence the reference to them as Pelasgians (Schol. in Apol. Arg. 1. 1024; Schol. in Xenophon *Anabasis Cyri* 4. 8. 5). The tendency to attribute the existence of peoples and places in the Black Sea to colonisation from homonymous or almost homonymous parts of Greece proper was not unusual, as we can deduce from the similar example of Trapezous, whose citizens, according to Pausanias (8. 27. 5–6), welcomed the inhabitants of the city of Arcadia called Trapezous – quitting it because they were unwilling to join Megalopolis – in their city 'as their kindred, as they bore their name and came from their mother-city'. The commentators of Apollonius occasionally also called the Macrones a Pontic nation (Schol. in Apol. Arg. 98. 1), a Scythian nation (like so many others in the eastern part of the littoral), or Euboean colonists (161. 7).[555]

One further issue arises from Anonymous' identification of the Macrones with the Macrocephaloi (*Periplus* 37), an identification which, however, is not unfounded. His work is based on four *Periploi*,[556] the earliest of which, the one by Pseudo-Scylax, not only makes no mention of the Macrones but places the Macrocephaloi between the Mossynoikoi and the Becheires (85–86; just like Pomponius Mela 1. 107, and Pliny *NH* 6. 11), and so actually in the Macrones' general region, and the temptation to identify the two peoples, with their similar names, must have been irresistible.

Nevertheless, Pliny places the Macrocephaloi in the area of Kerasous and mentions separately the Sannoi (6. 12), to the east of Trapezous, and specifically as *gens Sannorum Heniochorum*.[557] He also mentions the Machorones (6. 11),[558] in a position which seems impossible (to the east of Kordyle and the

[554] Bryer and Winfield (1985, 300) attribute to Herodotus the information that the Macrones were 'proto-Georgian Pontic groups that survived the demise of Urartu', and to Strabo the identification of the Chaldaioi who lived above Trapezous with the 'ancient Macrones', but both attributions are erroneous.

[555] W. Tomaschek (*RE* III.1 (1897), 181, *s.v.* Becheires) seems to consider the Macrones as indigenous peoples who spoke Caucasian dialects.

[556] See in the Introduction Section 2.2.

[557] For this specification offered by Pliny, see below.

[558] The same as the Machelones, according to Herrmann, *RE* XIV.1 (1928), *s.v.* Machelones (see below). For Pliny's specific references, see also Edwards 1988, 129–30.

Becheires, but to the west of Pharnakia, Tripolis and Philokaleia), in a rather problematic throwback in his description of the coast between Amisos and Trapezous.

Having mentioned the Macrocephaloi, it is worth dedicating a few lines to them. What else do we glean from the remaining sources dealing with this people? Was there indeed such a people to the west of Colchis? The first concrete preserved reference to them comes from Hippocrates, who relates (*De aere* 14) that 'their custom was to bind and compress the heads of their newborn infants, while they were still soft and pliable, to make them longer'. A bit later, Pseudo-Scylax's reference to them must have established the image of a real people, consolidating their place among the south Pontic peoples,[559] after their mention by other geographers as well (see above) and in combination with Hesiod's supposed reference to them (Apollodorus fragm. 160b, *apud* Strabo 1. 2. 35; Harpocrates *Lexicon* 197).

But then came Strabo's (1. 2. 35) disillusioning critique, in one of his numerous critical commentaries to Apollodorus, speaking of scholars who confound myth and history and adducing the example of several poets who invented mythical tales (*cf.* 7. 3. 6), among them Hesiod, who spoke of Hemicynai ('men who are half-dog'), concluding that we should keep in mind that we must not take poets' words literally when they compose stories on many subjects in the guise of history, 'for it is self-evident that they are weaving in myths intentionally, not through ignorance of the facts, but through an intentional invention of the impossible, to gratify the taste for the marvellous and the entertaining'. Later (11. 11. 8), presenting some customs of the 'barbarous' peoples in the area of Caucasus, he referred to the Sigginoi, who, among other things, were said to practise making their heads appear as long as possible (μακροκεφαλώτατοι) and their foreheads project beyond their chins.[560] Much harsher, Flavius Philostratus (*Vita Apollonii* 3. 47) presents the philosopher Apollonius of Tyana, who lived only a few years later than Strabo, as having doubted the very existence of men that have long heads, and 'other poetical fancies which the treatise of Scylax recounts about them'.

What might one deduce from all the above? Apparently, that the Macrocephaloi were not a specific people inhabiting a specific part of the Black Sea region. The Greek authors were always intrigued by the strange customs of distant barbarians, especially those concerning interventions to the human

[559] They were purportedly mentioned also by Antiphon and Palaephatus (Harpocratus *Lexicon* 197; Photius *Lexicon*, *s.v.* Μακροκεφάλους; the *Suda*, *s.v.* Μακροκέφαλοι). *Cf.* Aelius Herodianus 3. 1. 159. 19; Stephanus of Byzantium, *s.v.* Μακροκέφαλοι.

[560] On Strabo's criticism, see also Counillon 2004, 41–42.

body. And it seems that there were plentiful ones of that kind, as is the case with some tribes to this day. What better material could there be for both poets, such as Hesiod, and physicians, like Hippocrates, to deal with, inevitably giving them a Greek name that described their custom and appearance but was of course not theirs, as was the case with the Macrones? Nonetheless, neither of the two mentioned a specific place as the home of a specific Macrocephaloi, and there could have been more than one people who practised this custom indeed, Strabo (11. 11. 8) speaks about a Caucasian nation, not calling them Macrocephaloi, but describing their custom using the superlative degree of the qualifying adjective.

Were the Macrones one of those nations? Probably yes,[561] but we would expect Xenophon to have said something relevant in his account of them. And we should not exclude the possibility that the specific custom was practised by more than one nation in the south-eastern Black Sea region, hence its reference in different parts of it. The reasonable question here is why Pseudo-Scylax and later geographers confidently named the people between the Mossynoikoi and the Becheires Macrocephaloi. To impute it to an error, like the several detected in the preserved *Periplus*, attributable for example to copyists or editors,[562] would be the easy solution.

Nonetheless, one cannot fail to notice that Pseudo-Scylax's text is the first known geographic work to replace the Macrones by the Macrocephaloi, and that the latter are never again mentioned there in geographical texts of the pre-Christian period. The fact that Hecataeus (who spoke of Macrones) was more or less a contemporary of Scylax of Caryanda allows us with sufficient certainty to exclude the possibility that the replacement was made by the latter. Besides, the Macrocephaloi do not appear either in Xenophon or in Apollonius,[563] and thus this was probably also the case with Ephorus. Pseudo-Scylax's south-eastern part of the Black Sea is considered to have been based on the one hand on Hecataeus and Milesian tradition, and on the other on sources that predate Xenophon's work.[564] Notwithstanding our obvious ignorance of what the source for the Macrocephaloi might have been, the passage of Hippocrates which seems to place the Macrocephaloi not far from Colchis (but not in

[561] In the Byzantine literature some authors suggested that the Macrones had this name because 'most of them had long heads' (*Etymologicum Magnum*, *s.v.* Μακρῶι; Schol. in Apol. Arg. 1. 1024).

[562] See, for example, Counillon 2004, 27; Shipley 2011, 2–3.

[563] So, Apollonius does not seem to have relied directly on Pseudo-Scylax, at least for this particular area, as Counillon (2004, 28) generally assumed.

[564] Counillon 2004, 44–46. *Cf.* Shipley 2011, 157.

a specific area) could, if nothing else, provide us with us a *terminus ante quem* for this unknown source.

In the taxing effort to explain connections and confusions between the Macrones, Sannoi, Macrocephaloi, Machelones and Machorones, it is worth dwelling a little more on Pliny's puzzling references, especially the one to the Sannoi (6. 12), as *gens Sannorum Heniochorum*, living between Trapezous and the River Apsaros in the Colchis, an area that coincides with the territory of the Sannoi given by Arrian. It is the only time we read about the Sannoi as belonging to the Heniochoi, who, however, lived on the eastern Black Sea littoral, and this might throw further obscurity over the origins of all the peoples (that ended up) living in the eastern part of the southern littoral. More than a century ago, E. Kiessling, writing on the Heniochoi,[565] sketched out the movements of those peoples of the eastern coast that had not been subjugated by the Colchians around 400 BC,[566] among them the Heniochoi, to north Anatolia, with the help of geographical and historical sources of the Roman period. He placed the beginning of these movements in the 1st century BC and estimated its duration as more or less a century. Among these movements, those of the tribes of the Heniochoi were considered by Kiessling the most important. Among these tribes he included the Sannoi, following Pliny (6. 12).[567] Similar was the fate of the Machelones (Cassius Dio 68. 19. 2; Lucian *Toxaris* 44–55, as Machlyes),[568] whom Arrian (*Periplus* 11) placed between the Sannoi and the Heniochoi. After their establishment in the area, at the end of the 1st millennium BC, both the Machelones and the Heniochoi were ruled by king Anchialus. His palace was near the Prytanis river (Arrian *Periplus* 7, 11; *cf.* Anonymous *Periplus* 42, 8B).

In Strabo, however, who died when Pliny was just entering the world, the Heniochoi still live only on the Caucasian coast (11. 2. 1, 5. 6), and nothing is said about them or any part of their population living on the south coast or about any connection with the Sannoi. According to Kiessling,[569] this can be explained by the fact that Strabo lived only shortly after the movement of the Sannoi from the east to the south and rather ignored them, considering them as being identified with the Macrones.

[565] Kiessling 1912. On the Heniochoi, see more recently Asheri 1998.

[566] We might better say 'after Pseudo-Scylax's work', since the Heniochoi are mentioned there as still autonomous on the eastern coast (76).

[567] Where the initial homeland of these Sannoi might have been is indicated by Pliny again (6. 14), who mentions the *gens Sanicarum* between Phasis and Gyenos, not far from the Heniochoi who remained there.

[568] Who for Counillon (2004, 97) are undoubtedly the same as the Macrocephaloi.

[569] Kiessling 1912, 268.

If the Sannoi did indeed come from the Caucasian coast, then their identification with the Macrones (by Strabo), as with the Drilae (by Arrian), seems faulty. There is nothing to indicate that the Macrones originated from the eastern Black Sea littoral; there is only a very late reference to them as a Scythian nation (Schol. in Apol. Arg. 161. 7), which connects them with the northern part. Neither is there any such indication for the Drilae – according to the only author who mentions them (Xenophon); in this case, it would be Pliny who is to blame for the confusion. It is nonetheless noteworthy that in their vicinity Strabo mentions the Kerkitae or Appaetae,[570] who also seem to have migrated from the Caucasian area, where they were known as Kerketae or Toretae (Pseudo-Scylax 73), probably in the same period as the Heniochoi Sannoi.[571]

Consequently, it is not improbable that, as Kiessling implied, the key to the issue is the time of the southwards migrations of the 'Caucasian' peoples,[572] compared with the time of our geographic sources. It is more or less the time of Strabo. The geographer from Amaseia knew of the existence of the Sannoi;[573] indeed, he is the first known author to mention them, but not in the way he regularly uses to describe people and places familiar to him. His identification of them with the Macrones is not confident, as is the case with Arrian's identification of them with the Drilae.[574] Pliny, however, is aware of the movements, as is Arrian.

On the other hand, some 500 years had passed since Xenophon's unique account of the Drilae by the time Arrian connected them with the Sannoi, and this connection is nothing more than pure hypothesis, based on their warlike and riotous nature. It seems that a decisive element in the connection between these peoples – Macrones, Drilae, Sannoi, Macrocephaloi, Heniochoi – perceived by the Greeks was a similar idiosyncrasy: the Drilae were warlike and very hostile to the Trapezuntians (Xenophon *Anabasis* 5. 2), and so were the Sannoi (Arrian *Periplus* 11). The Macrones were autonomous people, free from any rulers (Xenophon *Anabasis* 7. 8. 25); similarly, the Sannoi did not live under a monarchical government (Arrian *Periplus* 11). The Macrocephaloi were also of unruly disposition (Mela 1. 107), while Pseudo-Scymnus, probably

[570] See Section 2.16.3.

[571] *Cf.* also Kiessling 1912, 265.

[572] For the features common to Anatolia and the Caucasus area concerning cultural, ethnic and linguistic elements, see Kavtaradze 2002.

[573] Later authors referred to the Sannoi as Tzannoi (Eustathius *Comm. in Hom. Od.* 2. 73; *Comm. in Dion. Per.* 765; Procopius *De bellis* 8. 1. 8–9, 2. 30), for which see also Kavtaradze 2002, 71–73, with earlier bibliography.

[574] See Section 2.12.1.

following Ephorus,[575] called the Heniochoi a people hostile to strangers (*cf.* Anonymous *Periplus* 9B).

Whether this is an adequate reason to strongly associate all of them with each other is extremely debatable. The Drilae might or might not have been related to the Macrones,[576] but by the time the Sannoi migrated from the east they had probably disappeared. The Sannoi appeared on the scene far too late to be identified with the Macrones, who were a very ancient nation on the southern Black Sea littoral and hinterland; neither is there any concrete reason to suppose that they were gradually assimilated by them. The origin of the Macrones remains in the dark. As for the Macrocephaloi, their existence as a specific nation and indeed identified with the Macrones is better credited to one of those charming influences of legend and poetry over some geographic works.

13.3. Some Social and Anthropological Notes

Judging from Xenophon's account, hostility was not the most representative characteristic of the Macrones. Upon arriving in their country the Greeks found a dense forest with such thick undergrowth that they could not pass through it. At the other end of this forest were the Macrones, who took the Ten Thousand for an invading army and drew up their battle lines, armed with wicker shields and lances and hair tunics (*Anabasis* 4. 8. 3). This description corroborates Herodotus' account of about a century earlier, concerning the Persian Wars, according to which the Macrones' equipment was similar to that of the Moschoi: 'they wore helmets made of wood, and carried shields and spears of a small size: their spear-heads, however, were long' (Herodotus 7. 78). What draws the attention in Xenophon's account, however, is their astonishingly low level of military organisation, for he presents them, in their rather primitive armour, 'cheering one another on and throwing stones, which fell into the stream; for they never reached the Greeks or did them any harm' (*Anabasis* 4. 8. 3).[577]

With the help of a peltast who knew their language, Xenophon explained to the Macrones that he merely wished to pass through their country. The Macrones accepted this, and not only helped the Greeks clear a passage for their baggage animals but also supplied them with food, helped them make their way easily through the country, and opened their markets to them

[575] Asheri 1998, 271.
[576] See Section 2.12.1.
[577] Lendle 1995, 282.

(*Anabasis* 4. 8. 4–9. *Cf.* also Diodorus 14. 29. 4–5). This certainly does not indicate a people hostile to strangers. Their initial reaction to a foreign army approaching their land was anything but unjustified; indeed, their quick change of mood and especially the opening of their market to the Greeks was something rather unexpected, which the latter would remember and emphasise to the Sinopeans in Cotyora, who, although Greeks, were very reserved towards them (5. 5. 18).

Of special interest is the incident with the soldier who knew the language of the Macrones: Xenophon recounts (4. 8. 4–5) that the man himself said that he had been a slave at Athens, and that he recognised the language the men were speaking. He said to Xenophon, 'I think that this is my native country, and if you have no objection, I would like to speak to them.' Xenophon assenting, the peltast asked them who they were, receiving the answer 'Macrones'. The first thing we obtain from this narrative is that the Macrones may have belonged to those Black Sea peoples that supplied classical Athens with slaves.[578] This particular slave must have come to Athens at a very young age, since he was not totally sure that he recognised his mother tongue.[579] And since it is not his name that could lead us to infer, probably erroneously,[580] his origin, but his language, we can be more confident about that origin. The second intriguing element is that we have a very rare (indeed, unique for the southern Black Sea) case of indigenous people who themselves tell the Greeks what their name is. Although a seemingly Greek word, Macrones is clearly the name stated by the people themselves in Xenophon's account. This further weakens the possibility that the Macrones had any connection with Euboea,[581] since in such a case the two peoples would not have appeared so strange to each other.[582]

Finally, Herodotus included the Macrones among the peoples who practised circumcision (Herodotus 2. 104; *cf.* also Josephus *Against Apion* 1. 170), clarifying that they learned this custom from the Colchians, one of the first

[578] *Cf.* Braund and Tsetskhladze 1989, 120 and 122, n. 46 for the possibility that the vase-painter of early 5th-century BC Athens, Makron, was a Macronian; Tsetskhladze 2008a, 310. For slaves from the southern Black Sea, see also Avram 2007, 245–46.

[579] Maksimova (1956, 137) assumed that he was captured during a war, most probably between the Macrones and the Colchians, and sold to Greek colonists, who sold him to Athens. For a wonderful and sentimental approach to the Macronian peltast's possible history, and his lost identity, see Ma 2004, 330–32.

[580] Since the ethnic names (Scythes, Brygos, etc.) might indicate the places where the slaves were sold, and not necessarily their native places. See, indicatively, the bibliography given above, in n. 578.

[581] See above in Section 2.13.2.

[582] Kavtaradze (2002, 74) saw in the name of the Macrones a Kartvelian origin.

nations to circumcise, as did also the Leucosyroi, and that this happened 'lately', so we would guess in the early 5th century BC.[583]

14. The Peoples of the Eastern Part

A number of peoples have been named as living in the eastern part of the southern Black Sea littoral (Fig. 10.3), between the Mossynoikoi (or the Macrones) and the Colchoi, though not all of them by the same authors. The first to mention almost the whole list was Apollonius, in his *Argonautica* (2. 393–396, 1242–1244). According to his narrative, after the land of the Mossynoikoi and the Island of Ares, the Argonauts reached successively the lands of the Philyres, the Macrones, the Becheires, the Sapeires and the Byzeres, before arriving finally at Colchis. Before him, only the Becheires and the Byzeres are mentioned, by Hecataeus and Pseudo-Scylax, and the Saspeires by Herodotus (see below), while he is the first known writer to mention the Philyres, whose name was afterwards repeated by authors influenced by or based on him. We have already spoken of the Macrones, and with the exception of the enigmatic Philyres all the other peoples examined here are reported to have lived to the east of them. Here, we shall focus on each of these peoples separately.

14.1. The Philyres

It is in Apollonius' *Argonautica* that we first hear of a people called Philyres on the southern Black Sea coast, between the Mossynoikoi and the Macrones (*Argonautica* 2. 393). Apart from their name, nothing more is said of them, not even a single characteristic expressed in an adjective, as is the case with the 'arrogant' Sapeires (2. 1243). All the important sources prior to Apollonius are totally silent: there is not a word of them in the works of Hecataeus, Herodotus, Xenophon or Pseudo-Scylax. Nonetheless, the Philyres managed to persuade some later authors of their existence as a southern Black Sea nation: these were Dionysius Periegetes (*Orbis descriptio* 767; *cf.* his *Paraphrases*; Schol. in Apol. Arg. 161), the author of the *Orphica Argonautica* (758), Ammianus Marcellinus (22. 8. 21) and the scholars Aelius Herodianus and Stephanus of Byzantium,[584] all of whom treat them as tersely as Apollonius. With Valerius Flaccus (*Argonautica* 5. 152), as we shall see, the case is somewhat different.

[583] See also in Sections 2.8 and 2.15.2.

[584] Aelius Herodianus *Peri pathon* and *De prosodia catholica*, *s.v.* Φίλυρες; Stephanus of Byzantium, *s.v.* Φίλυρες.

It should not escape our attention that all these later authors who mention the Philyres were influenced by Apollonius and borrowed material from him. Both the *Argonautica* of Valerius Flaccus and the *Orphica Argonautica* are clearly based on the Hellenistic epos, while Dionysius Periegetes is also considered to have gathered material from very different authorities of several periods, among them Apollonius.[585] A similar explanation must lie behind the specific geographical data furnished by Ammianus Marcellinus[586] for peoples of which he declares himself to be ignorant (22. 8. 21). One might, indeed, assume that the Philyres were actually an invention of Apollonius.

This being the case, his presentation of the Philyres and their land merits a thorough examination. In fact, apart from the bald reference to them, there are some further elements concerning the Philyres in the Hellenistic epic that might attract our attention. First of all, there is the delineation of their territory, which is not consistent in the two passages recounting it. In the first, where Phineus describes to the Argonauts the route they should follow from his land to reach Colchis, he says that the Philyres dwell beyond the Island (of Ares) and the opposite mainland (Apollonius *Argonautica* 2. 392–393).[587] In the second passage, however, where we read the more detailed account of their actual voyage, things are a bit different: the Philyres are not mentioned, and the first nation encountered after the Mossynoikoi are the Macrones (*Argonautica* 2. 1117–1142). In their place, however, another island appears: Philyreis (2. 1231), an island that is never mentioned again in the entire preserved body of ancient literature, a new island which also changes the geography of our area.

In contrast to the brevity that he has in store for the Philyres, Apollonius dedicates several verses to the island he calls Philyreis: the Argonauts had left the Island of Ares opposite the land of the Mossynoikoi (on which they had experienced an adventure narrated over 231 whole lines – *Argonautica* 2. 1030–1230), at dawn, and at nightfall they came to the island of Philyreis. On this island, 'Cronus, son of Uranus, at the time he reigned over the Titans on Mt Olympus, and Zeus was yet being nurtured in a Cretan cave by the Curetes of Ida, lay beside Philyra, having deceived Rhea; and the goddess found them in the midst of their dalliance; and Cronus leapt up from the couch with a rush in the form of a steed with flowing mane, but Ocean's daughter, Philyra, in shame left the spot and those haunts, and came to the long Pelasgian ridges, where by her union with the transfigured deity she brought forth huge Chiron,

585 Müller 1861, xxv, 152; Hoefer 1904, 559; Tozer 1971, 286.
586 *Cf.* Hoefer 1904, 559. See also below, for the Sapeires.
587 That it is the Island of Ares is made clear in the previous lines (2. 385–387).

half like a horse, half like a god' (2. 1231–1241).[588] Thence they sailed on to the land of the Macrones.

Apparently, while Apollonius was silent concerning the Philyres, he spent ten lines explaining the name of the island, but without even mentioning the people who were supposed to live there, as he did with the neighbouring Mossynoikoi and Macrones. Contrary to the first passage, reporting Phineus' words, here the Island of Ares clearly belongs to the territory of the Mossynoikoi, as we know also from Pseudo-Scylax (86),[589] and between them and the Macrones no other people intervenes. The island of Philyreis appears in the narrative instead of the Philyres, as if the poet wanted to 'correct' the sequence presented in the first passage. The myth of Philyra used by him was well known from the time of Pherecydes and Pindar at the latest (Schol. in Apol. Arg. 2. 1235; Pindar *Pythian* 4. 103; 6. 22; *cf.* Pseudo-Apollodorus 1. 9. 1–2), if not already by Hesiod (*Theogony* 1001–1002), but no source has ever said that Philyra's intercourse with Cronus, which resulted in the birth of Chiron, took place on this southern Black Sea island.[590] On the contrary, Apollonius' teacher, Callimachus (*In Delum* 118) cites Mt Pelion as the bridal chamber of Philyra (*cf.* Schol. in Callim. Hymnos 4. 118). It is only Aelius Herodianus (*s.v.* Φίλυρες) and Stephanus of Byzantium (*s.v.* Φίλυρες) who mention a nation in Pontus called Philyres which took its name from Philyra, who, however, is described as the mother of Cronus, not Chiron![591]

One question remains: were the Philyres of the southern Black Sea coast indeed a creation of Apollonius? This could explain Valerius Flaccus' reluctance (*Argonautica* 5. 152) to refer to Philyres, only to 'shores named after Philyra'.[592] In modern scholarship there is a tendency to attribute many unexpected elements in the works of some authors, especially poets, to earlier sources, and to search for these sources. This, however, leads to the entirely reasonable question of what might have been the source of those sources. To avoid possible causes of refutation, the works proposed as initial sources are usually ones that are not preserved but for which there is some evidence that they had influenced later scholars. In our case, U. Hoefer, in a thorough study on the appearance of the Pontic nations, especially of the eastern part of the southern coast, in the ancient literary sources, declared himself to be confident

[588] Translation by R.C. Seaton (Loeb ed.) (London 1912), amended slightly.

[589] See in Section 2.11.1.

[590] Apart from Valerius Flaccus (*Argonautica* 5. 153), also the commentators of Apollonius, of course (Schol. in Apol. Arg. *varia* 392–394a, 1231–1241a).

[591] A place called Φίλυρα is mentioned in Pseudo-Zonaras' *Lexicon*.

[592] So they were not mentioned by him as a people, as is usually said: e.g. Hoefer 1904, 559; W. Ruge, *RE* XX.1 (1941), 214 (*s.v.* Philyres).

that the Philyres were not an invention of Apollonius. Since Pseudo-Scylax, widely used by the poet, did not mention the Philyres, nor seemingly did Ephorus, one of his chief sources, Hoefer concluded that Apollonius must have borrowed the Philyres from Nymphodorus, who had written a *Periplus of Asia* and is presented by the commentators of Apollonius (Sch. Apol. Arg. 2. 1010, 1029; 3. 202–209a) as having been, together with Ephorus, one of the sources used by the epic poet.[593]

I cannot say that I am convinced by this theory. First of all, the commentators of Apollonius do not present Nymphodorus as a source used by Apollonius for the specific passage;[594] they just attribute to him (as well as to Ephorus) some of their knowledge when commenting on Apollonius' work. Besides, how would Nymphodorus know of the Philyres on the southern Black Sea coast if neither Ephorus nor Pseudo-Scylax nor any other before him was aware of them?[595] All the more so if Nymphodorus was a contemporary of Apollonius, and indeed his *Periplus of Asia* is considered to have been written at the end of the 3rd century BC, so probably even after the *Argonuatica*.[596]

Apollonius, a great poet and mythographer, might well have inserted the Philyres and their island (after all, would a fictional island first appear in a *periplus*?) in the southern Black Sea to present a variation of the myth narrated by his teacher, and by then his rival, Callimachus. If he did indeed use an earlier source, then I would suggest that this should be sought among the lost poetic works rather than the lost historical or geographical ones.

14.2. The Becheires

Unlike the Philyres, the Becheires seem to have been a people well known from early on, but yet in their case, too, very little information regarding them is furnished by the ancient literature. Hecataeus (*FGrH* 1a 1 F 207) mentions the nation of the Becheires, as well as their country, Becheirike, as neighbouring the Choi.[597] Pseudo-Scylax (84) places the Becheirike to the east of the

[593] Hoefer 1904, 556–64.

[594] This could be said only for the passage concerning the Colchians (Schol. in Apol. Arg. 3. 202–209a). *Cf.* Morrison 2020, 16, n. 82.

[595] To think that the Philyres as a nation appeared in the time between Ephorus and Nymphodorus, a time not known for the appearance of new peoples in the specific area, would be more improbable than to attribute the Philyres' existence to Apollonius.

[596] It is unclear whether it was the same Nymphodorus who wrote the 'Customs of the Barbarians' and the 'Periplus of Asia', as Hoefer thought. The uncertainty among scholars persists. *Cf.* F. Laqueur, *s.v.* Nymphodoros 1 and 2 in *RE* XVII.2 (1937), 1623–27; H.A. Gärtner, *s.v.* Nymphodorus 1 and 2 in *Brill's Neue Pauly*, 2006.

[597] See Section 2.16.2.

Macrocephaloi (probably meaning the Macrones)[598] and Trapezous, and to the west of the Ececheirieis (see below). The Becheires seem to have been a significant nation, or, better, a native group. There are several factors indicating this: first of all, they are the only people of the eastern part of the littoral who appear to have given their name not only to their territory but also to a harbour and a city, which was actually a *polis Hellenis* (Pseudo-Scylax 84).[599] Furthermore, several authors present the Becheires as being a group of nations, including – according to Herodorus and Philostephanus (Schol Apol. Arg. 2. 1024) – the Macrones.

While this latter detail is not corroborated by other sources, Apollonius affirms the existence of 'immense tribes of the Becheires' (*Argonautica* 2. 394), inhabiting a 'far-stretching land' (2. 1242) between the Macrones and the Sapeires. Although there is nothing to indicate that the territory of the Becheires was indeed large, they frequently appeared as a native group (Dionysius Periegetes 765 and its *Paraphrases*; *Orphica Argonautica* 741; Schol. in Apol. Arg. 161. 17), and this might have played its role in creating the image of a large area. In fact, not even the precise location of this area is determined, since the very neighbours of the Becheires vary from source to source (see above): for example, Dionysius Periegetes (765) places them between the Macrones and the Byzeres, skipping both Hecataeus' Choi and Apollonius' Sapeires.[600]

Another issue associated with the Becheires is their very name, since the first preserved text to mention them in the nominative and accusative cases, the one by Pseudo-Scylax, first mentions them as Becheiroi (84), but immediately after as Becheiras, in the accusative (85), which corresponds to the nominative Becheires. All the other sources until the Roman period mention them in the genitive form (*ἔθνος Βεχείρων*), which could fit both types. However, many grammarians from the 2nd century until the mediaeval period quote the nominative singular Becheir,[601] which clearly results in the plural form Becheires.

[598] See Section 2.13.2. Mela (1. 017) and Pliny (*NH* 6. 11) place them between the Macrocephaloi and the Buxeri (Byzeres).

[599] For these settlements, see Section 5.1

[600] In *Tabula Peutingeriana* the Becheires are placed somewhere between Polemonion and Kerasous, so more to the west than they appear in the rest of the sources. It is noteworthy, though, that they are depicted on a curve of the coast to the north and then west, while some commentators mention the Macrones as living to the south of the Becheires (Eustathius *Comm. in Dion. Per.* 765), or at the head of Pontus, near the Colchians (Schol. in Apol. Arg. 211. 10d). All this reminds us of the 'deep gulf of Pontus' mentioned by Apollonius (2. 1246) between the peoples examined in this section and Colchis, maybe also as a kind of a significant turning point, in geography as well as (especially) in the plot of the myth.

[601] Aelius Herodianus, *s.v.* Βέχειρ; *Peri orthographias* 3. 2. 431; *Peri kliseos onomaton* 3. 2. 748; Arcadius *De accentibus* 21; Stephanus of Byzantium, *s.v.* Βέχειρ; Theognostus *Canones*

Finally, there is the question of their origin. We have already seen that a significant majority of the peoples inhabiting the eastern part of the southern Black Sea littoral have been considered as being of Scythian origin by many scholars from the Roman period on.[602] The Becheires belong to these peoples. When their origin is mentioned, they are always presented as Scythians,[603] Eustathius (*Comm. in Dion. Per. 762*) giving a slight twist by calling them 'a Pontic and Scythian nation'. The first adjective might well carry a geographical meaning. An unexpected piece of anthropological information regarding the Becheires comes to us from Zenobius (*Epitome* 5. 25), according to whom they threw to the dogs anyone who was gravely ill. It is true that this custom is reminiscent of several Scythian ones,[604] but this is not sufficient evidence to label them a Scythian people. W. Tomaschek has counted them, based on their name, among the indigenous peoples who spoke Caucasian dialects, like the Macrones.[605]

14.3. The Ececheirieis

Pseudo-Scylax (83) mentions a people called Ececheirieis, between (from west to east) the Becheires and the Bouseres, within whose borders were two rivers (the Pordanis and the Arabis),[606] and two cities, Limne and Odeinios, the latter being a *polis Hellenis*. It is highly noteworthy that, although the Ececheirieis are for the rest unknown to ancient literature, they seem to have occupied a large area. Actually, Anonymous (*Periplus* 42), the only other geographer who seems to have known of them, obviously based on Pseudo-Scylax,[607] wrote that the *Ececheirieis*, as they appear in his work, used to inhabit the area between the rivers Archabis and Ophius, where in his time dwelt the Machelones and the Heniochoi, probably after their migration from the Caucasian region.[608] The area between the two rivers is a huge one, covering the most of

225; Georgius Choeroboscus *Prolegomena et scholia in Theod. Alexandr. can.* 321; *Epimerismoi in psalmos* 4. 3 (as Βέκχειρ); *Etymologicum Genuinum*, *s.v.* Βέχειρ; *Etymologicum Magnum*, *s.v.* Βέχειρ; *Etymologicum Symeonis*, *s.v.* Βέχειρ; Pseudo-Zonaras *Lexicon*, *s.v.* Βέχειρ.

602 See, for example, in Sections 2.9.4 and 2.11.4.

603 Aelius Herodianus *De prosodia catholica*, *s.v.* Βέχειρ; Stephanus of Byzantium, *s.v.* Βέχειρ; Georg. Choeroboscus *Epimerismoi in psalmos* 4. 3; *Etymologicum Magnum*, *s.v.* Βέχειρ; *Etymologicum Symeonis*, *s.v.* Βέχειρ; Pseudo-Zonaras *Lexicon*, *s.v.* Βέχειρ; Schol. in Apol. Arg. 161. 20d.

604 See Sections 2.9.4 and 2.11.4.

605 *RE* III.1 (1897), 181, *s.v.* Becheires.

606 Probably Prytanis and Archabis. See more below, in Section 2.16.5.

607 *Cf.* Diller 1952, 107–09. The Ececheirieis have been also mentioned by the 17th-century scholar Leo Allatius (*Hellas* 425).

608 See Section 2.13.2. *Cf.* Edwards 1988, 130.

the southern Black Sea coast to the east of Trapezous, which raises the logical question of why this nation is otherwise absent from the ancient texts. Could they have been one of those tribes of the Becheires mentioned above (actually, their names do not differ much),[609] who, specifically in Pseudo-Scylax's time, occupied such a big area that they would be mentioned separately? Do we have here another indication of the obscurity surrounding this specific part of the littoral, where peoples and places were sometimes named after anthropological characteristics,[610] but very few of them can actually be placed on the map with any precision?[611] Probably yes, but we should also note that Apollonius, who seems to present a geographic order of all these peoples that for the most part coincides with that given by Pseudo-Scylax,[612] omits the latter's Ececheirieis and replaces them (between the Becheires and the Byzeres) with the Sapeires (*Argonautica* 2. 395, 1243), who are absent from his text but were known in his time (4th century BC), since they were mentioned by Herodotus (see below).

14.4. The Sa(s)peires

The Saspeires were known to the Greeks from the time of Herodotus at the latest. The historian from Halicarnassus included them in the 18th Persian satrapy, together with the Matienoi and the Alarodians (3. 94), and added that, like these last, they were armed like the Colchians, namely with wooden helmets and small shields of raw oxhide, short spears and swords, and their commander was Masistius (7. 79). Elsewhere, he mentions the Saspeires as the only nation between the land of the Colchians and Medea (1. 104; 4. 37), and their territory as being 'very high and mountainous and covered with woods' (1. 110). That the Saspeires must have been a significant nation, or at least a populous one, might be deduced from the fact that Herodotus refers to them as one of the four peoples who, together with the Persians, Medes and Colchians, fill the whole space between the Red and the Black Seas (4. 37; 4. 40). Nothing in his words, however, indicates that the Saspeires bordered the Black Sea; on the contrary, they appear to be placed inland and closer to the Caspian (4. 37), while Strabo talks about the Syspititis land in Armenia (11. 4. 8, 14. 9, 14. 12). This image would be unassailable, were it not for Apollonius, who

[609] Counillon (2004, 98), referring to the Becheires, the Ececheirieis and the Bouseres, speaks of an 'ethnographic ensemble'.

[610] The Ececheirieis, for example, remind us of the word ἐκεχειρία (armistice). Counillon 2004, 98, n. 311; Shipley 2011, 157.

[611] *Cf.* Counillon 2004, 98.

[612] *Cf.* Hoefer 1904, 558.

clearly places them on the southern Black Sea coast, between the Becheires and the Byzeres (*Argonautica* 2. 395, 1243), followed, as usual, by the *Orphica Argonautica* (755), and Ammianus Marcellinus (22. 8. 21),[613] but not Valerius Flaccus.

None of the other authors who mention the Sapeires or Saspeires in Late Antiquity elaborate on their geographic location,[614] apart from Aelius Herodianus and Stephanus of Byzantium,[615] who mention them vaguely as a nation in the Pontic hinterland. Some later scholars also report that the Sapeires took their name from the sapphires that were supposed to be a product of their land (*Etymologicum Magnum*, *s.v.* Σάπειρες; *Etymologicum Symeonis*, *s.v.* Βέχειρ; Schol. in Apol. Arg. 161). The accuracy of this information is obviously debatable, as is the view that the Sapeires were a Scythian nation (Schol. in Apol. Arg. 161), which might have been created (only by the commentators of Apollonius) because of the relevant attribution by some late authors to the peoples that neighboured the Sapeires in the *Argonautica*.

The only help to be obtained comes from Herodotus. Combined with his description of the course of the Halys in 1. 72, his words indicate that the Matienoi did not live on the coast, but inland, to the south of the Leucosyroi, who bordered the Paphlagonians to their west, thus to the south-east of the latter, the Halys flowing between them.[616] And since the Alarodians, the third nation of the satrapy, are probably to be connected with the Urartians,[617] also far from the Black Sea, then the Saspeires too might be considered an inland people. Even if the Turkish town of Ispir, more than 70 km from the sea and high in the mountains (about 1200 m altitude), should be associated with the Saspeires,[618] which is possible but far from being proven, there is still nothing to indicate that their territory extended to the Black Sea. Nothing, except for Apollonius' account. Apparently, there are two choices open to us: either to agree that the Saspeires did indeed expand towards the sea sometime in the late 4th or the 3rd century BC,[619] or to suggest that Apollonius' placement of them did not reflect what we call historical reality. Taking into account the case of the Philyres, examined above, in combination with the Saspeires'

[613] Also Schol. in Apol. Arg. 161, 211. Compare the case of the Philyres, above.

[614] Including the passage 7. 8. 25 of the *Anabasis* (where they are mentioned as Hesperitae), which, however, is probably a later interpolation. For Kavtaradze (2002, 81), the Hesperitae were the western Armenians.

[615] *S.v.* Σάπειρ and Σάπειρες respectively.

[616] See Section 2.6.1.

[617] Barnett 1982, 365.

[618] Rayfield 2012, 18.

[619] *Cf.* Herrmann, *RE* 02, II A.1 (1921), 57 (*s.v.* Saspeires).

absence from the coast in all geographical works, I would have no hesitation in favouring the second.[620]

14.5. The Byzeres

The easternmost people in this group on the eastern seaboard to be encountered before the Colchians were the Byzeres. Their particularity does not pertain to the question of their existence, but to their name, which appears in many different forms in the ancient literature. The first to have mentioned them under this name was Pseudo-Scylax (82),[621] followed by Apollonius (*Argonautica* 2. 396, 1244; *cf.* the *Scholia*), Strabo (12. 3. 18), Valerius Flaccus (5. 153), Dionysius (*Orbis descriptio* 765; *cf. Paraphrases*), the *Orphica Argonautica* (756) and Eustathius (*Comm. in Dion. Per.* 762). Before all of these, however, Hecataeus called them Dizeres (*FGrH* 1a 1 F 207). In Mela (1. 107) and Pliny (*NH* 6. 11) we encounter them as Buxeri, while Ammianus Marcellinus calls them Buxari (22. 8. 21).

Almost everyone agreed that to the east of the Byzeres lived the Colchians; the exceptions are the *Orphica Argonautica* (756), where to their east are said to live the 'inhospitable tribes' of the obscure Sigynnae,[622] and Pliny, who includes them in his messy passage describing the coast between Amisos and Trapezous.[623] This general unanimity is not the case when it comes to their western neighbours, as might have become obvious from the previous subsections: as such we occasionally meet the Choi (Hecataeus), the Ececheirieis (Pseudo-Scylax), the Sapeires (Apollonius and Ammianus Marcellinus), the Macrones (Valerius Flaccus) and the Becheires (Mela, Dionysius and Eustathius).

Proximity to the Colchians is the best element we have for imagining where, more or less, the Byzeres might have lived, since very few place-names are cited for their territory. These few are the rivers Daraanon and Arion, mentioned by Pseudo-Scylax (82), and Byzerikos limen, mentioned only by Stephanus of Byzantium (*s.v.* Βύζηρες; *cf.* Eustathius *Comm. in Dion. Per.* 762) but with no indication of its location,[624] no more than for the people

[620] Apollonius (*Argonautica* 1243) allowed them only a single word of description, namely that they were 'arrogant', without explanation or implication, so perhaps for purely poetic reasons (to create an impression or even just for metric requirements).

[621] Bouseres in some codices. *Cf.* Leo Allatius (*Hellas* 424).

[622] Who have been variously located in Scythia, near the Istros (Herodotus 5. 9; Apollonius *Argonautica* 4. 320), or near the Caucasus and having Persian customs (Strabo 11. 11. 8).

[623] See Section 2.13.2.

[624] So probably an invention of Stephanus, equivalent to Becheirikos Limen (Pseudo-Scylax 84 – see above)?

themselves who are just called a nation in Pontus. Equally obscure is the location of the two rivers; the Daraanon has sometimes been identified with the Archabis,[625] but this is not certain.[626]

Let me return, however, to the names of this people.[627] Among those that appear in the sources, it is Hecataeus' version – Dizeres – that differs the most from the others; in fact, he seems to have been the only one to use this particular name. This would be of little importance were it not for a passage of Lycophron, a couple of centuries later. The poet (*Alexander* 1021–1026), speaking of the city of Polae established by the Colchians who pursued Jason and Medea in Illyria – an event narrated also by Callimachus, as Strabo informs us (1. 2. 39; 5. 1. 9) – mentions that the city was established 'by the deep stream of Dizerus'.[628] One cannot easily overlook the striking similarity of the names, especially since the Dizeres were supposed to be neighbours of the Colchians. It seems very likely that we are dealing here with one of those cases (and they are not few) in which names of places, persons or peoples that appear in a mythical tradition are encountered in different periods of time in various parts of the world that are mentioned in those traditions, for example in supposed colonies of these peoples.

[625] Tomaschek, *RE* II 1 (1895), 431 (*s.v.* Archabis); *RE* III 1 (1897), 1159 (*s.v.* Byzeres); *RE* IV 2 (1901), 2150 (*s.v.* Daraanon), proposing that the name in Pseudo-Scylax should be Archauon.

[626] There is clearly a problem concerning the rivers of the easternmost part of the littoral. Pseudo-Scylax mentions river names that we don't meet later and ignores names known to us from the later, Roman sources. Tomaschek's identification of the Daraanon with the Archabis is problematic, since Pseudo-Scylax mentions separately the Arabis (Section 2.16.3), which sounds almost the same as Archabis. Similarly, his Pordanis (Section 2.16.3) could be the Prytanis, as Bryer and Winfield (1985, 336) suggested. But they identify him with the Büyük Dere, which is far to the west, towards Rize. In the *Barrington Atlas* (1230) the Arion is identified with the Abu, thus placed to the west of the Archabis (the Daraanon is just ignored), while Counillon (2004, 92, fig. 1) suggests the opposite, based on Pseudo-Scylax's order. The truth is that the latter seems to always move from east to west, but this should not necessarily be taken for granted for all the places, especially if the Arabis and the Pordanis are indeed the Archabis and the Prytanis respectively; thus, Counillon's placements are speculative. Given the difference in river names between Pseudo-Scylax and the Roman sources, it might be that the name of some rivers simply changed. Given also the general geographic order of peoples and territories, which is indeed followed by Pseudo-Scylax, then the Arion and the Daraanon must be more to the east than the Archabis (modern Arhavi), as Counillon suggests. In this case, they could be identified with the Çamlı and the Karaosmaniye Creek, to the west and east of the Hopa stream, which is rather the Kissa, if one of them is not the (later renamed) Kissa himself. See also Section 1.3, Table 1.

[627] Only once do we meet a comment on the Byzeres' origins, and this is in the Scholia in Apollonius (161. 21), where they are mentioned as a Scythian nation, like most of their western neighbours.

[628] *Cf.* also Schol. in Lycophr. 1021. 20, 1026. See Šašel Kos 2004, 500–01, with proposals for the identification of the Dizerus, and more bibliography.

In our case the Argonautic tradition, and particularly the variant presenting the Argonauts' return to Greece via the Istros and the Adriatic Sea, as is fully known to us through Apollonius' work, offered just such an opportunity to attribute to the Colchians who found themselves in Illyria in pursuit of Medea the founding of a city by a river.[629] That the name of the river bears such a similarity to the name of the founders' neighbours would thus be easily explained. However, there is another explanation for this name, provided by Aelius Herodianus and Stephanus of Byzantium (both *s.v.* Δίζηρος): namely that it was given because the Colchians were there seeking for (δίζησθαι in ancient Greek) Medea. If this was indeed the case, could we surmise that Hecataeus changed the name of the Byzeres, consciously or not, affected by the phenomenon described above? It might sound a bit strange for one of his authority, but we should not forget the comparable aforementioned case of the Sigynnae, whose land is mentioned in the same myth by Apollonius as being encountered by the Argonauts right after their departure from Colchis, at the Istros; while in the *Orphica Argonautica* (756) they lived in the Colchis – and both places were accepted as their homelands by other authors (Herodotus and Strabo respectively).[630] In any case, we should note that Aelius Herodianus and Stephanus of Byzantium mention both the Dizeres, citing Hecataeus (*s.v.* Χοί), and the Byzeres, as a 'nation in Pontus' (*s.v.* Βύζηρες).[631]

15. The Colchians

15.1. The Colchian Question

The Colchians have undoubtedly always been among the best known peoples of the Black Sea region, owing their fame *inter alia* to the celebrated Argonautic myth. Although they were indisputably the dominant people of the eastern Black Sea littoral, it seems that they had occasionally spread their tentacles across the eastern part of northern Anatolia as well. However, there is a vagueness attending this expansion, since it is attested essentially only by Xenophon, though he is one of the most credible sources we have on the region. According to the *Anabasis*, not only was Trapezous clearly situated in the territory of the Colchians (4. 8. 22), the first people whom the Greek army met when they reached the coast, but this people seems to have occupied the littoral as far to

629 *Cf.* also Braund 1994, 34.

630 One might also compare the name Arion, which accompanies a river both in Illyria (Pseudo-Scylax 24) and in our area (Pseudo-Scylax 82).

631 The issue seems to have confused Jacoby, who initially rejected the identification between Dizeres and Byzeres (1912, 2723), later changing his mind (1957, 358).

the west as Kerasous (5. 3. 2),[632] at least at the time of the Ten Thousand's march through the area, namely in 400 BC, while to the south they bordered the territory of the Macrones (4. 8. 8–9).[633]

Conversely, the evidence from the rest of our sources for the geography of the littoral leads to a quite different conclusion: as we have seen in the previous sections, Trapezous and far less Kerasous were never presented as being situated within Colchian territory. The peoples occasionally mentioned around the two Sinopean colonies are the Macrones,[634] the Macrocephaloi, the Tibarenoi, even the Chalybes (e.g. Ctesias in Schol. in Apol. Arg. 2. 1015; Pseudo-Scylax 85; Strabo 7. 4. 3; *cf.* 11. 14. 1; 12. 1. 3; 12. 3. 1; 12. 5. 18–19, 28–29; Pliny *NH* 6. 12), but surprisingly not the Mossynoikoi, who, according to Xenophon, were the western neighbours of the Colchians, but lived more to the east of the Tibarenoi and the Chalybes in all the sources.

There is, therefore, an obvious disagreement regarding Colchian expansion along the southern Black Sea coast, which becomes even more significant since one of the sources contradicting Xenophon was written not long after the *Anabasis* and is indeed a *Periplus*: the one attributed to Pseudo-Scylax. In this geographical work we read that Kerasous belonged to the territory of the Assyrians, then followed the Chalybes, the Tibarenoi, the Mossynoikoi, the Macrocephaloi, whose land included Trapezous, the Becheires, the Ececheirieis, the Bouseres/Byzeres, and only then came the Colchians, near the River Apsaros (81–89). Their territory included, from north to south, three Greek cities (Dioscurias, Gyenos, Phasis),[635] a barbarian one (Aia, the birthplace of Medea) and nine rivers, from the Gyenos to the Apsaros (81). To the north of them were the Gelones and the Melanchlaenoi (80).[636]

Apart from Pseudo-Scylax (82), the Byzeres[637] would be mentioned as neighbours of the Colchians by Apollonius (*Argonautica* 2. 396–397, 1242–

[632] It was only after Kerasous, on their way west, that Xenophon's army reached the borders between the Colchians and the Mossynoikoi (5. 4. 1), before arriving at Cotyora (5. 5. 3). See Sections 2.11.1 and 5.1, also on the question of which Kerasous is the one mentioned by Xenophon.

[633] See also Section 2.13.1. Influenced by or citing Xenophon: Diodorus 14. 30. 3; Arrian *Periplus* 11, citing Xenophon, despite the confusing reference to the Ophis, which separates the land of the Colchians from the land of Thianinike, mentioned only by him (7); Procopius *De bellis* 8. 1. 8, 2. 15; Eustathius *Comm. in Dion. Per.* 772.

[634] In Xenophon these are just an inland people.

[635] The first of these is not called Greek by the author.

[636] As far as the southern borders are concerned, Hecataeus' reference to the Moschoi as Colchians (*FGrH* 1a 1 F 288) and neighbours of the Matienoi brings us to a wide area between north-east Anatolia and the Caspian. The Moschian Mountains (Strabo 11. 2. 1, 17) are usually identified with the Meskheti Range in Georgia. For the Matienoi, see Section 2.6.1.

[637] See Section 2.14.5.

1280; *cf.* the *Scholia*), Strabo (12. 3. 18), Valerius Flaccus (*Argonautica* 5. 153–180), Mela (1. 108); Pliny (*NH* 6. 11–12); Dionysius (*Orbis descriptio* 765; *cf. Paraphrases*), the *Orphica Argonautica* (756) and Eustathius (*Comm. in Dion. Per.* 762), most of whom also mention the rest of the peoples we have examined between the Mossynoikoi and the Colchians – Philyres, Becheires, Ecechirieis, Sapeires, to which we could add the Mares and the Macrones.[638] Besides, the rest of the ancient sources refer to the Colchians as the people who lived around the Caucasus and the Phasis,[639] with varying levels of generality.[640]

Trying to solve the problem of this serious discrepancy, M. Lipka proposed that the first author who mentioned the peoples between the Mossynoikoi and the Colchians (hereafter: the 'eastern peoples') must have been Hecataeus,[641] who was the source for all the others except Xenophon and probably also, like him, an eyewitness. If this is the case, then we are led to the conclusion, always according to Lipka, that a westward Colchian expansion took place sometime between the age of Hecataeus and the passage of the Ten Thousand, so somewhere in the 5th century BC. This expansion resulted in the displacement of the nations between Kerasous and Colchis to the south or west, while Lipka also left open the possibility of an eastwards displacement, saying however that such a displacement is not perceived in the written sources.[642] According to him, this Colchian expansion was responsible also for the division of the Chalybes' land into two parts, as we have seen, a coastal and an inland one.[643] Lipka's solution of the problem is the only clear one proposed so far, but it raises some new questions.

Did Hecataeus really visit the southern Black Sea littoral? The truth is that there is nothing to indicate that he 'certainly did',[644] and apart from Egypt (Herodotus 2. 143) we cannot be positive that he visited any other part of the world he described. Besides, we should not forget that the first after Hecataeus

638 For the last two, see Sections 2.16.1 and 2.13.

639 The name of this river was sometimes used to denote Colchis in general. See Braund 1994, 25.

640 E.g. Hecataeus *FGrH* 1a 1 F 17, 18a, 210, 288; Herodotus 1. 2. 2, 104. 1, 3. 97. 4; Ctesias *FGrH* 3c 688 F 57; Herodorus fragm. 52; Timaeus *FGrH* 3b 566 F 84; Pseudo-Scymnus 937; Strabo 1. 3. 2, 7, 21, 11. 2. 1, 14, 16–17, 5. 6, 14. 4, etc.; Appian *Mithridatica* 282, 556; Dionysius Periegetes 688–689, 762; Aelius Herodianus 3. 1. 96: Pyenis, city of the Colchians (Stephanus of Byzantium, *s.v.* Πυῆνις, mentioning also another city called Tyenis: *s.v.* Τυῆνις), 3. 1. 102, 271, 332; Hesychius *Lexicon*, *s.v.* Φᾶσις; Stephanus of Byzantium, *s.v.* Αἶα, Φᾶσις; Procopius *De bellis* 3. 1. 11; 8. 1. 8; Eustathius *Comm. in Dion. Per.* 19. 14–15, 689. 21.

641 *Contra* Hoefer (1904), who had suggested that it was Ephorus.

642 Lipka 1995, 66 and n. 9. For the matter, see also Braund 1994, 133–35.

643 Lipka 1995, 67. See more in Section 2.9.3.

644 As Jacoby (1912, 2689) suggested.

and Xenophon to have added the 'eastern peoples' on the coast was Pseudo-Scylax. The latter worked on the *Periplus* of Scylax of Caryanda, who not only preceded Hecataeus by a few years but is also supposed to have influenced him.[645] So why would one look for the source of the 4th-century compiler and the later authors in Hecataeus and not in Scylax himself? In other words, if both Hecataeus and Xenophon were describing the situation of their day, as Lipka suggests, then the specific part of Pseudo-Scylax's *Periplus* could have belonged to the original work of Scylax of Caryanda,[646] who lived in the same era as Hecataeus. But then, if the situation had indeed changed by the late 5th century BC, why did Pseudo-Scylax not correct the information given in the 6th century BC, since he was almost contemporary with Xenophon, who described the situation of their era?

Furthermore, Lipka is correct in saying that an eastwards displacement of the 'eastern peoples' is not perceived in the written sources. But neither is a southwards or westwards displacement. On the contrary, these are the directions from which, respectively, Xenophon's army came to Trapezous and in which they headed after Trapezous. Thus, one would expect to read about the nations in the *Anabasis*, which is not the case, since to the south of the Colchians we encounter the Macrones and the Drilae and to their west the Mossynoikoi, the Chalybes and the Tibarenoi. Thus of all these it is only the Macrones who fit Lipka's theory, while all the other 'eastern peoples' have never been mentioned inland or farther west on the coast in the ancient literature, apart from the Sapeires, who, however, appeared rather in the eastern Black Sea hinterland.[647]

Returning to Pseudo-Scylax: as we saw, to the north of the Colchians he placed the Melanchlaenoi and the Gelones (79–81). Herodotus, however, more than clearly places the Melanchlaenoi and the Gelonoi in Scythia, in the northern Black Sea region, and indeed in the hinterland (4. 20, 100–102, 108–109, 119, 125).[648] Similar is the placement of the Melanchlaenoi by Hecataeus (*FGrH* 1a 1 F 185), the supposed source of Pseudo-Scylax, who seems to have committed a serious inaccuracy here.

[645] Gisinger 1927, 634.

[646] The veracity of Scylax's journey has also been doubted, although not unanimously. See the bibliography in Parker (2008, 14–18), who does not share the doubts. More recently, Shipley 2011, 6, who does. See also in in the Introduction Section 2.2.

[647] So, an 'eastwards displacement'? For the Sa(s)peires, see Section 2.16.4.

[648] *Cf.* Aristotle *Mirabil.* 832b; Ephorus *FGrH* 2a 70 F 158; Pseudo-Scymnus 863, only for the Gelonoi, while he places the Melanchlaenoi near Colchis (937); Dionysius Periegetes 309–310. See also Counillon 2004, 88–90.

On the other hand, if we accept that Apollonius and the later authors of geographical works were really based on Pseudo-Scylax as well as Ephorus,[649] irrespective of the possible dependence of the former on the latter,[650] then we run the risk of ascribing the error to Ephorus. But Ephorus is in turn considered to have been influenced by, among others, Xenophon, who had written his *Anabasis* just a couple of decades before him.[651] Consequently, a plausible explanation seems to be that Pseudo-Scylax's carelessness is one of the causes of the whole problem. As already stated, his work contains several errors, which are attributed to multiple factors.[652] Here we have already spotted the misplacement of the Melanchlaenoi and the Gelones,[653] which is a rather serious mistake. Apparently, the Athenian compiler was relatively unconcerned with furnishing an accurate account of the real situation in all the parts of the world presented by the work he had worked on; otherwise, he would have taken Xenophon's *Anabasis* into consideration, which after all had been written in his home city only some years before.[654] Instead, he preferred to concentrate his attention on amending some specific parts of the original *Periplus*,[655] while most probably he had never been to the places mentioned in the work.[656]

Pseudo-Scylax's responsibility is, of course, just a plausible but unproved eventuality. In any case, there can be no doubt that accurate knowledge of the specific part of the Black Sea littoral was not among the strongest points of the ancient Greek pre-Christian authors, geographers or not.[657] Nevertheless, before abandoning the issue as impossible to be solved, it is worth directing our attention to another particularity regarding the Colchians, which might prove to be of some help. For it is not only the demarcation of their territory

[649] Even if indirectly, as Hoefer (1904, 558–64) suggested. On the contrary, Shipley (2011, 21) recognises a very limited, if any, influence of Pseudo-Scylax on later authors, including Apollonius, but I find this view rather exaggerated.

[650] See the bibliography on the relevant debate in Peretti 1961.

[651] Hoefer 1904, 544–54; Stylianou 2004, 80.

[652] Counillon 2004, 27–28 and 41, for the practical uselessness of the work especially as regards the Black Sea; Shipley 2011, 2–4.

[653] Compare also the case of the Macrocephaloi in Section 2.13.2.

[654] Especially since he is considered to have worked based mainly on the Athenian geographical and in general intellectual perspective of his era. Counillon 2004, 46; Shipley 2011, 6, 16–17.

[655] Shipley (2011, 11–12) doubts whether the original *Periplus* was the basic source of Pseudo-Scylax.

[656] Shipley 2011, 9–10.

[657] Especially as regards Pseudo-Scylax, see Counillon 2004, 38–39, 97. Sometimes a general description would be preferred, to overcome the difficulty. For example, Strabo (11. 2. 14) mentions the coast of Colchis as adjacent to the coast of Trapezous (which lies *before* Colchis: 12. 3. 17), adding that the two coasts form a considerable bend. On the other hand, the Island of Ares is presented by Stephanus of Byzantium (*s.v.* Ἄρεως νῆσος) as being situated 'towards the Colchians'.

that is marked by obscurity, but also their very status: were they one distinct nation, divided or not into different tribes, like for example the Mossynoikoi, or should we consider the term 'Colchians' as denoting a wider group of different peoples, similar to the case of the Scythians and the Thracians? There is evidence that argues for the latter. Already Hecataeus is reported to have mentioned at least two peoples as 'Colchian nations': the Coraxoi (Hecataeus *FGrH* 1a 1 F 210) and the Moschoi (Hecataeus *FGrH* 1a 1 F 288). The second instance is indeed noteworthy, since the Moschoi are otherwise encountered as an independent people in the ancient literature, starting from Herodotus, who clearly distinguished between the Moschoi and the Colchians (3. 94; 7. 78).[658] 'Colchian nations' are also mentioned by Pseudo-Scymnus (937; *cf.* Bessarion *Ἐγκώμιον εἰς Τραπεζοῦντα* 91. 2).

On the other hand, one cannot fail but notice that it is usually in the mythical narrations of the Argonautic expedition that the Colchians are encountered under this general term. As corresponding cases in Hecataeus we could indicate the Thracians and the Scythians. In his fragments we read about several Thracian and Scythian nations, but Thracians and Scythians are mentioned only in general geographical or historical contexts, just like Thrace and Scythia as wider areas (e.g. *FGrH* 1a 1 F 18b, 119). The significance of this observation is apparent.

In the general geographical description given by Herodotus, which we have already mentioned, the Saspeires lived to the north of the Medes and the Colchians to the north of the Saspeires, at the Phasis (1. 104; 4. 37, 40, 45),[659] while the historian also talks about the neighbouring tribes of the Colchians, who dwell between them and the Caucasus (3. 97). One of these could be the Coloi, who were mentioned by Hecataeus as neighbours of the Coraxoi 'inhabiting the foot of the Caucasus', but not as Colchians like the latter (*FGrH* 1a 1 F 209–210).[660]

Thus, apparently there has always been confusion regarding which of the tribes in the Caucasus area were actually Colchian and which not, notwithstanding the Colchians' general prominence from such early times.[661] As a result,

[658] In fact, the Colchians appear as having paid a special tribute to the Persians and not belonging to one of the 20 regular satrapies of the Persian Empire, like the Moschoi (Herodotus 3. 94 and 97). *Cf.* Braund 1994, 122–23.

[659] See Section 2.16.4.

[660] But how sure are we in fact that Hecataeus used the ethnic adjective in all the cases where it was required, or, even more, that Aelius and Stephanus transcribed his passages without the slightest alteration? Let us not forget that we have no words of Hecataeus on the Colchians themselves except those provided by the two ancient scholars.

[661] *Cf.* also Section 2.13.2.

the modern bibliography on them seems to be equally hesitant.[662] Generally, the name Colchis has prevailed to denote the area that more or less coincides with modern Georgia,[663] while the people(s) who inhabited this or a slightly larger area are called Colchians, albeit without further specification.[664]

Taking all this into consideration, we could think of a possibility which would practically confirm both Xenophon and the authors mentioning the 'eastern peoples': that the latter or most of them were actually Colchian. Xenophon could not have been mistaken. He and his soldiers were there; they did not just pass through the eastern part of the southern Black Sea littoral, they stayed there for many days, were in contact with the locals, discussed, negotiated and fought with them, witnessed their customs and way of life. If Xenophon had met other peoples there, he would have mentioned them. The same would be expected from Herodotus' presentation of the Persian provinces, where of all the 'eastern peoples' only the Saspeires are mentioned. On the other hand, Hecataeus and his supposed followers would not necessarily be mistaken in numbering the 'eastern peoples' among those living on the coast. An element that could enhance this theory is the lack of any anthropological data concerning any of them in any of the texts where they are mentioned, contrary to almost all the other peoples of the littoral. Moreover, Strabo's words potentially point also in this direction: the geographer, who generally places Colchis 'after' Trapezous, near the Mossynoikoi, the Moschian Mountains and Mt Scydises (12. 3. 17–18), referring to the savage peoples of the mountains who poisoned Pompey's soldiers with the locally made 'crazing honey',[665] states that 'some of these barbarians were also called Byzeres'.

If this is the case,[666] then we should imagine the eastern part of the southern Black Sea inhabited by the Mossynoikoi, the Macrones and the Colchians (Fig. 10.3). Their borders moved from time to time during the 1st millennium BC, leaving Trapezous in the territory of the one or the other. Of these three, the Macrones were sometimes totally absent from the coast, as is indicated by

662 Striking is the fact that there is no entry for the Colchians, not even for Colchis, in the *Paulys Realencyclopädie der classischen Altertumswissenschaft* (*RE*), where all the other peoples we have been dealing with in this part of the book, without exception, are treated.

663 One notes that, for example, in Braund's thorough monograph on ancient Georgia (1994), where the terms 'Colchis' and 'ancient Georgia' or 'Georgia in Antiquity' are alternated with each other, denoting practically the same area.

664 As Braund (1994, 92) correctly observes.

665 See Section 2.11.3.

666 That the Becheires have been mentioned as a group of nations by Apollonius (*Arg*. 2. 394) does not really mean much; after all, according to Herodorus and Philostephanus (Schol Apol. Arg. 2. 1024), the Macrones were supposed to belong to this group, which would confuse things even more. *Cf.* Counillon (2004, 98), who, referring to the Macrocephaloi, the Becheires, the Ececheirieis and the Bouseres, speaks of an 'ethnographic ensemble' (see Section 2.14.3).

Xenophon's narrative.[667] We could also remark that, contrary to the case of Paphlagonia, which ended up denoting an area larger than the actual place inhabited by the Paphlagonians and including other peoples as well,[668] the name of Colchis seems to have been established as denoting an area (more or less modern Georgia) actually smaller than that inhabited by all the (possibly) Colchian nations in antiquity.[669] Should we then list among these Colchian peoples also the Drilae and the Macrones, who after all, especially the former, were encountered by Xenophon in the same area, in the light as well of our previous remarks on the two peoples?[670] Even if Xenophon does not clearly say so, his words on the location of the Drilae in reference to the Colchians (*Anabasis* 5. 2) might lead to such a suspicion.

15.2. Historical, Economic and Anthropological Data

It is remarkable how much else besides the extent of their territory is vague about the otherwise well-known Colchians. Not least is their very origin. Again, we have in effect only one author's testimony in this regard: in his Second Book, Herodotus claims that

> there can be no doubt that the Colchians are an Egyptian race. Before I heard any mention of the fact from others, I had remarked it myself. After the thought had struck me, I made inquiries on the subject both in Colchis and in Egypt, and I found that the Colchians had a more distinct recollection of the Egyptians, than the Egyptians had of them. Still, the Egyptians said that they believed the Colchians to be descended from the army of Sesostris. My own conjectures were founded, first, on the fact that they are black-skinned and have woolly hair, which certainly amounts to but little, since several other nations are so too; but further and more especially, on the circumstance that the Colchians, the Egyptians, and the Ethiopians, are the only nations who have practised circumcision from the earliest times [...]. I will add a further proof to the identity of the Egyptians and the Colchians. These two nations weave their linen in exactly the same way, and this is a way entirely unknown to the rest of the world; they also in their whole mode of life and in their language resemble one another. The Colchian

[667] The proximity between the territory of the Mossynoikoi and that of the Colchians might be further indicated by the references to the famous 'mad honey' of the wider area, narrated in contexts including both peoples (Xenophon *Anabasis* 4. 8. 20–21; Strabo 12. 3. 18; Pliny *NH* 21. 770). See also Section 1.1.3.

[668] See Section 2.6.1.

[669] Conversely, this 'variety' of Colchian nations has later resulted in confusion and the presention of several other non Colchian peoples of the wider area as Colchian: see, for example, Tacitus *Annals* 6. 34; Joan. Laur. Lydus *De mensibus* 4. 146; Procopius *De bellis* 2. 15. *Cf.* Braund 1994, 14, 132.

[670] See Sections 2.12.1 and 2.13.2.

> linen is called by the Greeks Sardinian, while that which comes from Egypt is known as Egyptian (2. 104–105).[671]

Herodotus' theory was adopted or merely cited by other ancient authors.[672]

As noteworthy as Herodotus' information about the Egyptian origin of the Colchians is the certainty with which he proffers it, from the very first sentence.[673] A certainty which he justifies on the basis of both what he has heard from the two peoples and his own anthropological observations, mainly relating to physical appearance and customs. The connection of the Colchians with the army of Sesostris (irrespective of which one, probably the third) takes us back to the early 2nd millennium BC, and thus would present them as one of the most ancient indigenous peoples of the southern Black Sea littoral, even if just of a small stretch of it in the east. And yet, we know so little about these people, setting aside anything deriving from the world of mythology.

How reliable, though, is Herodotus' information? As we have seen, it is doubtful that he ever visited Colchis. When he says that he asked both the Colchians and the Egyptians about their possible connection to each other, this does not mean that he asked them in their homelands. He could have met people from both nations anywhere. As for his own conjectures, which were founded 'on the fact that they are black-skinned and have woolly hair', this is also an observation that could have taken place anywhere where Colchians could be encountered. The similarity of the two peoples' customs is a good indication, but not enough to be considered as proof.

Once again, it is Xenophon's *Anabasis* that is our safe recourse text of choice. In fact, Xenophon nowhere implies anything that can confirm any of Herodotus' statements. As we have seen, in his narrative he frequently dwells on anthropological features of peoples he encountered that drew his or his soldiers' attention, especially of peoples with whom they remained in contact for a lengthy period of time, like the Colchians (see below). One may recall, for example, the fattened children of the wealthy Mossynoikoi, who were soft and white to an extraordinary degree, and pretty nearly equal in length and breadth, and decorated their bodies with tattoos (*Anabasis* 5. 4. 32).

[671] Translation by G. Rawlinson, New York 1942.

[672] Diodorus 1. 28. 2, 55. 4–5; Strabo 1. 3. 21, 11. 2. 17; Josephus *Contra Apionem* 1. 168–70; Aelius Theon *Progymnasmata* 83. 5–13; Dionysius Periegetes 689; Celsus *Alethes logos* 3. 41. 13; Origenes *Contra Celsum* 5. 41. 17, 47. 2, 48. 2; Eustathius *Comm. in Dion. Per.* 689. 8. The Colchians are mentioned as descendants of Japheth by Hippolytus (*Chronicon* 84. 8) and Epiphanius (*Ancoratus* 113. 5. 4).

[673] The Colchians φαίνονται ἐόντες Egyptians, not just φαίνονται εἶναι: the participle following the verb φαίνομαι indicates a certainty (there is no doubt that they are), contrary to the infinitive, which would indicate uncertainty (they seem to be). *Cf.* also Pliny *NH* 33. 15.

If the Colchians' physical appearance was as noteworthy as Herodotus suggested, we would expect Xenophon to have mentioned it. Nevertheless, this omission cannot be considered as an argument against the validity of Herodotus' information.[674] But why would the Colchians that Herodotus met spread the tradition that they were of Egyptian descent[675] – if their 'recollection' of the Egyptians really meant that they originated from them? In absence of further evidence from other sources of any kind, one cannot elaborate more on the matter, which rather seems to be rooted in mythical tradition.[676] This is supported by Hippocrates' description of the dwellers on the Phasis, in the centre of Colchis, who 'in colour they are sallow, as if affected with jaundice' (*On Airs, Waters and Places* 15), and so totally contrary to the black skin claimed by Herodotus.[677]

There are, however, more tangible facts to be deliberated on concerning the Colchians. First of all, their military organisation, which was obviously on a higher level than that of, for example, the Macrones,[678] for as soon as they noticed Xenophon's army approaching the southern, mountainous part of their territory they immediately drew up in line of battle (Xenophon *Anabasis* 4. 8. 9). It seems that this preparedness, combined with the difficulty of the terrain, startled the Greek soldiers, who spent some time deciding how to face the Colchians most effectively and taking up their positions for the battle (4. 8. 9–16), which they eventually won (4. 8. 17–19). The Colchians having abandoned the mountain, the way was open to the Greeks to reach the Colchian villages, which were numerous and contained provisions in abundance. The only surprising item was the local honey, which turned out to be toxic and caused terrible symptoms in those who tasted it (4. 8. 20–21).[679]

From these mountains the Greek soldiers reached Trapezous, 'a Greek city in the territory of Colchis', as Xenophon called it. Thus, the description following concerns the coastal part of the Colchian land, where the Greeks spent as many as 43 days, the first 30 in Trapezous (4. 8. 22). This part was also wealthy and full of settlements, where the Greeks stayed and which they used as a base to plunder Colchis (4. 8. 23–24; 5. 1. 6). Again, Xenophon speaks of villages, and there is nothing in his account to indicate the existence of bigger

[674] The circumcision (see also Sections 2.8 and 2.13.3), as well as the similarities in language are strong indications, but not enough to prove common origins.

[675] For the tradition spread by the Egyptians that the Colchians were among the emigrants who left Egypt, see Diodorus 1. 282; 1. 55. 4–5.

[676] *Cf.* Braund 1994, 17–18.

[677] Braund (1994, 16) criticises the whole passage on the inhabitants of Phasis as 'negative to the point of misinterpretation'.

[678] See Section 2.13.3.

[679] For the 'mad honey', see also Sections 1.3.3 and 2.11.3.

towns or of villages that could be considered of some administrative importance compared with the others, as was, for example, the case with the metropolises of the Mossynoikoi and the Drilae.[680] However, such centres cannot be excluded, since they could have lain outside the areas visited by Xenophon.

Of special interest is the apparent division of the Colchians into two categories, those who were friendly to the Greeks and those who were not, which reminds us of the similar division among the Mossynoikoi.[681] With, however, a slight difference, for here the friendly attitude was mainly towards the Greek citizens of Trapezous, with whom it seems that the Colchians had developed a good relationship (which encouraged some of the coastal Colchians to give gifts to the Trapezuntians for the Greek soldiers – 4. 8. 24); while their hostility was concentrated on the Greek soldiers of Xenophon's army, which, as he admits, was only natural since they had driven them out of their houses.[682] This is why the Trapezuntians would not lead the Greeks to Colchian districts, from which provisions could be secured easily, but only to the Drilae, at whose hands they were continually suffering losses (5. 2. 1–2).[683] Finally, the Colchians were persuaded by Xenophon to repair the coastal roads in order to get rid of the Greeks as soon as possible (5. 1. 13–14; 5. 3. 2). By these roads Xenophon's army marched for three days until they reached Kerasous, which was still in Colchian territory. In this city they remained for another ten days, which makes a total of 43 days in the Colchians' land (5. 3. 2–3).

As far as the local economy is concerned, apart from the honey, the gifts of hospitality that the citizens of Trapezous gave to the Greek army – oxen, barley-meal and wine (4. 8. 23–24) – can be considered indicative of the area's resources. Additionally, Herodotus' information about Colchian linen, which was well-known to the Greeks, and how it was woven (Herodotus 2. 105), is also significant, as well as unique regarding the southern Black Sea littoral. Wood was also worked by the Colchians, who in the Persian Wars were armed with wooden helmets, small shields of raw hide and short spears, besides which they had swords (Herodotus 7. 79).

Apart from circumcision and the way in which they wove their linen, another piece of information concerning Colchian customs, at least in the area around the Phasis and so not necessarily applying to the southern Black Sea littoral as well, comes to us from Apollonius (*Argonautica* 3. 200–209): 'it is an abomination with the Colchians to burn dead men with fire; nor is it lawful

[680] See Sections 2.11.1 and 2.12.2.

[681] See Section 2.11.2.

[682] *Cf.* Maksimova (1956, 122–23), who sees in this division between the Colchians the absence of a united tribal society. See also Lendle 1995, 288, 299; Braund 1994, 132.

[683] *Cf.* also Section 2.12.1.

to place them in the earth and raise a mound above, but to wrap them in untanned oxhides and suspend them from trees far from the city. And so earth has an equal portion with air, seeing that they bury the women; for that is the custom of their land'.[684] From Apollonius' commentator we learn that the poet derived this information from Nymphodorus (Schol. in Apol. Arg. 3. 202).

Finally, as far as the Colchians' relations with the Persians is concerned, it seems that, being at the edges of the Empire, the status of the Colchians occasionally varied between full and semi-dependence. Herodotus pointed out that:

> the Colchians, and the neighbouring tribes who dwell between them and the Caucasus – for so far the Persian rule reaches, while north of the Caucasus no one fears them any longer – undertook to furnish a gift, which in Herodotus' day was still brought every fifth year, consisting of a hundred boys, and the same number of maidens (Herodotus 3. 97).

The image created is that of a peripheral ally of Persia, although the distinction between the regular tributes of the peoples in the provinces and the Colchians' regular contributions 'is both fine and uncertain'. It may suggest the absence of a Persian satrap, but still a measure of Persian control.[685] In the *Anabasis*' last – and probably interpolated – passage (7. 8. 25), the Colchians are presented as autonomous.

16. Other Peoples

16.1. The Mares

The Mares belong to those enigmatic cases of peoples who, although mentioned early in two significant and reliable sources, Hecataeus and Herodotus, would never again be encountered in the ancient literature. The Milesian historian and geographer must have alluded to the Mares as a people that were adjoining the Mossynoikoi (*FGrH* 1a 1 F 205 = Stephanus of Byzantium, *s.v.* Μᾶρες). Neither his nor Herodotus' two references to the Mares necessarily confirm a coastal placement, although they clearly point to the general location.[686] In Herodotus, together with the Moschoi, the Tibarenoi, the Macrones and the Mossynoikoi, the Mares formed the 19th Persian satrapy, and had to pay 300 talents (Herodotus 3. 94). Nevertheless, in the Persian Wars they had the same commander as the Colchians, namely Pharandates (7. 79).[687] It has

684 *Cf.* Claudius Aelianus *Varia historia* 4. 1. 30–31.

685 Braund 1994, 122–23.

686 Hecataeus sometimes mentioned also peoples of the hinterland. *Cf.* Jacoby 1912, 2722.

687 In these wars the Mares wore on their heads local plaited helmets peculiar to their country, and used small leathern bucklers and javelins (Herodotus 7. 79).

sometimes been argued that, if they were indeed a coastal nation, the Mares could be identified with Apollonius' Philyres,[688] but there is nothing to substantiate such a theory. Of the peoples examined in Section 2.14, the Becheires, the Ececheirieis and the Byzeres are also omitted by Herodotus, so, by the same logic, why could the Mares not be connected to one of them? Such a dearth of information about the Mares affords no opportunity for anything more than mere speculation: for example, that after the early classical period they migrated or were assimilated by neighbouring peoples, whom they might have even been kindred to from the beginning.

16.2. The Choi

A people called Choi are mentioned by Hecataeus (*FGrH* 1a 1 F 207 = Aelius Herodianus and Stephanus of Byzantium, *s.v.* Χοί) as living near the Becheires, with the Dizeres as their neighbours to the east, and therefore probably on the Black Sea coast, but not necessarily. The Choi are otherwise unknown in the ancient sources, at least by that name. If indeed they lived between the Becheires and the Byzeres, then they appear in the place where Pseudo-Scylax would later mention the Ececheirieis and Apollonius the Sapeires, two peoples also accompanied by serious elements of, at least geographical, obscurity.[689]

W. Tomaschek tended to identify the Choi with the Taochoi (Xenophon *Anabasis* 4. 4. 18; 4. 6. 5; 4. 7. 1, 17; 5. 5. 17; *cf.* Pseudo-Zonaras *s.v.* Τάοχοι), considering the name as an erroneous script for Taoi, which some used for the Taochoi, according to Sophaenetus (Aelius Herodianus and Stephanus of Byzantium, *s.v.* Τάοχοι), just like the Chaoi mentioned by Diodorus (14. 29. 1).[690] F. Jacoby added to the discussion the Koitoi mentioned by Xenophon (*Anabasis* 7. 8. 25) and suggested that the correct form of Choi should be Choitoi, adding that they must have belonged to the 'immense tribes of the Becheires' (Apollonius *Argonautica* 2. 394).[691] If this is the case, and the Choi were actually no other than the Taochoi, which is not necessarily correct,[692] then they cannot have lived on the coast, since the Taochoi have always and clearly been described as living in the Pontic interior, even further inland than

[688] Herrmann, *RE* XIV.2 (1930), 1680 (*s.v.* Mares); Jacoby 1957, 357.

[689] See Sections 2.14.3 and 4.

[690] *RE* III.1 (1897), 181 (*s.v.* Βέχειρες); III.2 (1899), 2356 (*s.v.* Choi); V.1 (1903), 1248 (*s.v.* Δίζηρες).

[691] Jacoby 1957, 358.

[692] Aelius Herodianus (3. 1. 226) and Stephanus of Byzantium (*s.v.* Τάοχοι) make separate reference to the Taoi/Taochoi and the Choi, but this may not mean much.

the Macrones and the Scythenoi (Xenophon *Anabasis* 4. 7),[693] and never on the coast.[694]

16.3. Kerkitae and Appaetae

When describing the hinterland of the coast between Trapezous and Pharnakia, Strabo says:

> Above Trapezous and Pharnakia are situated the Tibaranoi, the Chaldaeoi and the Sannoi, in early times called Macrones and Lesser Armenia; and the Appaetae, in earlier times called Kerkitae, are fairly close to these regions. Their country is crossed by the Scydises, a very rugged mountain, which joins the Moschian Mountains above Colchis and which is occupied by the Heptacomitae [by which he means the Mossynoikoi]. From here passes also the Paryadres, which extends from the region of Sidene and Themiscyra to Lesser Armenia and forms the eastern side of Pontus. Now all these peoples who live in the mountains are utterly savage (12. 3. 8).

This passage creates the impression that the Appaetae were another people that dwelt on the southern Black Sea littoral; however the area defined is vast, and the peoples mentioned are not given in a clear geographical sequence. Therefore, the place where the Appaetae lived is completely uncertain, as is their distance from the sea, which could have been considerable. Nonetheless, it was the earlier name of these people that attracted the interest of some modern scholars: Kerkitae.

One immediately thinks of the Kerketae, a well-known people occupying part of the north-eastern Black Sea coast, on the northernmost slopes of the Caucasus and mentioned by many authors (e.g. Pseudo-Scylax 73; Pseudo-Scymnus 913; Mela 1. 110; Pliny *NH* 6. 16; Dionysius Periegetes 628), including Strabo himself (11. 2. 1, 14). Thus, it has been proposed that the Kerketae belonged to those peoples of the Caucasian area that are supposed to have migrated to the eastern part of the southern Black Sea littoral, like the Heniochoi Sannoi, beginning in the 1st century BC.[695] To this we could add Curtius Rufus' passage (6. 4. 16–17), in which the Kerketae are mentioned together with the Mossynoikoi and the Chalybes, even though the passage in question is a very confusing one.[696]

[693] See Section 2.9.3.

[694] Tomaschek's words also indicate an inland location, since he says that the Choi lived 'to the south of' or 'behind' the Becheires (*RE* III.1 (1897), 181; III.2 (1899), 2356).

[695] Kiessling 1912, 265–68. See Section 2.13.2.

[696] See more in Section 2.8.1. Later authors mention an Indian nation called Kerkitae (Pausanias *Att. nam.*, *s.v.* Κερκῖται; Photius *Lexicon*, *s.v.* Κερκῖται).

Finally, in the *Orphica Argonautica* (755), a people named Napatae is mentioned in the accusative form (Napatas) between the Philyres and the Sapeires, namely in the place where we usually meet the Becheires, who are not mentioned in the poem. The name is totally alien to the ethnic names of the southern Black Sea littoral, and one can only think of Strabo's Appaetae, which sounds much the same.[697]

16.4. Generality is Safer: The Ponticoi

One thing that must have become clear after the last few sections is that the eastern part of the southern Black Sea littoral has been inhabited by many peoples, totally or somewhat different, who periodically moved or even changed territories with each other, or were forced to move by others. One reason for this situation that has been suggested is a migration of peoples from the Caucasus region to north-eastern Anatolia;[698] another could be a Colchian expansion in the same direction.[699] The fact remains that a gradual confusion characterises the written sources concerning the indigenous peoples of the area and their exact territories, especially since most of their authors did not visit it themselves but relied on others' accounts. This confusion culminated in the Roman period, at the beginning of which the above-mentioned migration is supposed to have taken place, and seems to have affected even authors who, for the rest, were fully conversant with the peoples of Asia Minor, such as Strabo, since the latter happened to live at the very time of this migration. The emergence of a new people's name that encapsulated a geographical determination was inevitable.

Strabo, referring to the raids of the Sacae, wrote that 'they advanced as far as the country of the Cappadocians, particularly those situated close to the Euxine, who are now called Ponticoi' (11. 8. 4). The name would be repeated later, on the occasion of the reference to the city of Zela and its temple of Anaïtis, where 'all Ponticoi make their oaths concerning their matters of greatest importance' (12. 3. 37). The geographical delineation of the name is made clear in the former passage, where the Ponticoi appear as those Cappadocians who lived on the Pontic coast. In the latter passage, the Ponticoi could be generally the people of Pontus in a smaller or larger area around Zela, but the reference to Pythodoris some lines before could also imply that it is the people

[697] For the rest, Napata(e) was a well-known Nubian city on the west bank of the Nile (Strabo 17. 1. 54, 41. 3; Ptolemy 4. 7. 19; Stephanus of Byzantium, *s.v.* Ναπάται; the *Suda*, *s.v.* Ναπάται).

[698] Kiessling 1912, 265–68. See Section 2.13.2.

[699] See Section 2.15.1.

of the client kingdom of Pontus that are meant. In fact, from the 1st century AD at the latest, the term Ponticoi would frequently be used also for the people or soldiers of the Pontic kingdom, in historical narratives, mainly of the Mithridatic Wars (e.g. Plutarch *Lucullus* 13. 5; Memnon, *passim*), or even their kings (e.g. Appian *Mithridatica* 83).

Ridiculing the people who were influenced by Alexander, the false prophet of Abonouteichos, Lucian (*Alexander* 17) referred to the uneducated 'fatheads' Paphlagonians and Ponticoi, the latter most probably meaning the peoples of Pontus other than the Paphlagonians, without specifying exactly which. Later, defining the southern Black Sea, Procopius said that 'it begins from Byzantium and ends at the land of the Colchians, inhabited from west to east by the Bithynians, the Honoriatae,[700] the Paphlagonians and the Ponticoi' (*De bello gothico* 8. 2. 1–3; *cf.* Hippolytus *Chronicon* 209; *Chronicon paschale* 58). The area of the Ponticoi in these cases coincides in general terms with the one on which we read the name Pontici inscribed in red on the *Tabula Peutingeriana* – a purely geographical work – along the coast between Sinope and Polemonion, as well as the area that in all later times, and starting from the fluid borders of the Pontic kingdom, would be defined as Pontus. Consequently, one could claim that the name Ponticoi in those instances has a geographical and/or a political, but not an ethnic character – they might be some of the peoples of northern Asia Minor, or the people of the Pontic kingdom, but not a specific nation bearing this name. When Eustathius (*Comm. in Dion. Per.* 762) calls the Becheires 'a Pontic and Scythian nation', he seems to mean that they were a nation of Scythian origin (irrespective of whether this is correct or not) that lived in Pontus, the geographical area; otherwise, he would have said 'a Pontic or Scythian nation'.

Even though we might suspect that the name Ponticoi first appeared with its political meaning, there is nothing to negate that it became a timely and well-suited tool for later authors who wished to refer to places and peoples in the eastern part of the littoral while avoiding the pitfall of having to identify and correctly locate each of the numerous peoples in that confusing area.

We will complete our tour of the peoples of the southern Black Sea littoral that have been mentioned in the ancient literature with the solitary case of Pliny's reference (*NH* 6. 11) to 'the *gentes Caenarum*[701] and the Chalybes', between Themiscyra and Cotyora. Unlike the Chalybes, the former people are

[700] Another case of a name of political/geographical nature: the people of a new province (Honorias) created in the late 4th century BC by Theodosius I, who gave it the name of his son, Honorius.

[701] See the Loeb edition of Pliny (by H. Rackham, London 1961), which translates as: the Caenares.

totally unknown, and their name has sometimes been corrected to *gentes Genetarum*,[702] since they are placed in the area of Cape Genetes, to the east of Cape Iasonion, namely the eastern cape of the peninsula between Polemonion and Cotyora. In the ancient sources Genetes is mentioned there as a cape, especially of Genetaios Zeus (Apollonius *Argonautica* 2. 378, 1009; Schol. in Apol. Arg. 159. 11; Strabo 12. 3. 17; Valerius Flaccus *Argonautica* 5. 148),[703] a harbour (Pseudo-Scylax 88) and a river (Anonymous *Periplus* 32).[704] However, a nation by that name is equally unknown, and the view that the cape was named after a local people, obviously based on Pliny's unique passage,[705] is highly speculative and unsubstantiated. As we have seen in the previous sections, the territory around Cape Genetes has at various times been reported as belonging to the Paphlagonians, the Leucosyroi, the Chalybes and the Tibarenoi.[706] But we never meet a people called Genetae elsewhere than in Pliny, who did not omit any of the other four peoples. Not being one of our most reliable geographic sources in any case, Pliny could have mentioned such a people either in error or, perhaps bewildered by the different situations described by different authors, considering a name giving a geographical determination safer than anything else – an instance similar to the Ponticoi.

[702] E.g. the Teubner edition of Pliny (by K.F.T. Mayhoff, Leipzig 1906).
[703] Which, according to Bryer and Wienfield (1985, 120), indicates that a temple of Zeus stood there.
[704] *Cf.* also Stephanus of Byzantium, *s.v.* Γενήτης, with all options.
[705] Bryer and Wienfield 1985, 120.
[706] See Sections 2.6.1, 2.8.1, 2.9.1, 2.10.1.

PART 3

INVADERS, CONQUERORS AND OTHER SOJOURNERS

1. The Cimmerians

In a geographical area as blessed by nature as northern Anatolia and at a time of huge population movements in Eurasia, the indigenous peoples inhabiting the southern Black Sea littoral would inevitably not long remain undisturbed. One of the first peoples for whom there is evidence of overland movement into Anatolia in the first half of the 1st millennium BC are the Cimmerians. If we take into consideration the words of two Greek authors – Herodotus and Pseudo-Scymnus – as well as a passage by Strabo,[1] it seems that this invasion also affected the southern Black Sea coast. However, we shall first devote some paragraphs to the Cimmerian invasion of Anatolia in general, in order to provide some idea of the difficulty of the issue.

Over recent decades much has been written about this invasion – when it took place, what results it might have had, and what kind of evidence, written or archaeological, there is for it. The conclusions of this research are admirably summarised by J. Olbrycht: 'Of all the nomadic peoples who were present in the Caspian steppes and in Western Asia in the 1st millennium BC, none has probably caused historians and archaeologists more trouble than the Cimmerians.'[2] It seems that this trouble springs *inter alia* from the impossibility of confirming archaeologically the data provided by the written sources.

In this regard, let me cite the distinction made by A. Ivantchik, to whom we probably owe the most thorough investigation of the Cimmerians to date:

> The written sources on the Cimmerian question make up two totally dissimilar groups. The first of these consists of data in the Classical tradition. They are fairly numerous and varied, yet most of the data are second- or third-hand and have survived in the works of authors, who lived at a time far later than that when actual contacts between Greeks and Cimmerians took place. [...] The second group of sources – markedly different in character from the first – consists of Akkadian cuneiform texts written at the time when Assyrians and Cimmerians were in direct contact with each other. There are also a few biblical ones. These sources enable us to acquire very reliable information first hand. Many Akkadian documents are well dated, often with the actual year, sometimes even the month and the day: this makes it possible for us to put together a reliable chronology of Cimmerian history and to include it in the general context of the history of Asia Minor.[3]

[1] See all of them below.
[2] Olbrycht 2000a, 71.
[3] Ivantchik 2001b, 307–08.

What do we learn from these sources? In brief, the Cimmerians lived in the north-northeast Black Sea region until they were driven out by the Scythians in the 8th century BC, and forced to head south, over the Caucasus, and then west, into Asia Minor. About 715–714 BC the Cimmerians attacked the Urartu and in the first half of the next century they conquered Phrygia. In the mid-7th century BC they attacked the Lydians, killed their king, Gyges, and carried out raids against Greek cities in Ionia and Aeolis. In the 620s, the Cimmerians were defeated by the Scythian king Madyes in Anatolia and at the end of the 7th century BC by the Lydian king Alyattes. After that nothing was heard of them and their remnants are supposed to have been assimilated by the local populations in Anatolia.[4]

Thus far things do not appear especially obscure, as there is general agreement between the Akkadian and Greek sources. The trouble begins when we attempt to trace the Cimmerian presence in Anatolia archaeologically, as might initially be considered sensible, since this presence lasted for almost a century.

One of the major problems that emerged from the very beginning of this effort has been the insurmountable difficulty in differentiating between Cimmerian and Scythian objects. Until the first decade of the present century, identifications of tumuli of the *kurgan* type, horse burials and other archaeological finds in Anatolia as Cimmerian (as well as Scythian) appeared a bit more casually in the bibliography than in recent years.[5] Besides, a broad destruction layer detected during the excavations at Gordion had been connected to the Cimmerian attacks mentioned by the written sources.

[4] Tokhtas'ev 1991; Olbrycht 2000a, 77–83, 92; Kõiv 2007, 160; Ivantchik 2012, 323, with all the sources and critical approaches to them. The Greek ones are mainly Herodotus (1. 6, 15–16, 103; 4. 11–12), Pseudo-Scymnus (986–997) and Strabo (1. 3. 21; 11. 2. 5). According to Parker (1998), there were earlier sources that Herodotus did not make full use of (probably for lack of interest), as is apparent from examination of other authors' texts, like that of Pseudo-Scymnus, who gives more details than Herodotus. Among them are said to have been Aristeas of Proconnesus, Callinus and Archilochus (Tokhtas'ev 1996, 1–11; Olbrycht 2000a, 75–77; Kõiv 2007, 155; Xydopoulos 2018, 66, 72–76). The first to have mentioned the Cimmerians was Homer (*Odyssey* 11. 14–19); for this passage, see Olbrycht 2000a, 72–75; Ivantchik 2005a; 2005b; Manoledakis 2015b, 302–06, with bibliography. For the Cimmerians in the Near East in general, see Bouzek 1983; Ivantchik 1993. For the disputable Treres, who are repeatedly mentioned by Strabo as connected to the Cimmerians, indeed even as a Cimmerian tribe (1. 3. 21; 14. 1. 40), see Leaf 1923, 45; Tokhtas'ev 1991; Olbrycht 2000a, 92. Contrary to most scholars (see above), Tsetskhladze (1994a, 116, as well as de Boer 2006, 44) considered Strabo 1. 3. 21, where the geographer wrote that the Cimmerians and the Treres advanced to Anatolia along the western shore of the Black Sea, as 'deserving our confidence'. He even claimed that this route is supported by archaeological material, an assertion he would later indirectly refute (see his views on 'Cimmerian' archaeology below).

[5] See, for example, Akurgal 1967; Bouzek 1983, 148–60; Ivantchik 2001a, 55–56; Dönmez 2007a, 60; 2011; 2012, 169.

However, things have now changed. In any case, in most instances there was nothing to substantiate such identifications, but a decisive factor seems to have been the new chronologies proposed for the destruction level at Gordion, based on radiocarbon and dendrochronological data from the site, which shift it from *ca.* 700 to 830–800 BC, thus disconnecting it from the Cimmerian invasion.[6] Yet the apparent need to speak of Cimmerian archaeological finds essentially never perished. Other terms would emerge to smooth over the difficulty, cardinal among them 'Pre-Scythian' and 'Early Scythian'. Their devising was based on the observation that there is 'enough evidence for a cultural similarity between the Cimmerians and the Scythians',[7] or that 'the Cimmerians cannot be differentiated archeologically from the Scythians'.[8]

But even this theory is not free of problems. Is an archaeological similarity enough to prove a cultural similarity? And, if so, is a cultural similarity based on an archaeological one enough to demonstrate an ethnic kinship, which is actually implied by a term like 'Early Scythian' used for the Cimmerians?[9] Indeed, the question is a difficult one, as is apparent from the words of those adopting this theory. Even in the most cautious cases we read that 'both Cimmerians and Scythians appeared in the Near East under the same cultural image – same weapons, etc., which the Greeks would call "Scythian", but were first brought to the East by the Cimmerians'.[10] Elsewhere we find that some monuments in Anatolia 'are enough to leave no doubt that they belong to the Cimmerian ("early Scythian") archaeological culture. Without doubt, these monuments belonged not to the Scythians, but to the Cimmerians', especially when they are situated in areas through which the Cimmerians are reported by the written sources to have passed.[11] What can be deduced from them is difficult to decide.[12]

[6] Voigt 2009; 2011, especially 1082; Genz 2011, 333–43, with similar chronologies from other sites in Anatolia. For more bibliography as well as the opposite view see Hellmuth 2008, 109–10; Tsetskhladze 2012a, 331.

[7] Ivantchik 2001a, especially 16–56, with earlier bibliography, and 283 with a map of the relevant find-spots. *Cf.* also Ivantchik 2001b; 2001c.

[8] Tokhtas'ev 1991.

[9] See, for example, Ivantchik 2001b, 339. More recently, see Hellmuth 2008.

[10] Ivantchik 2001a, 16, 67–68, 279.

[11] Ivantchik 2001a, 55–56; 2001c, 329; 2010; 2012, 65. A bit different in Ivantchik 2001b.

[12] Ivantchik 2001b, 339: 'What we are confronted with are two groups which were very close to each other from both the cultural point of view and also probably the linguistic and ethnic one as well, but which could nevertheless be quite clearly distinguished one from the other'; 2001a, 280; Sulimirski and Taylor 1991, where the Cimmerians (555–60) are examined as a more or less early Scythian people (547–90). *Cf.* also Pogrebova, Raevskii and Yatsenko 1998, as regards even the written sources.

The truth is that there is no safe archaeological evidence for a Cimmerian invasion or any other Cimmerian presence in Anatolia. Even worse, we know nothing of the Cimmerian culture in general, and archaeological identification of them is still being sought. Thus, applying terms like 'Cimmerian' and 'Thraco-Cimmerian' to archaeological finds is misleading, as has been correctly pointed out, and find categories like those mentioned above (*kurgan*-type and horse burials, arrowheads, etc.) may have belonged to any nomadic people.[13] This, indeed, is how another term for all these peoples emerged: the vague catch-all 'Eurasian Horse-Riding Nomads'.[14] Nevertheless, the absence of safe archaeological evidence for the Cimmerians does not refute their very existence in Anatolia, as has sometimes been implied.[15] To deny the existence of the Cimmerians (in Anatolia or generally) for want of clear archaeological evidence is excessive. The written sources mentioning them are too many and too important to be ignored. By the same logic, after all, one should also doubt the existence of most of the indigenous peoples examined in the previous part.

But let us now focus on our area of interest. As already noted at the beginning of this section, there are three passages in ancient Greek literature that mention a Cimmerian presence on the southern Black Sea littoral, and more specifically in its central section. The earliest of these belongs to Herodotus, according to whom 'it appears that the Cimmerians, when they fled into Asia to escape the Scythians, established a settlement [κτίσαντες] on the peninsula where today the Greek city of Sinope is built' (Herodotus 4. 12). Some centuries later, Pseudo-Scymnus would provide us with the most detailed description of the founding of Sinope and its several phases, though in a mythical context (*Periplus* 986–997):

> Then Sinope, a city named from one of the Amazons, whose village was nearby. Once, as they say, it was settled by noble Syrians, but after them by Greeks who went over against the Amazons, Autolycus and Phlogius with Deileon, who were Thessalians, and then Habrondas, a Milesian by race. He seems to have been killed by the Cimmerians. And after the Cimmerians, again Coos and Cretines, who were fugitives from the Milesian borders. These rebuilt [συνοικίζουσι] it when the Cimmerian army was campaigning in Asia.

Both passages clearly indicate Cimmerian activity on the Sinop peninsula, and indeed connected with the founding of Sinope, one of the most important

[13] Tsetskhladze 1999, 471, 483–86; 2011, 116–17; 2012b, 237. *Contra* his earlier view (1994a, 116) and de Boer (2006, 44) – see above.

[14] Dönmez 2011, 135; 2019a, who abandoned the term 'Cimmerian' after criticism by other authors, like Tsetskhladze 1999, 483–86.

[15] Tsetskhladze 1999, 483–86; 2011, 115–17.

Greek colonies on the Black Sea coast.[16] In addition to these two passages, there is also a piece of information provided by Strabo (12. 3. 8), referencing other, unspecified, sources, that the Enetoi were a people neighbouring the Cappadocians who campaigned with the Cimmerians and ended up on the Adriatic.[17] Strabo mentions this in the context of his examination of the Enetoi and how from Paphlagonia they found themselves in the Veneto; and despite the fact that the information is confirmed by no other source,[18] it is worth noting that the location of the peoples more or less coincides with the area around Sinope. Irrespective of the degree of its correspondence with historical reality, this tradition was apparently inspired by a passage of the Cimmerians through the general area sometime in the past, and the fact that the Cimmerians are mentioned as partners of the Enetoi in an expedition might reflect how they were perceived by the creators of this tradition.

Of what nature was the Cimmerian sojourn in the central part of the southern Black Sea littoral, and when did it take place? Considering the approximate times estimated for their movements in Anatolia (see above), the Cimmerians must have entered the peninsula from the east at the end of the 8th century BC. If they had already conquered Phrygia in the first half of the 7th century (some scholars say at its very beginning), they would have passed along the northern coast and through the Sinop region earlier, so sometime between the last years of the 8th and the first of the 7th century. Whether it was just one branch that headed there is another question, as is the route they followed – for example, did they go along the narrow coastland starting from the east, or up through the Halys valley?

Once again, the lack of archaeological finds plays a critical role, and controversial views predominate in the irresolute landscape of modern scholarship. The sole object that has been attributed to the Cimmerians is a silver bowl found at Unie, to the east of Amisos,[19] but it should rather be considered a work of Persian art.[20] Thus, we are forced to confine ourselves to the written evidence and must endeavour to look behind the words, acknowledging that this cannot produce the kind of conclusions we would wish.

The crucial word in Herodotus' account is κτίσαντες. It is the participle of a verb clearly signifying that the Cimmerians built a settlement. Pseudo-Scymnus is less clear, and focuses more on the impetuous nature of the Cimmerians, who killed the Milesian settlers they found at Sinope. After they left

[16] On its founding, see Section 4.3.2 in detail.
[17] On the Enetoi, see Section 2.7.
[18] See other traditions on the migration of the Enetoi in Section 2.7.
[19] Akurgal 1967.
[20] Ivantchik 2001a, 96.

the region to continue their campaign in Asia, other Milesians rebuilt the city. The verb used here is συνοικίζουσι, and it has been rather convincingly argued that its meaning is to rebuild a city after its destruction.[21] We may therefore assume that the Cimmerians destroyed the Greek settlement, which had to be rebuilt after their departure, but can draw no conclusions as to how low they remained there, although the implication is that they did not settle there for any length of time, for otherwise we would not hear of the rebuilding of an older, destroyed, settlement.

In any case all this belongs to speculations of a philological nature, since the choice of words might have been somewhat random and we are unaware of the formulation of the source text used by the two authors, if it was indeed the same one.[22] However, it must be noted that the two accounts seem to be at variance with one another. The same variance marks the modern literature: J. Hind relies more on Herodotus, proposing a long Cimmerian settlement in Sinope,[23] while Ivantchik finds such a settlement highly improbable, preferring to follow Pseudo-Scymnus, whose version indicates a short presence or merely a destructive raid.[24] Either could be right, but I incline towards the latter case. Not so much because the information given by Pseudo-Scymnus for the founding of cities is generally considered reliable[25] whereas the validity of Herodotus' information on the Cimmerians is problematic,[26] but rather because of the nature of the Cimmerian presence in Anatolia as it is described by most sources, which present a nomadic people invading a huge land and trying to proceed through it as far as possible, showing no interest in settling anywhere along their way. Apart from that, there is nothing to prove a short Cimmerian sojourn in the Sinope area, and certainly no archaeological evidence, as has already been said. Therefore, the absence of any such evidence – in an in any case practically unexcavated region – is not the most apposite argument in this debate.[27] Neither is the nature of the natural landscape around the Sinop peninsula, which is cut off from the hinterland.[28]

[21] Ivantchik 1998, 298–99, n. 3, with all the previous bibliography. *Cf.* Müller 1855, 236.

[22] See more above.

[23] Hind 1988, 214.

[24] Ivantchik 1998, 311, n. 42; 2012, 65–66.

[25] See, for example, Roebuck 1959, 119; Hind 1999, especially 34; Summerer 2005a, 147, for the case of Amisos; Saprykin 1997, 21, for the case of Heraclea.

[26] Tsetskhladze 1999, 483–84.

[27] As, for example, by Ivantchik (1998, 311, n. 42; 2012, 65–66).

[28] Ivantchik 1998, 311, n. 42; 2010, 66; 2012, 65–68. In the latter publication Ivantchik appears a bit confused as regards both archaeological finds and the landscape. On the one hand, he recognises Cimmerian objects immediately to the south of the Pontic Mountains (as opposed to the coast), and specifically at Imirler, to the north-west of Amasya, some 150 km to the south of Sinope (*cf.* also Ivantchik 2001a, 42–49), while, as we have seen, we have absolutely no finds

To sum up, the only evidence we have so far for a possible Cimmerian presence on the southern Black Sea littoral are two Ancient Greek passages, by Herodotus and Pseudo-Scymnus, to which one could add a rumour recorded by Strabo – that is, three texts composed long after the events they mention. Nothing more, neither archaeological evidence nor Akkadian texts. From this scant evidence we might deduce that this Cimmerian presence was limited to the central part of the littoral, around the Sinop peninsula and lasted for a rather short time somewhere between the late 8th and the early 7th centuries BC. What the Cimmerians seem to have achieved during this sojourn is what they usually did in Anatolia: they caused trouble, probably plundering or even destroying a Greek settlement on the Sinop peninsula,[29] before carrying on with their invasion of Anatolia.

2. The Scythians

The question of the Scythian presence in Anatolia, and subsequently on the southern Black Sea littoral, is interrelated with that of the Cimmerians, examined in the previous section, since it was in pursuit of the latter that they are supposed to have entered the peninsula. Despite our greater general knowledge of the Scythians, which compared with the case of the Cimmerians is considerably fuller and additionally enriched through several objects of their material culture, their activity or even presence on the Black Sea coast is almost equally obscure. We are again limited to scraps of written evidence, which provide data of a different type from that existing for the Cimmerians, and moreover do so indirectly, while once again there is no help to be had from archaeology.

that can certainly be considered Cimmerian. On the other hand, while he initially says that the founding of a village in Sinope by nomads is difficult to imagine because of the huge mountains, whose passes would be controlled by the local populations, he soon suggests that the nomads could nonetheless ransack the region, which was after all scarcely inaccessible to those who had succeeded in crossing the Caucasus (2012, 67–68), citing in confirmation the archaeological finds at Imirler... However, the finds at Imirler were later dated to the early 8th or even the late 9th century BC, like other similar finds from elsewhere, which would automatically exclude any connection with the Cimmerians (Hellmuth 2008, 105–10). Dönmez (2019a, 465) does not seem to be disturbed by this dating, and indeed claims that it points to an even earlier presence of the 'Proto-Scythians' or 'Early Scythians' in Anatolia, as far back as the Middle Bronze Age. Specialists in Cimmerian and Scythian scholarship will assess this theory in the future.

[29] More thoughts about the kind of this settlement are provided in Section 4.3. According to a version, Mariandynus, the eponymous hero of the Mariandynoi, was son of Cimmerius, the eponymous hero of the Cimmerians (Domitius Callistratus in Schol. in Apol. Arg. 1. 1126 – *cf.* Section 2.4.2). However, this version was created within the frame of later Heraclean traditions, and cannot considered an indication of Cimmerian presence in the area of Heraclea. Ivantchik 2005a, 129–34.

In general, our corpus of written information on the Scythians is comprised of Akkadian cuneiform texts together with Greek and Latin works. Through the former, it happens that the earliest data we have concerning the Scythians concern events connected with their campaigns in the Near East. Early in the third decade of the 7th century BC the Scythians were reportedly at the Assyrian border, while some scholars suggest that the Scythians might have started raiding Transcaucasia and Media even before 680 BC.[30] As far as the Greek sources are concerned, most data come from the usual suspect, Herodotus, but we must bear in mind that his information is not always reliable, since it is often mixed with folk tradition and exaggeration.[31] Nor is it confirmed by other sources – on the contrary, other authors sometimes give conflicting data: the most typical example concerns the duration of the Scythian presence in Asia Minor, which is estimated by Herodotus at 28 years (Herodotus 4. 1; 1. 106) and by Pompeius Trogus at only eight (Justin 2. 5. 1).

But let us first see what the historian from Halicarnassus has to say on the subject, in the Books 1 and 4 of his *Histories*: in the 7th century BC the Scythians burst into Asia in pursuit of the Cimmerians whom they had driven out of Europe,[32] and entered Median territory, when their king was Madyes, son of Protothyes (Herodotus 1. 103; 4. 1). The way they followed was not along the eastern Black Sea coast, as is considered to have been the case with the Cimmerians; rather, they took 'the upper route, which is much longer, keeping the Caucasus upon their right' (1. 104; 4. 12). They invaded Media and, after defeating the Medes, became masters of Asia (1. 104; *cf.* 1. 73). This dominion lasted for 28 years, during which time their insolence and oppression spread ruin on every side. 'For besides the regular tribute, they exacted from the several nations additional imposts, which they fixed at pleasure; and further, they scoured the country and plundered every one of whatever they could'[33] (1. 106; 4. 1). Herodotus ends by assuring his readers that 'this is another account that is common both to Greeks and barbarians' (4. 12). The word 'other' (ἄλλος) indicates that this was not the sole version of their history.

Confirmation of this comes in Justin's epitome of Pompeius Trogus, where, as already noted, the Scythian presence in Asia Minor is reported to have lasted for eight years (Justin 2. 5. 1). Apparently drawing on several different sources,[34] the author also provided his readers with another two narratives of

[30] Most recently, see Ivantchik 2018, with bibliography.
[31] See, for example, Ivantchik 2018.
[32] See Section 3.1.
[33] Translation by George Rawlinson, New York 1942.
[34] See Ivantchik 1999, 497–500.

Scythian incursions into Asia. One of these being most probably fictitious,[35] we are left with two versions from Pompeius Trogus plus one from Herodotus, for a total of three. One of the former differs from Herodotus chiefly in the duration of the Scythian presence in Asia (eight years as opposed to 28) and a few other details.[36] The other, which in Justin's epitome appears just one chapter earlier (2. 41–48), speaks of two Scythian youths of royal lineage who, being driven from their country by a faction of the nobility, took with them a numerous company of young men and founded a settlement on the coast of Cappadocia, near the Thermodon, occupying the plains of Themiscyra. Here, having made it their practice for several years to rob their neighbours, they were eventually, by a joint force of the surrounding people, cut to pieces in an ambuscade. Diodorus, writing in the same period (1st century BC), elaborated on this version, adding that at some later time the descendants of those two youths (here called kings) subdued much of the territory beyond the Tanaïs river as far as Thrace and, advancing with their armies to the other side, extended their dominion as far as the Nile in Egypt. Thus, after enslaving many great peoples they pushed the boundaries of the Scythian empire on the one side as far as the ocean to the east and on the other to the Caspian Sea and Lake Maeotis.

> It was by these kings that many of the conquered peoples were removed to other homes, and two of these became very great colonies: the one was composed of Assyrians and was removed to the land between Paphlagonia and Pontus, and the other was drawn from Media and planted along the Tanaïs, its people receiving the name Sauromatae (Diodorus 2. 43. 3–7).[37]

Diodorus' passage is significant, because it begins by revealing the provenance of the account: all these events, he says, took place 'as the Scythians recount the myth'. Scythian tradition clearly lies behind this version, and it has been suggested that a similar source was used by Herodotus, so that in practice our Greek sources reproduce (multiple) Scythian traditions concerning the presence of the Scythians in Asia.[38] These would contain many exaggerations and other inaccuracies, this time introduced by the Scythians, and so we may imagine a case equivalent to the issues arising from the one-sided presentation of events narrated by the Greek authors that we have been used to take into account when examining the presence of the Greeks and indigenous peoples on the littoral.[39] Be that as it may, we should not exclude interference in these

[35] The events it recounts are dated to the 2nd millennium BC.
[36] For which see Ivantchik 1999, 499.
[37] I shall return to this version later in this section.
[38] Ivantchik 1999, 500–09.
[39] See, for example, Part 2.

traditions from the Greek authors dealing with the Scythians,[40] or additional versions from the Asiatic peoples who encountered the Scythians in their lands.[41]

The result has been that modern bibliography is marked by severe scepticism towards many aspects of the Scythian presence in the Near East, including the very nature of that presence, at least as presented by Herodotus.[42] For example, it has been argued that it was 'not a rule, but one or several successful and lengthy raids, during which the Scythians never lost touch with their main territory, the steppes of the North Pontic region and the northern Caucasus'.[43] Another demonstrable problem with Herodotus' account is that while the Scythians are reported to have invaded Asia in pursuit of the Cimmerians, the two peoples appear to have followed different routes, passing respectively east and west of the Caucasus.[44] Furthermore, there is no certainty about whether the Scythians continued to the west of the Halys in Asia Minor or not.[45]

Finally, there is the crucial matter of precisely when in the 7th century BC the Scythians invaded Asia. With the inability of the Greek sources to help establish an accurate time taken as given, at least two different theories have been formulated: one, that the Scythians entered the Near East in the first half of the century, specifically about 680–670 BC, and left towards its end,[46] the other that the Scythians were present in Asia Minor between 626 and 616 BC.[47] Consequently, there is also a vagueness regarding the time span between the Scythian and the Cimmerian presence in Asia Minor, which might thus be longer than indicated by Herodotus' statement that the Scythians burst into Asia in pursuit of the Cimmerians.[48]

As with the case of the Cimmerians, one cannot resort to archaeology in search of more concrete answers. Several attempts have been made to identify archaeological remains in the Near East as Scythian or at least connected to

[40] To which we should add among others Polyaenus (*Strategemata* 7. 44. 2), Ammianus Marcellinus (22. 8. 41) and Curtius Rufus (7. 8. 17–18), all with brief accounts.

[41] *Cf.* Ivantchik 1999, 507.

[42] See the relevant bibliography in Ivantchik 1999, 497, n. 2.

[43] Ivantchik 1999; 2018.

[44] Tokhtas'ev 1996, 23–24; Olbrycht 2000a, 83.

[45] For what exactly Herodotus might have meant by 'Asia', see Vaggione 1973; *cf.* Ivantchik 1999, 507.

[46] See indicative bibliography in Tsetskhladze 2011, 96. *Cf.* Tsetskhladze 2019a, 10.

[47] Ivantchik 1999, 511–17. An even later period for the end of the Scythian presence in Anatolia is proposed by Dönmez (2019a, 471–73), who suggests that the Scythians were still there in the early 5th century BC, albeit subject to the Persians. Tsetskhladze (2019a, 11) finds this rather unlikely.

[48] See, for example, the several views in Goetze 1957, 193; Olbrycht 2000a, 83; Kõiv 2007.

the Scythians,[49] but we are still far from being able to prove such identifications. Almost all these finds are of the same categories (*kurgan*-type burials, horse burials, arrowheads, etc.) as those that have been linked with the Cimmerians, who as we have seen come under the umbrella term 'Early Scythian culture', so that the main problem is actually the same as with the identification of Cimmerian finds:[50] such find categories may have belonged to any nomadic people.[51] It is, therefore, extremely difficult to distinguish between 'Scythian' and 'Scythian-type' objects, especially those found in areas that have been traversed by several nomadic peoples apart from the Scythians, with the result that modern literature is characteristically hesitant when it comes to using these terms, which are therefore generally used together.[52] To do otherwise, it seems, might be misleading, since it can almost never be proven that a certain object in Anatolia was made by Scythians.[53] Besides, one should also take into consideration that the Scythians were probably influenced by the Near Eastern cultures they encountered, which makes it difficult to identify the actual creators of the objects that have been considered Scythian.[54]

But it is not only objects found in Anatolia that have been associated with the Scythian invasion. Many of these objects come from sites in the Halys valley, such as Akalan, Pazarli and Maşat Höyük, where a number of architectural remains have been also preserved. It appears that some settlements, like Akalan and Pazarli, were fortified, and this fortification has sometimes been attributed to the invasion of the Scythians (and probably also the Cimmerians).[55] The problem here seems to one of chronology and arises out of the dating of the decorated architectural terracottas from these sites. Despite the several related theories, including those concerning their origin, it is now widely accepted that these terracottas should be dated to the 6th century BC[56] and the

[49] See, indicatively, Olbrycht 2000b, 108; Ivantchik 2001a, 21–96, especially 57–70; Dönmez 2006, 20; 2007a; 2007b, 1211; 2011, 138; 2012, 169.

[50] See Section 3.1. *Cf.* Ivantchik 2001a. For the difficulty in ascribing a particular archaeological culture to the Scythians, see Tsetskhladze 2011, 120–21.

[51] Tsetskhladze 1999, 471, 483–86; 2011, 105, 116–18 (where he notes that some arrowheads found in Anatolia could have belonged even to the Achaemenids); 2012b, 237; Hellmuth 2008. For the so-called 'Scythian-type arrowheads', see Dönmez 2007a.

[52] Or just the latter. For Scythian and/or Scythian-type finds in Georgia, see Tsetskhladze 2011, 96–115.

[53] *Cf.* Tsetskhladze 2011, 117, with bibliography.

[54] Tsetskhladze 1999, 476–78.

[55] Dönmez 2006, 20; 2007a, 61–62; 2007b, 1208–11; 2012, 169.

[56] Summerer 2005a, 148; 2005b; 2007, 31–34; 2008, 267–78, with bibliography for other views. Even Dönmez (Dönmez and Ulugergerli 2008, 253–54) seems to have later accepted the dating of the architectural elements of these sites to the 6th century BC. See more recently Manoledakis 2018b, 199, n. 106, with more bibliography, also as regards the origins of the terracottas.

appearance of fortified settlements in general to the latter part of the 7th.[57] As has been shown, the Scythian presence in Anatolia had already ended by that time, or at least was about to end, making it probable that 'other explanations may need to be sought for the appearance of fortified settlements there than as a defence against Scythian nomads'.[58]

It is time now to turn our attention to the southern Black Sea littoral. One cannot help but be struck by the fact that in all the written sources mentioning the Scythian presence in the Near East (which are presented above), the only Anatolian place names that appear belong to our area of interest: that is, the land east of the Halys. In the version featuring the two Scythian youths of royal extraction, we are told that they took with them a numerous company of young men and founded a settlement on the coast of Cappadocia, near the Thermodon, occupying the plains of Themiscyra. There they lived for some time by robbing the surrounding peoples, until the latter eventually ambushed and killed them (Justin 2. 4. 1–8). In a variant of this version, we hear that sometime later the descendants of these two youths subjugated a vast territory as far as Egypt, conquering many great peoples and resettling them in new homes, some of which became great colonies; among these were the Assyrians, who were removed to the land between Paphlagonia and Pontus (Diodorus 2. 43. 3–7).[59] The territory in both accounts is actually the same, and one cannot help but suspect that, notwithstanding the obvious time difference between their protagonists, they both belong to the same tradition, which, as Diodorus says, was related by the Scythians (see above).

Although the Scythian tradition was obviously characterised by exaggerations, it is noteworthy that, as far as the Scythian presence (irrespective of what kind) in the specific area is concerned, further written evidence exists. Strabo, for example, records that the Sacae, who were Scythians (*cf.* Herodotus 7. 64), advanced as far as the country of the Cappadocians,

> particularly those situated close to the Euxine, who are now called the Ponticoi.[60] But when they were holding a general festival and enjoying their booty, they were attacked by night by the Persian generals who were then in that region and utterly wiped out. And these generals, heaping up a mound of earth over a certain rock in the plain, completed it in the form of a hill, and erected on it a wall, and established the temple of Anaïtis and the gods who share her altar – Omanus and Anadatus, Persian deities; and they instituted an annual sacred festival, the

[57] Tsetskhladze 2011, 119.

[58] Tsetskhladze 2011, 119. See also the similar case with the finds at Imirler, mentioned in Section 3.1, n. 28, as well as the bronze hilt found in Sinope, which has been attributed to the 'Eurasian horse-riding nomads' (Dönmez 2012, 171; 2019a, 465).

[59] See also Section 2.8.3C.

[60] See Section 2.16.4.

> Sacaea, which the inhabitants of Zela (for thus the place is called) continue to celebrate to the present day (Strabo 11. 8. 4; *cf.* 12. 3. 37).

And he continues:

> Now this is the account which some writers give of the Sacae. Others say that Cyrus made an expedition against the Sacae, was defeated in the battle, and fled; but that he encamped in the place where he had left behind his supplies, which consisted of an abundance of everything and especially of wine, rested his army a short time, and set out at nightfall, as though he were in flight, leaving the tents full of supplies; and that he proceeded as far as he thought best and halted; and that the Sacae pursued, found the camp empty of men but full of things conducive to enjoyment, and filled themselves to the full; and that Cyrus turned back, and found them drunk and crazed, so that some were slain while lying stupefied and asleep, whereas others fell victims to the arms of the enemy while dancing and revelling naked, and almost all perished; and Cyrus, regarding the happy issue as of divine origin, consecrated that day to the goddess of his fathers and called it Sacaea; and that wherever there is a temple of this goddess, there the festival of the Sacaea, a kind of Bacchic festival, is the custom, at which men, dressed in the Scythian garb, pass day and night drinking and playing wantonly with one another, and also with the women who drink with them (11. 8. 5).

These accounts are unlikely to be based on Scythian tradition, since they refer to the inglorious end of that people in the region, as well as the establishment of the temple of Persian deities.[61] But they do affirm a Scythian presence there, and actually in the time of Cyrus, the 6th century BC. Let us recall here Elderkin's theory that the cult of Apollo must have come to Scythia from the land of the Leucosyroi, as suggested by the Scythian name of the god, (G)oetosyrus (Herodotus. 4. 59. 2), which has been considered 'a loan word, which came with the god from the region south of the Black Sea; a compound adjective, the first part meaning *white* and the second *Syrian*'.[62] This would also indicate cultural interaction between the two areas.

But there is further evidence, which could be considered even more significant. As we realised when examining the indigenous peoples of the littoral in the previous part of the book, many of them, especially to the east of the Halys – the Leucosyroi, the Chalybes, the Tibarenoi, the Mossynoikoi, the Macrones, the Becheires, the Sapeires, even the Caucones, who lived more to the west, in Paphlagonia – were connected to the Scythians, even described as Scythian nations, by later authors.[63] We should bear in mind that most of these were scholars belonging to the period of Late Antiquity, and that the vast majority of the works in question are either dictionaries, such as Hesychius' *Synagogē*,

[61] It could well have been derived from a Persian tradition.

[62] Elderkin 1935, 342. See more in Section 2.8.3C.

[63] See Sections 2.5, 2.8.3C, 2.9.4, 2.10.3, 2.11.4, 2.13.2, 2.14.2 and 2.14.4, as well as 2.1.

the *Suda* and the *Etymologicum Magnum*, or commentaries on ancient works, like those on Homer, Aeschylus, Callimachus, Apollonius and Dionysius Periegetes (Scholiasts and Eustathius). Besides, in some cases there are also other theories on the origins of the same peoples presented in the same commentaries, as, for example, the Macrones, who appear in the *Scholia* on Apollonius both as a Scythian nation and as Euboean colonists.

Such observations make us unavoidably wary of these attributions, since they might have been based on several different traditions created by different peoples at different times. We have, therefore, been trying to find common anthropological features shared by the Scythians and specific indigenous peoples of the littoral. In some cases such similarities were indeed traced: metalworking, consumption of dolphin oil, ferocious behaviour against enemies, and outrageous customs such as devouring human flesh or throwing the ill to the dogs (Chalybes, Mossynoikoi and Becheires),[64] but would these similarities be enough to lead to conclusions about ethnic kinships or common origins? And what about the other peoples, for which nothing resembling a Scythian trait was ever recounted?

The truth is that these similarities are not enough for such conclusions, although this does not mean that the peoples in question necessarily had no connection with the Scythians. The Leucosyroi are one intriguing case; the Chalybes, who – be it noted in passing – lived in the same neighbourhood, are another, for here too there is a classical author, about a generation earlier than Herodotus, who associates the southern Black Sea nation with the Scythians. In Aeschylus' *Seven Against Thebes* (728), produced in 467 BC, a character called Chalybos is said to come from Scythia, while in *Prometheus Bound* (709–735) the Chalybes are presented as neighbouring Scythia, but somewhere to the north of the Caucasus, in a peculiar and fictional geographical description of the whole eastern part of the Black Sea area.[65] The scholiasts of the two plays also mention a Scythian city called Chalybos (Schol. in Aesch. Theb. 728), as well as the 'Chalybian land of Scythia' (Schol. in Aesch. Prom. 301). Furthermore, Eustathius (*Comm. in Dion. Per.* 767) conveys the information, provided by others, that the Thermodon, which flows close to the land of the Chalybes, was considered a Scythian river.[66] Whether following a Scythian tradition, as in Diodorus' account, or not, these references to the Scythians indicate a perception of the Chalybes as kindred of the Scythians from as early as the 5th century BC.

[64] See Sections 2.9.4, 2.11.4 and 2.14.2.
[65] See Section 2.9.1.
[66] For all this, see Section 2.9.4.

Such cases seem to be of particular importance, and there are also some reverse instances. Several decades ago, the existence of a city called Scythopolis in Palestine, first mentioned by Polybius (5. 70. 4), had been attributed to the Scythian raids in Asia in the 7th century BC.[67] However, more careful studies rejected this attribution, especially since the excavations there (modern Beth Shean) demonstrate destruction of the Iron Age settlement in the 8th century BC and reoccupation in the Hellenistic period, when the name Scythopolis appears. It is therefore now suggested that the name Scythopolis had nothing to do with the Scythian raids in the Near East, which took place after the destruction of the site and in any case could not have accounted for the foundation of any permanent settlement, but should rather be connected with the installation of 'Scythian' veterans of Ptolemy II Philadelphus' army there, as cleruchs.[68] These were not necessarily pure-blooded Scythians; many were natives of the Greek cities of the Bosporan kingdom.[69]

There is, however, another name, this time of a people and much closer to our area of interest, which also hints at a Scythian presence and which has not yet been given the attention it merits in the bibliography: the Scythenoi (*Σκυθηνοί*). Recounting the return of the Ten Thousand, and in particular the part covering the march towards the (Black) sea, Xenophon describes passing successively through the lands of the Taochoi, the Chalybes,[70] the Scythenoi, the Macrones and finally the Colchians, on the coast, around Trapezous (Xenophon *Anabasis* 4. 7–8). His narrative is quite detailed and may be summarised as follows: leaving the country of the Chalybes, the Greeks passed the Harpasus river and the land of the Scythenoi (4. 7. 18), where they reached the city Gymnias (4. 7. 19). After five days marching from this city they reached Mt Theches, from where they could see the sea (4. 7. 20–21). Descending this mountain they reached the country of the Macrones, which was separated from that of the Scythenoi by a river (4. 8. 1).

As we have already noted,[71] given the distances recorded by Xenophon and the reported sequence of the lands they traversed, it is hard to say whether the Greeks followed a continuous northwards route, which would place the Scythenoi to the north of the Chalybes who were north of the Taochoi, or made

[67] See, for example, Sulimirski and Taylor 1991, 567, n. 68.

[68] For Ptolemy's practice of keeping mercenary soldiers in his service by endowing them with land, see Chamoux 2002, 249.

[69] Avi-Yonah 1962, 123–27, with the written sources; Ivantchik 1999, 509–11; Schoville 2005, with the previous bibliography. Jones (1971, 232–33, 241) saw no connection of the site with the Scythians at all.

[70] Those of the hinterland. See more in Section 2.9.3.

[71] In our examination of the Chalybes: Section 2.9.3.

a rather long curve to the east before turning north and west, in which case the Chalybes and the Scythenoi would be almost east–west neighbours. The two places mentioned as borders of Scythenian territory, namely the Harpasus and the river near Mt Theches, are almost on the same latitude, which would seem to indicate a westward or at least a north-westward route; but that does not concern us here. Whatever the case may have been, Xenophon refers clearly to a people called Scythenoi who lived just to the south of our well-known Macrones.[72] The name is indeed striking and it is impossible not to suspect a relation with the Scythians.

Poring over Xenophon's narrative, we realise that the Scythenoi were not – or, at least, are presented as not being – a negligible populace encountered by chance, even if they would never again be mentioned in the written sources.[73] Their territory included several villages on a level plain, as well as a large and prosperous inhabited city called Gymnias. From the villages the Greeks were able to collect provisions and thus remained there for three days (4. 7. 18), while from Gymnias the ruler of the area, and probably of the Scythenoi, sent the Greeks a guide who led them through hostile territory to Mt Theches (4. 7. 19–21). It was from this mountain that the Greeks saw the sea, in one of the most celebrated narrative incidents in ancient Greek literature, which thus took place in the land of the Scythenoi. And it was the Scythenoi who helped the Greeks reach this destination, even if it was in order to secure their aid in destroying a hostile territory rather than from any desire to be helpful (4. 7. 20).

We acquire the image of a land that stretched from the Harpasus in the south to Mt Theches in the north, and a nearby river which separated it from the land of the Macrones.[74] The Harpasus is almost unanimously identified with the Çoruh Neri, and Mt Theches with the Zigana Pass, which is about 160 km directly west of the part of the Harpasus supposedly reached by the Greeks.[75] Gymnias is generally placed in the area of Bayburt, a bit further south, about 76 km from the coast. The river that formed the border between

[72] See also Section 2.13.1.

[73] Apart from Diodorus (14. 29), who repeated the account, probably of Ephorus, distorting the names (River Harpagus, Scytinoi, Gymnasia, Mt Chenion). See also Aeli Herodianus *De prosodia catholica*, *s.v.* Σκυθηνός; Stephanus of Byzantium, *s.v.* *Σκυθηνοί*; Pseudo-Zonaras *Lexicon*, *s.v.* Ἄβαρις.

[74] This river cannot have been a major one, since, as Lendle (1995, 281–82) points out, all the big rivers of the littoral flow naturally northwards, into the Black Sea, while the river in question must have flowed east–west if it separated two countries lying one to the north of the other.

[75] For the several related opinions, see Lendle 1995, 270–72, 276–82. As regards the average east–west distance of the 160 km, which indeed coincides with the roughly 7 stades and 30 parasangs estimated from Xenophon's account, compare the similar route presented in Section 2.9.3, and see also Rood 2010, 52.

the Scythenoi and the Macrones could have flowed near Hamisköy.[76] Thus, the land of the Scythenoi covered an area of at least 40 km from north to south, and 160 km from east to west, making it larger than the lands of several of the indigenous peoples of the littoral.

Moreover, the land of the Scythenoi was actually no great distance from the coast, only about 36–76 km as the crow flies, and also very close to the peoples inhabiting the eastern part of the littoral, who have been considered 'Scythian nations'. Unlike Scythopolis, whose name can be explained by its founding and the installation of 'Scythian' veterans in the Hellenistic period, the Scythenoi are a people who were well settled in the large country outlined above in the 5th century BC. Looking at the history before this time,[77] I can see no recorded event that could explain the name of this people other than the Scythian raids in the Near East. Who these Scythenoi were, how they came to be established there and how they became known by that name[78] are matters we can scarcely even guess at. But it would have been extremely difficult to ignore the Scythian presence in the area as the most probable explanation for their existence.[79] After that, the way was open for the emergence of traditions associating several coastal peoples near the Scythenoi with the Scythians, traditions that remained alive until very late in antiquity, even if some of them seem far-fetched, since they refer to people living even west of the Halys, like the Caucones.[80] The degree to which all these traditions reflected reality is obviously indeterminable, but their variety, their reference to the same general area, and the fact that they were spread by both Scythian and Greek sources, are elements that cannot pass unnoticed. In one way or the other, the Scythians had left their mark on the southern Black Sea littoral.

[76] Lendle 1995, 272–73. For the region between Gymnias and Trapezous, see also Bryer and Winfield 1985, 48–53; Edwards 1988, 128, focusing especially on marshlands in north-east Anatolia.

[77] And considering the form and the ending of the name *Σκυθηνοί*.

[78] Had, for example, a group of invading Scythians decided to settle there while en route for or returning from the Near East? Was it a case of intermarriages with a local population? It is worth noting that the only hypothesis concerning the Scythenoi that has been advanced so far is that they might have been immigrants (Lendle 1995, 272). Their later destiny is also unclear, since there is no mention of them in the written sources after Xenophon and Ephorus (Diodorus), who are quoted by the later authors.

[79] See also the tradition about the battles between Scythians and Persians in the 6th century BC mentioned above.

[80] Strabo (12. 3. 5) says that, according to 'some people', the Caucones were Scythians. He does not, unfortunately, say who these people were or when they lived, but he takes care to record all the other versions concerning their origin, which in fact are not few (see Section 2.5, in detail). The attribution of Scythian origin to the Caucones was also known to the Scholiasts of Homer (Schol. in Hom. Il. 10. 429).

3. Under the Persians

Croesus, king of Lydia, was the first ruler reported to have subjugated many of the peoples inhabiting the southern Black Sea (in the mid-6th century BC), and is thus in fact the first known historical person involved with a large part of the littoral.[81] According to Herodotus (1. 6, 28), he subjugated almost all the nations west of the Halys, among them the Bithynians, the Mariandynoi, the Paphlagonians and the Chalybes.[82] This domination cannot have been of long duration, because soon after the mid-6th century BC Asia Minor had come completely under Achaemenid rule under Cyrus II (559–530 BC).[83] Consolidated under Darius I (522–486 BC),[84] the Achaemenid Empire was organised and divided into administrative regions, for which the term 'satrapy' has prevailed. An idea of this division is provided by Old Persian inscriptions attributed to Darius I,[85] two passages from Herodotus' *Histories* (3. 90–94; 7. 72–79),[86] and, for a later period, the lists of the satrapies that were shaped by the Successors of Alexander the Great, with the Partitions of Babylon (323 BC) and Triparadeisos (321/0 BC).[87]

There are, however, a number of problems with these sources. The first two, which are supposed to refer to the same period, are not in full agreement with each other; either as regards the number of provinces or as regards their geographical territories. The very term 'satrapy' seems problematic, since Herodotus, who uses it, does not seem quite confident of it, using other terms as well, like *arche* and *nomos*. It also appears that the number of these regions was not firmly fixed but varied over time. Moreover, it is highly probable that the Old Persian inscriptions refer to groups of peoples rather than to provinces,[88] as seems to be the case with Herodotus, who, however, describes the provinces by naming the peoples forming them. For these and other reasons, both the Old Persian inscriptions and Herodotus' data have bother scholarship for many decades; both have been considered influential but misleading or unreliable,

[81] For the first locals known by name, see below.

[82] For the perplexing inclusion of the Chalybes among the peoples west of the Halys by Herodotus, see below and Sections 2.9.1 and 2.9.3.

[83] Cook 1985, 209–12, 292–96; Briant 2002, 62; Kuhrt 2007, 135; Dusinberre 2013, 1; Thonemann 2013, 11, 37.

[84] Cook 1985, 217–25; Briant 2002, 62–63; Kuhrt 2007, 135; Dusinberre 2013, 1.

[85] See, for example, Kuhrt 2007, 11, 136–57.

[86] The first refers to the division into 'satrapies' under Darius, and the second to the peoples (in our area of interest) who joined Xerxes in the Persian Wars, around 481/0 BC, and their equipment.

[87] See Klinkott 2000.

[88] See, indicatively, Cameron 1973.

and several theories have been put forward[89] in favour of one or another view, or even the one or the other source.[90] In the end, the only firm conclusion to date is that no firm conclusion can be drawn about the matter.

The divergence between the sources is also evident in our area of interest. Following the Old Persian inscriptions, there are only two provinces that seem to have included the southern Black Sea littoral: Cappadocia and Armenia, the former embracing a much larger part of it.[91] In Herodotus, however, things are somewhat different. Of his twenty 'satrapies', the third one included the Asiatic Thracians (Bithynians), the Paphlagonians, the Mariandynoi and the Syroi, together with the Hellespontians, 'of the right coast as one enters the straits', and the Phrygians (Herodotus 3. 90). The 19th 'satrapy' included the Moschoi, the Tibarenoi, the Macrones, the Mossynoikoi and the Mares (3. 94). The two provinces are separated more or less by the Thermodon, according to the estimations of the several peoples' territories made in Part 2; they thus differ from the data provided by the Old Persian inscriptions.[92]

But there are further differences. First of all, Herodotus' provinces do not extend as far inland as, for example, the Cappadocia of the Persian inscriptions. We read, for instance, that the 18th 'satrapy' contained the Matienoi, the Saspeires and the Alarodioi (3. 94). The first of these peoples must have been situated somewhere to the south of the Leucosyroi, according to Herodotus 1. 72,[93] but the other two are rather to be placed close to the Caspian Sea,[94] which gives the impression of a somewhat elongated province. There is also the very ambiguous 13th 'satrapy', which includes the Pactyic country, Armenia, 'and the countries adjoining as far as the Euxine sea' (3. 93). Which part of the Black Sea is actually meant here that is not covered by the 19th 'satrapy' or

[89] The bibliography is huge. Merely indicatively, the reader might see Herzfeld 1968, 288–97; Cameron 1973; Kimball Armayor 1978b; Cook 1985, 244–77; Tuplin 1987; Debord 1999; Klinkott 2000, 11–13; Briant 2002, 62–106; Tuplin 2004; Kuhrt 2007, 11, 136–57; Dan 2013; Dusinberre 2013; Michels 2017; Tsetskhladze 2019e, 135.

[90] Sometimes quite obsessively and losing the crux of the matter. This is, for example, the case with Kimball Armayor (1978b), who ended up dwelling on how bad a Father of History Herodotus was in his opinion.

[91] One could entertain the possibility that another province mentioned in the inscriptions, Lydia, included the westernmost part of the littoral instead of Cappadocia, even if Herodotus' division (see just below) does not mention it.

[92] The border between Cappadocia and Armenia would probably have been further east, but how much further? Note also that the Chalybes are absent from Herodotus' 'satrapies', while he had mentioned them among those that had been subjugated to Croesus of Lydia some years before the Persian conquest (1. 28). See also Sections 2.9.1 and 2.9.3.

[93] See also Section 2.6.1.

[94] See Section 2.14.4.

the Colchians, who are named as paying tribute to the Persian king,[95] but not belonging to the 'satrapies' (3. 97)?[96]

Whatever the case may have been, one thing is undeniable: that the southern Black Sea littoral came under Persian rule and remained so for more than two centuries. This was the first time that the area had become part of a great empire, since the Hittites had not managed to annex it to theirs.[97] We must also bear in mind that from the very outset of this Persian occupation there were already Greek colonies planted on the coast: at least the first four primary ones, Sinope, Tieion, Amisos and Heraclea, the first two of these long since.[98] The question pertaining to the scope of this book is: what, in practical terms, did this status mean for both the local peoples and the Greek cities of the littoral – to what degree and in what aspects did the Persian rule interfere with and alter their political, economic and cultural affairs?

As far as politics and administration are concerned, for the first century or more of Achaemenid rule we have virtual silence. There are only two indirect pieces of evidence of Persian presence in the littoral, from the worlds of art and of later tradition respectively. A marble head 0.27 m high found in Heraclea,

[95] And also joined the Persians in the Persian Wars (7. 79). For the position of Colchis in the Persian Empire, see Braund 1994, 122–32; Brosius 2010.

[96] In any case, when studying the history of the littoral under the Persians one needs for the first time to examine different regions separately, but this time regions that were not necessarily occupied by one people nor even belonged to the same 'satrapy' according to one or the other source, which indicates how difficult it must have been to rule a province of such a size as those known as satrapies, and some of the solutions that were therefore invented (see Michels 2017). The ancient texts referring to the Persian administration of the littoral make us follow a different division into provinces, since they mention, for instance, rulers of Paphlagonia or satraps of Cappadocia. In fact, these officials are the first historical persons from the littoral we get to know, figures like Doedalses of Bithynia (Memnon 12. 3–4; Strabo 12. 4. 2), and Corylas and Otys of Paphlagonia (for which see Manoledakis 2019), centuries after the mythical Pylaemenes. In the previous part of the book we have several times pointed out that some of the indigenous peoples of the littoral managed to give their name not only to the territory they inhabited but also to larger one including some of their neighbours, whether because of their cultural or political influence or their domination over some inhabitants of the broader region or for other reasons. Most of the southern Black Sea coast, the part we include in our western and central zones, was belonged to Bithynia, Paphlagonia and Cappadocia. It should nonetheless be stressed that these names did not correspond to territories with fixed borders through time. For example, Bithynia and Paphlagonia are not known as names of 'satrapies' in any of the relevant sources (see above), but are reported to have had different rulers, local or Persian, especially after Alexander's conquest. Later, however, they were names of Roman administrative provinces, again not with the same territories throughout the whole Roman period. As for the part of the littoral east of the Thermodon, with its various indigenous peoples, known widely as Pontus after the establishment of the homonymous kingdom in the 3rd century BC, under the Romans it became initially part of the province of Cappadocia, to be later divided into Pontus Galaticus, Pontus Polemoniacus and Pontus Cappadocicus.

[97] See Section 2.1.

[98] See Section 4.4.1.

which depicts a man wearing a Persian tiara, has been dated to 520–510 BC and is a work not of Persian but of Greek art.[99] This find has been sometimes used as an argument substantiating the good relations between Heraclea and Persia in the late 6th century BC, soon after the founding of that city.[100] S. Saprykin, who put forward this view, pointed out that these relations were responsible for the stability of the oligarchic regime in Heraclea, a regime that was imposed by the Persians, who generally favoured tyrannies.[101] Thus he justified also Justin's information that

> when the Athenians were at the height of power, and, after the overthrow of the Persians, had imposed a tax on Greece and Asia for the support of a fleet, and when all were promptly contributing to the maintenance of their safety, the Heracleans alone, from friendship for the kings of Persia, refused to pay (16. 3. 9).[102]

Accepting this view, Summerer further suggests that Heraclea was among the cities that were involved in Darius' preparations for his Scythian expedition in 513/2 BC, precisely the period from which the marble head dates. Even if there is no source to confirm this, as she admits, the city's geographical position would be a good reason to accept this possibility. Moreover, she cites the narrative of Ctesias (*FGrH* 3c 688 F 13), who relates that Ariaramnes, satrap of Cappadocia,[103] crossed over into Scythia with 30 penteconters, following Darius' order, and carried off a number of prisoners, and names Heraclea and Sinope as the likeliest candidates for the satrap's departure.[104] For all these reasons, Summerer suggests that the marble head form Heraclea belonged to the statue of a Persian ruler, created by the Heracleans within the context of their good relations with Persia in the specific period of time.[105]

Stimulating as these views may be, the truth is that there is no confirmation for them in the ancient literature. There is no positive evidence to testify that tyranny in Heraclea was imposed by the Persians; more probably, it was simply the outcome of social turbulence among the Heracleans themselves.[106] A marble head alone is not enough to attest to good relations between Heraclea

[99] Summerer 2005c, with a thorough examination and full bibliography.

[100] Saprykin 1997, 49. *Cf.* Summerer 2005c, 242–44.

[101] Saprykin (1997, 49–50, 133–34), according to whom the Persians fostered the establishment of tyrannies in the Greek cities in order to bring them under their control. *Cf.* Summerer 2005c, 243.

[102] Eventually, Athens managed to make Heraclea a tribute-paying city, as indicated by the latter's name in the tribute assessment decree of 425/4 BC (*IG* I³ 71. IV. 127; *cf.* Justin 13. 3. 10, on Lamachus' relevant efforts). See below.

[103] The first Persian official, apart from kings, known in relation to the littoral.

[104] Summerer 2005c, 242–44, with bibliography on other candidate cities.

[105] Summerer 2005c, 244–45.

[106] *Cf.* Aristotle *Politics* 1304b, 1305b. Franke 1966, 130; Manoledakis 2021f. Note also that Heraclea struck coins on the Aiginetan standard (Avram *et al.* 2004, 958), like Sinope, even in

and Persia, although such relations might indeed be a good occasion for its creation. As regards Ctesias' account, it would obviously be better if he had mentioned specific cities, to avoid our having to speculate. But we cannot let pass without remark the fact that Herodotus, who recounts Darius' preparations for the Scythian expedition, does not mention any city of the southern Black Sea having come to the king's aid, only cities from the Aegean and the Hellespont (4. 137–138).[107] Finally, the first – and indeed the only – author to speak directly of these relations, Justin, is referring to a much later period than that of the Scythian expedition and the sculpted head.

Nonetheless, whoever the person depicted in the marble statue may have been, the fact is that we have a very early depiction of a Persian in Greek art, and indeed the first one in a statue.[108] This is obviously interesting, even if there is only one such work, especially since the first concrete evidence of Persian political presence in the littoral comes from much later periods (see below).

The second piece of evidence comes from the written sources and pertains to the temple dedicated to Anaïtis at Zela by the Persians, which was mentioned in the previous section on the Scythians (Strabo 11. 8. 4; 12. 3. 37). Even if Strabo's account of this temple may belong to a tradition of unknown time as well as provenance, one of the variants he mentions refers to the period of Cyrus II, thus the very beginning of the Persian occupation. Zela is, of course, far from the littoral, but it is still important that the whole conflict between Persians and Scythians followed a Scythian invasion in 'the country of the Cappadocians, particularly those situated close to the Euxine, who are now called the Ponticoi'. Indeed, the temple of Anaïtis is supposed to have been depicted on later coins of the wider area, where the existence of similar temples in antiquity is also suggested.[109]

Apart from these two pieces of evidence, which cannot prove anything tangible, it is only in the late 5th century BC that we begin to acquire some concrete testimonies, once again thanks to Xenophon. His relevant accounts may not be abundant, but they are enough to provide an idea of the Achaemenid presence and interest in the southern Black Sea. In brief, this presence could be described as discreet or permissive, both towards the local communities and towards the Greek colonies, as long as both were prompt in their tax and

the period of Clearchus, and not on the Persian standard, as did, for example, Amisos (see Section 4.4.1).

[107] In any case, it worth bearing in mind that neither Ctesias nor Herodotus is the most accurate source we could have. See Fol and Hammond 1988, 240.

[108] Summerer 2005c, 145.

[109] Dönmez 2007c, 109.

tribute obligations.[110] Only thus can one comprehend the highly telling incident narrated by the Sinopean envoy Hecatonymus, that 'very recently', i.e. before 400 BC, some Paphlagonians refused to present themselves when summoned by the Persian king, because their archon – who must have been Corylas – was too proud to obey and because they regarded their cavalry as superior to the whole of the king's cavalry (Xenophon *Anabasis* 5. 6. 8). I cannot imagine many empires where such an official attitude towards the king would not have caused immediate and severe repercussions.

Similarly, some years later another Paphlagonian official, Otys,[111] refused to go up to the Persian king when he was summoned by him (Xenophon *Hellenica* 4. 1. 3). Otys seems to have had the habit of ignoring the king's commands, for there is more about him: he made an alliance with Spithridates, a Persian nobleman and official in the court of Pharnabazus II, satrap of Hellespontine Phrygia, against whom he revolted and joined forces with Agesilaus in his war against Persia in 396 BC. Spithridates went with Agesilaus to Paphlagonia, where Otys provided him with three thousand men and married his daughter (Xenophon *Hellenica* 4. 1. 1–14; *Agesilaus* 3. 4; *Hellenica Oxyrhynchia* 21. 6–22. 2; Plutarch *Agesilaus* 11. 2). An explanation for Otys' habit of disregarding the king's commands might lie in the fact that he was a near relative of Datames (they were sons of a brother and a sister), who was the satrap of Cappadocia (Nepos *Datames* 2), to which Paphlagonia must have belonged; Otys might thus have felt that he could afford to be a bit more relaxed toward the commands of the Persian king. However, he eventually (*ca.* 380 BC) paid for this attitude in a rather humiliating way (Nepos *Datames* 2–3; Theopompus 115 F 179), while later Datames managed to take possession of Paphlagonia, and probably even of the powerful city of Sinope (see below).[112]

It must have been extremely difficult to keep a 'satrapy' of such vast size and with so many different peoples under full control and its peoples suppressed,[113] especially those who seem to have stood out as experts in diplomacy and political manoeuvring throughout history, like the very self-confident

[110] For the taxes and tributes exacted by the Achaemenid Empire from its subjugated peoples, see Dusinberre 2013, 35–42. For gold and silver tax in the Achaemenid Empire, see Zournatzi 2000.

[111] For Corylas and Otys, the different official titles of Paphlagonia that are found in the ancient literature, and the territory of Paphlagonia over which these officials ruled, see Manoledakis 2019, with full bibliography.

[112] See more details in Manoledakis 2019, 218; 2021f; Tsetskhladze 2019e, 135.

[113] For the interesting question of possible differences in the status of satrapies, as well as a ranking among the different peoples within a satrapy, see Brosius 2010, 30.

Paphlagonians,[114] for whom, incidentally, more evidence is available as regards our topic here than for any other people of the littoral. Paphlagonia's subjection to Persia was not what an autocratic monarch would wish,[115] and the fact that the Paphlagonians had sent soldiers to Cyrus in Cunaxa (*Anabasis* 1. 8. 5) does not necessarily imply a general Paphlagonian 'obligation to serve in satrapal forces':[116] after all, the Greek mercenaries did exactly the same. The king at that time was Artaxerxes, Cyrus' enemy, and the peoples subjected to him would be expected to send soldiers to Artaxerxes. The Paphlagonians serving with Cyrus should not be considered as the official Paphlagonian forces in the war: they were just one thousand horsemen, who may have been be the only 'barbarians' mentioned by their ethnic name by Xenophon but were still very few compared with the 'more than 120,000 Paphlagonian infantry' (*Anabasis* 5. 6. 9), or the 100,000 barbarians serving under Cyrus (*Anabasis* 1. 7. 10), and even fewer in comparison to the 1,200,000 (*Anabasis* 1. 7. 11) or 400,000 (Plutarch *Artaxerxes* 13. 3, citing Ctesias) soldiers and 6000 horsemen (*Anabasis* 1. 7. 11) serving with Artaxerxes, whose nationalities are not recorded but certainly included Paphlagonians.

In any case, after Otys the Paphlagonians seem to have returned to a greater loyalty towards the Persians, since in *ca*. 361–360 BC Artabazus tried to draw supplies from the '*friendly* lands of Upper Phrygia, Lydia and Paphlagonia' (Demosthenes *Against Aristocrates* 155) for his war against Charidemus in Hellespontine Phrygia.[117] But this does not change the image of the elusive Persian presence in the littoral.[118] After all, Paphlagonia and Paphlagonians do not refer only to the littoral, where the Greek element was by this time as strong as the Paphlagonian or any other local one. Therefore, we must also examine the Persian attitude towards the Greeks and their cities there, as well as their intervention in issues between Greeks and locals.

As far as the latter issue is concerned, Xenophon's account, and especially what he does not say in it, implies a quite loose attitude, at least in the period around 400 BC. Notwithstanding the several instances of friction between Greeks and locals (be it the Colchians, the Drilae, or the Mossynoikoi) or between neighbouring locals, let alone between two factions of the same

[114] On this matter, see in detail Manoledakis 2019.

[115] *Contra* Tuplin 2004, 177. *Cf.* Manoledakis 2019, 217–19.

[116] Tuplin 2004, 177 and n. 74.

[117] For the dating and location of these events, see Heskel 1997, 119. More about Otys in Manoledakis 2019, 218–19.

[118] Some decades later, when Alexander the Great invaded Asia Minor, Curtius (3. 1. 23) mentions that the Paphlagonians did not even pay tribute to the Persians. But this is not confirmed by any other author. For the doubts about Curtius' historical accuracy, see, indicatively, Baynham 1998, 1–2, 57–67.

people, as was the case with the Mossynoikoi,[119] no Persian intervention is ever reported, as if the Persians were completely indifferent to what was happening in their administrative territory.[120] In 400 BC, Hecatonymus, the Sinopean envoy to Xenophon's army, threatens the latter in Cotyora that he will join forces with the Paphlagonians of Corylas to drive them out of his city's colony (*Anabasis* 5. 5. 12). Corylas appears to have expansionist views over Sinope (5. 5. 22–23),[121] but no Persian attitude is mentioned towards these views. The only remarkable mobilisation from the Persian side mentioned by Xenophon took place in Bithynia. There, Pharnabazus, the Persian satrap of Hellespontine Phrygia, sent forces to aid the Bithynians attacking the Ten Thousand, in order to prevent them from entering their province (*Anabasis* 6. 4–5), though unsuccessfully.[122] Nonetheless, this incident is not enough to indicate a steadily peaceful relationship between the Bithynians and the Persians, and Pharnabazus specifically, since Xenophon elsewhere mentions several conflicts between the two sides (e.g. *Hellenica* 3. 2. 2).[123]

As for the progress of Greek colonisation, which had begun before the Persian conquest, not only did the latter not disrupt it but, on the contrary, most of the colonies (both primary and secondary) seem to have been established during this occupation – indeed two of the most important, Amisos and Heraclea, must have been founded only a few years before the conquest but still experienced stable development over time.[124] It is noteworthy that in 400 BC the Greek colonies of Sinope are mentioned as paying tribute to their metropolis (*Anabasis* 5. 5. 7, 10), and one might wonder whether they were also paying tribute to the Persians. Even towards the plans of Xenophon's newly arrived Greeks to establish a colony in the vicinity of Cotyora it is only the aggressive attitude of merchants from Heraclea and Sinope that is mentioned, nothing from the Persian side (*Anabasis* 5. 6. 15–21). In fact, no Persians at all are mentioned by Xenophon anywhere in the littoral.

But there is another strong pointer, irrespective of the events narrated by Xenophon and in fact predating them: namely, the Athenian presence in the Black Sea. There is multiple evidence of this – albeit short-lived – presence in the southern littoral in the third quarter of the 5th century BC. Plutarch

[119] All of them described in the relevant sections of Part 2.

[120] See also Lane Fox 2004, 30–31; Tuplin 2004, 176–78; 2007a, 13, 25–28. *Cf.* Briant 2002, 498.

[121] In fact, Corylas appears at the same time to be a friend of the Sinopean Hecatonymus, who is his official representative at Sinope (5. 6. 11). See Sections 2.6.3 and 2.6.4A.

[122] See also Section 2.2.3.

[123] Michels 2017, 41–42.

[124] See Section 4.4.

(*Pericles* 20. 1) reports an expedition of Pericles to the southern Black Sea, and specifically to Sinope, in the 430s BC or a bit earlier,[125] where he left 13 ships of war and soldiers under the command of Lamachus to aid the Sinopeans against the tyrant Timesileos. Plutarch's account of Pericles' expedition to the Black Sea been considered by some scholars as unreliable, or even an Athenian invention of the 4th century BC.[126] However, we cannot prove that the reference to his visit in Sinope (specifically) is a product of fiction.[127]

Secondly, we have the Athenian re-colonisation and renaming (to Peiraieus) of Amisos in about 437/6 BC, which is mentioned by Strabo (12. 3. 14) and is linked to Pericles' expedition.[128] And finally, there is the presence, with more or less certainty, of some cities of the littoral – Heraclea, Caroussa (less possible) and Kerasous (*IG* I^3 71. IV. 127, 129, 169)[129] – in the tribute assessment decree of 425/4 BC, which indicates that Athens had managed to enrol several cities of the coast among the tribute-paying members of the Delian League.[130] Here too no Persian attitude is mentioned, and the Athenians of Pericles and Lamachus seem to have been able to act undisturbed by the Persians. Had the latter demonstrated a forceful policy towards the Greek cities of the coast at the beginning of their occupation, Athens might not have been able to be so active in the area.

Nevertheless, this situation did not last forever. Shortly before 362 BC Datames, the satrap of Cappadocia, attempted not only to conquer Paphlagonia, but also to attack and besiege Sinope,[131] misleading its citizens as to his alleged desire to help them in their conflict against Sestus. But when the Persian king Artaxerxes II ordered him to stop the invasion immediately, he obeyed, although displeased (Polyaenus 7. 21. 2–5; Aeneas Tacticus 40. 4).[132] A possible reason for this turnaround, if indeed such there was,[133] could be the rapid prosperity that Sinope and other Greek cities of the coast began to know from the 4th century BC, as manifested through archaeological finds in or

[125] The two main theories proposed talk about *ca.* 450 BC and 438–36 BC. See the bibliography in Tsetskhladze 1997 and de Boer 2005, 168. See also Meritt *et al.* 1950, 114–18.

[126] Ferrarese 1974; Tsetskhladze 1997, with bibliography. De Boer (2005, 167) seems to consider this view exaggerated.

[127] *Cf.* de Boer 2005, 167.

[128] Atasoy 2003, 1342–47; Avram *et al.* 2004, 954–55; Summerer 2005a, 149.

[129] For Heraclea, see above.

[130] For the Athenian intervention in the Black Sea, see Braund 2005b.

[131] According to Doonan (2004, 76): 'the walls of Sinope were sufficiently established by the early 4th century BC to frustrate the siege of the breakaway Persian governor Datames'.

[132] Datames is also reported to have had the intention of using Amisos as a mint (Polyaenus 7. 21. 1).

[133] See below.

from the southern Black Sea,[134] including amphorae fragments and coins. Such prosperity would hardly leave a Persian satrap of the area indifferent.[135]

Interestingly, on certain 4th-century coins of Sinope we observe that, although they adhere to the standard type for this century,[136] the legend with the abbreviated name of the city/citizens (ΣΙΝΩ) is replaced by Greek or Aramaic inscriptions with abbreviated names of Persian officials. Two of these are identified with Datames, the satrap of Cappadocia (385/4–362) mentioned above, and Ariarathes, satrap and later king of Cappadocia (350–322),[137] or alternatively with Ariobarzanes, as proposed by other scholars,[138] while the identification of the rest remains difficult. In an interesting attempt to identify the officials and place the coins in historical context C.M. Harrison,[139] countering the earlier views of J.P. Six,[140] identified Orobantes, Hydarnes and Mithropastes, son of the satrap of Hellespontine Phrygia Arsites, and suggested that the coins in question should be dated to the period between 360 and 330/320 BC. Acknowledging that, while not provable, her identifications are quite possible, she attributed the striking of the coins of the three officials to their activities in 340–330 or to the short-lived efforts of the Persians immediately after the battle of the Granicus to repossess areas that had been conquered by Alexander (*cf.* Curtius 4. 1. 34; 4. 5. 13), efforts that took place in Cappadocia and Paphlagonia, which had not totally fallen to the Macedonian king (*cf.* Diodorus 18. 16. 1) and for which the striking of coins would be required.[141]

According to E.S.G. Robinson, however, these issues indicate that Sinope eventually fell into the hands not only of Datames – apparently in a second attempt, probably when he conquered Paphlagonia (according to Nepos *Datames* 5. 6) – but also of other Persian officials as well, until Ariarathes' defeat and death at the hands of Eumenes and Perdiccas in 322 BC and the Macedonians' final conquest of the area.[142] Nevertheless, it must be noted that even during the period of those coins (370–325 BC), coinage with the name of

[134] See in detail Section 4.4.3.
[135] See more in Manoledakis 2021f.
[136] Price 1993, pls. LI–LV.
[137] Harrison 1982, 181, with bibliography; Price 1993, pl. LIII.
[138] Avram *et al.* 2004, 961.
[139] Harrison 1982.
[140] Six 1885, 15–65; 1895, 169–210.
[141] Harrison 1982, 193.
[142] This is the period in which the inscription ΣΙΝΩ reappears on the legend of the coins, which, always according to Robinson, implies that the city had recovered some measure of autonomy. Robinson 1920, 10–16.

Sinope and Greek personal names was still being struck,[143] which makes us wonder whether Sinope was indeed conquered by the Persians or whether we have here just an effort by Persian officials to arrogate its commercial success.[144]

Similarly, with regard to Heraclea, its economic prosperity might explain why, as recorded by Memnon (1. 4), Clearchus, the tyrant of the city between 364 and 352 BC, sent many embassies to Artaxerxes II and III, while from a reference in Justin (16. 3. 9) we conclude that earlier, too, in the mid-5th century BC, the city had friendly relations with the Persian king, so that it could refuse to pay tribute to Athens (see above). These two cities, Sinope and Heraclea, were both admirably located for the development of sea trade. Amisos, by contrast, not having a good natural harbour but situated at the entrance to the only natural passage from the coast into the Anatolian hinterland, turned mainly to contacts with local populations there. This can be inferred from the use of the Persian standard in its coinage, which circulated in both the Black Sea region and Asia, which was under Persian occupation.[145]

Finally, on an inscription of 353/2–346/5 BC[146] we read that the citizens of Sinope concluded an alliance with the tyrants of Heraclea, according to which each side would assist the other in case of attack on the territory or city by anyone except for the Persian king. Should the aggressor claim that his attacks were sanctioned by the king, the two contracting parties should send envoys to the king and the aggressor and ask the latter to leave their land. And if the aggressor should refuse, then the one contracting party should support the other in every possible way. The citizens of Cromna and Sesamos (situated between the other two)[147] could also join the alliance if they wished.

This inscription, which is safely dated by the names of the tyrants of Heraclea recorded in it,[148] shows that these two important cities had taken measures in case of aggression, for example from local peoples such as the Bithynians and the Paphlagonians, or from Persian officials, but still recognised the suzerainty of the Persian king.[149] Obviously, the inspiration for such an alliance was supplied at least in part by Datames, who a little earlier had attempted to invade Sinope (see above), while the attitude of king Artaxerxes at the time

[143] Price 1993, pls. LIII–LIV.

[144] As Avram *et al.* (2004, 961) also suggest. See more on this issue in Manoledakis 2021f.

[145] Malloy 1970; Price 1993, pl. XLI; Avram *et al.* 2004, 955.

[146] French 2004 = *IK Sinope* 1.

[147] On these cities, see Part 4.

[148] French 2004, 4.

[149] *Cf.* Avram *et al.* 2004, 957; Tuplin 2007, 24. According to one theory, the big Greek cities of the coast were subject directly to the king and not to the local satrap. Langella 1989. *Cf.* Summerer 2005c, 244. However, the coins mentioned above do not confirm this.

might also have played its role in the composition of the text. Indeed, the differentiation between the Persian king and anyone else, even Persian, is noteworthy; and what is again indicated is the state of balance between him and the Greek cities of the coast, which, although under his sovereignty, were able to conclude such alliances with each other, each maintaining its regime, and thus functioning autonomously.[150]

Let us now move on to the sphere of culture. The natural isolation of the littoral, and perhaps other factors as well, may have discouraged Iranian settlement there (except for officials and soldiers), as seems to have been the case with some parts of the Anatolian hinterland;[151] at least, there is no relevant evidence. However, the cultural impact of the Achaemenids on the whole of Anatolia, including the Black Sea littoral, is evident, and indeed from quite early on. The archaeological evidence implies that the area (like others) had become a melting pot of local, Greek and Persian cultural elements, by the 5th century BC at the latest. This is apparent, for example, in several rock-cut tombs in Paphlagonia, where Persian architectural and iconographic features can be detected, together with Greek ones.[152] Furthermore, the area has yielded some 'Achaemenid-type' pottery samples and metal objects dating from the late 6th century BC on,[153] as well as grave-reliefs of Persian influence,[154] while even the existence of a satrapal manufacturing centre for luxury goods in Sinope has been suggested, especially in the light of the several silver vessels found in the wider region.[155] Finally, there are the Greek objects that suggest Persian influence or presence in the littoral, such as the Persian names on amphora handles and coins, especially those of Sinope that were mentioned above.[156]

Nevertheless, finds that are either Persian or demonstrate Persian influence are not as numerous as one would expect in an area that was under Persian

[150] *Cf.* Marek 2003; Lane Fox 2004, 30.

[151] Klingenberg 2014, with bibliography for the different views on the topic; Michels 2017, 44–45.

[152] Marek 1993, 14–15; Summerer 2003, 19–20; Dönmez 2007c, 108; Summerer and von Kienlin 2010, who date the monuments to the late 5th and 4th centuries BC, but mention also other dating proposals that have been put forward.

[153] Summerer 2003, with more bibliography; Dönmez 2007c, 108–11; 2015; 2019b; Tuplin 2007a, 25; Tsetskhladze 2014, 321; 2019e, 134.

[154] Summerer 2003, 20.

[155] Tsetskhladze 2014, 321. For the silver vessels, see Summerer 2003. Persian influences are seen in the Colchian culture as well. Braund 1994, 126–30; Tsetskhladze 2014, 316, 321; 2017, 29; 2019e, 138–42. *Cf.* Brosius 2010.

[156] On the coins see above. See also Summerer 2003, 19; Erciyas 2006, 56; Keleş 2006.

domination for two whole centuries.[157] And they are mainly related to more official aspects of the Achaemenid culture, rather than the everyday Persian one.[158] It is also noteworthy that we are not even quite sure about how Persian some of these objects are, so that we often read about 'Persian' and not Persian finds.[159] In fact, many of them seem to have been produced locally, indicating contact with Persian craftsmen rather than manufacture by them.[160] The oft-quoted lack of archaeological research in the littoral could be an adequate explanation. However, it has several times been stressed that the weakness of any discernible impact of Persian culture concerns a much bigger part of the conquered areas, including the rest of Asia Minor.[161] Thus, there seems to be a 'consistency' of written and archaeological evidence as far as a 'silent' Persian presence in our area of study is concerned. As a result: 'it is undoubtedly difficult to detect archaeological evidence for an Achaemenid impact on many levels of society and culture across Asia Minor following its political absorption into that empire.'[162]

What other explanations could be imagined for such a phenomenon? Among the several that have been put forward,[163] one that deserves some attention is M.C. Root's opinion that:

> the apparently unimposing impact of Persian culture on the western empire was a reflection of the success of a deliberate, assertive central policy – as opposed to a vaguely defined 'tolerant attitude'. This policy might have sought to play down the conspicuous presence of Persian power in the provinces on a variety of social/cultural levels.[164]

Such a policy would not have been followed without a reason. As E.R.M. Dusinberre has argued, having to incorporate a large number of peoples with very different traditions, into their empire, the Achaemenid kings 'had to

[157] *Cf.* Dönmez 2007c, 108. However, Summerer (2003, 19) stated that the Persian culture influenced the northern part of Anatolia more than the western. *Cf.* Brosius 2010, 29.

[158] Francis 1980, 68. For exceptions elsewhere in Anatolia, see Dusinberre 2013, 268.

[159] Tuplin 2007a, 25; Tsetskhladze 2014, 321; 2017, 29; 2019e, 128–29, 134. For an effort to classify objects from the Black Sea that were made or influenced by the Persians, see Rehm 2010.

[160] Brosius 2010, 34–35, even if referring mainly to the western and eastern sides of the Black Sea. Brosius sees in the 'fine Achaemenizing luxury metalware produced by Greek craftsmen' an indirect Persian influence on the Greek cities of the Black Sea, which, although politically autonomous, were economically dependent on the local hinterland.

[161] Root 1991, 1; Erciyas 2006, 56; Matthews 2009b, 155; Tsetskhladze 2019e, 128–29.

[162] Matthews 2009b, 155.

[163] For these see Root 1991. Herodotus' remark that 'of all men the Persians most welcome foreign customs' (1. 135) has also been invoked.

[164] Root 1991, 3; *cf.* Matthews 2009b, 155–56. See Thonemann 2013, for a similar but not identical case regarding the Phrygian civilisation after the Persian conquest.

implement an imperial system that would consolidate and keep the empire, one that would allow for diverse populations and cultural backgrounds but still unite the people into a stable whole'. Such a system should achieve a flexible combination of authority and autonomy. Indeed, Anatolia became a region where 'the Achaemenid imperial authority was expressed powerfully', but at the same time 'individuals were able to exercise some type of autonomy, mostly in ways that allowed them to express their individual identity':[165] for example, aspects of culture such as religion and language were not extensively exploited by the central authority as avenues for imposing power.[166] This policy of the Achaemenids was one of the most significant reasons for the longevity of their empire.

Root also proposed five additional factors, which, according to her, 'have combined to create and perpetuate the cliché of low Persian presence in/cultural impact on the western imperial territories': problems in the archaeological retrieval of Persian material in the west; negatively presented quantification of the Persian material that has been retrieved; miscategorised works of portable art of Achaemenid type in numerous collections; misappropriated monumental art claimed as major examples of a purely Greek repertoire even though exhibiting important features reflecting the impact of Achaemenid society and/or art forms; and uncritical categorisation of large numbers of art works as 'Graeco-Persian'.[167]

If these suggestions are correct, then, on the one hand, it becomes apparent how crucial the interpretation of the archaeological evidence is,[168] while on the other it appears that one should be very cautious when talking about an overall 'silent' or 'elusive' Persian presence in the southern Black Sea littoral, even around 400 BC. After all, this is based only on Xenophon's accounts in his *Anabasis* and refers only to the political presence of a conqueror in the subjugated land. In the first place, however, Xenophon was not primarily seeking to describe the Persian presence in the littoral in this work;[169] there may have been many elements that escaped his attention or interest. And, secondly and most importantly, there is much more than the strictly imperialistic sense in the political presence of the Persian, and specifically Achaemenid, civilisation. Apart from the division into administrative provinces, the logic of which was maintained in practice, even if with different borders, by the next conquerors of Anatolia, it is noteworthy that, notwithstanding the earnest efforts of their

[165] Dusinberre 2013, 270–71.
[166] Dusinberre 2013, 63–64, 266–71.
[167] Root 1991, 7.
[168] See more on that subject in Section 4.3.
[169] See, for example, the incident narrated in the *Hellenica* (3. 2. 2 – see above).

local predecessors to avoid being integrated into the Persian Empire, the later dynasties in some – not all – regions of northern Anatolia aspired to legitimise their rule by linking it with the Achaemenids.[170]

But possible expediency could be ascribed to the other side as well, the side of the local inhabitants, which from the Classical period on included both the indigenous peoples and the Greeks of the colonies.[171] Some events from the period during which Alexander the Great was conquering Asia and declaring himself lord of it might be indicative. After 333 BC he met the Paphlagonians at Ankara, the closest place to the littoral he ever reached, where he accepted their request not to invade and not to require them to pay tribute and just ordered them to obey Kalas, the satrap he had appointed in Hellespontine Phrygia (Plutarch *Alexander* 18. 5; Arrian 2.4; Curtius 3. 1. 22–24). In fact, as soon as Alexander moved away from the region the Paphlagonians could not only continue their normal life and activity in their land, but also allow it to be an arena for Persian efforts to recover areas of the empire that had been conquered by the Macedonians (Curtius 4. 1. 34, 5. 13).

Some few years later, in about 330 BC, returning to his camp in Hyrcania after having conquered the Mardians, Alexander found there Greek envoys, some of whom were from Sinope, and Lacedaemonian envoys to Darius. He arrested and imprisoned the latter but released the Greeks from Sinope 'because these people had no share in the commonwealth of the Greeks; and as they were in subjection to the Persians, they did not seem to be doing anything unreasonable in going on an embassy to their own king' (Arrian *Alexandri Anabasis* 3. 24. 4).

Generally, Alexander proved to be rather indifferent to the area and its Greek cities and local peoples (*cf.* Justin 38. 7. 2), some of which managed to keep the Macedonians away even after his death. He might have been convinced that the specific populations were so well integrated into the Persian state that they had little in common with the Greeks of the Aegean coast, towards whom he demonstrated a quite different attitude. This could have been a result of a deliberate, adaptive policy of the population of the littoral, who did not react against the Persians as long as they could continue their basic activities and maintain their prosperity (as seen above), just as the Persians were rather discreet towards them, at least until the 4th century BC, as long as they received their taxes. A similar kind of policy has been proposed for the Phrygians during the period of Persian sovereignty, as we shall see in the next section. What we may conclude from the above examination is that the Persian

[170] Michels 2017, 42, 48, 50.
[171] See Section 4.4.3.

presence in and impact on the southern Black Sea littoral remains surprisingly nebulous, contrary to what one would expect given the amount of both written and archaeological evidence from and for the two centuries in question.[172]

4. What about the Phrygians?

Having migrated from the Balkans in the period of the dramatic changes that marked the last centuries of the 2nd millennium BC,[173] the Phrygians achieved uncontested supremacy over a vast expanse of central Anatolia, extending at least as far eastward as the Iris.[174] This dominance lasted for almost the whole first half of the 1st millennium BC and was due to a remarkable variety of elements: the construction of a complex state with admirable levels of urbanism, social complexity, craft specialisation, monumental architecture, art, cults that soon spread to other cultures and – unlike most contemporary peoples, for example in the Black Sea littoral – a language with a script, which is known to us through two categories of preserved Phrygian inscriptions from as early as the 8th century BC.[175]

Inevitably, a reasonable question arises: did the people with such great political and cultural influence manage to spread as far as the southern Black Sea coast, which, if anything, would allow them to further develop their economy? In other words, should this section rather appear in the previous part of the book? Based on the written evidence, the answer would appear to be negative. In the ancient literature there is not the slightest hint of Phrygian occupation on the coast, neither in the geographies nor in any other works. The

[172] For the littoral in the era of Alexander and his successors, see Manoledakis 2021f. See also Manoledakis 2019, about Paphlagonia specifically. For the littoral in the Hellenistic period, especially under the Mithridatids, see, indicatively, Højte 2009; Marek 2010, 180–308. See also Sartre 2003. For the littoral in the Roman period, see Magie 1950; Sartre 1991; Marek 1993; 2003; 2010.

[173] The route that the Phrygians are traditionally held to have followed from the Balkans to Anatolia is via the Hellespont. Recent research on the Neo-Phrygian inscriptions has shown that they belong to the same Achaean dialect group as Arcado-Cypriot and Pamphylian. This, in combination with the known change of the people's name from Bhryges to Phryges – based on a development of Greek (the shift from voiced fricatives *bh* to unvoiced *ph*) that is considered to have taken place in the Achaean dialect – permits the hypothesis that another route may also have been followed, via Greece proper, the Peloponnese, Cyprus and Pamphylia. See Manoledakis 2016 in detail, with the relevant bibliography.

[174] For the eastern borders of Phrygia, see most recently Ökse 2019. Few scholars (Summers 1994, 245; Dönmez 2019a, especially 458, *contra* 2006, 17) disagree with including the Halys valley in the Phrygian territory. See additionally below.

[175] For an overview of the history and the above-mentioned aspects of the Phrygian culture and civilisation, see Roller 2011; Thonemann 2013, both with bibliography. Most recently, see Tsetskhladze 2019b.

archaeological evidence, however, does not seem to be as categorical as the literary, especially in the last few decades.

Nonetheless, for a long period of time, beginning about a century ago, the view that part of the southern Black Sea coast belonged to Phrygia appeared every now and then in the literature. The most striking characteristics of these – admittedly rare – instances are, first of all, that if combined they result in associating the Phrygians with a very large area – from Bithynia to the Iris, almost two-thirds of the total length of the coast – and, secondly, that none of their presumptions of Phrygian habitation of an area is substantiated by hard evidence: on the contrary, it was sometimes even based on ancient written sources, none of which confirm any such thing.

Perhaps the most emphatic formulation of a theory clearly placing the Phrygians as far north as the Black Sea coast, and based mainly on ancient texts at that, is the one proposed by O. Haas in his justifiably influential monograph on the Phrygian language.[176] Haas invokes the references to Phrygia and the Phrygians by Homer and Herodotus; he claims that Phrygia as mentioned in the Trojan Catalogue (*Iliad* 2. 862–863), in its first appearance in Greek literature, is to be placed 'near Lake Ascania, in the Black Sea littoral, close to the Sangarius river'.[177] He continues by quoting the passage in Herodotus according to which the Phrygians belonged to the third Persian satrapy, together with the Hellespontians of the right-hand coast as one enters the straits, the Asiatic Thracians (namely the Bithynians), the Mariandynoi, the Paphlagonians and the Syroi (the Cappadocians) (Herodotus 3. 90. 2; *cf.* Xenophon *Cyropaedia* 8. 6. 8). Haas thinks that this passage also shows that the Phrygians mentioned by Herodotus 'certainly lived on the Black Sea coast'.[178] He distinguishes between the Phrygians of the 'Black Sea coast' (mentioned by Herodotus as those who came from the Balkans), whom he calls 'North-East Phrygians', and those of the hinterland, an elder group, in what is referred to as 'Greater Phrygia', a distinction based in part on linguistic differences in their dialects.[179]

Earlier, other elements had been alleged to support Phrygian expansion to the sea, to more or less the same area and a bit farther east, among them the name of the city Tieion/Tios, which has been held to be Phrygian,[180] thus leading to the view that the Phrygians possessed part of the 'land between

[176] Haas 1966.

[177] Haas 1966, 9–11, 241.

[178] Haas 1966, 9–11, 242.

[179] Haas 1966, 11–12, 16–17. He even calls the territory of the former 'Das Nord-Ost-Phrygische Reich' (established in the 8th century BC), a convention that is not supported by the sources.

[180] Kretschmer 1896, 183.

Heraclea and Bithynia'[181] – thus apparently a very small coastal area and in any case far from Tieion.[182]

This notwithstanding, other evidence was used to place the Phrygians much farther east along the coast, including by Haas himself, an integral element of whose theory is that the Phrygians and the Paphlagonians had much in common, specifically that the population of Paphlagonia must have belonged to the same 'folk' (*Volkstum*) as the elder group of Phrygians and that Paphlagonian and Phrygian were similar languages.[183]

In addition to citing some more written sources, such as Herodotus' passage on the similar dress of the two peoples (7. 73), Haas also highlighted the presence of rock-cut tombs (*Felsgräber*), which are often encountered in the Phrygian as well as the Paphlagonian landscapes, to which latter this type of monument was brought from 'North-East Phrygia'.[184] Thus, from the area around Heraclea and Tieion we have now been transferred to the environs of Sinope, where pottery described as Phrygian by E. Akurgal had led the latter to speak of Phrygian habitation of the city.[185] Haas attributed this pottery to his 'North-East Phrygians', and cited some Old Phrygian inscriptions, like the one mentioning the Paphlagonian ruler Otys and his Persian wife,[186] in justification of his theory.

This theory, however, is not based on solid evidence, not to mention its numerous inherent contradictions.[187] Despite the several similarities that do indeed exist between Paphlagonians and Phrygians, although not mentioned by Haas,[188] there is nothing to demonstrate either an ethnic or a linguistic kinship between them.

Even the geographical placement of the Phrygians on the Black Sea coast as cited by Haas is problematic. The fact that most of the rest of the peoples mentioned by Herodotus as belonging to the same (third) satrapy lived on the Black Sea littoral (Bithynians, Paphlagonians, Mariandynoi and Syroi) does not mean that everyone in that satrapy lived there. The 'Hellespontians of the

[181] Meyer 1925, 12, 149; Ruge 1941, 788, 796–97. For some New Phrygian inscriptions where the genitive form of Zeus, Dios, is converted to Tios, see recently Öztürk 2013, 149, with bibliography.

[182] Bithynia's eastern border was usually considered the Sangarius or the Hypius. See Section 2.2.1.

[183] Haas 1966, 9, 11, 14–17, 181, 231, 235, 242. See also Section 2.6.2.

[184] Haas 1966, 14–15. *Cf.* Leonhard 1903; von Gall 1966.

[185] Akurgal 1955a, 93, 126; 1956, 57. *Cf.* also Leonhard 1915, 298; Saprykin 1991, 249.

[186] Concerning this inscription, Haas went so far as to consider the early 4th-century BC Paphlagonian ruler Otys the beneficiary who may have taken advantage of the collapse of the Phrygian kingdom (Haas 1966, 16–18, 179–84).

[187] For a detailed criticism on Haas's suggestion, see Manoledakis 2021d.

[188] On this, see in detail Manoledakis 2021d.

right coast as one enters the straits' mentioned by Herodotus are obviously the Phrygians of Hellespontine Phrygia,[189] whose capital was Dascylium and which was located south-southeast of the Hellespont and the Propontis (Sea of Marmara) as far as the banks of the Sangarius (in its central/southern and not its northern part) and even included Ascania, thus has no connection with the Black Sea coast, the western part of which was inhabited by the Bithynians. This area corresponds to the territory mentioned in the *Iliad*. Herodotus then makes separate mention of the 'Phrygians' in the same satrapy; these are the Phrygians of Greater Phrygia, in the interior of Anatolia, and not the Hellespontine ones. In any event, neither group bordered the Black Sea, especially not the 'Phrygians' of Herodotus as Haas thought.[190]

Up to now, the only piece of evidence offered in support of a possible Phrygian presence on the southern Black Sea coast that is worthy of attention is the Phrygian pottery[191] found in Sinope[192] and in and around Amisos.[193] Curiously enough, it is only the latter that has led scholars to create maps of Phrygia including the corner of the Black Sea where Amisos – but not Sinope[194] – is located, while usually Phrygia is shown ending before the coast, at Akalan, about 18 km inland. Together, these theories cover a large area of the coast, from the Sangarius to the Iris, parts of which have occasionally been considered as belonging to Phrygia. But how credible would such a conclusion be?

From a geographical point of view, the easternmost part of this area, around Amisos, seems to unite the most probabilities for a Phrygian occupation. A crucial matter here is whether the Phrygians did in fact spread into the Halys valley or not. The vast majority of scholars agree that they did.[195] This view is based on both pottery, including monochrome grey ware,[196] and some terracotta revetment plaques found at several sites all along the valley, among them

[189] Thus the 'straits' are the Hellespont and not the Bosporus.

[190] In Haas 1966, 11, 242, it becomes clear that the author always refers only to them as his Black Sea Phrygians and not to Herodotus' 'Hellespontians of the right coast as one enters the straits', who would be a better geographical match for his North-East Phrygians at the Ascanian Lake.

[191] For the debate as to whether all this pottery called Phrygian is actually Phrygian, see below.

[192] Akurgal and Budde 1956, 49 and pl. IIIa–d; Akurgal 1956, 57; Budde 1956, 5, 7; Boardman 1999, 254–55.

[193] *Cf.* Boardman 1999, 92, 255; Summerer 2005a, 144; 2014, 203; Dönmez 2010, 523; Tsetskhladze 1998a, 47; 2007, 188; 2012a, 349; 2012b, 236. For the pottery found in Kurupelit (Atakum district), about 9 km from Amisos, together with Athenian black- and red-figure ware (see below), see most recently Şirin and Kolağasioğlu 2016, with bibliography.

[194] E.g. Sivas and Sivas 2012, 13, commented by Tsetskhladze (2019a, 6).

[195] For some exceptions, see n. 174 to this section.

[196] Which even Dönmez (2006, 17) had accepted as Phrygian, though later he revised his previous opinion, questioning even the Phrygian influence on the peoples of the Halys valley (2019, 461–63).

Akalan, Köyiçi Tepesi, Parali Tepe, Pazarli, Boğazköy, Kerkenes Dağ, Kaman-Kalehöyük and Suluca Karahöyük. Although neither of these find categories is unanimously considered Phrygian,[197] this is not, of course, the case with the Old Phrygian inscriptions found at some of these sites,[198] so few doubts remain concerning a strong Phrygian influence on the latter, or even a kind of Phrygian presence in some of them. Kerkenes Dağ, a site that has been systematically excavated over the last years,[199] has a cardinal place among them. The settlement there seems to have been established in the 7th century BC and destroyed in the mid-6th century BC.[200] Almost all scholars consider it an indisputably Phrygian settlement, or at the very least displaying strong Phrygian influence.[201] Usually the site is identified with Pteria, mentioned by Herodotus as having been demolished by Croesus during his campaigns against Cyrus in the 540s (Herodotus 1. 76),[202] although Akalan is another candidate for this identification.[203] If Kerkenes was indeed a Phrygian settlement, then I do not see how we can deny Phrygian expansion into the Halys valley,[204] at

[197] For the pottery, see below. As regards the architectural terracottas, which date from the 6th century BC, there are a variety of views as to their origin: Dönmez (2006, 13, 22–29, 38–39; 2007b, 1208; 2010, 517–22) ascribes them to cultures of the central Anatolia region with no Greek influence, while others associate many of them (excluding those from Akalan) with the Phrygians, for instance Summers (2006, with an interesting opinion as to the role of Lydia in the transfer of this artistic tradition to Phrygia and relevant bibliography) and Tsetskhladze (2007, 188–92, see detailed bibliography in Summerer 2005b, 125; 2008, 281, n. 54; more recently, see also Tsetskhladze 2012b, 237–38). Summerer (2005b; 2008, 267–78) has suggested that they bear decorative elements with a clear Greek influence and thinks that the moulds from which the reliefs were produced came from the Greek colonies on the coast, particularly Amisos, which was famous for its terracotta tradition and workshops, and indicate local familiarity with Greek culture. *Cf.* also Summerer 2007, 31–34. Similar views have been expressed by Boardman (1999, 92, 255), Atasoy (2003, 1347) and Roller (2011, 570). *Cf.* Barnett 1975, 430. Summerer also explains her 6th-century dating and her objections to the earlier datings proposed (7th and even 8th century BC): see Summerer 2005a, 148; 2008, 272, with bibliography. For a more detailed discussion of the Akalan terracottas, see Cummer 1976; Summerer 2005b, 125–39. *Cf.* also Maksimova 1956, 52–65. For Boğazköy, see Bossert 2000, 145.

[198] Haas 1966, 179–84; Brixhe and Lejeune 1984, 223–51.

[199] See https://sciences.ucf.edu/anthropology/kerkenes/ and http://www.kerkenes.metu.edu.tr.

[200] Roller 2011, 564; Thonemann 2013, 11; Branting *et al.* 2019.

[201] Draycott and Summers 2008; Roller 2011, 564; Tsetskhladze 2007, 188–92; 2012b, 238; Thonemann 2013, 11; Branting *et al.* 2019. Dönmez (2019a, 457), who, as already stated, denies Phrygian influence on the Halys valley settlements, claimed that Kerkenes Dağ and other settlements of the valley that have been deemed Phrygian (such as Pazarli and Maşat Höyük) have only minor similarities with Gordion but differ from it in several respects. Summers (2006, 686, n. 31) acknowledged the existence of some differences but pointed out that 'for the rest Phrygian culture is manifestly evident there'.

[202] Summers 1997; 2006, 685; Roller 2011, 564; Branting *et al.* 2019, 540.

[203] Tsetskhladze 2019a, 5, n. 13, citing also an earlier paper by Dönmez.

[204] Where, incidentally, Old Phrygian inscriptions have also been found (e.g. Haas 1966, 179–84; Brixhe and Lejeune 1984, 223–51).

least for a specific period of time (late 7th–mid 6th century BC),[205] before the Persian conquest.

Deciding whether the Halys valley belonged to the Phrygian territory or not is of great importance, for an obvious reason. It has already been pointed out that this valley offers the only significant natural access from the Black Sea coast to the Anatolian hinterland, which for the rest is cut off from the sea.[206] The northernmost site in the inland basin, Akalan, is only about 18 km from Amisos, which controlled the coastal end of the valley. Thus the way from the valley sites to the coast is open, and the pottery considered Phrygian that was found in Amisos would provide perfect confirmation.

But is this pottery indeed Phrygian? Over recent years several doubts have been expressed concerning the identity of this pottery, examples of which have been found in Sinope as well (see above). According to some scholars, this pottery is characteristic of the whole central Black Sea region of Turkey,[207] or was imported to the coast from central Anatolia.[208] However, this view cannot be archaeologically confirmed, at least for the time being; similarly, of course, there is no justification for the indiscriminate use of the term 'Phrygian' as synonymous for 'Iron Age' in central Anatolia.[209] Therefore, a hesitation in this regard, even a confusion at times, appears among scholars, veiled by newly invented terms such as 'Phrygian-type' or the rather awkward 'Central Anatolian type Phrygian culture pottery'.[210] The latter might seem an exaggerated case, but it shows that no clarity has yet emerged regarding what archaeologists used to call 'Phrygian pottery'. If we accept the term 'Phrygian-type', this means that we can certainly identify Phrygian pottery and differentiate it from Phrygian-type pottery,[211] while acknowledging similarities between them[212] or imitations of the former in the latter – for otherwise why risk

[205] Branting *et al.* 2019.

[206] See more in Part 1.

[207] Summers 1994, 245; Summerer 2007, 30; Tsetskhladze 2012b, 236. In the bibliography on Anatolia the term 'Central Black Sea region' is actually used to refer to the central part of the southern Black Sea littoral.

[208] Dönmez 2007a, 60. *Cf.* the discussion in Türker and Özdemir (2019), as regards the pottery from Suluca Karahöyük, in the southern end of the Halys valley.

[209] Summers 1994, 247, n. 1.

[210] Dönmez 2010, 523. *Cf.* Dönmez 2006, 39; 2007b, 1208; 2012, 167, with a general tendency to underestimate the Phrygian cultural influence on central Anatolia, including the Black Sea coast, claiming that the dominant culture there was the 'local central Anatolian' one.

[211] Sometimes we read about the finding of 'Phrygian and Phrygian-type pottery' in a place. See, for example, Tsetskhladze 2019a, with bibliography.

[212] Summers (1994, 244–45) distinguishes between 'two very distinct pottery traditions, western (Phrygian) and Central Anatolian (including Post-Hittite)', considering as Phrygian only the Phrygian Grey Ware pottery.

describing the latter in that way? But are these similarities enough to call that pottery 'Phrygian-type'? Looking at the published examples, it seems doubtful.[213] The use of this term therefore has no real meaning.

On the other hand, if we accept the term how can we distinguish 'Phrygian-type' within central Anatolian or 'local'? If we cannot, then the term 'Phrygian-type' is again meaningless, used only because of the strong political and cultural presence of the Phrygians in central Anatolia, which of course is not enough to characterise a pottery. Presumably, it would be much simpler to talk about local pottery, which could be Phrygian. Thus we pass to another issue, of a more methodological nature: in an attempt to avoid 'Phrygiocentrism'[214] we sometimes tend to deny any Phrygian presence in central Anatolia and emphatically differentiate between 'Phrygian' and 'local'. But here we may tend to forget that we are talking about central Anatolia in the Archaic period. Who could be considered local in that place and time[215] if not the Phrygians, who had been there for at least six centuries? Not them alone, of course, but them as well, and *par excellence*.[216] Whom else do we know there as established, not to say dominant, for such a long time? So what, in the end, would actually be the difference between 'local', 'Central Anatolian' and 'Phrygian'?[217]

It is a fact worthy of note that scholars, even those who are wary of the term 'Phrygian', continue to use the terms 'Phrygian (cultural) layers/phases' and 'Phrygian period', or the terms 'Phrygian', 'Iron Age' and 'Post-Hittite', to refer to the same thing in the archaeology of central Anatolia.[218] However confusing this may be, it certainly shows who remains the major candidate for identifying a specific period in the specific area. This should in no case and for no reason lead to the conclusion that everything in this area at this period is Phrygian and only Phrygian. As noted in the introduction to the previous part of the book, dedicated to the indigenous peoples of the littoral, no people has always lived entirely and only in the same place and, conversely, very few peoples totally cease to exist in an area. Moreover, it is more than natural for

[213] The existence of such similarities was, for example, mentioned by Summers (1994, 245), but without any evidence. Chemical analyses and comparisons, which would offer some answers, have not been made.

[214] Summers 1994, 245.

[215] *Cf.* the definition of indigenous given in Section 2.1.

[216] Türker and Özdemir (2019, 424) correctly remind us that the Phrygians were not the only people who inhabited central Anatolia in the Iron Age. Unfortunately, however, we know nothing about the others, not least of their pottery.

[217] Using the term 'post-Hittite' (Summers 1994, 245) or anything similar for something that existed six centuries previously is very awkward – much like calling Athenian black-figured pottery 'post-Mycenaean'.

[218] See more recently, for example, Türker and Özdemir 2019, 412–13, 424, with the relevant bibliography. *Cf.* Kealhofer and Grave 2011.

peoples who migrate to be influenced by the cultures they meet at their destination as well as by those they encounter on their way. The Phrygians are a fairly representative case, even in respect of their language, which was characterised by several modifications and by the adoption of Greek and Thracian linguistic elements, due to the routes they followed,[219] and so obviously in respect of their pottery and vase painting as well. Their adoption of ceramic or iconographic traditions pre-existing in the area[220] is entirely normal and does not mean that their art is in the end not Phrygian because of it; the same, *mutatis mutandis*, holds true for their language.

Nevertheless, and to close this parenthesis, how to characterise the specific pottery is not in the end the most crucial matter as far as the northwards expansion of the Phrygians is concerned. Even if we agree that the pottery in question from Sinope and Amisos *is* Phrygian, it remains undoubtedly true that the existence of some examples of Phrygian pottery in a place is not enough to prove that Phrygia extended to that place[221] – in our case, the Black Sea coast. The few Phrygian potsherds that have been found in Sinope or Amisos do not necessarily prove Phrygian habitation of the area,[222] but could easily be explained by relations of any sort between the two neighbouring areas.

More evidence is needed, and, indeed, some new evidence has emerged in the past few years. A rescue excavation undertaken in 2009 in Kurupelit, about 9 km from Amisos,[223] brought to light several multi-level structures, rooms and roads within a large area that may belong to a shrine. Among the finds are many terracotta figurines of Cybele (both standing and enthroned), pottery that could be characterised as local or Phrygian,[224] as well as fragments of Greek black-glazed pottery (mainly cups) displaying graffiti inscriptions with dedications to Kybebos (in the dative form: Κυβηβώι).[225] The name matches that of the priest of Kybebe (Hellenised form of the Neo-Hittite Kubaba), another name for Cybele, attested in the written sources from the 6th century BC.[226] All these finds have led to the conclusion that there was a temple dedicated

[219] See the first note in this section and Manoledakis 2016 with the relevant bibliography.

[220] As, for example, Türker and Özdemir (2019, 413–26) propose for the pottery from Suluca Karahöyük. See generally Kealhofer and Grave 2011, especially 421–24.

[221] Summers 1994, 245; Matthews 2009b, 153; Tsetskhladze 2012b, 236.

[222] For the well-known question of whether pots can always be equated with peoples, see Tsetskhladze 2012a; Szamalek 2014, with bibliography.

[223] For the pottery of which see above. See Şirin and Kolağasioğlu 2016, with the previous bibliography and many images.

[224] Together with Athenian pottery, as in Sinope. See above.

[225] Şirin and Kolağasioğlu 2016, 16–36, with colour illustrations. See also Dönmez 2010, 522–23; Akyüz 2013; Summerer 2014, 200–05; Tsetskhladze 2015, 19–20; 2019a, 6.

[226] Roller 1999, 124; Munn 2006, 120–25; Summerer 2014, 203.

to Cybele here.[227] It is roughly equidistant from Akalan and Amisos, and L. Summerer suggests that it could have been a common place of worship, used by both Greeks from Amisos and local people.[228] If this is the case, then the possibility that Akalan was inhabited by Phrygians (see above) is strengthened. The whole complex has been dated to the period between the mid-6th and the early 5th century BC,[229] although among the finds depicted in the latest publication one can clearly discern fragments of Athenian red-figure pottery of the late 5th or early 4th century BC.[230]

Now, in terms of the possible Phrygian occupation of the specific part of the coast, I do not think that the shrine in question is enough to lead to such a conclusion.[231] From the finds it is obvious that the Greek element (apparently from Amisos) is strongly represented. Apart from the black- and red-figure pottery,[232] the black-glazed pottery fragments with the votive (Greek) inscriptions that are published[233] seem of good quality and also more likely to be imported from Attica. Like the black-figure pottery, they belong to the late 6th or early 5th century BC. It is noteworthy that the earliest Athenian pottery in the shrine considerably predates the supposed Athenian presence and re-colonisation of neighbouring Amisos (Strabo 12. 3. 14),[234] corresponding almost to the time of the city's original founding.[235] That Phrygians might have been among the worshippers there is perfectly possible and would be very natural, given its equal proximity to Akalan and the valley, as well as the identity of the cult deity. But the evidence from the site does not appear adequate to suggest anything more than that.

Finally, the most recent find on the whole coast comes from Tieion, where during the 2019 excavation season a small pottery fragment bearing an Old Phrygian inscription was unearthed together with 7th- and 6th-century BC

[227] According to Roller (1999, 127–28), the cult of Cybele in the Greek cities of Ionia, Miletus included, might have come from the Ionian Greek colonies of the southern Black Sea coast, where the Greeks lived very close to the Phrygians, with all that this might imply for their cultural contacts.

[228] Summerer 2014, 203.

[229] Summerer 2014, 203; Şirin and Kolağasioğlu 2016; Tsetskhladze 2019a, 6.

[230] Şirin and Kolağasioğlu 2016, 17 fig. 17a, the last two fragments.

[231] See below on what is meant by 'occupation'.

[232] Şirin and Kolağasioğlu 2016, 17 fig. 17a.

[233] Şirin and Kolağasioğlu 2016, 33–35 figs. 34–38.

[234] Which is connected to Pericles' Black Sea expedition (Plutarch *Pericles* 20. 1–2), and thus dates from about 437/6 BC. For the doubts concerning the authenticity of that expedition, as well as the opposite view, see Section 3.3.

[235] For which see Atasoy 2003, 1342–47; Avram *et al.* 2004, 954–55; Summerer 2005a, 137–38, 148–49, 152.

pottery.[236] This is of great importance, since it is the first Old Phrygian inscription ever found on the Black Sea coast. But still, it is only one inscription, thus not enough to lead to the conclusion that the city was at some time systematically inhabited by Phrygians.[237] There are too many other ways in which it could have arrived there.

Whatever the case may be, permanent Phrygian possession of any part of the Black Sea coast cannot yet be proven.[238] The archaeological finds are scant and might indicate relations between the coastal communities – basically the Greeks of the colonies – and the Phrygians,[239] but nothing more than that. These relations were entirely normal and likely, since the Phrygians were politically and culturally dominant in their area, as were the Greeks in their cities' territories, in a region very close to Phrygia.[240] Besides, one must not forget the cultural closeness between Greeks and Phrygians, which has several times been highlighted,[241] and of which the shrine at Kurupelit could indeed have been an indication. Furthermore, if the Phrygians inhabited the coast in permanent Phrygian settlements, there would be some indication of this in the written sources, which is not the case in any historical or geographical text. Therefore, we should be very cautious – at least for the time being – about numbering the Phrygians among the indigenous peoples of the southern Black Sea littoral as we examined them above,[242] despite the evidence that some Phrygians were certainly visiting the area, or occasionally even living there.

In other words, while we can claim that Phrygians may have lived on the Pontic coast, we cannot claim that *the* Phrygians occupied part of it, that is, that the Phrygian territory expanded as far north as the coast. There is only one factor that could raise a reservation about this conclusion, as regards the period before the Persian conquest, and that is that our geographical sources start

[236] Yildirim 2022, fig. 13.

[237] Even less to lead to the certainty of such a habitation: Tsetskhladze 2019a, 3, quoting Ş. Yıldırım.

[238] *Cf.* also Dönmez 2006, 17. Huxely's argument (1960, 19, 22) that 'there had been a Phrygian settlement in the land of the Leukosyroi, for west of Sinope lay the little port of Armene. The name points to an Armenian settlement, and the Armenians were colonists of the Phrygians' is unsubstantiated.

[239] Yildirim 2022.

[240] Similar relations, at least of a commercial character, between the colonies and the Phrygians of Gordion can be detected later, in the 4th and 3rd centuries BC, when we find amphorae from Heraclea and some Sinopean fragments in the Phrygian capital. Lawall 2010, 162–63; Tsetskhladze 2013, 76; Manoledakis 2015a, 85–87.

[241] See more recently Manoledakis 2021d, with the previous bibliography. *Cf.* also Manoledakis 2016.

[242] One thing that might possibly be deduced from this is that the Phrygians were by nature non-violent, for at their peak they could certainly have attempted to seize part of the coast, especially the easily accessible area around Amisos, had they so wished.

from the second half of the 6th century BC and so belong to the periods from the Persian onwards. According to one thought-provoking theory, from 550 to 330 BC the powerful and well-organised Phrygian state ceased to exist, 'leaving a post-literate, post-urban, largely non-stratified, cellular agro-pastoral society across much of central Anatolia'. This was a deliberate decision on the part of the Phrygians, a 'rational adaptation to the circumstances of Persian rule', since the latter was essentially a mechanism for exacting tribute, which in a vast area like the conquered one required local centralised structures to help the Persians collect their tax and tribute money. Retreating from complexity, the Phrygians in practice made it very difficult for the Persians to extract their taxes.[243]

This also explains, according to the same view, the dramatic decrease (almost disappearance) in Phrygian inscriptions in the classical period and their reappearance in Roman times. If one accepts this view, then one might be able to give an explanation for the total absence of the Phrygians from the peoples that inhabited the Black Sea littoral in the ancient sources. Indeed, the so-called Phrygian pottery mentioned above dates only from the late 7th to the first half of the 6th century BC. This could mean that after the mid-6th century BC one should in any case not expect to find tangible evidence of a Phrygian presence on the littoral, although I am not sure whether this is an adequate reason. However that may be, for the preceding period the conclusion suggested above remains, unless future excavations provide different evidence.

[243] Thonemann 2013, 3, 8, 11–15, 37–38.

PART 4

THE GREEKS IN THE SOUTHERN BLACK SEA

1. Introduction

For better or worse, when people, including scholars, hear or talk about the Greek presence in the (southern) Black Sea (in antiquity), they mainly mean the Greek colonisation of the area, part of the huge movement designated the Second Greek Colonisation. Despite the fallacy of such a perception, which will be analysed in the following pages, no one can deny the momentousness of the colonisation in the ancient history of the region.

Although many peoples have inhabited, visited, invaded or conquered the region, as demonstrated in the preceding parts of the book, what we have here is quite different. It is not the emergence or evolution of an indigenous people from earlier local populations occupying a part of northern Anatolia bordering the sea, like the Paphlagonians. Neither is it the expansion of a neighbouring people, as was the case with the Thracians from the east and perhaps, according to one theory, with the Phrygians from the south. Nor is it the ephemeral but catastrophic invasion of a people from far off, as with the example of the Cimmerians, or even political occupation by a major power, as happened with the Persians and the Romans. Here we have a case of colonisation. Numerous small groups from separate cities of a faraway people – not the whole or a massive part of it – come and settle in specific places, all of them coastal, for specific reasons, without the intention of eliminating the existing peoples, and create new permanent communities there, eventually collaborating with local populations and gradually, to a greater or lesser degree, being integrated into them.

This happened only once in the southern Black Sea littoral, in the 1st millennium BC, and it changed its history dramatically and forever. The impact of the newcomers on the indigenous peoples[1] (and *vice versa*) was determinative in terms of culture[2] and politics, and eventually ethnography, as indicated by the emergence of, *inter alia*, several myths, even though the primary reasons for their coming seem to have been of an economic nature;[3] in fact, vestiges of this impact still exist in the area. Moreover, this coming took place via the sea, in contrast to all the other occasions mentioned above. So whoever the

[1] For which see Part 2, with a definition of 'indigenous' provided in the introduction.

[2] The role of the Greek language, which unlike all the local ones had a script, must have been decisive here. *Cf.* Tsetskhladze 2002b, 86. Looking at it from another viewpoint, it is through the products of this language that the indigenous southern Black Sea peoples have mainly become known, as already mentioned (see Part 2).

[3] See more on this below, Section 4.4.

'Sea Peoples' of Egyptian records might have been,[4] for those living in the southern Black Sea littoral in the Early Iron Age the Greeks were the first foreign 'sea people' to arrive on their shores with obvious long-range ambitions.

However, to consider the colonisation of the littoral as the beginning of the Greek presence in it would be utterly misleading, as was pointed out in the beginning. For an earlier phase had preceded, which unintentionally paved the way for this colonisation and facilitated its accomplishment. In this part of the book, I shall approach these two broad 'phases' separately, merely to emphasise the significance of the earlier and less prominent one for the Greek presence in the southern Black Sea littoral. Numerous essays having been published, especially over the last 30 years, on the Greek colonisation itself, covering almost every single aspect of it,[5] my intent here is not to scrutinise it once again. I shall, rather, confine myself to some selected facets of it and offer some personal comments. With a view to easing the bewilderment caused by the (at least) two broader 'phases' of the Greek presence in the southern Black Sea by looking at how they relate to each other, I shall start with a section dedicated to the relevant terminology that has been used in scholarship and the several theories that have emerged.

2. A Game of Terms

Our knowledge of the Greek presence in the area of this study is so dominated by the colonisation, mainly resulting from the plethora of ancient texts referring to it in stark contrast to what went before, that it has become difficult to approach any kind of earlier Greek presence independently of it. The crucial question as regards the emergence of the Greeks in the (southern) Black Sea has very often been expressed in terms of when the colonisation began. Attempting to answer this question has, in turn, led to a dispute over which category of evidence is more reliable, textual or archaeological, a dispute that for no real reason has assumed almost dramatic dimensions in modern scholarship, apparently between two sets of scholars serving disciplines absolutely interdependent: historians and archaeologists. For it has been persistently claimed that the answers gleaned from the written sources are different from the clues provided by the archaeological finds, at least as far as both have been interpreted by historians and archaeologists respectively. Thus, the historians date the beginning of colonisation to the 8th century BC, based on written

[4] For the 'Sea Peoples', see, indicatively, Oren 2000; Woudhuizen 2006; Bauer 2014.

[5] See more recently Tsetskhladze 2006a; 2018; Manoledakis 2018b, all with the previous bibliography.

references,[6] while most archaeologists – not all – refuse to see any Greek settlement before the second half of the 7th century BC,[7] the earliest period to which, in the majority view, Greek finds in the Black Sea region can be dated.[8]

There is no need for an extensive exposition on how naïve and unscientific such a confrontation is. Already in 1990 J. Bouzek aptly pointed out that:

> the sources themselves – both written and archaeological – only very rarely contradict each other: they just show the subject from different angles. The contradictions are mainly caused by our premature extrapolations, by hypotheses derived from limited or one-sided observations and by perpetuating these mistakes.[9]

In order even to think of claiming that textual and archaeological evidence are at odds with each other, one should first make sure that he has an adequacy of both. This is not in the slightest the case with the southern Black Sea. The written sources alluding to the outset of the Greek settlements there are scant, often ambiguous,[10] and come from periods later than the events they describe.[11] Even more meagre is the assistance provided by the archaeological evidence. The very promising excavations at Tieion since 2006 are still the only large-scale systematic excavations taking place in the littoral, and the latter is still by far the least archaeologically investigated part of the Black Sea.[12] The

[6] For example, Drews 1976, 18–31; Graham 1958, 25–42; 1971, 35; 1982, 123; 1989, 52–54; 1994, 4–5 (many of Graham's essays cited here are now collected in Graham 2001); Braund 2005a, especially 107.

[7] See, for example, R. Cook 1946, 67–98; Boardman 1991, 387–90; 1999, 245–67; Tsetskhladze 1998a, 19; 1999, 470; 2002b, 81; 2011, 96; 2012b, 237; Erciyas 2007, 1195; and others. However, other archaeologists, such as Minns (1913, 439), Miltner (1939), Bouzek (1990, 13–15; 2008) and Atasoy (2018, 127) take the opposite view. For this confrontation between historians and archaeologists, see more recently Tsetskhladze 2012a, 335; Manoledakis 2018b, 174, 182.

[8] Although some scholars supporting this view talk persistently about the last third or last quarter of the 7th century BC, several studies of the earliest Greek pottery in the Black Sea conclude that some of the fragments, for example from Nemirovo and Orgame, date from the third quarter of the century, sometimes from the middle of it: Vachtina 2007, 27–34; Handberg 2013, 2–3, both with bibliography; Tsetskhladze 2018, 41, 52. Handberg (2013, 3, 11) suggests that the second stage of Ionian colonisation in the Black Sea area must be dated to the late 7th–early 6th century BC. Although I am not a supporter of its division into stages (see below), I assume that, in this view, Greek colonisation must have started (first wave) in or before the mid-7th century BC.

[9] Bouzek 1990, 169–71, especially 171.

[10] The most characteristic example regarding the southern Black Sea is, of course, Eusebius' dating of the founding of Sinope in 631 BC and of Trapezous in 756 BC, which cannot be true, since the latter was a colony of the former.

[11] *Cf.* Braund 2005a, 101.

[12] See in the Introduction Section 2.

archaeological material from the littoral is, as a result, so disappointingly scant that no one can appeal to it to argue about either the beginning of Greek colonisation or many other aspects of it.

Besides, while archaeology might as a rule provide the most reliable evidence, it cannot tell us what cannot have happened in a place before a certain point in time – namely, the date of the earliest finds from that place; all it can tell us is what certainly happened from that point onwards. Moreover, we need to be sure that we have reached the earliest habitation levels of the place. Therefore, the few archaeological finds of the late 7th century BC, for example, which are considered the earliest archaeological evidence of Greek settlement in the region,[13] provide only a *terminus ante quem* for a settled Greek presence.[14] Many more finds would be needed, and from many sites along the littoral (at least those that are numbered among the primary Greek colonies), in order to reach safe conclusions, according to the methodological norm universally acknowledged among archaeologists.[15] I shall return to this later.

Be that as it may, even if we manage to identify the beginning of colonisation, we have not solved the whole problem. For the Greek presence in the Black Sea did not emerge with the colonisation. The next section examines some evidence that points to a Greek presence in the southern littoral that cannot be associated with organised colonisation and apparently precedes it. This fragmentary evidence has always plagued modern scholarship, which has attempted to bypass the stumbling block of its interpretation by treating it as the product of an historical phenomenon that has been called 'pre-colonial contacts', 'proto-colonial contacts', 'pre-colonisation', 'reconnaissance period' and the like; 'penetration' is another commonly used term, although seemingly not always in the same sense.[16] What do all these terms denote, what are the differences between them, and what do they have in common?

To start with the most popular, 'pre-colonial contacts': the first part, the adjective, is a general but not semantically erroneous term, which simply describes anything that could have taken place before colonisation.[17] The second part, however, is a noun that clearly denotes interaction between Greeks and locals. This conceals a certainty that not only was there a Greek presence in the Black Sea before the establishment of the colonies but this presence had

[13] See below, Section 4.3

[14] Graham 1958, 32–33; Manoledakis 2010b, 572.

[15] See, indicatively, Tsetskhladze 2006a, xxxiii; 2019d, 61.

[16] See the use of these terms in, for example, Graham 1958, 25–42; 1989, 45–60, especially 54; Asheri 1972, 17; Bouzek 1990, 172–73; 2013; Tsetskhladze 1994a, especially 113; 1998a, 10; Petropoulos 2005a; 2005b, 217; de Boer 2006, 48–52; Summerer 2007, 29, 35; 2008, 264.

[17] But how far before? It is as a general term as 'pre-classical'.

also developed into contacts with the locals; the next step would, of course, be to ask whether such contacts had resulted in any settlements. Seemingly similar is the term 'pre-colonisation', although in fact it is a delusive one, because it is not an adjective defining, for example, contacts, activity and the like, but a noun and as such seems to encompass a whole situation, like colonisation itself. However, it is highly doubtful whether all that preceded colonisation was, like the latter, a specific, concrete and homogenous situation, even though in our introduction we have generally distinguished it from colonisation as a separate 'phase'.

The term 'reconnaissance period' has been proposed as a designation for the period during which the Greeks 'collected information' about the regions they would later colonise, and so a period that 'served for the next move', the establishment of settlements.[18] This term suggests, therefore, that what we call here the early Greek presence in the Black Sea had occurred deliberately with a future colonisation in mind. Similar allusions could be traced in theories that speak of early journeys of exploration to regions known for their riches.[19] That the Greeks must have had notions of their new land when they decided to colonise it[20] seems logical. But this does not necessarily mean that in every visit to the area by Greeks centuries before the colonisation there had been underlying thoughts of possible future colonisation.

The term 'proto-colonial', although used in exactly the same contexts as 'pre-colonial', is not synonymous with it. Proto-colonial (proto = first) describes what belongs to the first stage of colonisation, which means that the colonisation has already begun, while in the pre-colonial period it has not. Finally, another popular term, especially in the previous century, is 'penetration'.[21] As apparent as its meaning seems to be, Greek 'penetration' has sometimes been used to designate the beginning of Greek colonisation. In reality, of course, 'penetration' refers to the 'pre-colonial' period[22] (before the 'proto-colonial'), for it implies an incursion into a territory and clearly

[18] Petropoulos 2005b, 216–17, even if referring just to the northern Black Sea. *Cf.* Petropoulos 2005a.

[19] For example, Huxley 1990, 200 (as mentioned by Tsetskhladze 1994a, 114); Boardman 1991, 387; 1999, 244.

[20] Braund 2005a, 107; de Boer 2006, 49.

[21] Carpenter 1948; Graham 1958; Danoff 1962; Tsetskhladze 1994a; 2018; Petropoulos 2005a.

[22] The two terms seem to have a different meaning in Tsetskhladze 1994a, for example pp. 113 and 125. Later, Tsetskhladze (1998a, 19) clarifies that he is referring to penetration by Greek colonists and not by the Greeks generally. Probably this is also the case with his recent paper (Tsetskhladze 2018).

suggests that this is happening for the first time.[23] When, therefore, we speak of Greek 'penetration' into the Black Sea, we can only mean the first ingress of the Greeks into that sea and not necessarily the founding of the first organised colonies. Unless, of course, we deny any Greek contact with the area before colonisation.[24]

The question is whether all these terms have consistently been used with absolute cognizance of their precise meanings. Some of them, like 'reconnaissance period' or even 'pre-colonisation', sound rather 'colonio-centric', at least judged by how they are attested, in the sense that they denote an intention to colonise. But this is utterly exaggerated, because it would mean that the Greeks had been preparing their colonisation for centuries – as many as the evidence presented in the next section indicates. It would be erroneous to claim that any Greek visits to the Black Sea, irrespective of how early or scattered, must only have had such a purpose.

On the other hand, while the term 'pre-colonial' seems right in terms of denoting simply a time before colonisation, it inevitably accompanies nouns like contacts, activities or presence, and it is therefore rather surprising to find it widely used by scholars who have firmly denied any Greek presence in the Black Sea before the colonisation – especially if those contacts would imply some form of settlement, as asked above. Carpenter's theory,[25] as absurd as it may seem, at least entails a degree of honesty, since he implies (but does not directly claim) that the first Greek appearance in the Black Sea came with colonisation and thus does not have to deal at all with any 'pre-colonial' matters.

In conclusion, the use of most of these terms in themselves – with the exception of 'penetration' when it is used with its proper meaning – betokens acknowledgment of the existence of a Greek presence in the (in our case southern) Black Sea before the systematic founding of colonies. What is really interesting is the fact that this acknowledgment is sometimes subconscious, or seems to emerge rather unwillingly. For example, J. Boardman, one of the

[23] The equivalent German term is 'Eindringen' (see, for example, Danoff 1962, 1051, 1055).

[24] As for example Carpenter (1948), in his outworn theory, according to which, the Greeks could not have reached the Black Sea before the development of the penteconter, in the first half of the 7th century BC. *Cf.* also S. West (2003), who also emphasised the rough climatic conditions in the Bosporus and the Black Sea, as if navigators before the 7th century BC were incapable. Kolb (2004, 594) also accepted this theory (mostly, it seems, because it fitted his own theory about the Mycenaean absence from the Black Sea – see below), although he placed the invention of the penteconter in the 8th century BC. Criticism of Carpenter's theory can be read in Labaree 1957, 29–33; Graham 1958, 26–31; Drews 1976, 25–42; Danoff 1962, 1053–54; etc. As I see it, the equation of penetration with colonisation is as risky as the equation of pots with people.

[25] See the previous note.

champions of the view that colonisation began in the 7th century BC,[26] nonetheless entertains serious suspicions that the Greeks had earlier contacts with the Black Sea. Although he emphasises the presence of archaeological finds in the region from the 7th century BC and places the beginning of colonisation at that time, several phrases clearly betray his conviction that contacts between the two, and specifically Greek visits to the Black Sea, had begun earlier. These contacts – these unknown 'pre-colonial' contacts – he sets down to 'exploration'[27] and to 'occasional voyages by sailors and merchants'.[28]

Recently it has been argued that, as more evidence comes to light in the regions of Greek colonisation, the term 'pre-colonial contacts', as it was 'once' used to explain finds which 'supposedly' predated the establishment of the first colonies, has gradually receded, and that the earliest Greek pots found may be treated as 'some form of diplomatic gift or prestige objects'.[29] The argument appeared in an essay concerning the northern Black Sea, where probably the earliest Greek finds happen indeed to be finds from the earliest Greek colonies. But this is not everywhere the case. Besides, when talking about a couple of Greek finds, we cannot be sure whether they were brought to the region where they were found by Greeks, as has reasonably been pointed out (the classic question of whether pots equal people).[30] But they still indicate a contact between the region and the Greeks, not to mention that such contacts are not necessarily implied only by finds from the region in question itself.[31] If we reject the term 'pre-colonial' for them, how then do we explain them? Certainly not as 'colonial', I trust.

Speaking of finds brings us back to historical and archaeological evidence. If both are as meagre as indicated above, then what exactly do they demonstrate? Visits? Contacts? Settlement? And of what kind? Obviously, the problem has been recycled. Archaeological evidence, it is said, does not prove an early Greek presence in the southern Black Sea, nor any Greek settlement there before the last third of the 7th century BC, which is the reported date of

[26] Boardman 1991, 387–90; 1999, 245–67.

[27] Boardman 1991, 387; 1999, 244.

[28] Boardman 1999, 240–41, 64–65, where he also admits that 'there does seem to be some connection between Caucasian and some Geometric Greek metal-work'. *Cf.* Bouzek 1990, 15, 123, 173, 184; 1997, 200–01; 2005, 65–66. De Boer (2006–07, 283) on the other hand, does not reveal how he imagines 'pre-colonial' contacts, which he places in the 8th century BC.

[29] Tsetskhladze 2012a, 352–53. For the theory about diplomatic gifts, see Tsetskhladze 2012a, 338–39, with further bibliography; 2012b, 238; 2013, 69, 72, 74; 2014, 220; and, in greater detail, 2010.

[30] See more recently Tsetskhladze 2012a; 2012b; 2014, 316.

[31] See, for example, Boardman 1991, 387; 1999, 64–65, 244; Braund 2005a. For such kind of finds, see in the next section.

the earliest Greek finds from Sinope and Tieion.[32] And historical evidence regarding these early periods, or at least its interpretation by historians, cannot be considered trustworthy. Neither can a couple of Greek finds prove much. Nevertheless, in the next section we shall see that each of these axioms might be or has been proved vulnerable, even by its own supporters; we shall observe this by examining some aspects of the early Greek presence and the Greek colonisation in our area of interest, as well as the quite significant case of Sinope, which might provide an intriguing link between these two stages.

3. Early Greek Presence

3.1. Indications and Thoughts

Tracing the outset of the Greek presence in the southern Black Sea is a tremendously difficult task – one might almost say a vain endeavour. First of all, what could be considered 'presence' apart from colonisation? Factors that might be considered determinant are repetition of visits, their duration, sufficient number of visitors, and some kind of settlement, even temporary or seasonal. Unlike the northern Black Sea, for the south there are several indications of people from Greece having been there before the colonisation and initially without such intention, or at least without such a result. These indications do not include specific archaeological finds from the littoral itself, and this constitutes the foremost reason for the aforementioned difficulty. They derive mostly from indirect evidence.

The legend of the Argonauts has always been treated as reflecting very early Greek visits to the Black Sea, and their outward route, hugging the south coast, might be highly indicative of real knowledge, especially since the Greek mythographers are virtually unanimous on the subject, in contrast to the various accounts given of the Argonauts' return from Colchis.[33] The initial core of the Argonautic myth was certainly older than Homer, as is clear from his reference to the 'famous *Argo*' (*Odyssey* 12. 69–72), and in the timeline of mythology the expedition must have taken place sometime in the 13th century BC.[34]

While the mythical narrative is too flimsy a basis for drawing conclusions, additional hints might be furnished by other evidence, mostly archaeological,

[32] See most recently Tsetskhladze 2006a, lxxii; 2018, 53. However, between these two papers, we read that 'there is no 7th-century BC (Greek) pottery from these two colonies (Sinope and Amisos) and the earliest pottery dates to the first half of the 6th century' (Tsetskhladze 2012a, 349). For Tieion, see below.

[33] Manoledakis 2018b, 187.

[34] Manoledakis 2018b, 187.

from the same time period. For example, the gold grave offerings found in a Mycenaean tholos-tomb at Kazanaki, near Volos, an area identified with the region of ancient Iolkos, the place from where the Argonauts set out on their expedition, were subjected to chemical analysis which showed that they were made in the Late Helladic period (LH III A 2–III B 1) of alluvial gold identical in composition to that of Georgia, the Argonaut's destination (Colchis).[35]

Moreover, several Linear B texts, mostly from Pylos, contain names bearing a striking resemblance to proper names in the Argonaut story, and have thus led, combined with Mycenaean finds, to a theory about Mycenaean activity in the Black Sea, even to consideration of the epic tale of the Argonautic expedition as Mycenaean in origin.[36] While on the subject of Mycenaean finds, we note that for the southern Black Sea region such finds come only from the hinterland, not the littoral, but are still contemporaneous with the evidence presented above. In the main they are Mycenaean vases found in the Halys valley, and specifically at Maşathöyük (about 130 km from the sea, together with Hittite pottery) and Boğazköy.[37] The finds from Maşathöyük include at least seven fragmentary vessels,[38] which have been generally dated to the LH III A2–B (1380–1190 BC),[39] and by association with other finds more specifically to the LH III B – the first half of the 13th century BC, or perhaps a little later, after the middle of that century.[40]

The Halys valley, that unique natural highway connecting the Black Sea coast with the Anatolian hinterland, has also yielded some fragments of

[35] Adrimi-Sismani, Guerra and Walter 2009, with bibliography.

[36] Hiller 1991, 213–15; *cf.* Nilsson 1932, 136–38. See more in Manoledakis 2018b, 208–10.

[37] For the sherds at Maşathöyük, see Mellink 1974, 105, 115; 1975, 204, 208; 1976, 265, 270; 1984, 445; 1985, 558; Bittel-Heidenheim 1976, 528–29; Mee 1978, 132–33; 1997, 141; Özgüç 1978, 65–66, 127–28, pls. 83–84 and D1; 1980, 309; 1982, 31, 102–03 and pl. 47.5–6; French 1982, 21–23; Cline 1991, 3; Summerer 2007, 30–31, with bibliography. The sherds at Boğazköy, which are nowhere depicted, are mentioned by Bittel 1976, 9; Summerer 2007, 31; Schachner 2011, 305–06. For the other Mycenaean finds elsewhere around the Black Sea (weapons, fibulae, double axes), see Hiller 1991; Cline 1996; Jablonka and Rose 2004, 626; Harding 2005, 299; Petropoulos 2005a, 16–17, 63; Vanschoonwinkel 2006, 77–78.

[38] Hiller 1991, 208, with more bibliography; Cline 1991, 3. Kacarava *et al.* (1999, 66) are more sceptical about the meaning of Mycenaean pottery at Maşat, but they do not seem to be aware of all the fragments found there.

[39] Mee 1978, 133; French 1982, 21; Özgüç 1978, 65–66, 127–28; 1980, 309; 1982, 31, 102–03.

[40] Özgüç 1980, 309; Cline 1991, 3. Mycenaean pottery has also been mentioned in Akalan, but it is not certain that it is really Mycenaean (see Mee 1978, 124; Cline 1991, 3). Although the Mycenaean finds in the southern Black Sea hinterland suggest that the Mycenaeans reached those regions, D. French questioned whether it was the Mycenaeans who took this pottery to Anatolia or the Anatolians who brought it from Greece (French 1982, 22; *cf.* Tsetskhladze 2012b, 239). It is worth noting that Mee (1978, 150) thinks the presence of Mycenaean pottery at Maşat 'surprising', because, as he says, there is no other evidence of Mycenaean trade with central Anatolia.

Geometric pottery. At Kaman-Kalehöyük (about 300 km inland), few sherds of the Protogeometric/Early Geometric periods were found in the Period IId level of the site, which dates from the late 11th to the mid-9th century BC.[41] They display motifs typical of Attic Geometric pottery of 1100–900 BC, such as pendant concentric semicircles, intersecting or otherwise, though they are certainly not Attic as is clear from the clay. They also bear strong similarities with Thessalo-Euboean parallels of the Late Protogeometric period.[42] The route that all these finds followed to arrive at the Halys valley settlements is a matter of speculation. Three main routes, both overland and by sea, have been proposed: overland, through East Greece, Sardis and Gordion; up from the south, through Cyprus, which corresponds to a major route followed by Achaean dialects like the Arcado-Cypriot, the Pamphylian and the Phrygian; and down from the north, through Troy and the north end of the Halys valley, where Amisos would later be founded.[43]

All three routes are plausible, but the last is based on weightier arguments, among them the role of Troy, which has yielded a significant quantity both of Mycenaean pottery and of amphorae with concentric semi-circles, as a very important node in the transport of Greek products from the Mycenaean period on.[44] Besides, this route is precisely the one supported by the legend of the expedition of the Argonauts, as already mentioned. Moreover, it is widely accepted that the sea routes were generally far preferred to the overland ones in these early periods.[45] And finally, the spatial distribution of archaeological sites in the Halys valley might also prove to be of the utmost importance. All

[41] Matsumura 2008; Matsumura and Omori 2010; Omura 2011, 1101.

[42] Manoledakis 2018b, 177–78.

[43] For the three routes, see, in detail, Manoledakis 2018b, 214–19, with the whole bibliography.

[44] Korfmann 1997, 94–95; 2001; Hiller 1991; Wright 1998. Cf. also Roebuck 1959, 116. That Troy was from the Mycenaean period a stopover for Aegean ships on their way to the Black Sea, and consequently that there had been a Mycenaean expansion to the latter, even to the Propontis, has been sometimes rejected (Kolb 2004; de Boer 2006–07, often almost repeating the former); here again, however, the authors, although seeming sure of their conclusion, leave open some possibility of the contrary (e.g. Kolb 2004, 592; de Boer 2006–07, 282, 296–97). They give much weight to the mythical data, which however are not the crucial element, and conclude that the Black Sea in the Bronze Age was a '*mare clausum*' for Mycenaeans (Kolb 2004, 594; de Boer 2006–07, 297), which is much more frivolous than trying to connect Mycenaean finds with some kind of Mycenaean presence. Actually, they claim that 'Mycenaean objects or impact do not necessarily mean Mycenaean trade of those objects [which they thus acknowledge]; they do mean cultural influence and presuppose contact, which was certainly connected with some exchange of goods.' Unfortunately, they do not enlighten their readers as to how they imagine such contacts and exchanges, which would be so different from trade that they could not entail what we call 'presence'. For more recent argumentation in favour of the significance of Troy as a trade station to the Black Sea (and elsewhere) in the Mycenaean period, see Easton *et al.* 2002, 101–06; Jablonka and Rose 2004; Vanschoonwinkel 2006, 77–78.

[45] See also below, Section 5.5.

along the valley, starting from the Amisos area and as far as Kaman in the south, there are numerous significant sites that have yielded finds belonging to several periods and cultures, but most importantly indicating an interaction between several cultures and civilisations that flourished in the central part of northern Anatolia in addition to the Greek.[46] An interaction that becomes especially visible as regards the 6th century BC, from which architectural terracottas and local pottery at many of these sites indicates a Greek influence.[47] Taking into account the importance of the valley's accessibility to the hinterland,[48] we cannot help but surmise that this third route was highly possibly the one followed in our case; the second one, from the south, demanded a long journey inland (from the Mediterranean to Kaman), over terrain entailing more time, effort and cost.

It seems, then, that the sea route along the southern Black Sea coast must have been used by people coming from the west from a very early period. Furthermore, we should not underestimate the existence of overland west-east routes in northern Anatolia, starting from the Bosporus or the Propontis and extending as far as the Caucasus and central Asia, just to the south of the big mountain ranges, for some of which it has been proposed – even if this is difficult to prove – that they were already in existence in the Bronze Age.[49]

The tale of the Argonauts is not the only textual basis for speculation on early Greek contacts with – and certainly knowledge of – the southern Black Sea littoral. In the Trojan Catalogue of the *Iliad* (2. 851–857), two peoples from north Anatolia are mentioned as having sided with the Trojans: the Paphlagonians and the Halizones. Irrespective of the latter's real identity,[50] the lines in question are believed to belong to the initial core of the Trojan Catalogue,[51] which the overwhelming majority of scholars today assign to the original core of the *Iliad*.[52] Thus, if Homer mentions north Anatolian peoples in the early 7th century BC,[53] it would seems that the Greeks were by that time

[46] Later, other significant cities, like Amaseia, the capital of the Pontic kingdom, would benefit by this route. See, indicatively, Ramsay 1890, 27–28; Munro 1901, 53; Burney 1956, 193; Saprykin 1997, 24; Summerer 2005a, 147; 2007, 29; Tsetskhladze 2013, 77; French 2016b.

[47] See below, Section 4.4.

[48] For the routes into the interior that this valley might have offered – among them the river road of the navigable Halys itself – see Section 5.5.

[49] For these routes, Section 5.5, with the relevant bibliography.

[50] For the Halizones, see Section 2.9.5.

[51] See the whole argumentation in Manoledakis 2013a.

[52] For the relevant argumentation, as well as all the theories that have been from time to time formulated on the relation of the Achaean and Trojan Catalogues to the *Iliad*, see more recently Weber 2011, 242–44; Kullmann 2011, 92–94; 2012.

[53] For the dating of Homer's works in the 7th and not the 8th century BC, see Manoledakis 2018b, 184–85, with the whole bibliography.

acquainted with the region; an acquaintance that is further confirmed by the information furnished by Apollodorus (Strabo 12. 3. 26) that Homer obtained his knowledge of the history of the Paphlagonians of the interior from persons who had travelled through that country on foot. Other early writers who seem to have been familiar with the region are Eumelus (*Corinthiaca* – Kinkel, *EGF* 8 and 17), who mentions the nymph Sinope and the Muse Borysthenis,[54] and Hesiod (*Theogony* 337–345), who names four rivers (Phasis, Aldescus, Sangarius and Parthenius) that flow into the Black Sea.[55] To assume that the familiarity evident in their works could indicate a knowledge that had been acquired earlier is, of course, reasonable, and the Geometric pottery mentioned above might help one who would consider the Mycenaean pottery far too early to imply such a connection with the Early Archaic period of the poets.

The question is whether all this indirect evidence from the Mycenaean to the Early Archaic periods might indicate Greek contacts with the southern Black Sea in these early periods, that is, before the organised colonisation. Or better, a kind of presence, in terms of the factors propounded in the beginning. In my opinion, they might. For they do not imply just individual one-off journeys; nor, of course, were these contacts necessarily developed with the objective of present or future colonisation. More importantly still, for the 8th century BC there are even strong indications of a specific settlement, and one moreover not established (at this specific time) by the Ionians of Miletus. That site, Sinope, will therefore be examined separately.

3.2. The Case of Sinope

The significance of Sinope becomes obvious from the diversity of the evidence referring to it, compared with a typical Greek Black Sea colony founded by Ionians in the 7th century BC or later. Let us start with the literary evidence, keeping in mind that Sinope (the Ionian colony) is unanimously considered one of the earliest if not the earliest Greek colony in the (southern) Black Sea.[56]

The most striking is undoubtedly the timeline for the city's population history given by Pseudo-Scymnus (986–997):

> the city of Sinope was named for one of the Amazons living in the area that was sometimes inhabited by the noble [or local?] Syrians, and afterwards, as they say,

[54] M.L. West 2002, 126; Braund 2005a.

[55] *Cf.* also Mimnermus (in Strabo 1. 2. 40). According to the new relative dating of the epics, all these texts are to be placed after the *Iliad*, which is held to be the oldest of them all, so after the mid-7th century BC: Janko 2012, 20–43.

[56] See, indicatively, Doonan 2007, 617; Doonan *et al.* 2017, 178.

> by those from the Greeks who went over against the Amazons, Autolycus and Phlogius with Deileon, Thessalians in origin, and then Habron(das) from Miletus; it is said that he was killed by the Cimmerians. And after the Cimmerians, again Coos and Cretines who were fugitives from Miletus. They built the city again, when the Cimmerian army was invading Asia.[57]

According to Pseudo-Scymnus,[58] Sinope was originally inhabited by the Leucosyroi,[59] while the first Greeks to live there were the Thessalians Autolycus, Phlogius and Deileon. Later came the Milesian (H)abron(das),[60] who is said to have been killed by the Cimmerians, and later still Coos and Cretines, exiled from Miletus, who rebuilt[61] the city when the Cimmerian army was campaigning in Asia. There is no need to repeat that mythical narratives may have an historical background but cannot be considered as corresponding to historical events. Nevertheless, the specific narrative contains some noteworthy elements.

First of all, the earliest Greeks to settle in Sinope are represented by Thessalian heroes, with the Milesians following. That the settlement appears to have belonged to the territory of the Leucosyroi is a fact that enhances the reliability of the source. Secondly, the Cimmerian campaign in Asia is mentioned, which we have placed somewhere between the late 8th and the early 7th century BC.[62] This seems to have been of rather short duration – again consonant with what has already been suggested – but brutal enough to cause

[57] Εἶτεν Σινώπη πόλις ἐπώνυμος μιᾶς
Ἀμαζόνων, ὧν πλησίον τὸ χωρίον,
ἥν ποτε μὲν ᾤκουν εὐγενεῖς ὄντες Σύροι,
μετὰ ταῦτα δ', ὡς λέγουσιν, Ἑλλήνων ὅσοι
ἐπ' Ἀμαζόνας διέβησαν, Αὐτόλυκός τε καὶ
σὺν Δηϊλέοντι Φλόγιος, ὄντες Θετταλοί,
ἔπειτα δ' Ἀβρώνδας γένει Μιλήσιος,
ὑπὸ Κιμμερίων οὗτος δ' ἀναιρεῖσθαι δοκεῖ·
μετὰ Κιμμερίους Κῷος πάλιν δὲ Κρητίνης,
οἱ γενόμενοι φυγάδες ὅρων Μιλησίων.
οὗτοι συνοικίζουσι δ'αὐτὴν ἡνίκα
ὁ Κιμμερίων κατέδραμε τὴν Ἀσίαν στρατός.

See Diller 1952, 175. The text is cited with slight differences in Müller 1855, 236 (Pseudo-Scymnus 941–952), but not as regards the content. *Cf.* Ivantchik 1998, 297–98.

[58] The story is also told by the anonymous author of the *Periplus* (22), one of whose sources was Pseudo-Scymnus. See in this regard Diller 1952, 102. For the sources of Pseudo-Scymnus on this particular point, one of whom must have been Demetrius of Callatis (late 3rd century BC), see Maksimova 1956, 50–51; Roebuck 1984, 119–20; Hind 1988, 210; Langella 1997, 98–103; Ivantchik 1998, 321.

[59] For the Leucosyroi, see Section 2.8.

[60] See Ivantchik 1998, 298, n. 2.

[61] The word *συνοικίζουσι* in Pseudo-Scymnus denotes that the city was rebuilt after a disaster. See Ivantchik 1998, 298–99, n. 3.

[62] See Section 3.1.

much trouble to the settlement of the Milesians, who – a new wave from Miletus this time – had to rebuild it after the Cimmerian withdrawal. If the narrative is trustworthy and the origin of the heroes faithfully reflects the origin of the settlers,[63] then the Thessalians must have settled the site in the 8th century BC, with the Milesians succeeding them in the late 8th or early 7th – but is it trustworthy?

Seemingly, there is no certain way to find out, but we should not deny the value of the verification provided by other evidence. The tradition regarding the Thessalian brothers was well enough known for them to be mentioned in the *Argonautica* (2. 955–961), where we learn that they were sons of the noble Deimachus of Tricca. Furthermore, Strabo says (12. 3. 11) that the people of Sinope held Autolycus to have been the founder of their city and that they worshipped him as a god, adding that there was an oracle of his in the city and also a statue made by the sculptor Sthenis (4th century BC), which Lucullus took away with him. We learn more about Lucullus and his part in the story from Plutarch (*Lucullus* 23. 4–5) and Appian (*Mithridates* 83), who relate that when Lucullus reached Sinope (in 70 BC) he saw in his sleep a figure that told him that Autolycus was there and wanted to speak to him. The next day, after capturing the city and driving out Mithridates' Cilician troops, he saw a statue lying on the shore. This, he was told, was the statue of Autolycus, one of the finest works of the sculptor Sthenis, which the soldiers of Mithridates had attempted to carry off but had had to abandon. Then Lucullus remembered the words of Sulla: that nothing is as reliable as what is revealed in dreams. After this he set Sinope free. These texts demonstrate that the Thessalian hero was venerated as the founder of the city at least until the Roman era.[64] Indeed, his statue and oracle might suggest the existence of a temple dedicated to him.[65] Also noteworthy is the existence of a votive inscription to his brother Phlogius (*CIG* 3. 4162), who is cited as co-founder of the city (Plutarch *Lucullus* 23. 4–5; Appian *Mithridates* 83).

It seems, then, that the data provided by Pseudo-Scymnus are not arbitrary. Should this shake the universally accepted view that Sinope was a colony of

[63] In general, information on the founding of cities given by Pseudo-Scymnus is considered reliable. See, for example, Roebuck 1959, 119–20; Hind 1999, especially 34; Summerer 2005a, 147, for the case of Amisos; Saprykin 1997, 21, for the case of Heraclea. For foundation stories, see Hall 2008.

[64] For all the details and arguments, plus the older literature on the founding of Sinope, see Manoledakis 2010b.

[65] Barat (2012, 54–60) has argued that the so-called 'Serapis temple' in Sinope, parts of which were revealed by Akurgal and Budde, was more probably a *heroon* dedicated to Autolycus. Her theory is rejected by Summerer (2014, 193–96). See the bibliography on this building in these two papers.

Miletus, as, among others, Strabo himself says earlier in 12. 3. 11? Not at all. For if we read Strabo carefully (at least the particular chapter) and without isolating selected passages, we see that he first describes the city of his own day (when he mentions it as a Milesian colony) but then gives a brief overview of its history, in which he mentions the tradition of the Thessalian settlement, adding that 'then *later* [ὕστερον] the Milesians, seeing the natural advantages of the place and the weakness of its inhabitants, appropriated it to themselves and sent forth colonists to it'. Thus, there is no controversy, just a sequence of events. The weakness of the inhabitants refers apparently to a rather small and sparsely populated settlement. This is how a Thessalian settlement there would have appeared when the Milesians came, if there had indeed been one.[66]

It is therefore, thoroughly astonishing that modern scholarship has almost unanimously denied the veracity of Pseudo-Scymnus' (mythical) order of events, but only selectively. Probably spurred by the fact that Aelius Herodianus and Stephanus of Byzantium (*s.v.* Σινώπη) name only the Milesians Coos and Cretines as founders of Sinope, many scholars have tried to establish an historical validation and a chronology only for a double Milesian founding of Sinope, once by Habrondas and once by Coos and Cretines,[67] treating it as self-evident that there was no previous Greek presence in the region and excluding any possible historical basis for the tradition that the city was founded by Autolycus,[68] even though in so doing they lapse into the controversy of considering Pseudo-Scymnus reliable in some lines but not in others.[69]

[66] On the issue of the Thessalian presence in Sinope, see in detail Manoledakis 2010b, with all the relevant sources and bibliography. Even O. Doonan, the excavator of Sinop province, seems to be leaning towards the existence there of a settlement earlier not only than the Milesian city but also than the destruction layer of Gordion (Doonan 2007). For Autolycus in Sinope, see also Malkin 1987, 207–08. For a thorough study of the foundation myths and their credibility, see Hall 2008. For Pseudo-Scymnus' narrative on the founding of Sinope, see also Bilabel 1920, 30–40.

[67] Robinson 1906, 149; Ehrhardt 1988, 55–57; Langella 1997, 96, 125; Ivantchik 1998, 311, 313; 2010; Doonan 2003, 1381; Højte 2008, 156. More bibliography in Tsetskhladze 2009, 230, n. 24. See also Boshnakov 2004, 114–22.

[68] E.g. Huxley 1960, 19; Hind 1988; 1999, 29; Langella 1997, 91. Højte 2008, 156. Huxley (1960, 22) argued that the Thessalian heroes were taken over from the earlier Phrygian settlers of Sinope. Langella (1997, 75), on the other hand, argued that the tradition about Autolycus could also have been belonged to the Milesians. See also Ivantchik 1997. More theories about this tradition, not always convincing, are mentioned by Langella (1997, 79–88). See more recently, in brief, Faulkner-Gentry 2021, 30–31.

[69] This, of course, is scientifically absurd, unless one can prove that some of the verses were interpolated or false, which is not at all the case with the passage in question. For this selective use of Pseudo-Scymnus, see Manoledakis 2010b, in detail, with bibliography.

Finally, Herodotus' statement that 'the Cimmerians, in their flight from the Scythians into Asia, also made a colony on the peninsula where now the Greek city of Sinope has been founded' (4. 12. 2) has been also interpreted as proof that the Cimmerians were the first to build a city at Sinope, before the Greeks.[70] Herodotus, however, does not say precisely that.[71] What he says is that the Cimmerians, driven by the Scythians out of Asia, settled on the peninsula where the Greek city of Sinope now stands. That does not mean that Sinope was originally founded by the Cimmerians and only later by the Greeks. It simply means that a Cimmerian settlement pre-existed the specific Greek settlement known to Herodotus, who may well have been unaware of the existence of any pre-Cimmerian settlement, Greek or other, for he makes no mention of a Leucosyrian settlement either. Besides, even if it is only Pseudo-Scymnus' Milesians who are to be taken into account with regard to the founding of Sinope, there still remains the question of a double Milesian founding at a greater or shorter remove one from the other, and again, the first of these would in any case have to be placed before the Cimmerian occupation and therefore in the late 8th or very early 7th century BC at the latest.[72]

To conclude, the evidence presented above might be considered enough to provide serious grounds for thinking that the site of Sinop, that by all accounts privileged location on the southern Black Sea coast, had been settled by Greeks prior to colonisation. Whether this settlement was town or *emporion*,[73] permanent or seasonal, with a mixed or purely colonial population, is very difficult to discover. But it is noteworthy that its Greek inhabitants seem unconnected with the Ionians, who later dominated the littoral, or the Megarians. They appear, rather, to be linked with central Greece, and there is further evidence to suggest that such a link is not unsubstantial.

Consider, for example, a decree of the Euboean city of Histiaea honouring the citizens of Sinope (*IG* XII. 9. 1186).[74] Although the inscription dates from the 3rd century BC, it states (line 22) that the Sinopeans were friends and brothers of the Histiaeans of old, and a highly probable explanation of this relation could be the common ancestry that the two cities shared. This ensues from traditions claiming that the Euboean Histiaea was founded by the

[70] See the bibliography in Manoledakis 2010b.

[71] *Cf.* Huxley 1960, 21. See more in Tokhtas'ev 1996.

[72] Maksimova 1956, 42–46; Hind 1988, 212–13; Langella 1997, 96; Højte 2008, 156.

[73] Doonan 2004, 78; 2007, 615, 617, with earlier bibliography. On the *emporia* see below. Summerer (2007, 29; 2008, 264) has suggested that an *emporion* could have preceded the colony at Amisos as well.

[74] For Sinopeans abroad and foreigners in Sinope, see Ruscu 2008.

Perrhaiboi, who ruled in Thessalian Histiaeotis,[75] while the alleged founder of Sinope is known to us as Autolycus of Tricca, a town in that same Thessalian Histiaeotis.[76] Even more abundant are the relevant mythical traditions connecting the southern Black Sea littoral with central Greece (mainly Thessaly and Boeotia),[77] which also happens to be the area the Argonauts departed from.

Keeping all this in mind, and at the same time pondering the exceptionally close connection between the Euboeans and the Thessalians from the Late Mycenaean to the Geometric period, as well as the role that both played in the overseas activities of the Greeks,[78] one can hardly avoid contemplating the possibility that the first Greeks who developed contacts with the littoral might have come from central Greece. After all, this is the area that seems to have forged early contacts with the northern Aegean,[79] and the similarities between the two cases, as evidenced by archaeological finds in central Macedonia and their dating, are remarkable.[80]

But what about archaeological finds form Sinope? Unfortunately, in contrast to the literary evidence, in the absence of light darkness prevails. Buried beneath the modern town of Sinop, the ancient city, and particularly the part near the sheltered harbour where the earliest Greek settlers must have dwelt,[81] is covered by modern buildings, which means that whatever archaeological evidence lies below is presumably irretrievably lost.[82] The only archaeological investigation that has taken place to date in this area is the excavation of some Archaic graves (in 1951–53) to the west of the isthmus, as well as of several parts of the city,[83] while in 2000 and since 2015 equally limited (so far) excavations have taken place close to the castle (*kalesi*) of Sinop.[84] To date it is mainly in the first that the earliest Greek finds have come to light. Although sparse indeed, they are nonetheless impressive for the way they have been so frequently and excessively invoked by some modern scholars, sometimes indeed even interpreted in unsubstantiated ways and thus coerced into misleading conclusions, as will be shown in the next section.

[75] For Histiaeotis, see, for example, Stählin 1967, 113–30.

[76] *Cf.* Asheri 1973, 74; Rigsby 1980, 245; Langella 1997, 56–57; Manoledakis 2010b, 570.

[77] See them in detail in Manoledakis 2018b, 210–11, with bibliography. See also Langella 1997, 60–64.

[78] The relevant bibliography in Manoledakis 2018b, 213–14.

[79] Tiverios 2008.

[80] Manoledakis 2018b, 223–25. For the Euboean interest in the area, as well the relations between Euboea and Miletus, see Langella 1997, 10–11.

[81] See, for example, Doonan 2004, 72, 74–75; 2007, 614; Erciyas 2007, 1195; Summerer 2007, 28; Manoledakis 2010b, 572–73.

[82] Manoledakis 2018b, 181, with earlier bibliography.

[83] Akurgal 1955b; 1956; Budde 1956; Akurgal and Budde 1956.

[84] Doonan 2016, 217; Doonan *et al.* 2015; 2017.

According to the excavators, the earliest finds are some fragments of Greek, and specifically East Greek and Corinthian, pottery found in the graves near the northern bay (Kumkapı). As read by the excavators, the East Greek pottery fragments were initially dated by one, E. Akurgal, with some reservations, to 'about the end of the 7th century BC',[85] while the other, L. Budde, spoke of the early 6th century BC[86] and attributed generally to the late 7th century only the earliest finds from the area around the temple, which, however, he did not name (some indeed were Phrygian).[87] In their joint publication, Akurgal avoided repeating this dating,[88] and dated only the earliest Corinthian pottery (some aryballoi) to the beginning of the 6th century BC, *ca.* 600 BC, clearly adding that 'our aryballos is the oldest Greek product unearthed until now in Sinope, and provides as such an approximate reference point for the dating of the upper time limit of the Greek colonisation in Sinope'.[89]

Three years later, Y. Boysal published his study of the earliest Greek finds from Sinope as presented by Akurgal and Budde, dividing them into those from the necropolis and those found in the city (see above). In the first group he included only the Corinthian aryballoi, which he described and dated. The earliest of these were dated to the first half of the 6th century BC, starting from 580,[90] while the finds[91] from the city, namely some 'orientalising' sherds and a large part of an Attic cup, he dated to 570–560 BC.[92] He therefore concluded that there was at that time no Greek find in Sinope earlier than 580 BC and consequently that nothing could be stated about the history of the settlement before that period.[93]

Some years later J.M. Cook and J. Boardman, two of the most experienced scholars of ancient Greek pottery, also dated the finds of Akurgal and Budde, correctly I believe, to the early 6th century BC.[94] It was only three decades after the excavations of Akurgal and Budde that J.G.F. Hind, talking about some 'chance finds' from Sinope – without referencing find location and date – published a drawing of a fragment of a Greek plate, which he dated to the

[85] Akurgal 1956, 55. *Cf.* also Hind 1964, 179.

[86] Budde 1956, 6.

[87] Budde 1956, 7; Akurgal and Budde 1956, 38.

[88] Akurgal and Budde 1956, 4.

[89] Akurgal and Budde 1956, 5, pl. I.

[90] Boysal 1959, 9–13, 18–20.

[91] The reference here is always to the earliest Greek ones.

[92] Boysal 1959, 13–14. The Attic cup had also been dated to the same period by the excavators (Akurgal 1956, 55; Akurgal and Budde 1956, pl. II).

[93] Boysal 1959, 15–16, 18.

[94] Cook 1960, 34; Boardman 1963, 51; 1999, 254–55.

late 7th century BC (Fig. 11).[95] That fragment, however, cannot provide enough indications for precise dating. Its decorative motifs are seen mainly in the early 6th century BC. Hind also mentioned fragments of East Greek pottery that were shown to him in Ankara,[96] which, however, have to date not been published; therefore, one cannot be sure about anything that has been said about them, and this applies to any find that is not published or at least not illustrated in a publication.[97]

The interpretation of the earliest Greek archaeological finds from Sinope is as astounding as modern scholarship's treatment of Pseudo-Scymnus' text on its settlement. For, despite the fact that these finds do not predate 600 BC, not only are they almost everywhere purported to belong to the last third of the 7th century BC (and indeed citing Boysal, who clearly dated them after 600 – see above), but they are also used as the most decisive evidence to argue that the Milesian colony of Sinope was founded in this period.[98] This is undoubtedly

[95] Hind 1984, 95, fig. 28.

[96] Hind 1964, 179; 1988, 216, 222, n. 40. It is to the Ankara Archaeological Museum (now Museum of Anatolian Civilizations) that the specific finds from Sinope were brought (Akurgal 1956, 59).

[97] This is another universally acknowledged methodological norm. A scholar's personal assessment is not enough (note that Akurgal and Budde clearly mention in their publications that what they depict are the earliest Greek pottery fragments from their investigations, and I suppose that if they had earlier ones, they would have depicted them). Finds should always be presented to the academic community; besides, the dating of small fragments from broad categories of pottery, like the East Greek, may vary slightly, but this slight variation might span parts of two different centuries. Consider, for example, apart from the plate fragment mentioned by Hind, some East Greek pottery fragments from Tieion: Atasoy and Erpehlivan date fragments Kat. 1–3 of their 2012 publication (here in Fig. 12, the two fragments above) to the last quarter of the 7th century BC. Kat. 1 and 2, however, might well be dated to the first half of the 6th century BC, while as for Kat. 3 (here in Fig. 12, below, left), I would say that this seems almost certain. Were they the only early East Greek finds from Tieion, the whole theory on its establishment in the 7th century would falter. But they are not. In his most recent publication, the excavator of the site, Ş. Yıldırım, talks about several other fragments from the 7th century, and depicts one of them (Yıldırım 2022, fig. 11), which actually seems to belong to its first half. It comes from a bird bowl, which resembles the one that was found in Mecitözü and is exhibited in the Museum of Amasya (Dönmez 2010, 546 fig. 38). In any case, the excavator is preparing an aggregated publication of the site's Archaic Greek pottery, which is eagerly awaited, since it will be the first from a Greek colony of the southern Black Sea littoral. Equally desirable is the publication of the Archaic pottery fragments from the recent excavations at Sinop Kalesi, which 'may' date from the 7th–6th centuries BC and have various provenances (Doonan *et al.* 2017, 193, 198; *cf.* Doonan *et al.* 2016, where he dates the earliest East Greek ones to the 6th century BC). I was recently informed that there is some hope for such a publication soon.

[98] See, just indicatively, Hind 1988, 216; Langella 1997, 103–05; Ivantchik 1998, 330; 2005a, 161; Doonan 2003, 1381–82; 2004, 70–71, 74–76; 2006, 52; 2009, 71; 2019, 105; Kacharava 2005, 12; de Boer 2006, 51; Erciyas 2007, 1195; Summerer 2007, 28; Tsetskhladze 1994a, 116 ('all pottery dates from the late-7th century to a little after 600 BC'); 2006, lxxii; 2007, 166; 2009, 232; 2018, 53; 2019c, 3. However, in 2012a, 349, he mentions that 'there is no 7th-century BC (Greek) pottery from these two colonies (Sinope and Amisos) and the earliest pottery dates to the first half of the 6th century'.

a rather misleading deduction, which, additionally, runs counter to some of the widely accepted axioms mentioned at the end of Section 4.2. If one intends to interpret the evidence, both literary and archaeological, in strictly hidebound fashion, then one would be forced to admit that the former points to an early Greek settlement of Sinope, in the 8th century BC, and the latter to only the 6th century BC. And that a 7th-century settlement is actually substantiated only through the former (e.g. Pseudo-Scymnus) and not the latter. Furthermore, the first Greek colony established in the southern Black Sea must have been Tieion, where several 7th-century pottery fragments have been unearthed (see below). How plausible would such conclusions sound?

3.3. Concluding Remarks

The reason for considering such conclusions awkward would be precisely that same hidebound manner of interpreting evidence, especially archaeological evidence, and the obsession with interpreting it in isolation from the literary evidence. Unfortunately, when we study 'grey' periods, where both kinds of evidence are vague and largely silent, as is the case with the beginnings of systematic Greek permanent settlement in our littoral, we are compelled to be less zealous in relying only on one or the other kind of evidence.[99] From what has been presented so far, it emerges that the period from the 8th (mid- or late)[100] to the late 7th century BC is indeed a 'grey' period in what concerns us here, for lack of solid evidence. It is for this reason that I have argued for examining the Greek presence in the southern Black Sea as a progressive evolution, a broader whole that encompasses and was shaped in at least two stages, the last or culminating phase being that of colonisation.[101]

Let me be more specific. The objections to a Greek presence (generally) on the southern Black Sea littoral before the 7th century BC are based on the grounds that there is no relevant archaeological material there. No one would deny that such a justification is reasonable, on the face of it and speaking strictly archaeologically. But there are two ways to respond to this, or even refute it. The first was put forward in the beginning of the previous section and concerns the universally acknowledged deficiency of archaeological finds

[99] Even de Boer (2006–07, 283–84) seems to accept 'pre-colonial' contacts of the Early Iron Age based on Homeric knowledge of the Black Sea, although the archaeological indications for such contacts in the specific period are even less than for the Mycenaean one, for which he excludes any contacts, invoking the fact that archaeological evidence does not prove them.

[100] When the Greek colonisation of other areas began. See Greco 2006, for southern Italy; Domínguez 2006, for Sicily; Tiverios 2008 for the North Aegean.

[101] Manoledakis 2018b, 225–27.

from the littoral, due to the lack of proper excavations there – even less of finds safely attributed to the earliest habitation levels of sites: one cannot appeal to archaeological material from an area that has not been archaeologically explored. Such an *argumentum ex silentio* would be irrational,[102] since the silence is not the result of an investigation but of the absence of investigation. What better area to confirm that absence of evidence is not evidence of absence than a practically unexcavated one?

Boardman was absolutely right to point out that the graves in Sinope that have been excavated to date may not be the most ancient ones at the site.[103] As was Hind, who noted that the first small Greek settlement of Sinope could hardly be expected to yield archaeological material.[104] In fact, neither do we have any traces of building construction from this period, which would be much more decisive for the substantiation of settlements. As Doonan stressed, 'intact settlement evidence of the Archaic town has not been found to date'.[105] On the same wavelength, D. Braund added another significant dimension as regards the interpretation of the earliest Greek pottery in the dating of colonisation:

> At what stage in the development of a colonial community do we imagine that fine Greek pottery came into use? [...] We may reasonably doubt (there is no evidence on the point) that the earliest settlers were greatly concerned to bring fine wares with them. [...] Indeed, most of the exchange between Greeks and non-Greeks in the early stages of what we consider colonisation will have been in much more perishable goods, which tend not to leave archaeological traces.[106]

In such a case, the earliest Greek pottery from a site would again define only a *terminus ante quem* for the founding of a Greek settlement there. Of course, this is just another conjecture, but it does, however, illustrate the relativity of the interpretation of early finds that has already been suggested.

The second argument derives from the archaeological evidence concerning the supposed beginning of colonisation and the 7th century BC itself. What about this evidence? Is it enough to document systematic settlement? The truth is that it is not. With the exception of Tieion, where excavations began in 2006, long after the whole debate had started, it was only Sinope that had

[102] *Cf.* Graham 1982, 123; 1994, 5.
[103] Boardman 1999, 254.
[104] Hind 1993, 110.
[105] Doonan 2004, 74. However, elsewhere (2007, 614–15) he did not exclude the possibility that a stone structure that has partially been revealed in the area of Sinop Kalesi might have belonged to the early colony – mainly because of the fact that stone architecture is not characteristic of non-Greek settlement in Sinop province. But this hypothesis cannot be confirmed, especially since the pottery found there was not Greek (p. 615).
[106] Braund 2005a, 100, 108–09, who adds that 'in fact, we have not the faintest clue about the date of the foundation of Sinope on archaeological grounds'.

allegedly yielded some East Greek pottery from the 7th century BC; this, however, as we have seen, is inaccurate.

But even if a handful of Greek potsherds from this century had indeed been found in Sinope, would they be enough to prove Greek settlement and in fact a founding of a Greek colony at Sinope by the Milesians? If we follow the widely – and correctly – accepted principles that (1) a few sherds are not enough to prove settlement in a place, since they might have arrived there on any occasion,[107] (2) pots, especially in such small quantities as in our case, do not necessarily equate to people; anyone could have brought them to the place where they were found,[108] and (3) archaeological evidence is in general more reliable that literary: then no, they would not.

Conversely, if we follow these principles strictly, then:

1. The earliest Greek pottery fragments from Sinope date from the early 6th century BC, so we should claim that Sinope was founded then and not in the late 7th century, although considering all we know of the history of Greek colonisation, this seems most unlikely.
2. The vast majority of these fragments are Corinthian, so the metropolis must have been Corinth and not Miletus – but again, there is nothing else to indicate a Corinthian colony in Sinope: the Corinthians were known to have been among the most active Greek traders in the Archaic period, in Asia Minor and elsewhere,[109] therefore they might have even brought the East Greek pottery with them.
3. If we accept that a couple of Corinthian aryballoi from the early 6th century BC are enough to establish Milesian colonisation in the 7th century BC, then why should seven Mycenaean vessels not suffice to establish (some kind of) Mycenaean presence in northern Anatolia in LH III?[110]

And finally, it has recently been argued, I suppose on the grounds of the paucity of finds, that Tieion was 'probably founded in the early 6th century BC',[111] although it has yielded Greek pottery of the late 7th century. So why should we not date the foundation of Sinope to the same period, on exactly the same grounds? It seems, therefore, that the very arguments used to reject Greek settlement in the littoral before the 7th century BC may be turned against settlement in that same century as well, and used to argue *ex concessis*.

[107] Tsetskhladze 2002b, 81, n. 3; 2006, xxxiv; de Boer 2006–07, 296–97.

[108] E.g. Tsetskhladze 2012a; 2012b; 2014, 316.

[109] Roebuck 1984, 47–48, 77–79, 126–27. For the role of Corinth in the Black Sea, see Boysal 1956, 16–17.

[110] See above, in Section 4.3.1.

[111] Tsetskhladze 2019c, 3.

One final note, now that Tieion has been mentioned: most noteworthy is the fact that all the earliest Greek pottery samples so far come from the highest part of the settlement, the 'Upper city', on which the acropolis would later be built.[112] One would certainly expect the earliest Greek settlement in the region (and thus the earliest Greek finds) to be on the coast, just as at Sinope and elsewhere, for obvious reasons regarding the nature of Greek colonisation. Apparently, patience is needed until systematic excavation extends to the 'Lower city' as well. Besides, should the discovery of finds from the first half of the 7th century BC continue,[113] then the dating of Tieion's founding could be moved back. This would affect the dating of Sinope's (Milesian) founding, if we insist on considering it the earliest Greek colony on the littoral (see above). In that case, Pseudo-Scymnus' report of the first Milesians there preceding the Cimmerians would be confirmed, and thus textual and archaeological evidence would no longer be at odds. In any case, thanks to the Tieion excavations, the scientific community finally has a very good opportunity to draw more conclusions about the beginnings of Greek colonisation on the southern Black Sea coast based on archaeological evidence, which will hopefully be able to shed more light on what is still a 'grey' period. The same applies to the Sinope excavations, which are being pursued in a limited but vital part of the site.

What I have been trying to point out is how volatile things are as regards the early stages of the Greek presence and activity in the southern Black Sea, how cautious we have to be when invoking the available evidence – of any kind – and how easily one can become inconsistent with what he had already claimed when he appears overconfident of having accurately interpreted ambiguous evidence. Moreover, that one should be rather more amenable to combining evidence, sometimes even from other areas that have known similar events. To conclude, the data presented in the two preceding sub-sections might be considered enough to imply a Greek presence in the southern Black Sea before systematic colonisation, and in fact a Greek presence not necessarily connected only with the Ionians from Miletus.[114] In the following pages we shall look into some essential aspects of the Greek colonisation in the southern Black Sea, irrespective of when exactly it began.

[112] At an elevation of *ca.* 75 m. Yıldırım 2022.

[113] Like the one mentioned above (n. 97).

[114] There is therefore no need at all to examine the Bronze Age history of Miletus, and Ionia generally, in an attempt to confirm or disprove any Mycenaean activity in the Black Sea (de Boer 2006–07, 278), which was colonised by Ionians only in the 7th century BC.

4. The Greek Colonisation

4.1. Generally

How can one specify a presence – Greek in our case – as colonisation? The bibliography on the efforts to define colonisation, distinguishing it from colonialism or even differentiating between different aspects of it, is rich.[115] It has even been doubted, not without logical argumentation, whether the term 'colonisation' actually characterises the process of events that it is used for. Nevertheless, fair as such concerns might be, 'we have to call this process something, and colonisation is as good a term as any',[116] keeping in mind and accepting that 'terminology can never reflect the full reality'.[117]

From my viewpoint, in order to be able to specify a presence as colonisation, we should have evidence for permanent, systematic and organised settlement. One could further add that we should also have evidence for the foundation of Greek cities, as mainly Greek-populated *poleis*, with all that a *polis* in its political and cultural sense denotes,[118] as was the case in Greece itself, thus excluding occasional settlements of mixed population which were used as production and trading centres with the locals. These were what are known as *emporia*,[119] and their existence, albeit continuing into the Roman era,[120] is one of the significant characteristics of what we have called the 'grey' period.

What causes a considerable part of a city's population to migrate and start a new life in a strange place? The reasons for colonisation that have been put forward include overpopulation, not enough land in the country or space in the cities, the need to find new agricultural land, desire to develop trade relations with new areas, banishment of political enemies or dissidents and socio-political turbulence in general, eagerness to explore faraway lands, even simply a craving for adventure.[121] These can be grouped into three main categories: economic, socio-political and adventurist. Reasonable as all of them may

[115] See Tsetskhladze 2006a, xxv–xxviii; 2009, 229; Tsetskhladze and Hargrave 2011. Lately, Tsetskhladze (2019d, 59) has favoured the term 'overseas settlement', affected by Boardman's seminal work (1999), though adding (correctly, I think) that the term is unenlightening.

[116] Whitley 2001, 125.

[117] Tsetskhladze 2006a, xxviii.

[118] For the political meaning of the term *polis*, see Hansen and Nielsen 2004, especially 27–36. *Cf.* Tsetskhladze 2006a, xxxviii–xli.

[119] See more on the *emporia* below.

[120] See Section 5.1.

[121] For a summary of the views, see more recently Descœudres 2008, 293–98, 360–64, who sees mainly political reasons for the colonisation. See also Roebuck 1984, 116–31; Tsetskhladze 1994a, 123–26; 2006a, xxviii–xxx; 2012a, 346; 2019d; Burstein 2006, 137; de Boer 2006, 46; Greaves 2007, 9; Tsetskhladze and Hargrave 2011, 169. A bit different in Figueira 2008, 427.

sound, the mass and frequency of this endeavour coupled with the great distance of the destination from the homeland, with all that that entails, point – undoubtedly in my view – to the cardinal role of the economic motives; it would seem that resources and production in the metropolises could no longer sustain their populations. This does not mean, of course, that people with other motivations might not have joined the migrant group.[122]

We should not forget that colonisation was considered a particularly hazardous undertaking, which, according to some scholars, based on ancient sources, was equivalent to a crime.[123] There was the length of the voyage itself, on a sea with often difficult weather conditions; the fact that the easternmost parts of our littoral, together with the eastern side of the Black Sea, were thought by the Greeks of the Archaic period to be the eastern edge of the world; the utterly strange peoples inhabiting these regions, making it seem like 'a barbarian ocean';[124] and piracy.[125] Consequently, the sets of reasons for the choice of the areas that were colonised must have been similarly compelling.

The champions of Greek colonisation in the southern as well as the other littorals of the Black Sea were indisputably the Ionians, but Dorians also played an important role. There is no need to refer again to the simplistic and utterly wrong perception, often encountered, that everything Greek in the (southern) Black Sea is Ionian and everything Ionian is Milesian – a perception that has even led to the absurdity of the use of the term 'Ionian colonisation' to denote Greek colonisation in general.[126] Information about the identity of the colonies' founders derives primarily, sometimes exclusively, from the written sources – further proof of their significance – since the archaeological material from the earliest periods of the colonies (in their vast majority not archaeologically investigated) is minimal, consisting only of a few fragments

[122] Manoledakis 2018b, 193–96. Scholars focusing more on the political reasons for colonisation invoke the case of Miletus, which was suffering from the Lydian policy against it in the Archaic period. At the end of the day, however, this policy had severe economic consequences for the Milesians, while at the same time we must not forget that the Ionian cities were not the only colonisers of the Black Sea. What about Megara and Boeotia, for example?

[123] Dougherty 1993a, 40–41 and 1993b, 187–88, based mainly on passages of Thucydides (6. 23), but also Mimnermus and Archilochus; Descœudres 2008, 295; Højte 2008, 151. Interesting is Dougherty's (1993a) interpretation of colonisation as a kind of purification.

[124] Jones 1971, 148.

[125] For piracy in the Black Sea, see Tsetskhladze 2002a.

[126] Detailed criticism of this perception in Manoledakis 2021e, with the relevant bibliography. This paper of mine was written and presented more than six years ago, and therefore the reader should not be surprised to observe some revisions or improvements to my then expressed views in this book.

of vases.[127] But even the ancient texts do not always accord with each other. A characteristic example of divergence is the case of Heraclea, the establishment of which has been attributed to both Dorian (Megara) and Ionian (Miletus) founders (see below). In the next part of the book, Table 3 presents the relevant versions provided by the 'geographical sources' for some of the colonies, as well as the terms that have been used to describe each one of the sites that were created during the colonisation, that is, in the Archaic and Classical periods.[128] The period of establishment for each one is in the vast majority of cases indicated only by the information contained in the written sources.

The possibility that Sinope may well have been the earliest site at which the Milesians established a colony, succeeding earlier settlements, has already been discussed. This took place in the 7th century BC, but one cannot be more specific than that.[129] Apart from Sinope, another three colonies were certainly established in the littoral by Greeks in the Archaic period: Tieion, Amisos and Heraclea (Fig. 13.2, 5). The first was founded by Milesians,[130] probably in the 7th century BC, as indicated by the earliest Greek pottery discovered there.[131] For Amisos, several cities, including Miletus and Phocaea, have been proposed as having had a hand in its founding, which occurred probably in the first half of the 6th century BC.[132] Heraclea presents a somewhat puzzling case, since all ancient authors apart from Strabo, who calls it a Milesian settlement, ascribe its establishment to the Megarians, probably

[127] East Greek pottery, to which most of the fragments belong, introduces an additional difficulty: it is not always clear which East Greece site they come from, and Miletus is not always the provenance. See, for example, Manoledakis 2018b, 220, n. 219.

[128] In consequence, references to ancient authors appear here only in special cases.

[129] According to one theory, the founding of Greek colonies in the southern Black Sea could only have been achieved after the final defeat of the Cimmerians (de Boer 2006; Dönmez 2007a, 62. Criticism of the view in Tsetskhladze 2011, 119). This, however, would point to the early 6th century BC. Furthermore, I do not see why the Cimmerian presence would be such a limiting factor for Greek colonisation: on the one hand, if we are to trace any evidence for Cimmerian presence in the Black Sea littoral, this refers only to the Sinope area; on the other, if the problem was in western Anatolia and the region of the Ionian metropolises, there is nothing to indicate that the Cimmerians ever managed to reach them. See Section 3.1.

[130] Perhaps with Phocaeans as well, according to Tsetskhladze (2018, 54).

[131] Atasoy and Erpehlivan 2012; Öztürk 2013; Atasoy and Yıldırım 2015a; Atasoy 2016; Aksoy and Yıldırım 2017; Yıldırım 2022, with more bibliography.

[132] Roebuck 1984, 120; Marek 1993, 19; Atasoy 2003; Avram *et al.* 2004, 954–55; Morel 2006, 361; Summerer 2005b, 129; 2007, 29; 2008, 264; Tsetskhladze 2007, 168–73. Roebuck (1984, 120, n. 27) suggested that 'Amisos was a Phocaean foundation of *ca.* 600, which was refounded by Milesian colonists *ca.* 560.' Note that the pottery often mentioned in the bibliography as having been found in Amisos (e.g. Boardman 1999, 253) was actually found in the local settlement of Akalan (Cook 1946, 82, n. 138). See more below, n. 177. For a possible Athenian settlement in the 5th century BC, which resulted in the name change from Amisos to Peiraieus, see in the Section 3.3.

jointly with Boeotians. The Dorian founding can be dated in the mid-6th century BC, probably around 554.[133]

Of particular significance is the fact that Sinope seems relatively rapidly to have developed to the point that her inhabitants were able to found new colonies of their own, namely Cotyora, Kerasous and Trapezous, all to the east of their metropolis.[134] This is the so-called secondary colonisation, another phenomenon which until now we have known only through the literary sources.[135] Although the time of their founding remains extremely difficult to estimate, the earliest of these three Sinopean colonies could have been established while the Milesians were still colonising the littoral.[136] Harmene, a port city just west of Sinope, is mentioned in a way suggesting that it was founded by Sinope, while Cytoron is reported by Strabo (12. 3. 10) to have been an *emporion* of that city at one time. The textual evidence ascribes secondary colonies to the Heracleans as well: Calpe, as an *emporion*,[137] and Thynias on the homonymous island (also called Apollonia). These four settlements appear for the first time in texts of the Classical period. There are also later references to settlements belonging to Heraclea, but their locations and establishment dates are unknown.[138]

Apart from these eleven (four primary and seven secondary) colonies, another 26 settlements are mentioned by name in the ancient literature as lying on the southern Black Sea coast in the Archaic and Classical periods. A detailed examination of the way these settlements are described in the sources[139] might reveal that probably all of them were created by Greek colonists. This betokens a fairly intense colonising activity, and subsequently a corresponding urban development in the littoral,[140] because the existence of

[133] Asheri 1972, 12, 23–28; Marek 1993, 16; Erciyas 2003; Avram *et al.* 2004, 955–58; Summerer 2005, 242; Vinogradov 2007; Tsetskhladze 2007, 164–65. See more on Heraclea below. For the theories on the fate of the local Mariandynoi during and after the founding of the city, see in Section 2.4.3.

[134] On these three secondary colonies of Sinope, see Tsetskhladze 2009, 230–39. Pausanias mentions that the citizens of Trapezous welcomed the inhabitants from the homonymous Arcadian city who went there being unwilling to move to Megalopolis 'as their kindred, since they bore their name and came from their mother-city' (8. 27. 6). This supposed kinship has never been confirmed by another source, and might have been owed to a tradition created much later than the founding of the Sinopean colonies.

[135] The division of the Greek colonisation into three successive waves or stages, from the late 7th to the late 6th century BC (Roebuck 1984, 121–24; Kacharava 2005, 11–14) is incomprehensible and misleading.

[136] See in Section 5.1, comment 16 of Table 3.

[137] On Calpe, see Aslan 2014.

[138] See Section 5.2.3 and Table 3, nos. 107–113.

[139] This is carried out in Section 5.2.3.

[140] This will be examined analytically in the next part of the book.

poleis, of which there were not a few among these 37 settlements,[141] might entail several other dependent settlements or installations: for instance, Xenophon mentions several χωρία in the territory of Cotyora (*Anabasis* 5. 5. 6–7, 10), as well as coastal χωρία (strongholds?) belonging to Sinope (5. 5. 23).

Which of these settlements were primary colonies and which secondary is extremely difficult to determine, although the latter must have been more numerous than the former. The cities mentioned by Hecataeus are more contentious, since they were established in the Archaic period.[142] Theoretically, they could equally be primary or secondary colonies, especially if we accept the founding of Cotyora, Kerasous and Trapezous in the same period. Equally unknown are the founders of all these settlements. One can only guess at Ionians from the Aegean coast of Asia Minor and Sinopeans respectively.

It is extremely noteworthy that the term *emporion* first appears in sources of the Roman era,[143] even though it refers to settlements that had clearly been founded earlier.[144] The type of settlement described as an *emporion* was one of the most fundamental elements of Greek colonisation. The *emporia* were production centres and local trading posts and had mixed populations. An *emporion* might have been either an institution within a settlement or the settlement itself. In the first case, what distinguished it from the (Greek) settlement's *agora* was that the *emporion* was a port or part of a port and was the hub of the settlement's foreign trade, while the agora was the centre of local trade. In the second case, the *emporion* constituted the entire settlement, which could also be, but was not necessarily, a *polis*.[145] Apparently, *emporia* existed from the beginning of Greek colonisation, and even earlier. The first Greek settlement at Sinope, for example, was in all likelihood an *emporion* of the second type (the entire settlement). And it would be entirely reasonable to suspect that many primary Greek colonies, at least the earlier ones, were founded as *emporia*[146] of the first type. In any case, exchanges with local inhabitants should rather be taken for granted, in view of the critical importance of the economic reasons for colonisation.

Being able to serve potentially as a purely commercial and not a political institution, thus being not necessarily tied to the existence or establishment of

[141] See Section 5.2.1 for discussion of the differentiation between the two meanings of this term.

[142] See Part 5 and especially Tables 3 and 7.

[143] See Section 5.2.1 and Table 4. *Cf.* Hansen 2006, 23; Hind 1995–96, 115.

[144] See, for example, Strabo (12. 3. 10) about Cytoron, which was already mentioned by Pseudo-Scylax (90), but only as a *polis*.

[145] On *emporia*, see Hansen 2006; Demetriou 2012, 16–23.

[146] See more below. *Cf.* also Manoledakis 2021e.

a *polis*, and with a population that included a local element, the *emporion* was a much easier type of settlement to establish. Therefore, the view that many of the so-called 'secondary colonies' were *emporia*[147] is very plausible; indeed, some of them were called emporia by ancient writers. Apart from Cytoron and Calpe, for which the founders are known, the term was applied to another six settlements,[148] half of which had, like them, been called towns in earlier periods (see Tables 3 and 4);[149] presumably, they were secondary colonies as well. Four of these eight *emporia* are described as Greek cities by Pseudo-Scylax,[150] and this could be considered to imply that they were *emporia* of the first type (institutions within a settlement).[151]

Might a secondary colony have established new settlements? I think that we should not rush to exclude the possibility. Not least because some of them developed into flourishing *poleis* in both senses of the term, striking coins and generally playing a significant role in the area. As regards Trapezous, for example, we know of the existence of Timesitheus the Trapezuntian, who was the official representative of the Mossynoikoi at Trapezous in the 5th century BC (Xenophon *Anabasis* 5. 4. 2), while the city struck coins on the Persian standard in the Classical period, which could indicate a pronounced interest in contacts with the local populations.[152] Even if Trapezous was still paying tribute to Sinope at this time (Xenophon *Anabasis* 5. 5. 10), and thus politically dependent on it, there was nothing to prevent it from establishing an *emporion* or installation to promote its own commercial activities. Such settlements or installations could, for instance, have been among those mentioned in the Roman period around the city, for example Hieron Oros, Cordyle, Ophious and Rhizeon (see Table 3 and Figs. 13.7–8).

Looking at the available information on the identity of the founders of the several colonies in conjunction with their dispersal over the littoral, it is apparent that the central and eastern parts were under Ionian control. It is only in the west that the Dorians seem to have managed to prevail, founding Heraclea,

[147] Tsetskhladze 2009, 229.

[148] Lillaion, Elaion, Cales, Psylla Limen, Kimolis, Caroussa.

[149] Psylla Limen, Kimolis, Caroussa.

[150] Psylla Limen, Cytoron, Kimolis, Caroussa.

[151] And that Pseudo-Scylax preferred to focus on the status of the settlements as Greek cities. After all, the term *emporion* was not yet widely used in his time. For the use of the term *polis Hellenis* by Pseudo-Scylax, see Section 5.2.3. The author might not necessarily have used the term *polis* only with its political meaning. For instance, it has been suggested that Cytoron was probably a *polis* only in the urban sense, still dependent on Sinope (Avram *et al.* 2004, 959). In any case, Cytoron has nothing to do with Cotyora (Tsetskhladze 2009, 232–33), and Strabo clearly differentiates between them. Nevertheless, there was nothing to prevent an *emporion* from gradually developing into a polis, with either of the two meanings. See Section 5.2.1.

[152] Avram *et al.* 2004, 964; Erciyas 2007, 1199.

and the Heracleans in their turn Calpe and Thynias. Probably the rest of the settlements and installations to the west of Heraclea belonged to the Doric sphere of influence too, and this must have been the case with some to the east of the city as well, such as Aconae (see Table 3 and Fig. 13.2).[153] After all, the Bosporus and the eastern end of the Propontis were also under Megarian control.

The first colony to the east that was certainly Ionian is Tieion. Nevertheless, even for the area to the west of Tieion, it has sometimes been suggested that it had been visited by the Ionians before the Dorian occupation. This assertion is based on Strabo's reference (12. 3. 4) to Heraclea as a Milesian colony, contrary to all the other ancient authors' claims. Several scholars have tried to explain this discrepancy, but failed to draw a firm conclusion. Although they did not consider it impossible for the Ionians to have preceded the Dorians in the region of Heraclea, they acknowledged that this could not be proven for the city itself, in the absence of relevant epigraphic, numismatic or other evidence. But the same does not apply to Heraclea's *chora*, which is supposed to have extended from the island Apollonia/Thynias and Calpe in the west to the rivers Parthenius and Billaeus in the east.[154] It is this area, from Heraclea to the Parthenius and a bit beyond, which is controversial.

According to D. Asheri, the Sinopeans were founding secondary colonies westwards (as they did undisturbed eastwards) until the mid-6th century BC, when the Dorians arrived, preventing their expansion beyond the Billaeus. Until then, this river may have been an early meeting point of Milesians and Mariandynoi, a century before the appearance of the Dorians in Heraclea. Indeed, he goes on to say that, as the enslavement of the Mariandynoi by the Milesians is described by Strabo (12. 3. 4) in words similar to those used by Poseidonius (according to Athenaeus *Deipnosophistae* 6. 84. *Cf.* Apollonius *Argonautica* 2. 747–748) in describing the enslavement (of the Megarians and Boeotians) by the Heracleans, it was probably not an invention of the latter but of the Milesians, which the Heracleans continued.[155] In addition, cities like Tieion and Cytoron, which had been founded by Ionians, had by the late Classical era passed into the control of Heraclea, whose sphere of influence reached

[153] Aconae was probably named after the poisonous herb aconitum, on which see in Part 1. *Cf.* Ivantchik 2005a, 127–29.

[154] Meyer 1893, 421; Hanell 1934, 132–36; Asheri 1972, 12–17; Burstein 1976a, 13–15; Saprykin 1997, 23.

[155] Asheri 1972, 16–23, where he also tries to find mythical genealogical data connecting the Milesians with the Mariandynoi. Generally, the mythic narratives about the Black Sea are considered products of the Ionian voyages. See Roebuck 1984, 117. The view of the enslavement of the Mariandynoi as a Milesian invention was not accepted by Saprykin (1997, 35–36).

Cytoron to the east.[156] Going one step further, Hanell and Saprykin proposed that Heraclea itself was first founded by the Milesians in the second half of the 7th century BC and afterwards by the Megarians and Boeotians, in the 6th century BC,[157] but this view did not prevail.[158]

In any case, all this seems to indicate that it was not unusual for settlements in the southern Black Sea to have more than one phase of Greek habitation, that is, with some Greeks first settling in a place and others coming later. This confirms the likelihood that the same happened with Sinope, where the Ionians may have not been the first Greeks to inhabit the city. The passing of an area from Ionian to Dorian political control, as might have been the case with the land between Heraclea and the Billaeus in the 4th century BC,[159] is something different.

Be that as it may, all settlements to the west of Heraclea (apart from Calpe and Thynias, already known as its colonies) must have been founded after the Classical period (see Table 3) and should therefore rather be considered as being controlled by that city. On the other hand, those of the later periods between Heraclea and Tieion could have belonged to either of the two cities, unless they were in close proximity to one of them (e.g. Aconae). Tieion might have not been excavated by Asheri's time, but the results of the archaeological investigation there indicate a rather prosperous city in the Roman period,[160] which, moreover, very soon after entering into the formation of Amastris had the confidence to withdraw.[161]

Not all the settlements that were founded by the Greeks in the littoral experienced the same evolution. As a matter of fact, some of them did not manage to survive for long. This seems to have been the case with a few of the cities established in the Archaic period and mentioned by Hecataeus, such as Stephanis, Teiria, Stamene and Choirades – that is, four of the twelve settlements founded in that era. A similar fate appears to have awaited seven or eight of the 24 settlements established in the Classical period (Table 7).[162] In both cases this is one in three of the settlements. There could be many reasons

[156] Vitucci 1953, 22–23; Asheri 1972, 13–15, 32–34 (his attribution of more than 40 settlements to the Heraclean territory sounds exaggerated; some of them appear only in the Roman period); Saprykin 1997, 33–35. *Cf.* Marek 1993, 16; Burstein 2006, 147.

[157] Saprykin 1997, 26–27. *Cf.* also Dunham 1915, 58. Saprykin's argument (1997, 25) that 'possible Milesian colonisation of Heracleia can be supposed from the fact that the most part of apoikiai on the Black Sea were founded by Miletus' is rather naïve.

[158] Erciyas 2003; Avram *et al.* 2004, 956.

[159] Asheri 1972, 15–16; Marek 1993, 16.

[160] See the bibliography on Tieion given above.

[161] Manoledakis 2013a, 31.

[162] See more on this matter in Section 5.2.2.

for the abandonment of a colony and it is hard to say whether this percentage is high or low. On the other hand, there was a general increase in the number of settlements in the following Hellenistic and Roman periods, by a factor of almost three. Even if the status of the new settlements and the reasons for their establishment were different from what applied in the Archaic and Classical periods,[163] this increase shows that the Greek colonists had managed to create an indisputably prosperous world throughout the littoral. Their success was decisively fostered by the participation of the local peoples, as demonstrated *par excellence* by the very existence of the *emporia* down the centuries. For that reason, then, this participation deserves more attention.

4.2. Greeks and Locals

As I argued some years ago, colonisation is a game for two, and in order to understand it more fully one must look at it from both sides.[164] So far, we have been dealing with the one; let us now take the other's viewpoint. An attempt has been made to exhaust the available evidence about the local peoples that inhabited the littoral and encountered the Greek newcomers in Part 2. One of the aspects on which stress was laid was that regarding their manners and customs. Although it was emphasised that our sources are almost exclusively Greek or Greek-influenced, thus probably presenting a one-sided, subjective image, one can hardly fail to observe that the vast majority of the littoral's indigenous peoples pursued a way of life that in Greek eyes was anything from curious to offensive – consider the savage Bithynians, the wretched Chalybes, the most barbarous Mossynoikoi and the most warlike Drilae, to name just some. Even if these descriptions appear in texts of the Classical period, there is nothing to suggest that things had been better when the colonists began to appear – obviously they could not have been. Moreover, many of the peoples already had numerous problems to deal with in their relations with their neighbours, thus the arrival of a new people with an eye on their in any case rather restricted land could not have been welcome.

And yet, colonisation succeeded, and indeed very soon gave birth to secondary colonisation.[165] If centuries after the beginning of colonisation most local populations had not yet displayed any great interest in advancing to a different

[163] See Section 5.2.2 and Tables 5 and 7.

[164] Manoledakis 2018b, 193.

[165] And, although the archaeological investigation in the southern littoral is very limited, the general image from the whole Black Sea, namely of unfortified Greek cities before the 5th century BC (with the exception of Histria), must apply to the southern side as well. Tsetskhladze 2002b, 83; 2014, 318; Højte 2008, 152; Manoledakis 2015c, 65.

(superior, according to the Greeks) cultural level, what might have spurred, or sometimes even forced, them to participate in this protracted colonial process? Judging by the available evidence, economic and political motives must certainly have been among the main reasons. As regards the former, the establishment of the *emporia* along the coast is an obvious attestation. Even if the indigenous peoples did not face the problems that were plaguing the Greeks in the mother-cities (overpopulation, lack of arable land, etc.), and even if they were economically autonomous, who would scorn the opportunity to improve their economy, either through trade or by enhancing their technical knowledge and ergo their productive forces? Some of the products the newcomers brought with them[166] may have helped in this direction,[167] and thus intrigued the local peoples.[168]

Familiarity with the new products would gradually lead to familiarity with the new culture; if not in terms of manners and customs, at least in terms of art. This familiarity is, for example, manifest in decorative architectural elements, particularly terracotta revetments on buildings in local settlements all along the Halys valley as far as Kaman-Kalehöyük.[169] According to the most convincing and widely accepted theory, these elements display a clear Greek influence and the moulds from which the revetment plaques were produced, probably in the 6th century BC, came from the Greek colonies on the coast, especially Amisos, which was famous for its terracotta tradition and workshops.[170]

Let us move on to the political reasons. Several passages of ancient texts, from both historical and mythical narratives, provide various indications that some of the indigenous peoples in the littoral suffered from conflicts with neighbouring peoples or even civil strife.[171] They also show that the coming of the Greeks might in some cases have been seen as a good opportunity to attract a potential ally in such clashes. In the second Book of his *Argonautica* (2. 752–814), Apollonius Rhodius presents Lycus, king of the Mariandynoi, warmly welcoming the Argonauts and bringing them gifts. He tells them of his

[166] Braund (2005a, 100–01) correctly suspects that these products were not necessarily of the finest quality (made of precious metals, for example), but probably perishable and of lower value, but necessary for the surviving of the colonists. Like those, for example, that Greaves (2007, 12) calls 'invisible archaeological materials'.

[167] We should not forget that *emporia* also functioned as manufacturing centres (see, for example, Tsetskhladze 2006a, xli; 2012a, 341).

[168] Manoledakis 2018b, 196–99.

[169] These are architectural terracottas found in Akalan, Köyiçi Tepesi, Parali Tepe, Pazarli, Boğazköy and Kaman-Kalehöyük.

[170] Summerer 2005a, 148; 2005b; 2007, 31–34; 2008, 267–78. For the bibliography about the different theories on the revetments and the whole debate, see Manoledakis 2018b, 199, n. 106.

[171] Examples may be found in several sections in Part 2.

admiration for Heracles, who, according to Pseudo-Apollodorus (2. 5. 9), had taken part in a battle between the Bebryces and the Mariandynoi, where he killed the brother of the former's king and then took much land from that people and gave it to Lycus (*cf.* Schol. Apol. Arg. 2. 786).[172] According to one theory, the narrative indicates that the Mariandynoi must have had an interest in the arrival of the Greeks, since they saw them as defenders against attacking neighbouring tribes, and furthermore that the Mariandynoi could have made a military and political agreement with the Greeks when Heraclea was founded.[173]

It is natural that such kinds of political and diplomatic relations would have strengthened considerably by the time Xenophon and the Ten Thousand marched across the littoral in 400 BC. The *Anabasis* contains several relevant references. Characteristic among them is the one referring to the Sinopean envoy Hecatonymus and his relation with the Paphlagonian ruler Corylas (Xenophon *Anabasis* 5. 5. 22–23; 5. 6. 11).[174] The case of the Paphlagonians and their relations with the Greeks, as well as the traditions that these relations had brought about regarding the origins of both peoples, have already been highlighted.[175]

Going further east, we hear about other local peoples that sought Greek support in their conflicts. For example, the Mossynoikoi were divided into at least two different groups, which were fighting among themselves for control of what was apparently their most important urban centre, the Metropolis. In fact, Xenophon took advantage of this civil war in order to pass through the country of the Mossynoikoi: he gained the aid of one group by promising assistance against the other (*Anabasis* 5. 4. 3, 15). The negotiations were mediated by Timesitheus the Trapezuntian (5. 4. 1–10), who was the official representative of the Mossynoikoi at Trapezous and acted as interpreter. His existence shows that the Greeks of the colonies maintained relations with the Mossynoikoi, while some of them also learned the local languages. Furthermore, some of the Colchians also seem to have developed good relations with the Greeks, since on the arrival of the Ten Thousand the Trapezuntians did not lead the latter to Colchian places, from which provisions could be obtained easily, because they were friendly towards their inhabitants. In fact, the Colchians who dwelt on the plain agreed to give hospitable gifts to the Ten Thousand (*Anabasis* 4. 8. 24; 5. 2. 2). Finally, Xenophon mentioned also the existence of some

[172] See Sections 2.3 and 2.4.
[173] Saprykin 1997, 24–39.
[174] See Sections 2.6.3–4 in detail.
[175] See Section 2.6.2; Manoledakis 2021c.

'barbarian' settlements in the mountainous area close to Kerasous, which were friendly towards the Greek inhabitants of the city (*Anabasis* 5. 7. 13–17).

The above-mentioned evidence reveals that the indigenous peoples of the littoral had reasons enough to let the Greek incomers settle in their privileged strip of land, even if they were culturally so dissimilar. This plainly explains their participation in the *emporia*, and probably in the creation of larger settlements as well. In fact, for the area of Amisos, there is an interesting if obscure passage in Strabo (12. 3. 14, citing Theopompus): Amisos, he says, was at some point in its existence expanded by a Cappadocian ruler named Timades, and it has been suggested that neighbouring native settlements were very likely absorbed in the process.[176]

This co-existence or co-operation in the early periods of colonisation is archaeologically attested as well, first and foremost in the Halys valley. Apart from the architectural terracottas that have already been mentioned, several local settlements along it, such as Akalan, Mecitözü, Pazarli, Maşat, Boğazköy, Alisar and Kaman-Kalehöyük, have yielded Greek pottery, together with local, mainly from the 7th and 6th centuries BC, indicating a Greek infiltration inland already from an early period.[177] This local pottery is, conversely, found together with Greek in Amisos.[178] The case of the Halys valley finds suggests that the almost exclusive interest of the littoral's Greek colonists in sea trade at the outset of colonisation was determined by the barrier to the hinterland posed by the long mountain ranges. For example, despite the fact that Sinope was one of the most significant ports and surely not unknown to peoples of the hinterland, as Phrygian pottery fragments found there indicate,[179] it seems that essential relations between the Sinopeans and the local inhabitants of its near hinterland did not develop before the 5th century BC, and until then the city

[176] Summerer 2005a, 137–38, 148–49, 152; 2007, 30, with bibliography. But this is just a theoretical view, because unfortunately Strabo's text has a lacuna at the most crucial point, where something about a leader of the Cappadocians is mentioned. Neither his name nor what he did is specified ([…] Καππαδόκων ἄρχοντα). Lasserre (1978, 212–20; 1981, 76) completes the passage as Τιμάδην δ' ὕστερον ἐπαυξῆσαι Καππαδόκων ἄρχοντα. However, this is just a proposal. We do not know of any Timades from any other source, and the original text was probably different. Olshausen (1994, 476) mentions only that a Cappadocian leader was at some time the governor of the city; Reinach (1975, 20) mentions a second founder, while Hirschfeld (*RE* 1.2, *s.v.* Amisos, 1839) refers to a conqueror or founder. No one, however, mentions a name. More bibliography in Summerer 1999, 24.

[177] See Summerer 2005a, 147, n. 103; 2005b, 129; 2007, 30–31; 2008, 262–67; 2009, 188; Kacharava 2005, 12–13; Tsetskhladze 2012a, 349; 2013, 74; Manoledakis 2015a; 2015c; 2018b, 175–82, all with bibliography. It has sometimes been suggested that this early Greek pottery might have arrived in the local settlements as diplomatic gifts. See Manoledakis 2018b, 198, n. 104, with the relevant bibliography, and comments on this theory.

[178] Summerer 2005a, 143–45, 152; 2005b.

[179] See Section 3.4.

was almost totally maritime oriented in its interests.[180] It was Amisos that seems to have been more cosmopolitan in the Archaic period, even though founded later than Sinope.[181]

Nonetheless, one should not be left with the impression that this assent and co-operation of the local peoples with the newcomers was universal and that the establishment of primary and secondary colonies and *emporia* was an easy, smooth and fast process. It seems that some parts of the littoral were still less receptive to such co-operation than others even as late as the Classical period. A characteristic example is the westernmost part, up to Heraclea. As recorded by Xenophon (*Anabasis* 6. 4. 2), between Byzantium and Heraclea there was no other city, either friendly or Greek, but the whole area was occupied only by Bithynians, at least until the 4th century BC.[182] In fact, these Bithynians were known in Xenophon's time to be ferocious towards any Greeks they encountered. This is confirmed by a look at Table 3 in the next part, where no other Greek settlement to the west of Heraclea (no. 12) is mentioned before the Hellenistic and Roman periods.

However, in addition to Calpe (no. 3), there were also a settlement and a port on the island of Thynias (nos. 5–6), mentioned by Pseudo-Scylax, which thus must have been established at the latest some decades after Xenophon's march. As a matter of fact, both these settlements were reported to have belonged to Heraclea.[183] Therefore, the view that the Heracleans had not established a strongpoint at Calpe or somewhere else on the coast to their west, because they were prevented by the Bithynians,[184] is not quite accurate. The Bithynians were indeed dominant on this part of the littoral, where they had plenty of settlements (*Anabasis* 6. 3. 1–3, 4. 6),[185] but the Heracleans had at least managed to create two settlements there, ports or *emporia*, and indeed at two very good points on the only island in the wider region, opposite its longest promontory, as well the best bay nearby.

To conclude, it is apparent that the Greek colonisation was of mutual benefit to both incomers and inhabitants.[186] For the Greek colonies, these economic relations with the locals and all that they entailed not only helped them survive

[180] Doonan 2004, 69, 71–72, 77–78, 88, 91–93; 2007, 616–17.

[181] For the economic relations between Greeks and locals from the Classical period onwards, see in the next section.

[182] *Cf.* Tsetskhladze 2007, 163; Højte 2008, 156–58.

[183] See above, Section 4 4.1, as well as in Part 5.

[184] Højte 2008, 158.

[185] See Section 5.3.

[186] Tsetskhladze 2006a, lii; Manoledakis 2018b, 196–203.

and prosper in a strange environment but also accelerated the process that transformed them into locals themselves.[187]

4.3. Becoming Locals

Although it is impossible to say exactly when the total integration of the Greeks into the ethno-cultural map of the southern Black Sea littoral took place, it must have been accomplished in the 5th century BC. Both the archaeological and the textual evidence argue for such a conclusion. One might also consider the founding of secondary colonies in an earlier period, but since the role of the local populations was utterly indispensable in that, as has been shown, I think that the establishment of the first secondary colonies cannot be regarded as an indication of this achievement, just a crucial stage in the process.

Nevertheless, from the 5th century BC onwards, there is considerable evidence to show that the Greeks felt and acted as true locals, that is to say, not just living in the littoral but being independent in many aspects, having interests there and being able to defend them on their own, even affecting other local peoples' life and culture.

To start with the archaeological evidence, nothing is more striking than the commercial activity of Sinope, and indeed other Greek cities, in the Classical and Hellenistic periods. The city's export trade peaked in the 4th and 3rd centuries BC, as can be deduced from, among other things, the presence of Sinopean amphorae in several places along the Black Sea coasts as well as in the eastern Mediterranean.[188] This circulation of Sinopean amphorae seems to have begun in the mid-4th century and lasted until at least the 2nd century BC. Specifically as far as the east coast of the Black Sea is concerned, large quantities of Sinopean coins of the mid-4th century BC have been found in Colchian settlements, such as Pichvnari, while amphorae made with local clay but imitating Sinopean types, also found there, are dated to the same period. Thus, it seems that potters from Sinope had settled in Colchis and established *emporia* in the region, where they both traded their wares and manufactured products using their own distinctive technique.[189]

[187] Manoledakis 2015c, 67.

[188] The relevant bibliography is huge. See, most recently and indicatively, Doonan 2004, 73, 148; 2007, 616; Conovici 2005, 101–04; Garlan 2007, 143–48; Monachov 2010, 23–27; Krapivina 2012, 467–77; de Boer 2013, 110–11; Kaba *et al.* 2019.

[189] Tsetskhladze 2009, 234, with bibliography. *Cf.* Hind 1993, 110–11; 1996, 120; Doonan 2007, 616; Kvirkvelia 2010, 130–31; Tsetskhladze 2010, 43.

It is certainly noteworthy that from this period a turnaround in Sinope's relations with the local communities is observed: the recent surveys in the valleys of Demirci and Karasu, to the west and south of Sinope, lead to the conclusion that, at least in the Hellenistic period, these two areas were inhabited harmoniously by both Greeks and locals; in fact, a specific burial type sparked a theory about the creation of a 'colonial middle ground' there.[190]

Heraclea Pontica seems to have enjoyed comparable growth in commercial activity in the 4th century, and at least until 330 BC, as indicated, for example, by the reference of Pseudo-Demosthenes (probably Apollodorus of Acharnae) to the Heraclean merchant Lycon in Athens (*Against Callipus*).[191] Other pointers to this same conclusion include inscriptions from Athens dated to 360 and 330 BC, one of which mentions the honouring of merchants from Heraclea in the city,[192] and the manufacture of stamped amphorae in Heraclea, which began in the late 5th century BC and continued until the first quarter of the 3rd, with the bulk of this activity in the middle and the second half of the 4th century.[193]

I can see no better explanation for the undoubtedly surprising attitude of the merchants from these two cities towards Xenophon's idea of establishing a new colony on the southern Black Sea coast in 400 BC, as so vividly narrated by the Athenian historian (*Anabasis* 5. 6. 15–21). On their way back from Cunaxa to Greece, the Greek troops found themselves in the vicinity of Cotyora and Xenophon thought it would be a good idea to found a new Greek colony there, given that there was already a large body of well-armed Greeks on the spot (roughly 8600 soldiers had survived to reach that point). Some of his soldiers who got wind of the plan and were anxious to return to Greece reacted and told merchants from Heraclea and Sinope, who in turn informed their cities. One might perhaps have expected that the Greeks of those two cities would have been pleased at the prospect of a stronger Greek presence in the region, but what in fact happened was exactly the reverse: the Heracleans and the Sinopeans were appalled at the idea and begged the soldiers who had passed on the information to do everything they could to ensure that the Greek army sailed away and returned home, even offering money to help speed them on their way.

Irrespective of the fact that in the end Xenophon dropped the idea, the reaction of the Black Sea Greeks to the possibility of other Greeks settling in their

[190] Doonan 2004, 84–88, 148; 2019, 104–05.
[191] Burstein 1976, 41–42, with more data.
[192] *IG* II^2 117, 360, 408.
[193] Saprykin 1997, 103–05, 108–19; Kac 2003, 261–78; Monachov 2010, 23, with previous bibliography.

territory is certainly noteworthy. On the one hand, it confirms the primacy of the economic factor over the cultural in the whole story of colonisation: the merchants of two important commercial port cities on the coast did not look favourably on the arrival of other fellow Greeks and the founding of another Greek city on the coast, since it would certainly have drawn away some of their trade. On the other hand, it demonstrates at the same time that the Greeks of the colonies had long become true locals in the littoral by 400 BC, in the way this was defined above.[194]

Remarkable as it may sound, this incident was not enough to make Xenophon abandon his desire to establish a new Greek colony on the coast, for a little later he outlined a similar plan, and indeed in the district least likely to be receptive to such a scheme – Bithynia (*Anabasis* 6. 4. 7, 14).[195] His narrative is more comprehensive this time, as he makes sure that he justifies his plan by describing in detail and praising that extremely privileged region, in terms of landscape, fertility, natural resources and room for 10,000 people (6. 4. 3–6).[196] Unfortunately for Xenophon, once again his plan was grounded by the opposition it met from most of the Greek soldiers. This time, however, word of it seems to have travelled farther afield before its abandonment: very soon 'market products came in from the Greek cities on all sides, and people coasting past were glad to put in, since they heard that a city was being founded and that there was a harbour. Even the hostile peoples who dwelt nearby, namely the Bithynians, began now to send envoys to Xenophon, for they heard that he was the man who was making a city of the place, to ask what they must do in order to be his friends' (6. 6. 3–4).

Apart from this turn of events, the two incidents are almost identical. In both cases Xenophon is more or less alone in desiring to found a Greek colony; most of his men react, because they want to return to Greece, and the idea is abandoned. And in both cases, the reaction of the locals, Greeks in Cotyora and indigenous peoples in Calpe, points to the same thing: the potential economic loss or profit outweighs any ethnic or cultural kinship or differences respectively. The Bithynians did not change their attitude only because Xenophon's army was large and difficult to confront[197] – this was the case in Cotyora as well – but also because, seeing merchants coming from different cities with so many products, they recognised an opportunity.[198] When Xenophon says that his men avoided encamping on the spot which might

[194] See Manoledakis 2015a, 86; 2018b, 200–01.
[195] See the previous section.
[196] See analytically in Part 1. *Cf.* Lendle 1995, 385–87; Højte 2008, 156–58.
[197] Højte 2008, 158.
[198] *Cf.* Lendle 1995, 402.

become a city for fear that this would imply an ulterior motive and cause opposition (6. 4. 7–8), he has in mind the opposition he had faced at Cotyora – but which opposition, that from his soldiers or that from the local Greeks? Probably both. The latter included Heracleans, as we have seen, and this time Xenophon was in a part of the coast that was supposedly under Heraclean control. However, the total absence of the Heracleans from Xenophon's specific narrative rather implies that this time they were not there; unless one of the cities from which merchants arrived was indeed Heraclea.[199]

Be that as it may, the Greeks of the colonies had apparently become locals. Another piece of relevant information comes from Justin (16. 3. 9), according to whom in the mid-5th century BC, the city of Heraclea had developed such relations with the Persian king that it could refuse to pay tribute to Athens. But even more striking are the relations of these Greeks with Alexander the Great, compared with those between the Macedonian king and the Greeks of the Aegean coast of Asia Minor.

During his dash through Asia, about seven decades after Xenophon's march, Alexander promised to liberate all the Greeks of the continent and proceeded to act on that undertaking in the Greek cities of western Asia Minor. His interest in the Greek cities of the north coast, however, was totally different. In fact, he did not even pass through the region, coming no nearer to it than Ankara. In stark contrast to the detail with which his activity in the western littoral of Anatolia is described by ancient authors, reports of any communication between Alexander and the Greeks of the northern littoral are minimal to non-existent. Only twice is he mentioned by the written sources as having met embassies of Greek cities, and both these meetings took place far away from that region.

The first occurred in 330 BC, when Alexander was campaigning against the Mardians, south of the Caspian Sea (Curtius 5. 6. 17–19; Arrian 3. 24). As we read in Arrian (3. 24. 4) (and only there), upon returning to his camp in Hyrcania after having conquered the Mardians, Alexander found there Greek mercenaries, some of whom were from Sinope, and Lacedaemonian envoys to Darius. He arrested and imprisoned the latter, while he released the Greeks from Sinope, 'because these people had no share in the commonwealth of the Greeks; and as they were in subjection to the Persians, they did not seem to be doing anything unreasonable in going on an embassy to their own king'.[200]

[199] It seems that the difference at Cotyora was that Xenophon's plan was immediately divulged. At Calpe, on the contrary, the thought seems to have been shared among many soldiers from the beginning and thus was easily spread. So it was too late for the Heracleans to react as at Cotyora, and therefore maybe unwise as well.

[200] For the relations of the Greek colonies with the Persian state, see Section 3.3.

Alexander's second meeting with people from a Greek southern Black Sea city concerns Heraclea. Here again, our information comes from a single author, in this instance Memnon. After Alexander's victory at the Granicus, the tyrant of Heraclea Dionysius, like other local rulers, had the opportunity to increase his power and he had become more oppressive towards his people.[201] According to Memnon (4. 1), at this time exiles from Heraclea sent envoys to Alexander, who by then had conquered the whole of Asia, asking him to help them return to the city and restore democracy there. However, we have no further information about the result of this meeting, and we know that Dionysius managed to remain tyrant of Heraclea for 32–33 years, through an impressive series of intelligent manoeuvres.[202] Memnon did not bother to inform his readers of Alexander's reply to the Heraclean exiles. Probably, Alexander did not bother with them either, and it seems that neither they nor democracy returned to their city.[203]

But how should one account for Alexander's indifference to this specific area, which is in such sharp contrast to his interest in the Aegean coast of Asia Minor? A similar question might be asked of the Greek side as well: it is really puzzling that not one Greek city of the southern Black Sea coast is recorded as having asked Alexander for its liberation. Unless they had already realised, from the examples of the Aegean coast of Asia Minor, that this famous liberation may have meant liberation from the Persians but not real liberty. One explanation could be that the southern Black Sea Greeks, after so many centuries of settlement there, were by then so thoroughly integrated that they were considered, by themselves as well as by others, as locals. One could then claim that Alexander was perhaps convinced that the specific Greek populations were so well integrated, culturally to the local element and politically to the Persian state, even if the latter was not always entirely the case,[204] that they had little in common with the Greeks of the Aegean coast.[205] Besides, the natural isolation of the region from the rest of Asia would cost him unwarranted delay in his final target, which the remote Greek element there did not deserve.

201 On the tyranny of Dionysius see Saprykin 1997, 144–46.

202 Manoledakis 2020.

203 A possible exception in this state of affairs is the laconic reference in Appian (*Mithridatica* 8. 24, 83. 374) to a decree with which Alexander restored democracy in Amisos, which was subject to the Persians, although it is highly possible that the story of such a decree was fabricated by the people of Amisos at some time before Appian's era. See more in Manoledakis 2020.

204 See Section 3.3.

205 Who, after all, belonged, geographically as well as politically, to the main, central Greek world. The Greeks of Pontus, on the contrary, living far from this world and surrounded by foreign peoples and territories, had an inherent need to integrate.

PART 5

THE URBAN DEVELOPMENT OF THE LITTORAL

The emergence of human settlements and installations constitutes a significant benchmark in any study of historical geography. It denotes a decisive moment not only in the history of a people, who pass from nomadic to permanent living, but also for the natural environment that hosts the settlements: these are the first extensive human interventions to the environment, and they usually modify it forever. As we shall see, from the ancient sources we know, by name, of more than 100 human settlements and installations over the more than 1000 km of the littoral, which means on average a settlement (with or without environs) every 10 km – not including those established by the indigenous peoples, which usually escaped the attention of Greek and Latin writers. It is these settlements and installations that we shall be examining in this part of the book.

1. Known Settlements and Installations of the Southern Black Sea Littoral

All the places mentioned by name are shown on the maps in Figs. 13.1–8, which thus provide an idea of their geographical dispersion. The settlements and installations that, according to the information provided by the ancient sources, were built or constructed on the southern Black Sea littoral during the 1st millennium BC are presented in the following table. The sites are listed in geographical order, from west to east.[1] In the few cases where the sources disagree on the geographical relation between two sites, I follow the most probable version (drawing on supplementary data) or insert a comment number as an exponent (e.g. in no. 27 – Erythinoi, with comment 6). Twenty-one such comment-indicators appear, for several reasons and in several columns, and the explanations are given immediately after the table, together with relevant bibliography if needed. These are followed, finally, by statistical observations and general conclusions that may be drawn from the table.

The first column of the table gives the names of the sites, with the original name in parentheses if it is Greek. In the next column, the sites are classified according to how they are characterised in the sources. If there is no specification, the word 'settlement' is used. The status of many sites apparently changed

[1] As in Table 1, with the borders between Turkey and Georgia always being our eastern boundary.

in later times,[2] but this is beyond our scope here. The sources are recorded in the third column, with their characterisation of the site in the original language, Greek or Latin, if it is not already part of the name (e.g. no. 21 – Psylla Limen). Only what I term 'geographical sources' are listed.[3] This does not imply that other texts do not refer to the more important of these sites; merely, sources other than the 'geographical' ones are given only if they add knowledge that these do not provide (e.g. in no. 13 – Aconae, where 'other' sources indicate an even earlier existence of the site). The sources appear in abbreviated form.[4] If a site name is rendered somewhat differently in any source, this is recorded in a parenthesis after that source.

In the next column the sites are located on the modern map and, where possible, identified with modern sites. For this identification I had recourse to several modern sources, chiefly the *Barrington Atlas*,[5] the *Pleiades* project,[6] the *Ancient Ports* project,[7] the *Digital Atlas of the Roman Empire* by the Centre for Digital Humanities, University of Gothenburg,[8] as well as Bryer and Winfield (1985). In several cases there are disagreements among them; in the few for which no identification is provided, I give those I consider most probable.[9] Some of the modern names may have changed in recent years or decades,

[2] See, for example, Rhizeon (no. 97) in the mediaeval period: Bryer and Winfield 1985, 331–34.

[3] For the term see the Introduction 2.2.

[4] The following abbreviations are used:

Anon. = Anonymous *Periplus Ponti Euxini*, following Diller 1952

Apollonius = Apollonius *Argonautica*, Book 2

Arrian = Lucius Flavius Arrianus *Periplus Ponti Euxini*

Dionysius P. = Dionysius Periegetes *Orbis descriptio*

Hecataeus = *FGrH* 1a 1 F

Mela = Pomponius Mela *De Chorographia*

Men./Marc. = Menippus *Periplus Ponti Euxini* in Marcian's edition (see Introduction Section 2.2), following Diller 1952

Pliny = Gaius Plinius Secundus *Naturalis Historia*

Pseudo-Scylax = Pseudo-Scylax *Periplus*

Pseudo-Scymnus = Pseudo-Scymnus *Periegesis (ad Nicomedem regem)*, following Diller 1952, unless otherwise indicated

Ptolemy = Claudius Ptolemy *Geography*

Strabo = Strabo *Geographica*

Xenophon = Xenophon *Anabasis*

[5] Talbert 2000.

[6] https://pleiades.stoa.org.

[7] https://www.ancientportsantiques.com.

[8] https://dh.gu.se/dare/.

[9] In a few other cases, the attributions of these sources are considered erroneous; for example, the Rhebas is only a river and no ancient source mentions it as a settlement as well, as we see in Talbert 2000, map 52, and vol. II, 791. Just Aelius Herodianus and Stephanus of Byzantium (*s.v.* Ῥῆβας) mention it also as a *chora*.

and thus do not appear in the earlier modern bibliography; here I use the names that are currently valid and appear on Google maps today (2021). It is essential to clarify that, for the vast majority of the settlements and almost all the port installations and facilities, no trace has ever been found. Their locations have therefore been surmised from a combination of several factors – literary descriptions, stated distances between them and securely located settlements, natural features, etc. Sites that have been archaeologically investigated (themselves or their wider area), even to a limited degree, are noted by an asterisk; the basic bibliography regarding these investigations appears in the Introduction Section 2.1.

The last column gives the period or periods (again abbreviated)[10] in which, always according to our sources, each site was in existence.[11] Here, it should be stressed that some of our sources – Menippus *par excellence* but also Strabo, Mela and Pliny – belong to the first years of the Roman period.[12] This is very important, since, although the sources clearly belong to the Roman period, one cannot exclude the possibility that a site mentioned in them already existed in the Hellenistic age. On the contrary, I would say, it is very difficult to imagine that Menippus, writing in the second half of the 1st century BC,[13] could have been aware of sites that had only just been established. This is why, especially in the cases where Menippus is our earliest source (e.g. nos. 1, 2, 8, 10, etc.), the Hellenistic period is also given as a very possible period of existence, in a parenthesis.[14] Parentheses are also used in cases where there are other reasons to assume that a site existed in a specific period – for example,

[10] The following abbreviations are used:
E = Early Iron Age
A = Archaic
C = Classical
H = Hellenistic
R = Roman

[11] Which means according to what the sources say, not necessarily according to their date.

[12] On the chronology of their works see the Introduction 2.2.

[13] And since Marcian's 'edition' of Menippus is considered to be mostly Menippus' own work. Diller 1952, 149. See more in the Introduction 2.2.

[14] There is a possibility that some of the settlements and especially the installations that first appear in Menippus had been created by the Romans soon after their conquest of the Bithynian and Pontic kingdoms earlier in the 1st century BC. This is when the Roman period effectively begins in our area of interest, although we shall adhere to the traditional general starting date of the Roman period (31 BC). Similarly, one may imagine that some of the cities mentioned only by Pseudo-Scylax (e.g. nos. 35 – Carambis, 43 – Coloussa, 47 – Tetrakis, etc.) would have continued to exist in the Hellenistic period, which began only a few years after the writing of his work. Here, however, there is the additional problem that we are not sure which of these were first mentioned by the 4th-century writer and which existed in the original work of Scylax himself, which would mean that they belonged only(?) to the Archaic period. Given the impossibility of resolving this problem, we consider the time of creation the time of the work that has survived,

when there are sources mentioning it in the Classical and in the Roman periods, which rather implies its existence in the intervening Hellenistic age as well (e.g. nos. 13, 62). Conversely, when archaeological evidence clearly indicates a period of existence not attested by the literature, the relevant abbreviation appears without a parenthesis but followed by a comment indicator (e.g. no. 23 with comment 3). If a site that has been investigated archaeologically is not followed by such a comment, this means that the archaeological results offer nothing new or different regarding its chronology from what is already known from the written sources.

For the rest, Roman geographical sources of the 2nd century AD – notably Ptolemy and Arrian – are included, although clearly later than the period covered by this book, namely the 1st millennium BC. This is because, first of all, they are two of the most, if not the two most, important geographical sources of the whole Roman period and even earlier for, respectively, the *Oecumene* and the Black Sea; secondly, because it is highly probable that some of the sites that first appear in them had been established earlier, for example in the 1st century BC; and thirdly, because, even if the latter is not the case, they provide us with a good image of how the urban environment of the southern Black Sea littoral might have changed in the final years of our period of interest – the Roman era covers parts of both millennia and in some cases there is no sense in drawing a line between the two. The much later anonymous *Periplus* is also taken into account, because it is essentially a compilation of four earlier works and may even provide lost parts of some of them; however, names that the author clearly identifies as entering into use in his time, for example 'Metroon, which is now called Aulia',[15] are not included here.

The Roman *Itineraria* and the *Tabula Peutingeriana*, on the other hand, are not used, because these are even later works with no clear influences from our 'geographical' sources.[16] Not only that, but most place names in them, from our area of interest at least, are so heavily distorted that in some cases it is impossible to comprehend their identity or to locate them, when compared with any other source: some examples are the sites Tycae, Cereas, Mileto and Abgabes.[17] Similarly, much later sources are also not taken into consideration,

in full awareness that we thereby risk drawing erroneous conclusions, especially in the statistical observations made later.

[15] Anonymous *Periplus* 12: Diller 1952, 120. The same applies to the roadstead of Trapezous called Daphnous (*Periplus* 36), which is mentioned only by this author.

[16] Moreover, the map of Agrippa, which seems to have been the source of inspiration for all the Roman itineraries, is lost; on it, see Arnaud 2015, with bibliography.

[17] Furthermore, it is sometimes unclear whether a name on the *Tabula Peutingeriana* refers to a settlement or a geographical feature, as is the case with Kissa, which is known from other

as was the case in Table 1. Thus, settlements such as Kelessa, Mocata and Gerna are missing, as is Rhebas.[18]

Table 3. Known settlements and installations in the southern Black Sea littoral (from west to east) (see also Fig. 13).

No.	Place-name	Classification	Sources	Location/ Identification	Period
1	Artake/Artane (Ἀρτάκη/ Ἀρτάνη)	town	Men./Marc. 5701 (χωρίον); Ptolemy 5. 1. 5 (χωρίον); Arrian 12 (ὅρμος); Anon. 3 (χωρίον).	Şile	(H) R
2	Psillion/Psillis (Ψίλλιον/ Ψίλλις)	town	Men./Marc. 5701 (χωρίον); Pliny 6. 4; Anon. 4 (χωρίον).	Ağva Merkez	(H) R
3	Calpe Limen (Κάλπη λιμήν)	town, port, *emporion*	Xenophon 6. 2. 17 (χωρίον, λιμήν); Men./Marc. 5701 (ἐμπόριον Ἡρακλεωτῶν); Pliny 6. 4 (*portus*); Arrian 12 (λιμήν, χωρίον, ὅρμος); Anon. 5 (λιμήν, ἐμπόριον Ἡρακλεωτῶν, χωρίον, ὅρμος).	Kerpe*	C H R
4	Rhoe (Ῥόη)	roadstead	Arrian 13 (ὅρμος); Anon. 6 (ὅρμος).	Kefken	R
5	Apollonia (limen)[1] (Ἀπολλωνία)	port	Apollonius 673 (λιμήν); Arrian 13 (λιμήν); Anon. 6 (λιμήν).	on Kefken Island*	H? R
6	Thynias/ Bithynis[1] (Θυνιάς/Βιθυνίς)	city	Pseudo-Scylax 92; Pseudo-Scymnus 1026 (πόλις, Ἡρακλεωτῶν ἄποικος); Mela 2. 98 (*urbs*); Anon. 6 (πόλις, Ἡρακλεωτῶν ἄποικος).	on Kefken Island	C H R
7	Chele/Chelae (Χηλή/Χηλαί)	settlement	Arrian 13; Anon. 6.	Dikili Sahili	R
8	Dio(s)polis/Dia (Διό(σ)πολις/ Δία)	city	Men./Marc. 5710 (πόλις); Ptolemy 5. 1. 7; Anon. 9 (πόλις).	Akçakoca	(H) R
9	Lillaion/Lilion (Λίλλαιον/ Λίλιον)	emporion	Arrian 13 (ἐμπόριον); Anon. 9 (ἐμπόριον).	Kemos Beach, east of Akçakoca	R

sources as a river but not also as a settlement. For the Roman *Itineraria* and the *Tabula Peutingeriana*, see most recently Rathmann 2018, and the bibliography in Manoledakis 2022a.

[18] Which in my opinion was not clearly a settlement (for the river see Table 1), since Aelius Herodianus and Stephanus of Byzantium mention a river and a *chora*, but not a city under that name (see above, n. 9).

No.	Place-name	Classification	Sources	Location/ Identification	Period
10	Elaion (Ἔλαιον)	emporion	Men./Marc. 5710 (ἐμπόριον); Arrian 13; Anon. 9 (ἐμπόριον).	Kocaman	(H) R
11	Cales (Κάλης)	emporion	Men./Marc. 5710 (ἐμπόριον); Arrian 13 (ἐμπόριον); Anon. 9 (ἐμπόριον).	Alaplı	(H) R
12	Heraclea (Ἡράκλεια)	city	Xenophon 6. 2. 1 (πόλις Ἑλληνίς Μεγαρέων ἄποικος, ἐν τῇ Μαριανδυνῶν χώρᾳ); Pseudo-Scylax 91 (πόλις Ἑλληνίς); Pseudo-Scymnus 1016 (πόλις, Βοιωτῶν κτίσις καὶ Μεγαρέων); Men./Marc. 5710 (πόλις); Strabo 12 .3. 4, 6 (πόλις, Μιλησίων κτίσμα, ἐν τοῖς Μαριανδυνοῖς); Mela 1.103 (*urbs*); Pliny 6. 4 (*oppidum*); Ptolemy 5. 1. 7, 11 (πόλις); Arrian 13 (πόλις Ἑλληνίς); Anon. 9 (πόλις Ἑλληνίς, Βοιωτῶν κτίσις καὶ Μεγαρέων).	Ereğli*	A C H R
13	Aconae	township/port	Pliny 6. 4 (*portus*); First mentioned by Theopompus (4th century BC), according to Athen. *Deipn.* 3. 2; Aelius Herodianus and Stephanus of Byzantium, *s.v.* Ἀκόναι (πολίχνιον).	Cığlik Beach	C (H) R
14	Metroon (Μητρῷον)	settlement	Arrian 13; Anon. 12.	Köseağzı Beach	R
15	Poseideion (Ποσείδειον)	roadstead	Men./Marc. 5710 (σάλος); Arrian 13; Anon. 12.	NE of Armutcuk, Alacaagzi	(H) R
16	Tyndarides/ Todarides (Τυνδαρίδες/ Τοδαρίδες)	settlement	Arrian 13; Anon. 12.	Kireclik, west of Çavuzağzı	R
17	Nymphaion (Νυμαφαῖον)	settlement	Arrian 13; Anon. 12.	Çavuzağzı	R
18	Oxines (Ὀξίνης)	roadstead	Men./Marc. 5710 (ὅρμος).	Ilıksu Beach	(H) R
19	Sandarake/ Sandarache (Σανδαράκη/ Σανδαράχη)	roadstead	Men./Marc. 5710 (ὅρμος); Arrian 13 (ὅρμος); Anon. 13 (ὅρμος).	Zonguldak	(H) R

No.	Place-name	Classification	Sources	Location/ Identification	Period
20	Crenides (Κρηνίδες)	roadstead	Men./Marc. 5801 (ὅρμος); Arrian 13; Anon. 13 (ὅρμος).	Kilimli	(H) R
21	Psylla Limen/ Psylleion (Ψύλλα Λιμήν/Ψύλλειον)	town, *emporion*	Pseudo-Scylax 90; Men./Marc. 5801 (χωρίον); Ptolemy 5. 1. 7; Arrian 13 (ἐμπόριον); Anon. 13 (χωρίον).	Eren Port	C H R
22	Stephanis (Στεφανίς)	city	Hecataeus 198 (πόλις Μαριανδυνῶν).	unknown[2]	A
23	Tieion/Tion (Τίειον/Τίον)	city	Pseudo-Scylax 90 (πόλις Ἑλληνίς); Men./Marc. 5801 (πόλις); Strabo 12. 3. 5 (πόλις), 12. 3. 8 (πολίχνιον); Mela 1. 104 (Tios *oppidum, Milesiorum colonia*); Pliny 6. 4 (Tium *oppidum*); Ptolemy 5. 1. 7; Arrian 13 (πόλις Ἑλληνίς); Anon. 13 (πόλις Ἑλληνίς, Μιλησίων ἄποικος).	near Filyos*	A[3] C H R
24	Mastya	town	Pliny 6. 5 (*oppidum Milesiorum*).	unknown[4]	R
25	Sesamos[5] (Σήσαμος)	city, acropolis	Pseudo-Scylax 90 (πόλις Ἑλληνίς); Apollonius 941; Strabo 12. 3. 10 (ἀκρόπολις Ἀμάστρεως); Mela 1. 104; Pliny 6. 5 (*oppidum*, ident. with Amastris); Anon. 15 (ident. with Amastris).	Amasra	C H R
26	Amastris[5] (Ἄμαστρις)	city	Pseudo-Scymnus 1001 (πόλις); Men./Marc. 5801 (πόλις); Strabo 12. 3. 10 (πόλις); Pliny 6. 5 (ident. with Sesamos); Ptolemy 5. 1. 7; Arrian 14 (πόλις Ἑλληνίς); Anon. 15 (πόλις Ἑλληνίς, ident. with Sesamos).	Amasra*	H R
27	Eryth(r)inoi (Ἐρυθ(ρ)ίνοι)	breakwater, settlement(?)	Apollonius 941; Arrian 14; Anon. 17 (Χηλὴ Ἐρυθίνων).	close to Çakraz[6]	H? R
28	Cromna (Κρῶμνα)	town	Apollonius 942; Men./Marc. 5801 (χωρίον); Strabo 12. 3. 10 (κατοικία); Mela 1. 104; Pliny 6. 5; Ptolemy 5. 1. 7; Arrian 14; Anon. 17 (χωρίον).	Tekkeönü or Kurucaşile	C[7] H R

No.	Place-name	Classification	Sources	Location/ Identification	Period
29	Cytoron/Cytoros/Cytoris (Κύτωρον/ Κύτωρος/ Κύτωρις)	city, *emporion*, roadstead	Pseudo-Scylax 90 (πόλις Ἑλληνίς); Apollonius 942; Men./Marc. 5801 (χωρίον); Strabo 12. 3. 10 (κατοικία/ἐμπόριον ἦν ποτε Σινωπέων); Mela 1. 104; Ptolemy 5. 1. 7, 11; 5. 4. 2 (πόλις); Arrian 14; Anon. 17 (ἐμπόριον, ὅρμος).	Gideros	C H R
30	Aegialos (Αἰγιαλός)	town	Men./Marc. 5801 (χωρίον); Strabo 12. 3. 10 (κώμη); Arrian 14; Anon. 17 (χωρίον).	Karaagac Limani, Cide*	(H) R
31	Crobialos (Κρωβίαλος)	settlement?[8]	Apollonius 942.	unknown	(H)
32	Climax (Κλίμαξ)	town	Men./Marc. 5801 (πόλις); Ptolemy 5. 4. 2 (χωρίον); Anon. 17 (κώμη).	Şehriban Beach, Sakallı	(H) R
33	Timolaion (Τιμολάϊον)	town	Men./Marc. 5801 (χωρίον); Anon. 17 (χωρίον).	Timne, Uğurlu	(H) R
34	Thymina/ Teuthrania (Θύμινα/ Τευθρανία)	settlement	Ptolemy 5. 4. 2; Arrian 14; Anon. 17.	Çayyaka	R
35	Carambis (Κάραμβις)	city	Pseudo-Scylax 90 (πόλις Ἑλληνίς). Otherwise, only as a cape.	Erse Beach, İlyasbey	C
36	Callistratis/ Callistratia (Καλλιστράτις/Καλλιστρατία)	village	Men./Marc. 5901 (κώμη); Ptolemy 5. 4. 2; Anon. 19 (κώμη).	Marçula Koyu	(H) R
37	Zephyrion (Ζεφύριον)	settlement	Ptolemy 5. 4. 2; Arrian 14; Anon. 19.	Doğanyurt	R
38	Garios (Γάριος τόπος)	settlement	Men./Marc. 5901; Anon. 19.	Özlüce	(H) R
39	Abonou Teichos (Ἀβώνου τεῖχος)	city, roadstead	Men./Marc. 5905 (πόλις); Strabo 12. 3. 10 (πολίχνιον); Ptolemy 5. 4. 2; Arrian 14 (πόλις σμικρά); Anon. 19 (πόλις, ὅρμος).	İnebolu	(H) R
40	Aeginetes Αἰγινήτης	township	Men./Marc. 5905 (πολίχνιον); Arrian 14; Anon. 20 (πολίχνιον).	İlişi, Yakaören	(H) R

No.	Place-name	Classification	Sources	Location/ Identification	Period
41	Kimolis/ Kinolis Κιμωλίς/ Κινωλίς/ Κίνωλις	city, *emporion*	Pseudo-Scylax 90 (πόλις Ἑλληνίς); Men./Marc. 5905 (κώμη); Strabo 12. 3. 10; Mela 1. 104; Pliny 6. 5 (*oppidum*); Ptolemy 5. 4. 2; Arrian 14 (ἐμπόριον); Anon. 20 (κώμη).	Ginolu West	C H R
42	Antikinolis (Ἀντικινωλίς)	anchorage	Men./Marc. 5905 (only ὕφορμος); Strabo 12. 3. 10; Anon. 20 (ὕφορμος).	Ginolu East	(H) R
43	Coloussa (Κολούσσα)	city	Pseudo-Scylax 90 (πόλις Ἑλληνίς).	Güllusu, Oymayaka	C
44	Collyris[9]	settlement	Mela 1. 104.	unknown	R
45	Stephane (Στεφάνη)	town/village, port	Pseudo-Scylax 90 (λιμήν); Men./Marc. 5905 (κώμη); Pliny 6. 5 (*oppidum*); Ptolemy 5. 4. 2 (κώμη); Arrian 14 (ὅρμος); Anon. 20 (κώμη καὶ λιμήν).	Istifan, Çaylıoğlu	C H R
46	Potamoi (Ποταμοί)	town	Men./Marc. 5905 (χωρίον); Arrian 14; Anon. 20 (χωρίον).	Çamurca Beach or Gebelit	(H) R
47	Tetrakis (Τετράκις)	city	Pseudo-Scylax 89 (πόλις Ἑλληνίς).	Sarıkum?	C
48	Harmene (Ἁρμένη)	city, port	Xenophon 6. 1. 15 (τῆς Σινώπης); Pseudo-Scylax 89 (πόλις Ἑλληνὶς καὶ λιμήν); Men./ Marc. 5905 (κώμη καὶ λιμὴν μέγας); Strabo 12. 3. 10 (κώμη Σινωπέων); Mela 1. 104; Pliny 6. 6 (*oppidum*); Ptolemy 5. 4. 2; Arrian 14 (λιμήν); Anon. 20 (κώμη καὶ λιμὴν μέγας).	Hamsilos, Akliman	C H R
49	Kerasous[10] (Κερασοῦς)	city	Pseudo-Scylax 89 (πόλις Ἑλληνίς).	close to the Sinop airport?	C
50	Sinope (Σινώπη)	city	Hecataeus 34; Xenophon 6. 1. 15 (ἐν τῇ Παφλαγονικῇ, Μιλησίων ἄποικοι); Pseudo-Scylax 89 (πόλις Ἑλληνίς); Apollonius 946 (indirectly); Pseudo-Scymnus 986 (πόλις); Men./Marc. 5905 (πόλις); Strabo 12. 3. 11 (πόλις); Mela 1. 105; Pliny 6. 5 (*colonia*); Dionysius P. 775; Ptolemy 5. 4. 3; Arrian 14; Anon. 21 (πόλις, Μιλησίων ἄποικος). First mentioned by Eumelus of Corinth (*FGrH* 3b 451 F5).	Sinop*	E[11] A C H R

No.	Place-name	Classification	Sources	Location/ Identification	Period
51	Cyptasia (Κυπτασία)	settlement	Ptolemy 5. 4. 3.	Demirciköy Limani	R
52	Gamge	town	Pliny 6. 7 (*oppidum*).	unknown[12]	R
53	Caroussa (Κάρουσσα)	city, port, *emporion*, roadstead	Pseudo-Scylax 89 (πόλις Ἑλληνίς); Men./Marc. 6001 (χωρίον); Pliny 6. 7 (*oppidum*); Ptolemy 5. 4. 3 (as Gazoron?); Arrian 14 (σάλος); Anon. 24 (ἐμπόριον καὶ λιμήν, πρώην Πολίχνιον-Polichnion).	Gerze	C H R
54	Gourzoubathe (Γουρζουβάθη)	town	Men./Marc. 6001 (χωρίον); Anon. 24.	Kurzuvet	(H) R
55	Zagora/ Zagoron (Ζάγωρα/ Ζάγωρον)	town	Men./Marc. 6001 (χωρίον); Pliny 6. 6?; Ptolemy 5. 4. 3 (as Gazoron?); Arrian 14; Anon. 24 (χωρίον, ἤτοι Καλλίπους-Callipous).	Çayağzı	(H) R
56	Zalic(h)os (Ζάλικος/ Ζάλιχος)	village	Men./Marc. 6001 (κώμη); Anon. 24 (κώμη).	north of Alaçam	(H) R
57	Naustathmos (Ναύσταθμος)	roadstead	Men./Marc. 6001 (σάλος); Arrian 15; Anon. 26.	close to Dereköy Limanı	(H) R
58	Eusene (Εὐσήνη)	settlement	Arrian 15; Anon. 26.	Kurupelit	R
59	Eupatoria	town	Pliny 6. 7 (*oppidum*).	north of Amisos Hill, Samsun	R
60	Amisos/Enete/ Peiraieus (Ἀμισός/Ἐνετή/Πειραιεύς)	city	Hecataeus 199 (Ἐνετή); Pseudo-Scymnus 956 (πόλις ἐν τῇ Λευκοσύρων γῇ, Μιλησίων καὶ Φωκαέων ἀποικία); Men./ Marc. 6001 (πόλις); Strabo 12. 3. 14 (πόλις, Roman colony); Mela 1. 105; Pliny 6. 7 (*oppidum liberum*); Ptolemy 5. 4. 3; Arrian 15 (πόλις Ἑλληνίς); Anon. 26–27 (πόλις Ἑλληνίς).	Samsun*	A C H R
61	Lycastos (Λύκαστος)	city	Pseudo-Scylax 89 (πόλις Ἑλληνίς); Mela 1. 105 (*urbs*); Pliny 6. 9 (*oppidum*).	south of Samsun, at the outlet of the Mert River	C H R
62	Chadisia (Χαδισία)	city	Hecataeus 200 (πόλις Λευκοσύρων); Men./Marc. 6001 (κώμη); Pliny 6. 9 (*oppidum*); Anon. 28 (κώμη).	close to Karabahçe and the outlet of the Abdal Steram	A (C H) R

No.	Place-name	Classification	Sources	Location/ Identification	Period
63	Ancon (Ἀγκών)	port	Ptolemy 5. 6. 2; Arrian 15 (λιμήν); Anon. 28 (λιμήν).	at the outlet of the Yesil River	R
64	Teiria (Τείρια)	city	Hecataeus 201 (πόλις Λευκοσύρων); Sotira *oppidum* of Pliny 6. 10?	unknown	A (C H) R?
65	Heracleon/ Lamyron (Ἡράκλειον ἱερόν/Λαμυρῶν λιμήν)	port, roadstead	Men./Marc. 9r9 (λιμὴν μέγας); Arrian 15 (ὅρμος ναυσίν); Anon. 29 (λιμὴν μέγας).	Sivaslılar Köyü	(H) R
66	Themiscyra (Θεμίσκυρα)	city	Pseudo-Scylax 89 (πόλις Ἑλληνίς); Men./Marc. 9r9 (πόλις); Mela 1. 105 (*oppidum*); Pliny 6. 10 (*oppidum*); Ptolemy 5. 6. 3; Anon. 29.	Terme, or close to it[13]	C H R
67	Phanoria	fortress	Pliny 6. 10 (*castellum*).	unknown, probably very close to Terme	R
68	Oenios/Oenoe (Οἴνιος/Οἰνόη)	anchorage	Men./Marc. 9r9 (ὕφορμος); Arrian 16; Anon. 30 (ὕφορμος).	Ünye	(H) R
69	Amelitos (Ἀμηλιτός)	settlement	Men./Marc. 9r9 (τόπος); Anon. 30 (τόπος).	Yüceler	(H) R
70	Chabax (Χάβαξ)	fortified town	Strabo 12. 3. 16 (χωρίον ἐρυμνόν).	unknown, close to Fatsa	(H) R
71	Phadissa(ne)/ Phabda(?) Φαδισάνη/ Φάδισσα/ Φάβδα(?)	port, fortress, fortified town	Men./Marc. 9r9 (λιμήν); Strabo 12. 3. 16 (χωρίον ἐρυμνόν); Arrian 16 (φρούριον); Anon. 30 (λιμήν).	Fatsa*	(H) R
72	Polemonion/ Side (Πολεμώνιον/ Σίδη)	city, fortified town	Men./Marc. 9r9 (πόλις); Strabo 12. 3. 16 (χωρίον ἐρυμνόν); Pliny 6. 11 (*oppidum*); Ptolemy 5. 6. 4; Arrian 16 (πόλις); Anon. 30 (πόλις).	Bolaman	(H) R
73	Iasonia (Ἰασονία)	city	Pseudo-Scylax 88 (πόλις Ἑλληνίς). Later only as a cape.	close to Yason	C
74	Stamene/ Ameneia (Σταμένη/ Ἀμένεια)	city	Hecataeus 202 (πόλις Χαλύβων); Pseudo-Scylax 88 (πόλις Ἑλληνίς).	close to Aziziye	A C
75	Boon (Βοῶν)	port, roadstead	Men./Marc. 9r9 (λιμήν); Arrian 16 (ὅρμος); Anon. 32 (λιμὴν καὶ ὅρμος).	Perşembe	(H) R

No.	Place-name	Classification	Sources	Location/ Identification	Period
76	Gene(sin)tes (Γενέσιντις/ Γενήτης)	port	Pseudo-Scylax 88 (λιμήν); Strabo 12. 3. 17 (or the cape?).	south-east of Perşembe	C (H?) R?
77	Cotyora (Κοτύωρα)	city	Xenophon 5. 5. 3 (πόλις Ἑλληνίς, Σινωπέων ἄποικος, ἐν τῇ Τιβαρηνῶν χώρᾳ); Men./Marc. 9r9; Strabo 12. 3. 17 (πολίχνη); Pliny 6. 11 (*oppidum*); Ptolemy 5. 6. 4; Arrian 16 (κώμη οὐδὲ μεγάλη); Anon. 32 (κώμη οὐδὲ μεγάλη).	Ordu*	(A?)[16] C H R
78	Ischopolis (Ἰσχόπολις)	settlement	Strabo 12. 3. 17 (destroyed); Ptolemy 5 .6. 5.	Bulancak	H? R
79	Kerasous (Κερασοῦς)	city	Xenophon 5. 3. 2 (πόλις Ἑλληνίς, Σινωπέων ἄποικος, ἐν τῇ Κολχίδι χώρᾳ);[14] Pseudo-Scymnus 911 (in Müller 1855, as Σινωπέων ἄποικος); Strabo 12. 3. 17 (κατοικία μέτρια)?;[14] Mela 1.107 (*urbs*)?;[14] Pliny 6. 11 (*oppidum*)?;[14] Ptolemy 5. 6. 5; Arrian 16 (Σινωπέων. ἄποικος, ident. with Pharnakia); Anon. 34 (Σινωπέων ἄποικος, ident. with Pharnakia)	Giresun*	(A?)[16] C H R
80	Pharnakia (Φαρνακία)	city	Pseudo-Scymnus 950 (πόλις); Strabo 12. 3. 17 (ἐρυμνόν πόλισμα); Pliny 6. 11 (*oppidum*); Ptolemy 5. 6. 5; Arrian 16 (ident. with Kerasous); Anon. 34 (ident. with Kerasous).	close to Giresun, or Giresun[15]	H R
81	Choirades (Χοιράδες)	city	Hecataeus 204 (πόλις Μοσσυνοίκων); Pseudo-Scylax 86 (πόλις Ἑλληνίς).	Gedikkaya, Giresun	A C
82	Zephyrios Limen (Ζεφύριος λιμήν)	town	Pseudo-Scylax 86; Men./Marc. 9r9 (χωρίον); Arrian 16; Anon. 36 (χωρίον).	Gülburnu	C H R
83	Tripolis (Τρίπολις)	settlement, fortress	Pliny 6. 11 (*castellum*); Arrian 16; Anon. 36.	Tirebolu	R
84	Argyria (Ἀργύρια)	settlement	Arrian 16; Anon. 36.	quarry at Halkavala, at outlet of the Doğankent stream Harsit	R

No.	Place-name	Classification	Sources	Location/ Identification	Period
85	Philokaleia (Φιλοκάλεια)	settlement, fortress	Pliny 6. 11 (*castellum*); Arrian 16; Anon. 36.	Görele	R
86	Coralla (Κόραλλα)	settlement	Arrian 16; Anon. 36.	on the cape north-east of Eynesil	R
87	Liviopolis	fortress	Pliny 6. 11 (*castellum*).	Adacık	R
88	Kerasous (Κερασοῦς)	settlement	Men./Marc. 9r9 (πόλις); Strabo 12. 3. 17 (κατοικία μέτρια)?[14]; Mela 1. 107 (*urbs*)?[14]; Pliny 6. 11 (*oppidum*)?[14]; Anon. 36.	Vakfıkebir	R
89	Hieron Oros (Ἱερόν ὄρος)	city, anchorage	Men./Marc. 9r9 (πόλις καὶ ὕφορμος); Arrian 16; Anon. 36 (πόλις καὶ ὕφορμος).	Yoroz	(H) R
90	Cordyle (Κορδύλη)	town, port, roadstead	Men./Marc. 9r9 (χωρίον); Pliny 6. 11 (*portus*); Ptolemy 5. 6. 11 (but as inland) or Ptolemy 5. 6. 6?; Arrian 16 (ὅρμος); Anon. 36 (χωρίον καὶ ὅρμος).	Akçakale	(H) R
91	Hermonassa (Ἑρμώνασσα)	city, roadstead	Hecataeus 208 (πόλις);[17] Men./ Marc. 9r9 (χωρίον); Strabo 12. 3. 17 (κατοικία μέτρια); Ptolemy 5. 6. 4; Arrian 16 (ὅρμος); Anon. 36 (χωρίον καὶ ὅρμος).	Akçaabat	A[17] (C H) R
92	Trapezous (Τραπεζοῦς)	city	Xenophon 4. 8. 22 (πόλις Ἑλληνίς, Σινωπέων ἀποικία, ἐν τῇ Κόλχων χώρᾳ); Pseudo-Scylax 85 (πόλις Ἑλληνίς); Men./Marc. 9r9 (πόλις); Strabo 12. 3. 17 (πόλις Ἑλληνίς); Mela 1. 107 (*urbs*); Pliny 6. 12 (*liberum*); Ptolemy 5. 6. 5; Arrian 1, 16 (πόλις Ἑλληνίς); Anon. 36 (πόλις Ἑλληνίς, Σινωπέων ἄποικος).	Trabzon*	(A?)[16] C H R
93	Psoron Limen/ Hyssos (Ψωρῶν λιμήν/Ὕσσος)	settlement, port	Pseudo-Scylax 85; Ptolemy 5. 6. 5; Arrian 3, 7; Anon. 38.	Araklı	C H R
94	Ophious/ Pityous (Ὀφιοῦς/ Πιτυοῦς)	roadstead	Men./Marc. 9r9 (σάλος); Ptolemy 5. 6. 6; Anon. 38 (σάλος).	Of	(H) R
95	Becheiras (Βεχειράς)	city	Pseudo-Scylax 84 (πόλις Ἑλληνίς).	Rize, or a bit more to the west	C

No.	Place-name	Classification	Sources	Location/ Identification	Period
96	Becheirikos Limen[18] (Βεχειρικός Λιμήν)	port	Pseudo-Scylax 84.	Rize, or a bit more to the west	C
97	Rhizeon/ Rhizous (Ῥίζεον/ Ῥιζοῦς)	port	Men./Marc. 9r9 (λιμήν); Ptolemy 5. 6. 6 (λιμήν); Anon. 39 (λιμήν).	Rize	(H) R
98	Odeinios/ Adienos/ Adinaios (Ὠδείνιος/ Ἀδιηνός/ Ἀδιναῖος)[19]	city, roadstead	Pseudo-Scylax 83 (πόλις Ἑλληνίς); Men./Marc. 9r9 (σάλος); Anon. 39 (σάλος).	Çayeli	C H R
99	Limne (Λίμνη)	city	Pseudo-Scylax 83 (πόλις).	unknown. Between Çayeli and Pazar?	C
100	Cordyle/ Cordyla (Κορδύλη/ Κόρδυλα)	town	Men./Marc. 9r9 (χωρίον); Ptolemy 5. 6. 6; Anon. 39 (χωρίον).	unknown.[20] Between Çayeli and Pazar?	(H) R
101	Athenai (Ἀθῆναι)	town, roadstead	Men./Marc. 9r9 (χωρίον καὶ ὅρμος); Arrian 3; Anon. 39 (χωρίον καὶ ὅρμος).	Pazar	(H) R
102	Morthoula (Μόρθουλα)	settlement, probably town	Ptolemy 5. 6. 6.	Findikli	R
103	Xyline (Ξυλίνη)	settlement	Ptolemy 5. 6. 6.	Sugören or Hopa	R

Another two settlements are mentioned as having been situated in the littoral, but not on the coast:

No.	Place-name	Classification	Sources	Location/ Identification	Period
104	Parthenia (Παρθενία κώμη)	town	Only in Aelius Herodianus and Stephanus of Byzantium, *s.v.* Παρθένιος, maybe Parthenopolis of Pliny 5. 148, but as non-existent any more.	Bartın? (between 23 and 25)	R
105	Gadelon (Γαδηλῶν)	settlement	Strabo 12. 3. 14, whence Γαζηλωνῖτις/ Γαδιλωνῖτις χώρα (12. 3. 13, 25); Pliny 6. 6?	Bafra*?[21] (between 56 and 57)	(H) R

Finally, another nine settlements are mentioned, but with no clarification as to their exact location, which could be either inland or coastal. The first of them is recorded by Strabo, who himself was uncertain about its exact location. The others have been mentioned by Aelius Herodianus and Stephanus of Byzantium. The reasons they are added here is that the authors either quote Roman sources, like Domitius Callistratus, the 1st-century BC historian who wrote on Heraclea, and Arrian, or imply with their words the possibility of an ancient establishment.

No.	Place-name	Classification	Sources	Location/ Identification	Period
106	Zygopolis (Ζυγόπολις)	settlement, probably town	Strabo 12. 3. 17 (κατοικία).	unknown, in the eastern end of the littoral	(H) R
107	Acrioessa/ Arciroessa (Ἀκριόεσσα/ Ἀρκιρόεσσα)	city	Domitius Callistratus (πόλις ἐν Πόντῳ, Ἡρακλείας ὑποτελής) in Aelius Herodianus, *s.v.* Ἀκριόεσσα, and Stephanus of Byzantium, *s.v.* Ἀρκιρόεσσα.	unknown, close to 12	R
108	Hodioupolis (Ὁδιούπολις)	town	Domitius Callistratus (χωρίον Ἡρακλείας πρὸς τῷ Πόντῳ) in Aelius Herodianus and Stephanus of Byzantium, *s.v.* Ὁδιούπολις.	unknown, close to 12	R
109	Agamos (Ἀγαμός)	city	Stephanus of Byzantium, *s.v.* Ἀγαμός (πόλις περὶ Ἡράκλειαν τὴν Ποντικήν).	unknown, close to 12	?
110	Cynna (Κύννα)	township	Aelius Herodianus and Stephanus of Byzantium, *s.v.* Κύννα (πολίχνιον πλησίον Ἡρακλείας).	unknown, close to 12	?
111	Panelos (Πάνελος)	city	Aelius Herodianus and Stephanus of Byzantium, *s.v.* Πάνελος (πόλις περὶ τὸν Πόντον. ἐκλήθη ἀπό τινος τῶν Ἡρακλεωτῶν, ὃς ἀφίκετο ἐκ Βοιωτίας).	unknown, close to 12	?
112	Priola (Πριόλα)	city	Aelius Herodianus and Stephanus of Byzantium, *s.v.* Πριόλα (πόλις πλησίον Ἡρακλείας). Relevant to the mythical Mariandynian Priolas (see chapter MR.2)?	unknown, close to 12	?

No.	Place-name	Classification	Sources	Location/ Identification	Period
113	Saggaros (Σάγγαρος)	town	Domitius Callistratus (ἐγκυριεύσαντος ἀπὸ τοῦ Σαγγάρου χωρίου ἐντὸς τοῦ Ψιλίου ποταμοῦ) in Aelius Herodianus and Stephanus of Byzantium, *s.v.* Ψίλιον.	unknown, close to 2	R?
114	Thibais (Θιβαΐς)	settlement	Arrian (τόπος πρὸς τῷ Πόντῳ) in Aelius Herodianus and Stephanus of Byzantium, *s.v.* Θιβαΐς.	unknown, probably in the wider area around 66, since they are connected to the Amazons.	R

Recent surveys and excavations on the Island of Ares (Giresun Adasi), opposite Kerasous (no. 79) have revealed part of its history, mainly of the Byzantine period.[19] From the outset the scholars who carried out the survey were willing to identify a classical settlement, based on the date of Kerasous as well as the reference in Apollonius (*Argonautica* 2. 385–386, 1169–1173) to a stone temple and altar of Ares, created by the Amazons (the only relevant reference in ancient literature), even if there was no evidence for such a conclusion. They even identified 'an open air temple of Cybele' on the island.[20] Later, during the excavations, some fragments of Attic black-glazed and other pottery were found, which were considered evidence of habitation of the island from the Archaic to the Roman periods.[21] The pottery fragments depicted in the later publications indeed indicate a use of the place from the Late Archaic, and more intensively form the Late Classical period,[22] of what kind, however, remains undetermined. Since we cannot, for the time being, be sure about the existence of, for example, a temple, we do not yet include the site in our catalogue, hoping that further excavations will clarify the issue.

Comments on the Table

1. Whether the so-called port of Apollonia was another harbour of the city of Thynias or not, and therefore whether we have one or two different

[19] Doksanaltı and Mimiroğlu 2010; Doksanaltı 2012; Doksanaltı and Ekici 2017; Temür and Iltar 2018. See the Introduction 2.1.

[20] Doksanaltı and Mimiroğlu 2010, 85–86, 95; Doksanaltı 2012, 65–67.

[21] Doksanaltı and Ekici 2017; Temür and Iltar 2018; Kaymakçi 2021, 95–97.

[22] Doksanaltı and Ekici 2017.

installations, is uncertain. Pseudo-Scylax, the earliest to mention a settlement on the island, does not characterise it, but makes clear that it was inhabited by the Heracleans. According to Diller (1952), Pseudo-Scymnus was the source of Anonymous for calling it a city, colony of the Heracleans. But the latter also mentions the port Apollonia separately; this is probably due to the different sources he used, and the port of Apollonia may well be the same settlement as the city of Thynias, or at least have belonged to the latter, on this in any case very small island (about 11 ha). Mela's information that the city on the island was called Bithynis because Bithynians lived there further confirms the existence of a city there; Apollonius' characterisation of the island as 'desert' must either be based on a very early source or, more probably, a product of artistic licence.

2. It is usually considered to be the same as Stephane (no. 45). The latter, however, is clearly not in the territory of the Mariandynoi, as Hecataeus says of Stephanis.

3. The first surviving written source to mention Tieion is Pseudo-Scylax. However, systematic archaeological excavations that have been carried out since 2006 have revealed many finds dating from the Archaic period, starting from the 7th century BC.[23] See more in the Introduction 2.1 and Section 4.4.1, with the bibliography on the site.

4. See Ramsay 1890, 126, 432.

5. Sesamos, one of the four cities that contributed to the founding of Amastris, is mentioned by Strabo (12. 3. 10) as that city's acropolis. Moreover, both Pliny (6. 5) and Anonymous (15) say that Amastris was built on the place where Sesamos had been.[24]

6. On the location of Erythinoi, as well as on the issue of whether this was a settlement or just a pair of rocky outcrops, see Manoledakis 2013a, 24–25, 28.

7. While the earliest surviving written source mentioning Cromna may be the Hellenistic epic of Apollonius,[25] there exist both coins struck by the city and inscriptions mentioning it dating from the 4th century BC.[26]

8. Apollonius, the only writer to mention Crobialos, provides no further information. The name may be in error for Aegialos or a blend of Cromna and Aegialos, as Strabo's words (12. 3. 10) might suggest. It must be noted that Apollonius is also the only author who places Aegialos to the east and not west

[23] Which were, naturally, unknown to Avram *et al.* (2004, 964: 'nothing of Classical date is known from the site').

[24] See more in Manoledakis 2013a.

[25] On the reference to the city in the *Iliad* and its possible dating, see Manoledakis 2013a.

[26] See, for example, Avram *et al.* 2004, 959.

of Cape Carambis.[27] Valerius Flaccus, in his *Argonautica* (5. 102), which is an imitation and sometimes a pure transfer of Apollonius' poem into Latin (which is why he is not cited in the Table as a separate source), mentions Crobilaus as a beach (*latus*).

9. Mentioned only by Mela. Perhaps the same as Coloussa (?), which, however, is again only mentioned once, by Pseudo-Scylax.

10. Perhaps placed there by mistake by Pseudo-Scylax, as most scholars agree.[28] There is nothing else to indicate the existence of a city of that name so close to Sinope, less still one described as a *polis Hellenis*. Only P. Counillon entertained the possibility of Pseudo-Scylax not having been mistaken, assuming in that case that the term *πόλις* was not used in its political sense.[29] Even then, I have deep reservations about the existence of two Greek cities so close to one another (and to Harmene) in the Classical period. The question is why Pseudo-Scylax made such an error. One might just accept that he misplaced Kerasous, the Sinopean colony, since he in any case paid little attention to the southern Black Sea and may have confused it with the mother city's name and location. But then we remember that it is not only Kerasous which he omitted, but also Cotyora, another colony of Sinope. According to Counillon, this double omission indicates that the two cities were not in Pseudo-Scylax's source and that they probably were not the kind of settlement that the author was interested in.[30] But why, then, did he not also omit Trapezous, which had exactly the same status (*polis Hellenis* and colony of Sinope) as the other two? A solid answer is very far from being reached. In any case, we shall consider no. 49 in the way it is mentioned as actually non-existent.

11. For the early history of Sinope, see more in Section 4.3.2.

12. Probably confused with the inland city of Gangra/Germanicopolis, although mentioned with the coastal sites.

13. Ptolemy clearly refers to Themiscyra as a coastal city, and the same conclusion may be drawn from its appearance in the *Periploi*. Terme is inland, so Themiscyra could have been situated a little farther east, unless the area was coastal in antiquity.[31]

14. The question of Kerasous is one of the most puzzling in the study of the southern Black Sea littoral. This is because the ancient literature mentions cities bearing that name in as many as three different places (nos. 49, 79, 88). I have already dealt with the matter elsewhere,[32] so I shall confine myself here

[27] See more in Manoledakis 2013a, 24; Rengakos 1993, 128–29.

[28] E.g. Avram *et al.* 2004, 958–59. See also Manoledakis 2010a, 142–47, with the earlier bibliography on the various opinions that have been expressed about the Kerasous of Pseudo-Scylax.

[29] Counillon 2004, 120.

[30] Counillon 2004, 45, 111.

[31] *Cf.* the case of Bafra mentioned in Section 1.3.

[32] Manoledakis 2010a; 2022b.

to the most important points for our examination. With regard to no. 49, see above, in Comment 10. As for nos. 79 and 88, the main question is which of the two was meant by Strabo, Mela and Pliny, whose descriptions are utterly unclear, and especially which one was meant by Xenophon as the colony of Sinope that he visited (*cf.* Diodorus 14. 30. 5; the *Suda*, *s.v.* Κερασοῦς). While no secure answer can be given for the three Roman authors, the question in relation to Xenophon has been troubling historical research for more than two centuries. Two spots on the southern Black Sea littoral have been proposed as the site of this city, Giresun and Vakfıkebir. Neither has ever yielded any archaeological evidence to support its identification with Xenophon's Kerasous, and so scholars have sought to draw conclusions based solely on Xenophon's words.

Despite the fact that Arrian, another ancient author and one moreover who had studied the *Anabasis*, offers valuable help in the matter, since he places the colony of Sinope called Kerasous opposite the Island of Ares and thus at Giresun, most scholars have opted for Vakfıkebir. The main argument has always been the fact that the three days which, according to Xenophon (*Anabasis* 5. 3. 2), the Greek soldiers took to march from Trapezous to Kerasous, is too short a time to cover the distance between Trabzon and Giresun; and indeed, none of the rest of the ancient sources mentioning cities called Kerasous is able to clarify the matter.[33]

Now, however, a new dimension has been added to the discussion with the application of a GIS technique called Cost Surface Analysis. Setting Trapezous as the starting point of a walking route to the west, it appears that Giresun could have been reached on foot in about three days, or a bit less, while marching to Vakfıkebir would require no more than ten hours, definitely closer to one than three days, given the fact that, as Xenophon clearly states (*Anabasis* 5. 3. 1), the soldiers doing the marching were young and healthy and unencumbered by any factor that could delay them. Notwithstanding the possible variances that any such method used in spatial analysis may entail, the difference between the two sites is still very great, and the coincidence between the time indicated by Cost Surface Analysis as needed from Trabzon to Giresun and the time recorded by Xenophon for the march from Trapezous to Kerasous is so noteworthy that it cannot pass unnoticed. It seems to me, therefore, that there can be little doubt about identifying the Sinopean colony visited by Xenophon's army with Giresun.[34] At all events, if there was indeed a second Kerasous, at Vakfıkebir, it first appears in the literature of the Roman period.

[33] For them, see in Manoledakis 2022b.

[34] For a detailed argumentation, see Manoledakis 2022b, with the relevant images.

15. For the three cities, Kerasous, Pharnakia and Choirades, which have sometimes been identified with each other, see Manoledakis 2010a, with bibliography. They were probably three different cities, very close to one another. Choirades might already have been abandoned when Pharnakia was founded.

16. The three cities, Cotyora, Kerasous and Trapezous, are mentioned by Xenophon as colonies of Sinope. He is the earliest known author to mention them. However, a version of Eusebius erroneously giving 757/6 BC as the founding date of Trapezous has caused a lot of discussion in modern literature.[35] Some scholars hold that Sinope established the three cities not long after her own founding,[36] but there is nothing to positively confirm such a view. Unfortunately, no excavations that could shed more light on the matter have yet been carried out in any of the three cities. Some limited rescue excavations took place during construction work in the centre of Trabzon in 1997, but these yielded no finds from the periods in question.[37]

What we are left with are Xenophon's words relating to the three cities, and the possible restoration of *Κερ* in the Athenian Tribute List of 425/4 BC as Κερασός.[38] Both the latter, if this restoration is indeed correct, and Xenophon's references suggest that the cities had been long established by the late 5th century BC. The Sinopean ambassadors who met the Ten Thousand at Cotyora told them that the Cotyorites were their colonists, and that 'it was we who gave over to the Cotyorites this land, after we had taken it from the barbarians; therefore they pay us a stated tribute, as do the people of Cerasus and Trapezous' (*Anabasis* 5. 5. 7–10). Furthermore, the city certainly had walls, an *agora* and private houses in 400 BC (*Anabasis* 5. 5. 5–11). In the same period Trapezous was striking silver coins on the Persian standard.[39] Consequently, it is very probable, although not provable, that the three cities were founded – initially as small settlements or *emporia* – by Sinope before 480 BC, and so in the Archaic period.

17. There has been discussion on whether Hecataeus was actually referring to this Hermonassa or to the one in the Cimmerian Bosporus. The debate arose from the vagueness of Stephanus' citation, who mentions both cities in the same entry (*FGrH* 1a 1 F 208: Ἑρμώνασσα· νῆσος μικρὰ πόλιν ἔχουσα ἐν τῶι Κιμμερίωι Βοσπόρωι, Ἰώνων ἄποικον, ὡς ὁ Περιηγητής. ταύτην ὁ

[35] See, for example, Avram *et al.* 2004, 964; Manoledakis 2010b, with bibliography.

[36] E.g. Erciyas 2007, 1196; de Boer 2015, 75. Tsetskhladze (2009, 233–34) suggested the 6th century BC. Wojan (2003, 264) is right to claim to be unable to determine the difference in the time of the establishment of the three cities.

[37] Kaya 1999; *cf.* Summerer 2014, 190–93.

[38] Avram *et al.* 2004, 958.

[39] Avram *et al.* 2004, 964.

Σκύμνος καὶ Ἑρμώνειαν καλεῖ. Στράβων ἐν ζ Ἑρμώνακτος κώμην φησί. Μένιππος δὲ χωρίον Τραπεζοῦντος ἐν Περίπλωι τῶν δύο Πόντων. Ἑκαταῖος δὲ καὶ Θεόπομπος πόλιν αὐτήν φασιν.). Stephanus begins with the Bosporan city, mentioned by Dionysius Periegetes (552), Pseudo-Scymnus (886) and Strabo (7. 3. 16), and then moves on to the 'town of Trapezous', citing Menippus (9r9), Hecataeus and Theopompus.

F. Jacoby[40] gave the impression of having considered the whole entry as referring to the southern Black Sea settlement, based also on W. Ruge,[41] who, however, erroneously attributed Pseudo-Scymnus' reference to this settlement although it is obvious that he is referring to the one on the Cimmerian Bosporus. M.H. Hansen, on the other hand, concluded that Stephanus erroneously considered the Bosporan Hermonassa a Trapezuntian possession, that the reference to Menippus was a mistake, and that it was the Bosporan colony which both Hecataeus and Theopompus described as a *polis*, and thus that Stephanus' reference is only to this city, since there is no indication in his words that he is talking about two cities.[42] This last argument is indeed well founded, but we cannot so easily accept the three preceding ones, especially the mistakes made by Stephanus. There is nothing to indicate that either Menippus or Stephanus considered the Bosporan Hermonassa a Trapezuntian possession.

The truth is that Stephanus seems unwilling to make a clear distinction between the two cities, but he does mention them separately, and such carelessness is not unprecedented in his work. As we read the entry, he mentions the Bosporan Hermonassa first (especially since the quotation to Strabo is only to his Book 7, and not also to Book 12 (12. 3. 17), where he mentions our Hermonassa), and then the one on the southern coast, quoting Menippus. Thus, Hecataeus and Theopompus, who follow, may well have referred to this city. I cannot claim to be certain about this conclusion but, being unable to prove the contrary, have decided to let it stand here.

18. The connection between this port and the city of Becheiras (no. 95) is unclear, as is its location in relation to the latter. Sometimes the city of Becheiras is considered to have been the predecessor of Rhizeon (no. 97). In any case, since none of these things is clear and we still have three different place names, we present them here separately. Nevertheless, the possibility that no. 96 was a settlement like no. 95 is small, since the name is actually the same. It seems more likely that it was a port connected with the city of Becheiras.

[40] Jacoby 1957, 358.
[41] Ruge, *RE* VIII 1 (1912), 899, *s.v.* Hermonassa 2.
[42] Hansen 1997, 23.

19. Probably Odeinios/Ὠδείνιος rather than Hodeinios/Ὡδείνιος.[43]

20. Sometimes identified with no. 90. However, the places given in the two cases are clearly far from each other.

21. *Cf.* Roller 2018, 702. The settlement in question might well have been coastal in antiquity.[44]

2. Statistical Observations and General Remarks

The preceding table, together with the accompanying comments, encapsulates a large part of the human building activity and intervention in the natural environment of the southern Black Sea littoral, which is visually illustrated on the maps in Figs. 13.1–8. There are, naturally, several conclusions that may be drawn from the table and the maps; here, I present the most noteworthy.

First of all, I should clarify the distinction between settlements and installations. Any product of extensive building activity is defined as a significant human intervention to the environment. This may refer not only to settlements, i.e. places where people have come to live and have built houses and other buildings, but also to places that contain equipment and machinery used for a particular purpose, such as roadsteads, breakwaters, etc. These are what are here called installations, and in the main they are port facilities. There are also two fortresses (nos. 67 and 87) and a temple (ἱερόν) (no. 65) mentioned on the coast, which might form separate categories, although the last is mentioned together with a port.

In total, 114 sites are mentioned in the written sources, the vast majority of them (103) coastal. This in itself manifests the significance of the sea in the communities' life and activities, as well the limited opportunities provided by such a narrow strip of land. Another two settlements are mentioned as lying inland, while there are a further nine settlements that must have been situated on the littoral but for which no additional specification concerning their exact location is given; two of these (nos. 111 and 114), in fact, might have been situated elsewhere in the Black Sea region. From these twelve settlements, eight were created in the Roman period (two of them perhaps in the Hellenistic), while the date of the others' establishment remains unknown.

On the relevant maps (Figs. 13.1–8) the settlements are indicated by a circle, the installations by an anchor.[45] This does not, of course, mean that the settlements on the coast did not have ports or port facilities. On the contrary:

[43] See Müller 1855, 63; Counillon 2004, 18, 44, 93; Shipley 2011, 157.

[44] *Cf.* Tsetskhladze 2007, 178. See also Part 1.

[45] A tower stands for a fortress and a temple for the temple.

as stressed in Section 4.4, the main reason for their creation and survival, whether they were *poleis*, *emporia* or anything else, was maritime communications and trade. One may thus be quite certain that almost every one of them contained a port facility of some description.

Of the 103 coastal sites, I decided to omit nos. 31 and 49 from the following statistical observations, since there is good reason to believe that they did not actually exist, as analysed in, respectively, comments 8 and 10 to the Table. Furthermore, no. 52, for the reason presented in Comment 12. We are thus left with 100 sites, which chances to be a convenient number for the statistics.

2.1. Site Classification Terms

The ancient authors used several different classification terms to denote the settlements and installations enumerated in our table. The former are described in Greek texts as πόλις, πολίχνιον, πόλισμα, κώμη, χωρίον, ἐμπόριον and κατοικία, and in Latin sources as *urbs* or *oppidum*. Installations are designated by the terms ὅρμος, ὕφορμος, σάλος and χηλή, while λιμήν and *portus* are also encountered. For the fortresses, the terms φρούριον, ἐρυμνόν (χωρίον) and *castellum* are found, along with one ἀκρόπολις. Three of the sites (nos. 38, 69, 114) are simply labelled τόπος. Finally, we frequently find combinations of terms, specifically κώμη καὶ λιμήν, ἐμπόριον καὶ λιμήν, χωρίον καὶ ὅρμος, λιμήν καὶ ὅρμος and πόλις καὶ ὕφορμος, or longer strings, such as λιμήν, χωρίον, ὅρμος (e.g. no. 3).

The term λιμήν appears to be one of the trickiest, like distinguishing between 'port' and 'harbour'. Does it describe a settlement with a port facility or just the latter? In our sources, as in the rest of ancient literature, the word is used for both categories of sites. Nonetheless, we should always bear in mind that the same term may be used differently by different authors (not necessarily of different periods), and sometimes even variously by the same author, as is the case with other similar terms. All we can do here is make some statistical observations. Thus, in our sources the term λιμήν is encountered 17 times, used by several authors.[46] In three of the cases one of those mentioning the site is Pliny, who uses the term *portus*.[47] In all three of these cases the site appears in other sources accompanied by settlement descriptors (χωρίον, ἐμπόριον, πολίχνιον), as in another six cases of this set of sites, for a total of nine cases out of the 17.[48] In five of the remaining eight cases, all the authors mentioning

[46] Nos. 3, 5, 13, 21, 45, 48, 53, 63, 65, 71, 75, 76, 82, 90, 93, 96 and 97.
[47] Nos. 3, 13 and 90.
[48] Nos. 3, 13, 21, 45, 48, 53, 71, 82 and 90.

the site call it λιμήν or ὅρμος.[49] As for the last three cases, we know the site either only from Pseudo-Scylax, who calls it *λιμήν*, in two instances as part of its name (nos. 76 – Genesintes Limen, and 96 – Becheirikos Limen), or from other authors, all of whom appear to use the term as part of the name, as seems to be indicated by the genitive (no. 93 – Psoron or Hyssou limen).[50] This suggests that the term could have been applied to both a settlement with a port and just a port facility. We therefore decided to assign the nine sites where it is also accompanied by a settlement term to the settlements and the five sites without such designation to the installations. As for the three 'special' cases, we assessed each of them separately, fully aware that we might be mistaken:[51] Genesintes Limen (no. 76) is considered more probably an installation, since the other author who mentions the name (Strabo) does not make clear whether he is speaking of a settlement of any sort or just the cape of that name. For Psoron or Hyssou limen (no. 93), we tend more to imagine a small settlement on the site, but only because of Arrian's words (*Periplus* 3) that his team found foot-soldiers there, whom they exercised. Finally, in the case of Becheirikos Limen (no. 96) we lean more towards an installation, because of the existence of the neighbouring homonymous city. In fact, all three could equally have been just installations or settlements as well.

Another intriguing case is that of Ancon (no. 63), which has been assigned to the installations. Of interest here is the name itself, which denotes a roadstead, natural or artificial, but occurs elsewhere (e.g. in Italy) as a settlement name; here, no author uses a settlement term for the specific site. Finally, there is the difficult case of Erythinoi (no. 27). Of all the authors mentioning the place, only Anonymous uses the term χηλή, which denotes an artificial breakwater. Apollonius describes Erythinoi as αἰπεινούς, which means steep, craggy. Strabo and Ptolemy clearly call them rocks (see Table 1), while Arrian, a contemporary of Ptolemy, uses no designation at all. The only thing suggesting an installation, therefore, would be Anonymous' term χηλή, were it not for the fact that Erythinoi is mentioned in the Trojan Catalogue (*Iliad* 2. 853–855), although in a passage that has been considered a Hellenistic interpolation, known to both Apollonius and Strabo.[52]

[49] Nos. 5, 63, 65, 75 and 97.

[50] It has been claimed that the word *λιμήν* might well have been used as a place name already in the Mycenaean period: Chadwick and Baumbach 1963, 218, *s.v.* λιμην.

[51] It is not accidental that in some of the statistical conclusions that follow the reader may see results like '34 or 35 sites', or sentences such as 'if the identification proposed in Table 3 is correct'. *Cf.* n. 14.

[52] See Section 2.5, with bibliography.

This reference to Erythinoi in the *Iliad* together with other towns has sometimes led to the view that Erythinoi was a town as well. Notwithstanding the fact that no source characterises the site as a settlement, the interpolation is indeed noteworthy, but equally noteworthy is the fact that it describes Erythinoi as ὑψηλούς, as Strabo (12. 3. 10) repeats, which means lofty and could certainly be applied to a settlement, while Apollonius, as we saw (2. 941), uses the adjective αἰπεινός, precipitous, which rather suggests a rock or a coast. But then Strabo himself adds the information that the Erythinoi were two rocks, at least in his time. If Anonymous had derived the term χηλή from Arrian, we might conclude that there was an installation there from the Roman period at the latest; but this is not the case, and the term seems to be an addition of the compiler.[53] All things considered, we decided to include Erythinoi in this list (and of course in Table 1, as there were certainly rocks known by this name – Strabo, Ptolemy) because of the term χηλή, but only as an installation since there is not the slightest evidence for Erythinoi being a settlement. All five of these sites (nos. 27, 63, 76, 93, 96) are problematic and are treated here with considerable reservation.

But even if we could solve this problem we are immediately faced with another, which affects not only the term λιμήν but also the other related ones, namely ὅρμος, σάλος and ὕφορμος: do these necessarily denote artificial roadsteads (the first two) and anchorages (the third) or just natural ones? For in the latter case the corresponding sites should not be included in our list.[54] In order to decide, we must consider the more general usage of the three terms in ancient literature, whilst additional light may be shed on specific places by other sources that mention it. After applying both procedures, we conclude that in most cases the installation described is an artificial one.[55] First of all, a considerable number of sites have been characterised variously as settlements and installations. If the differentiation occurs in different periods (e.g. no. 91) one might infer a gradual decline; otherwise (e.g. no. 1), it seems more plausible that the specific installation term describes an artificial rather than a natural feature.

Besides, looking at similar cases in the wider body of ancient Greek literature, especially the works we consider 'geographical', it becomes apparent that a λιμήν is almost always a man-made port facility, as, indicated for example,

[53] Diller 1952, 121.

[54] We are speaking here of sites described only by an installation term (e.g. no. 4); those described, for example, as ὅρμοι by some authors and χωρία by others (e.g. no. 1) are in any case classed as settlements.

[55] See also Manoledakis 2022a.

by several of Strabo's descriptions of sites outside the Black Sea.[56] An ὅρμος, an ὕφορμος, or a σάλος, on the other hand, [57] might be either a natural place or an artificial installation, but usually the latter.[58] In any case, references to these in the sources suggest that they were smaller installations than the *limenes* and could sometimes be parts of a *limen*.[59] So in the few uncertain cases in our area of interest where we have no other sources to crosscheck a place-name (e.g. no. 18) the site is added to our list, since we observe that the vast majority of such sites mentioned by geographers and authors of *Periploi* are man-made constructions. Noteworthy is the fact that the terms ὅρμος, σάλος and ὕφορμος appear only in the Roman period, while λιμήν was also used in the Classical age (Table 4).

Table 4. Site classification terms used in the 'geographical sources'.
The additional one in the last column is the settlement called Χηλή (no. 7 on Table 3).
The last row gives the number of different settlements mentioned under the specific term, while the previous row the number of references with each term.

	πόλις	ἐμπόριον	χωρίον	κώμη	πολίχνιον	πόλισμα	κατοικία	urbs	oppidum	λιμήν/portus	ὅρμος	σάλος	ὕφορμος	χηλή
Hecataeus	6													
Xenophon	5									1				
Pseudo-Scylax	20									6				
Apollonius										1				
Pseudo-Scymnus	6													
Menippus	12	3	16	6	1					4	3	4	3	
Strabo	6	1	3	2	2	1	4							
Mela								5	2					

[56] 1: Strabo 3. 4. 7: Ταρράκων πόλις, ἀλίμενος μὲν ἐν κόλπῳ δὲ ἱδρυμένη: so here λιμήν clearly means a man-made construction, which is absent in the (natural) bay where the city is built. Similarly 2: Strabo 4. 6. 2: ὅλως δὲ ἡ παραλία αὕτη πᾶσα μέχρι Τυρρηνίας ἐκ Μονοίκου λιμένος προσεχής τέ ἐστι καὶ ἀλίμενος πλὴν βραχέων ὅρμων καὶ ἀγκυροβολίων. 3: Strabo 7. 6. 1: Σαλμυδησσῷ. ἔστι δ' οὗτος ἔρημος αἰγιαλὸς καὶ λιθώδης, ἀλίμενος, ἀναπεπταμένος πολὺς πρὸς τοὺς βορέας.

[57] Pseudo-Scylax, who frequently uses the term λιμήν (eight of the 17 times in our list), never uses ὅρμος, σάλος and ὕφορμος, terms which we encounter only in the works of Menippus and Arrian (and thus, of course, 'Anonymous'). This could well indicate an interest in more limited kinds of sites, compared to the other two and later authors.

[58] E.g. Pausanias 6. 19. 9: ὁ δὲ ὅρμος ταῖς ναυσὶ χειροποίητος καὶ Ἀδριανοῦ βασιλέως ἐστὶν ἔργον. *Cf.* Arrian *Anabasis* 7. 19. 4.

[59] E.g. Arrian *Periplus* 24. 2: ἀλίμενος μὲν εὔσαλον δὲ καὶ εὔορμον.

Pliny									19	3				
Dionysius P														
Ptolemy	2		2	1						2				
Arrian	7	4		1						4	9	1		(1)
Anon.	11	6	13	8	1					6	6	2	3	1(+1)
Total mentions	**75**	**14**	**35**	**18**	**4**	**1**	**4**	**5**	**21**	**24/3**	**18**	**7**	**6**	**1(+1)**
Total different	**36**	**8**	**23**	**9**	**3**	**1**	**4**	**5**	**19**	**15/3**	**13**	**5**	**3**	**1(+1)**

Let me turn now to the terms that describe settlements: πόλις, πολίχνιον, πόλισμα, κώμη, χωρίον, ἐμπόριον, κατοικία, *urbs* and *oppidum*. The variety here is larger and the question of the differences in their meaning inevitable. Cardinal among them is, of course, the term *polis*. In recent decades many studies have sought to comprehend how this term was used by ancient authors, mainly of the Archaic and the Classical periods – in the urban or political sense (or both), consistently or not, etc. – and much debate has ensued;[60] there is, therefore, no point in returning to this issue here.[61] It seems, however, that, despite sometimes almost obsessive efforts to reach a conclusion, one may safely determine only that no consistent definition of the term as it is used in any ancient text is possible; each author may have used the word in more than one sense within the same work, depending *inter alia* on factors such as his sources and how accurately he interpreted them. Especially as far as the authors who are our main sources and have been studied here are concerned, we note that they may have used the term *polis* in both the urban and the political sense[62] and to describe 'barbarian' as well Greek cities.[63]

[60] See Hansen 2000; Hansen and Nielsen 2004, with the whole bibliography, also on the reservation expressed about the Copenhagen Polis Centre's research results. The most insightful view expressed in the publications of the Centre is that 'each occurrence of *polis* must be studied in context' (Hansen 2000, 177).

[61] We should note, however, that apart from the 114 sites known by name there are also a few cases of general references to minor Greek settlements or installations that belonged to important Greek cities but are not mentioned by name. In Xenophon's *Anabasis* we read about the χώρα (land, territory) of Cotyora (5. 5. 7, 10), in which were several χωρία belonging to the Cotyorites (5. 5. 6). Similar conclusions are drawn from the passage (5. 5. 23) referring to the expansionist views over the territory of Sinope entertained by the Paphlagonian ruler Corylas (see Manoledakis 2019, 216–17), where we read that the latter had a desire not only for Sinope, but also for its coastal χωρία (strongholds?). Sinope is also reported by Strabo (12. 3. 11) to have had several harbours, roadsteads and fisheries on either side of its isthmus.

[62] For two examples of the latter case, according to Xenophon's descriptions, see below, in Section 5.3: In the *Anabasis* we read about *choria* of the Cotyorites and the Sinopeans (5. 5. 6, 23).

[63] Hansen 1997 on Hecataeus; Nielsen 2000 on Xenophon; Flensted-Jensen and Hansen 1996 on Pseudo-Scylax, considered not the most reliable source as regards our area of interest; Hansen 2000. In some of these studies, some of the southern Black Sea cities that have been described as

All the more, then, in the Roman authors, the use of the term *polis* can give no reliable information about the status of the city, since it is often simply repeated from earlier sources; the term *polis Hellenis* in particular – encountered frequently in Anonymous, for example – has no meaning in his time or generally in the Roman period. In many other cases, too, it is obvious that the author has adopted a phrase unchanged from an earlier source, especially when he also lists additional historical information, citing his source by name (e.g. Xenophon or Pseudo-Scymnus).

It is noteworthy that the term appears 64 times in our sources, 31 in Archaic and Classical works and 33 in Hellenistic and Roman ones (see Table 4).[64] Of the 36 settlements characterised as *poleis*, 26 are *first* encountered in Archaic and Classical texts[65] and just ten in Hellenistic and Roman ones.[66] What concerns us here, however, is not the political status of the settlements described as *poleis* but the fact that they were cities (our focus is on the urban development of the coast), irrespective of whether they were city-states as well, which is certainly not the case for every one of them, or of whether they were Greek or 'barbarian'. In Table 3 all of them are classified as cities.

Of secondary significance are, similarly, the possible differences in meaning of the terms κώμη, χωρίον, πολίχνιον and πόλισμα, which appear here as villages, towns and townships. It is obvious that no firm conclusions can be drawn concerning their size[67] or status, for reasons that have been already discussed in the case of *πόλις*. In cases where more than one settlement term is given by different authors for the same site, we retain in the relevant column of Table 3 the one that denotes a larger settlement. If the site has been also accompanied by an installation term or the word *emporion*, these are also added, since they imply a different status for the site (e.g. no. 29). If no classification

poleis are missing from the relevant accompanying lists, or are even treated erroneously. See, for example, the case of Choirades, mentioned in Manoledakis 2010a, 140, 148, n. 55.

[64] We do not count Anonymous' references to those of the Roman or any other period. His work was a compilation of earlier works, so each time a term is mentioned there it is actually a repetition from an earlier work. The only exception is the term χηλή (see above).

[65] See Avram *et al.* 2004, 929, 954–64, who, however omit Stephanis (no. 22), Chadisia (no. 62), Teiria (no. 64) and Hermonassa (no. 91). This might be because of the fact that Hecataeus calls them cities of indigenous peoples, but in that case we would expect them also to omit Stamene (no. 74). If the reason is that Stamene is called a *polis* of the Chalybes by Hecataeus but *polis Hellenis* by Pseudo-Scylax, then we have also the similar case of Choirades (no. 81), which is called *polis* of the Mossynoikoi by the former and *Hellenis* by the latter. For what this discrepancy among the Greek sources might mean, see Manoledakis 2010a.

[66] Nos. 6, 8, 26, 32, 39, 60, 72, 80, 89 and 107. The cases of settlements mentioned only by Aelius Herodianus and Stephanus of Byzantium are not taken into consideration.

[67] Of all the southern Black Sea settlements the only one for which some estimate of size has been made is Tieion (about 60 ha): Yıldırım 2022. For the rest, the territories of Heraclea and Sinope have been estimated at about or more than 500 km^2. See Avram *et al.* 2004, 955, 960.

term appears in the sources, we just call the site a settlement (e.g. no. 34). What is noteworthy is that, unlike πόλις, which is used throughout our period (Table 4), the terms κώμη, χωρίον, πολίχνιον and πόλισμα were used only in the Roman era.[68] What we mean here by 'use' is: 'accompany specific settlements mentioned by name', since terms like χωρίον, for example, might be widely used in general contexts, as for instance when an author describes the urban environment of a region without necessarily knowing the names of its settlements. This is regularly encountered in descriptions of indigenous people's lands, and the question is dealt with in the corresponding section.[69]

Nonetheless, in general one may say that a *kome* is usually an unfortified village,[70] while several *komai* may be clustered around a *polis*, especially in the Greek world.[71] As regards the *chorion*, diversity seems to be its most salient feature.[72] It is worth noting that Strabo seems to care more than any other author to distinguish between the terms, since he is the only one who uses them all (Table 4). It is interesting to note his specification that the three *choria* he mentions were fortified (χωρίον ἐρυμνόν: nos. 70, 71 and 72), as was his *polisma* (ἐρυμνόν πόλισμα: no. 80). As for his two *komai* (30 and 48), they are not described as fortified, probably because this would be implicit in the term; however, if Harmene (no. 48) was indeed a *polis* in the political sense one would expect it to be fortified, unless we have here a case of a settlement that had declined since the days of Pseudo-Scylax or if Strabo wished to indicate its dependence on Sinope.[73]

A unique case is Strabo's use of the term κατοικία. Five settlements (nos. 28, 29, 79, 91 and 106), three of which had been called πόλεις by earlier sources, are described by him in this way, using a word that denotes a settlement, possibly a colony, without further elaboration. Moreover, we have the Latin words *urbs* and *oppidum*. The first appears to be Mela's favourite, for he used it five times. Pliny, on the other hand, preferred *oppidum*, which he used

[68] Thus the conclusion that in ancient literature the term *kome* was not used for any Greek colony (Hansen 1995, 69) is correct only when speaking of Archaic and Classical literature. From the early Roman period on, we find several instances of such use, as is evident in Table 3, some of which in fact refer to cities that had been called *poleis* by earlier sources (e.g. nos. 41, 48, 62, 77, the last being the most indicative case).

[69] See Section 5.3.

[70] *Cf.* Nielsen 2000, 135–37. However, when the terms are used for Greek settlements, the difference may also concern political status. *Cf.* Hansen 1995, 61–62.

[71] The term *kome* frequently appears in the plural form, to denote villages scattered over a large territory. See Hansen 1995, in detail; Nielsen 2000. For the use of the term *polis* for 'barbarian' settlements, see also Hansen 2000, 180–82; Hansen and Nielsen 2004, 36–37.

[72] See Section 5.3, in detail.

[73] *Cf.* Flensted-Jensen and Hansen 1996, 149; Hansen 2000, 193, n. 41. For Harmene, see Stoop 1977–78.

for 19 settlements, two of which (nos. 23 and 66) were also given this label by Mela. It should be noted that almost all these *urbes* and *oppida* were called πόλεις by earlier Greek authors.

Finally, there is the term ἐμπόριον. The meaning of the term and the significance of the *emporia* in the history of Greek colonisation and the relations between colonists and locals have already been highlighted.[74] However, it is worth stressing here that it is only in the Roman period that the term is applied to places on the southern Black Sea coast.[75] The geographers of this period report eight *emporia*,[76] against none in earlier geographical texts. Three of them (Lillion, Elaion, Cales) first appear as place-names in the Roman era. The other five (Calpes, Psylla, Cytoron, Kimolis, Caroussa) had appeared in the Classical period – four in Pseudo-Scylax and the other (Calpe) in Xenophon – although none of these was then designated an *emporion*, but instead called either λιμήν or πόλις. It is worth noting that even other geographical texts of the first centuries of the Roman era do not mention these settlements as *emporia*, but rather as *urbes* (Cytoros in Mela), or as *oppida* (Kinolis and Carusa in Pliny), or as λιμένες (Calpas in Pliny, as in Xenophon), or as πόλεις (Cytoron in Ptolemy), or with no classification (several in Strabo and Ptolemy). Moreover, Calpe (an *emporion* in Menippus) is mentioned as a λιμήν by Arrian (following Xenophon), although he generally uses the term *emporion*. Conversely, Psylla and Kinolis (*emporia* in Arrian) are mentioned by Menippus as χωρίον and κώμη respectively. Finally, Anonymous generally follows Menippus concerning the classification of settlements as *emporia*. Furthermore, we know of settlements that, like Harmene (no. 48), must almost certainly have been *emporia* although they are never so described in any of the many sources mentioning them. What conclusions may be drawn from these observations on the use of the term *emporion* in ancient Greek literature is a matter not to be attempted here.

A case of special interest is that of Cytoron (no. 29). Anonymous calls it an *emporion*. This is important, because, as already mentioned, he is considered to have been influenced mainly by Marcian's edition of Menippus' *Periplus* and the *Periplus* of Arrian, and secondarily by the *Periploi* of Pseudo-Scylax and Pseudo-Scymnus.[77] None of these four, however, refers to Cytoron as an *emporion*, but instead as a χωρίον (Menippus), or as a πόλις Ἑλληνίς (Pseudo-Scylax), or with no site classification term, merely the information

[74] See Section 4.4.1.

[75] Hansen 2006, 23, who mentions that 'securely attested examples of *emporia* which were not *poleis* belong only in the Hellenistic and Roman periods'. *Cf.* also Hind 1995–96, 115.

[76] Nos. 3, 9, 10, 11, 21, 29, 41, 53.

[77] Diller 1952, 1, 102, 106–07.

that there was an ὅρμος for ships (Arrian), or does not mention it at all (Pseudo-Scymnus, at least in the preserved text). The rest of the sources mention Cytoron either as πόλις (Mela; Ptolemy), or as πόλις Ἑλληνίς (Pseudo-Scylax), as a mountain (Pliny), or with no site classification term (Apollonius, who at least describes it as woody, as does Valerius Flaccus, by whom Pliny was probably influenced). The only one who records that Cytoron was once an *emporion* (of the Sinopeans) is Strabo.[78] This could mean that Anonymous was mainly influenced by the authors of the *Periploi* but also drew on other sources, which a geographer from Asia Minor could not have been without.

We also observe that, with the exception of Menippus (1st century BC), the frequency of the term *emporion* increases from the 2nd century AD on. One might therefore be led to suspect that Marcian may have made more interpolations in the work of Menippus than has been thought up to now,[79] such as for example the ascription of the term *emporion* to the settlements Cales and Elaion, which are not called *emporia* by other writers of the 1st century BC–1st century AD. Most *emporia* are only referred to by the latest source, namely Anonymous, while it is also worth mentioning that Eustathius in the 12th century is the first to apply the term to Trapezous (Schol. in Dion. Per. 687).

These observations on the use of the term *emporion* in the Roman texts raise further questions: for example, were the *emporia* that appeared in the Roman period newly-founded trading posts and, if so, what trade did they serve? In the age of Greek colonisation the *emporia* mainly served commercial contacts between the locals and the Greeks. The Romans, as a new power in the region, had reasons enough to create *emporia*, in the sense of the term accepted for the Archaic and Classical periods. The local populations, however, no longer appear to play an important role in the Roman era, and indeed tend to disappear in its last centuries. One might assume, therefore, that the *emporia* served either contacts between the Romans and the Greek cities or contacts between the Romans (and possibly the Greeks) and the foreigners, in the general sense,[80] who sailed the Black Sea. It is equally possible, however, that after a period of time the term *emporion* acquired a broader sense, encompassing not only generally the coexistence of locals and foreigners in a settlement, but even more generally the installation that also had the character of a port.[81]

[78] For the term *emporion* in Strabo, see Étienne 1993; Rouillard 1993.

[79] Two such interpolations have been recognised in the area we are dealing with: Marcian's reference to the newer name of Abonou Teichos (Ionopolis) and to *Duo Ponti*. See Diller 1952, 149.

[80] *Cf.* also the phrase 'ἐμπόρια βαρβαρικά' in Anonymous *Periplus* 7.

[81] This does not, of course, mean that the *emporia* were necessarily ports. *Cf.* Counillon 1993, 50–51.

Arrian's description of Kinolis as an *emporion* where ships may anchor during the summer might be characteristic.[82]

Finally, as far as the combinations of terms are concerned, Anonymous is the undoubted champion, using ten of them, and a very possible reason for this might be the fact that he was essentially a compiler synthesising from several different sources, which often used different terms for the same site. For the rest, three combinations are encountered in Menippus and only one in Pseudo-Scylax, in the intriguing case of Harmene. Sometimes, multiple terms may appear together, like λιμήν, χωρίον, ὅρμος (e.g. no. 3). The sites for which the most different terms (more than three) are used are Calpe (no. 3: ἐμπό-ριον, λιμήν, χωρίον, ὅρμος), Cytoron (no. 29: πόλις, χωρίον, κατοικία, ἐμπόριον, ὅρμος) and Caroussa (no. 53: πόλις, χωρίον, σάλος, ἐμπόριον καὶ λιμήν).

All told, then, of our 100 coastal sites 81 were settlements, 17 installations, specifically port facilities, and two are simply called fortresses (Table 5).[83] Of the 17 sites classified here as installations, some might have been settlements, as already noted: nos. 76 and 96, for example, are the only two that are mentioned in the Classical period, while the other 15 were created later; besides, Pseudo-Scylax, who mentions them, was generally interested in settlements, but for reasons that have been analysed we include them among the installations.

Table 5. Number of settlements, installations and fortresses mentioned in each period.

	Early Iron Age	**Archaic**	**Classical**	**Hellenistic**	**Roman**
Settlements	1	12	22	28	18
Installations			2	12	3
Fortresses					2

[82] See also Manoledakis 2022a.

[83] Nos. 67 and 87. This does not, of course, mean that there were no other fortifications. Strabo mentions four towns as fortified (nos. 70–72, 80) and Sesamos (no. 25) as the acropolis of Amastris, while the big cities were obviously fortified. He also mentions 75 strongholds on Mt Paryadres, built by Mithridates Eupator, but apparently far from the littoral. Of special interest is the unique temple that is recorded on the coast (no. 65), together with an installation, while the very name of the city Hieron Oros (no. 89) might also point to a sacred place.

2.2. *Diffusion over Time*

An indisputably significant matter is the chronological time-frame of these sites: when did each one appear and how long did they survive? Once again we must recall that, save for a couple of sites, archaeology is unable to provide us with any answers. And the written sources, in the vast majority of cases, furnish only indirect hints, for they usually just name a site, listing it among others, which does not necessarily tell us when it was established. Most of these sites are recorded only by our 'geographical' sources; very rarely do we find additional data about a site's early history. Interestingly enough, however, this happens to be the case with the earliest of them all, Sinope (no. 50). From a late source, Pseudo-Scymnus, we obtain information concerning its early existence that, as extensively presented in Section 4.3.3, enables us to know that a settlement had been created there, in one of the best protected and most strategic points on the southern Black Sea coast, in the 8th century BC at the latest, a settlement that was used, destroyed, and rebuilt by the local Leucosyroi, the Cimmerians and the Greeks well before it became a *polis*. This is the only site that we may confidently claim was settled in the Early Iron Age, that is, before 700 BC (Table 4).

In the Archaic period twelve new settlements appear on the coast,[84] most of them, we observe, in its eastern half. In the ensuing Classical period, the literature records 22 new settlements, plus the first two installations. Most of our sites – 63% – emerged in the Hellenistic and Roman periods (Fig. 14). In the foreword to Table 3 at the beginning of this section, we highlighted the difficulty in deciding in which of the two periods some of the 63 sites were established, as well as the reasons for preferring the Hellenistic in most cases. We now observe that the Hellenistic period was the one with the most new settlements (28) and installations (12), 40 in all, followed by the Roman period, with 23 new sites (18 settlements, three installations, two fortresses). Finally, all the inland sites and those of imprecisely determined location were settlements first recorded in the Roman era, although three were probably Hellenistic.

The population of these sites (Greek or 'barbarian'?) is a question that will occupy us later. For the moment, we merely note that the proportion of sites between the two earlier (Archaic and Classical) and the two later (Hellenistic and Roman) periods roughly corresponds to that of our sources from those periods and the number of sites mentioned by each of them (Table 6). We have

[84] Or eleven, if Hecataeus meant the Bosporan Hermonassa and not the one of no. 91. See comment 17 to the Table.

one Archaic, two Classical, two Hellenistic and seven Roman. In the first two periods there are 43 site references,[85] compared with 238 in the last two, chiefly in the Roman age, when most of the *Periploi* were produced. Anonymous, compiling a text from all of them, most probably in the 6th century AD, mentions a total of 75 sites.

Table 6. The numbers of sites mentioned in each 'geographical source'.

Author	**Coastal**	**Inland**	**Of unknown location**
Hecataeus	8		
Xenophon	7		
Pseudo-Scylax	28		
Apollonius	7		
Pseudo-Scymnus	7		
Menippus	55		
Strabo	24	1	1
Pomponius Mela	16		
Pliny	31	1	
Dionysius Periegetes	1		
Ptolemy	35		
Arrian	56		
Anonymous	75		

Another matter that may be of some interest is the survival of the sites in time, irrespective of any change in their status or even population.[86] Sinope is the champion, since despite changing hands several times it was inhabited uninterruptedly from the Early Iron Age, if not earlier, and throughout the whole of antiquity – and still is (Table 7). Of the twelve sites that seem to have appeared in the Archaic period, eight continued to exist until Roman times, while one (no. 22) soon disappeared. Two were recorded in the Classical period as well but not later (nos. 74 and 81),[87] and one may have reappeared in the Roman period (no. 64), if the identification proposed in Table 3 is correct; in that case, it might have existed uninterruptedly from the Archaic period.

[85] Not 43 different ones, but 43 times a site is named in all the sources of the period.

[86] On this matter, see directly below.

[87] The latter, Choirades, might, according to one theory, have been assimilated to Kerasous or, later, Pharnakia. See comment 15 to Table 3.

Table 7. The survival of the known by name sites in time.

	Archaic	**Classical**	**Hellenistic**	**Roman**
Sinope				

Archaic sites

Heraclea				
Stephanis				
Tieion				
Amisos				
Chadisia				
Teiria (Sotira?)		?	?	?
Stamene				
Cotyora				
Kerasous				
Choirades				
Hermonassa				
Trapezous				

Classical sites

Calpe				
Thynias				
Aconae				
Psylla				
Sesamos				
Cromna				
Cytoron				
Carambis				
Kimolis				
Coloussa				
Stephane				
Tetrakis				
Harmene				
Caroussa				
Lycastos				
Themiscyra				
Iasonia				
Genesintes			?	?

Zephyrios				
Psoron Limen				
Becheiras				
Becheirikos Limen				
Odeinios				
Limne				

Of the 24 sites that seem to have appeared in the Classical period, 16 continued to exist in the Hellenistic and Roman periods, while seven disappeared. For one of them (no. 76) we cannot be sure, since it depends on whether Strabo meant the port or the homonymous cape. Finally, all the sites that were established in the Hellenistic period are also recorded in Roman sources.

Hence, and by combining Tables 5 and 7, we can estimate how many known sites were existent in the littoral in each of the five periods of the 1st millennium BC: 1 in the Early Iron Age, 13 in the Archaic, 34 or 35 in the Classical (it depends on the identification of no. 64 in Table 3), 64 or 66 in the Hellenistic (it depends on nos. 64 and 76) and 90 or 91 in the Roman period.

One cannot, of course, assume that a site with a long history always maintained the same status. We should not take it for granted that all the settlements which are known to have been colonies were necessarily established as colonies from the outset. Sinope is a prime example: initially the site of a settlement created by the local Leucosyrians, it was later inhabited by Greeks, most probably from central Greece, before passing into the hands of the Milesians; it is only these last who established the colony we speak of today. In the previous phase the settlement was most probably an *emporion*, while in its original one it was just a simple local settlement.[88] No one can claim to be able to say how many of all the coastal settlements known today by name had succeeded earlier local settlements, even from as early as the prehistoric period.[89]

Other instances of change of status may underlie more puzzling cases, such as that of Ischopolis (no. 78), which was in ruins in Strabo's day (12. 3. 17) but is later mentioned by Ptolemy (5. 6. 5) as if it were a proper settlement; this suggests a city that had declined for a time but later prospered again. One could also mention Sesamos (no. 25), which after the establishment of Amastris became that city's acropolis (Strabo), although it had previously been a *polis*

[88] See more in Part 4.

[89] As must have been the case with Sinope, for example. Doonan 2004, 147; 2007, 615–16, 618; Doonan *et al.* 2016; 2017, 198; Demirel 2019. *Cf.* also Işın 1998.

(Pseudo-Scylax), or Cotyora (no. 77), which appears as a *polis* in the Classical age (Xenophon) but as a small village in the Roman period (Arrian).

To conclude, the only site classification terms encountered in the Archaic and Classical sources (and persisting later) are πόλις and λιμήν (Table 4). All the others first appear in the Roman texts. If we take into consideration that the earliest among our later texts may also refer to sites created in the Hellenistic period, as previously noted, this observation might well indicate that we can generally divide the urban history of the southern Black Sea littoral in antiquity into two main phases: an Archaic and Classical phase and a Hellenistic and Roman phase.

Thinking in terms of historical context, one cannot but think of the transition from the period of the *poleis* and their people to the period of the rulers, as so distinctly reflected in the change in numismatic inscriptions. In the first phase we have the founding of (relatively) large cities – mostly Greek colonies or secondary colonies – with ports, and perhaps just two installations. In the second, Hellenistic and Roman rulers seem to be interested in creating more numerous, and generally smaller, towns and many installations, attempting among other things to increase the exploitation of the already commercially successful spots on the littoral (and establish new ones) and expand their personal political influence (Fig. 14).[90] Despite the dramatic political changes in the last centuries of the 1st millennium BC, there was a constant escalation of urban activity, so that by the first centuries of the next millennium the littoral would have become densely built and populated, a region bustling with commercial activity, based on the foundations laid by the Greeks through their colonies and their relations with the indigenous peoples several centuries earlier.

2.3. *Greek or 'Barbarian'?*

As regards the 37 sites established in the Archaic and Classical periods, namely in the age of Greek colonisation, can we claim that all of them were products of that activity, that is to say, created primarily or entirely by Greeks? We have only four settlements that are clearly mentioned in the ancient literature as primary Greek colonies: Sinope, Tieion and Amisos, founded by Ionians, and Heraclea, founded by Dorians.[91] Another four are recorded as secondary

[90] I would find it more difficult to talk about an extension of the Greek colonial process in the Hellenistic period (Doonan 2004, 72). For the wider area of Cide (which includes several significant Greek settlements, like Cromna, Aegialos, etc.), see Şerifoğlu and Bakan 2015, 246–59.

[91] For the authors mentioning each of them, see Table 3. See more on Greek colonisation in Section 4.4.

colonies: Cotyora, Kerasous and Trapezous, founded by Sinope, and Thynias, founded by Heraclea. All eight were established as such in Archaic or Classical times. In the latter period two more settlements appear, which could be described as secondary colonies but in slightly different terms: Harmene, as a settlement of Sinope (Xenophon; Strabo), and Calpe Limen as an *emporion* of the Heracleans (Menippus; Anon.). It is usually Xenophon and Pseudo-Scymnus who provide the relevant information. Furthermore, Strabo mentioned that Cytoron was 'once' an *emporion* of Sinope. Finally, later scholars mention towns as belonging to Heraclea (Acrioessa and Hodioupolis), quoting Domitius Callistratus, but we know neither when they were they established nor, in the case of the former (no. 107), what precisely would have been meant by ὑποτελής (subject).

What about the remaining 26 settlements of the Archaic and Classical periods that are not described, directly or indirectly, as Greek? For most of them we may readily assume that they, too, were colonies, for although they are not clearly so labelled in the sources, they are frequently characterised as Greek cities. The phrase *polis Hellenis* is a term regularly used by Pseudo-Scylax, who applies it to all the colonies named as such by other authors, apart from Amisos, Kerasous and Cotyora, which he surprisingly omits, and Thynias, of which he speaks indirectly. What else could such a term mean in the Classical period if not a Greek colony? The best confirmation is provided by Xenophon, who also used this term, together with the term colony, when referring to Heraclea, Sinope, Cotyora, Kerasous and Trapezous.[92]

We may, therefore, be almost certain that the rest of the places called Greek cities by Pseudo-Scylax were also Greek colonies;[93] the only thing we cannot say is whether each of these 'Greek cities' was a primary or a secondary colony. We have, then, 22 Greek colonies, plus the two (Calpe and Thynias) indirectly described as such, so that of the 37 settlements on the coast in the Archaic and Classical periods a total of 24 seem to have been primary or secondary Greek colonies.

What about the other 13 settlements? Nine of them are first mentioned in the Classical period, and it is noteworthy that seven of them are recorded in a way denoting that their primary role was that of a port or *emporion*.[94] In fact, some of them have the word Limen as part of their very name, and some we

[92] The term is sometimes used also by Arrian (for Heraclea, Tieion, Amisos and Trapezous), perhaps influenced by Xenophon.

[93] These are Ameneia (Stamene), Choirades, Sesamos, Cytoron, Carambis, Kimolis, Coloussa, Tetrakis, Harmene, Caroussa, Lycastos, Themiscyra, Iasonia, Becheiras and Hodineios.

[94] Aconae portus, Psylla Limen, Stephane, Genesintes Limen, Zephyrios Limen, Psoros Limen, Becheirikos Limen. The special case of Cytoron could be counted here as well.

have characterised as installations. The leading role of the Greeks in the establishment of these settlements seems almost certain, even if in many cases local participation could have been vital. The other two are Cromna and Limne. For the first, the evidence of texts and inscriptions, and of coins from the 4th century BC,[95] leaves little doubt that it was a Greek city. As far as Limne is concerned, while Pseudo-Scylax, the only author to mention it, simply calls it a *polis*, it is not unlikely that he meant *polis Hellenis*, since this is the only time he uses the word *polis* without that epithet.[96]

We are left with the four settlements of the Archaic period: Stephanis, Chadisia, Teiria and Hermonassa. It is noteworthy that, prior to the Roman period, all four are mentioned only by Hecataeus; Stephanis indeed, is mentioned only by him in all of ancient literature, and perhaps also Teiria, if the identification with Sotira is erroneous. All four are called *poleis*, although not Greek. And here we come to another significant issue. Hecataeus calls Stephanis a city of the Mariandynoi, Chadisia and Teiria cities of the Leucosyroi, and Hermonassa just a city. Setting aside the last-named, can we take this as the earliest evidence of indigenous peoples' cities mentioned by name in Greek literature?

The truth is modern scholarship appears at least undecided, if not confused, about the matter. This is because some of these cities would later be called Greek (*polis Hellenis*) by Pseudo-Scylax, and if the specific passages come from the original Scylax's work, then the problem is even thornier, since he was a contemporary of Hecataeus. The cities in question and Hecataeus' description of them are as follows: Stephanis, city of the Mariandynoi (πόλις Μαριανδυνῶν); Chadisia, city of the Leucosyroi (Λευκοσύρων); Teiria, city of the Leucosyroi; Stamene, city of the Chalybes (Χαλύβων); and Choirades, city of the Mossynoikoi (Μοσσυνοίκων). Three of these, Stephanis, Chadisia and Teiria, are not encountered again before the Roman period, or ever. But

95 See, for example, in Avram *et al.* 2004, 959.

96 Counillon (2004, 44) considers it certain that he meant *polis Hellenis*. Flensted-Jensen and Hansen (1996, 138, 147) claim that when Pseudo-Scylax refers only to a *polis* (without the epithet *Hellenis*), he usually means a barbarian city, though with some exceptions. The authors also suggest (1997, 149 and n. 37) that there are some *poleis* that may have been so designated by Pseudo-Scylax in error. Since one of these is Tieion, which was undoubtedly a *polis*, their suggestion should be regarded with some reserve. The same applies to the classification of the city as a barbarian one (type A:γ) by Avram *et al.* (2004, 960). It has also been held that Pseudo-Scylax is an unreliable source for the south coast of the Black Sea because he sometimes applies the term *polis* to toponyms that apparently denote a river or a mountain – Limne being one such (Hansen 2000, 193; Hansen and Nielsen 2004, 27, with examples in n. 20). But how can the authors know that there was not a homonymous *polis* as well, as was so often the case (e.g. Artanes, Psillis, Elaeus, Cales, Oxines and many others)? For the term *polis Hellenis* in the ancient Greek literature, see also Kacharava 2005, 9–11.

the other two, Stamene and Choirades, are recorded by Pseudo-Scylax as Greek cities.[97]

Indicative of the confusion is, for example, the fact that the same scholars give different interpretations for Hecataeus' characterisation of these two cities within the same essay. They interpret his words regarding Choirades as 'a *polis* among the Mossynoikoi',[98] while a bit later Hecataeus' words cause them to wonder whether Stamene was 'a barbarian *polis*'.[99] But Hecataeus refers to both of them in the same way, as does Pseudo-Scylax, so why should more reliance be placed on the former in the one case and on the latter in the other?[100]

Hansen stressed the fact that Hecataeus' work is preserved only through excerpts quoted by Stephanus of Byzantium (and Aelius Herodianus) and, in an attempt to see how reliable these quotations might be, differentiated between the three methods of citation: placing Hecataeus' exact words in quotation marks ('Hecataeus said that "Choirades is a city in...".'), using indirect speech ('Hecataeus says that...'), and giving the name of the city followed immediately by that of Hecataeus as the reference.[101] As regards our five cities, four of them belong to the last category. Choirades, however, occupies a category of its own, a blend of the first and third, for Stephanus writes: Χοιράδες· πόλις Μοσσυνοίκων. Ἑκαταῖος Εὐρώπῃ·[102] «Τιβαρηνοῖσι δὲ πρὸς ἥλιον ἀνίσχοντα Μοσσύνοικοι ὁμουρέουσιν. ἐν δ' αὐτοῖσι Χοιράδες πόλις.». According to Stephanus, Hecataeus mentions the city in these precise words: 'and in their land (of the Mossynoikoi) (lies) the city of Choirades'. Stephanus interprets Hecataeus' words as meaning 'Choirades was a city of the Mossynoikoi'. But this might be misleading:[103] 'in the land of the Mossynoiki' does not necessarily mean 'of the Mossynoikoi'.

Let us recall that not everything in Pseudo-Scylax, who called Choirades a *polis Hellenis*, belongs to the unknown 4th-century BC author. There are specific, concrete reasons for dating the work to that period: primarily, the reference to some towns that did not yet exist in Scylax's day, while some

[97] Stamene as Ameneia. For Chadisia and Teiria, see also Summerer 2005a, 130; Dan 2012, 75. There is nothing to prove an identification of Teiria with Pteria.

[98] Avram *et al.* 2004, 955.

[99] Avram *et al.* 2004, 963.

[100] The authors eventually include Stamene in their list of Greek *poleis*, so their question seems rather redundant, as does the same question asked by them (Avram *et al.* 2004, 955) about Becheiras, which is mentioned only by Pseudo-Scylax and clearly as a *polis Hellenis*. If Pseudo-Scylax's description is insufficient, then why do they rely on him in the case of Choirades?

[101] Hansen 1997, 17.

[102] Ἀσίαι, in other codices. See *FGrH* 1a 1 F 204.

[103] We are not even sure about the book of Hecataeus cited by Stephanus as containing the specific passage. See the previous note.

characteristics of a political nature suggest that the author might have lived in Athens.[104] Choirades is not one of those cases, and no one can claim with certainty that it does not appear in the original 6th-century BC work; indeed, quite the contrary.[105] It might well, therefore, belong to Hecataeus' era, in which case there is additional reason to interpret his words as 'in the land of the Mossynoikoi', not 'of the Mossynoikoi'.

Xenophon's *Anabasis*, that invaluable source, appears once again to be of vital assistance. Let us see how he describes the Greek cities he encountered: Trapezous, Greek city, colony of the Sinopeans, in the land of the Colchians (*Anabasis* 4. 8. 22: πόλις Ἑλληνίς, Σινωπέων ἀποικία, ἐν τῇ Κόλχων χώρᾳ); Kerasous, Greek city, colony of the Sinopeans, in the land of the Colchians (5. 3. 2: πόλις Ἑλληνίς, Σινωπέων ἄποικος, ἐν τῇ Κολχίδι χώρᾳ); Cotyora, Greek city, colony of the Sinopeans, in the land of the Tibarenoi (5. 5. 3: πόλις Ἑλληνίς, Σινωπέων ἄποικος, ἐν τῇ Τιβαρηνῶν χώρᾳ); Heraclea, Greek city, colony of the Megarians, in the land of the Mariandynoi (6. 2. 1: πόλις Ἑλληνίς Μεγαρέων ἄποικος, ἐν τῇ Μαριανδυνῶν χώρᾳ). This is exactly how both Hecataeus and Pseudo-Scylax describe them; in fact, it is a combination of both. Xenophon wanted to be precise: for each of these cities he made sure to specify that it was a Greek *polis* established in the land of an indigenous people. The preposition used for the latter part is ἐν, the same used by Hecataeus. After all, this was the quintessence of Greek colonisation: the founding of Greek settlements in territories inhabited by indigenous peoples, often with their assent.[106]

It seems clear, therefore, that Choirades was a Greek city;[107] and the same holds true for Stamene (Ameneia). Let me, then, move on to Stephanis, Chadisia, Teiria and Hermonassa, the only four cities of the Archaic and Classical period whose status as Greek or indigenous remains to be determined. The procedure followed for the case of Choirades in Hecataeus on the basis of Hansen's differentiation among the forms of citation in Stephanus of Byzantium is decisive.[108] Stephanus presents Hecataeus' references to Stephanis, Chadisia and Teiria only in the third manner.[109] But since in that same way he

[104] See the Introduction 2.2; Counillon 2004, 24–27.

[105] Counillon 2004, 11.

[106] See Section 4.4.

[107] See also Hansen 1997, 19, 22; Manoledakis 2010a, 138–39.

[108] Even if Hansen made the differentiation to decide about Hecataeus' use of the term *polis* and not about all the information provided by him, which would include the attribution of a city to a specific people.

[109] Στεφανίς· πόλις Μαριανδυνῶν. Ἑκαταῖος Ἀσίαι (*FGrH* 1a 1 F 198); Χαδισία· πόλις Λευκοσύρων. Ἑκαταῖος ἐν Ἀσίαι (*FGrH* 1a 1 F 200); Τείρια· πόλις Λευκοσύρων. Ἑκαταῖος Ἀσίαι (*FGrH* 1a 1 F 201).

had erroneously conveyed Hecataeus' information about Choirades, which he also cited in the first way, I am afraid that he may have made the same error for these three cities as well.[110] In which case, all three of them could also have been Greek.

We are left with Hermonassa (no. 91), a very difficult case, as indicated in comment no. 17 to the Table. The additional problem here is that the city is not ascribed to any people. Stephanus uses Hansens's second manner of citation, namely: 'Hecataeus and Theopompus say that this (Hermonassa) is a city.' Nevertheless, if the reference to Hecataeus is correct, I do not see what else a city by that name in that place in the Archaic period could have been if not Greek.

Consequently, it seems that all the Archaic and Classical settlements of the littoral that are mentioned by name in the ancient literature were either Greek colonies or settlements created by Greek colonies. The greater or lesser presence of local populations in some of them, especially of the latter category, is at the very least likely and expected, for reasons that have already been analysed,[111] while little doubt can remain as regards the ascription of their establishment to Greek colonisation. Equally normal and expected would be the phenomenon of Greek colonies founded in locations previously inhabited by local populations, as the example of Sinope shows.

3. The Anonymous Settlements of the Indigenous Peoples

But what about the settlements of the indigenous peoples in the southern Black Sea littoral? One should not have the impression that the 114 sites reviewed above, plus the several smaller ones belonging to the larger of them, were the only ones existing there in antiquity. They were just the ones that chanced to be recorded by Greek and Roman writers, because they were created in the main by Greeks or by Hellenistic and Roman rulers and they dominated the coast, thus attracting the interest of the authors of the *Periploi*. The littoral, however, was much more densely built and populated, for it was also dotted with settlements of the several indigenous peoples that had inhabited it from a time before the Greek colonisation.[112] But since none of these settlements happened to be mentioned by name by any ancient author and their inhabitants

[110] Which thus might have been, respectively: Στεφανίς· πόλις ἐν τοῖς Μαριανδυνοῖς, Χαδισία· πόλις ἐν τοῖς Λευκοσύροις and Τείρια· πόλις ἐν τοῖς Λευκοσύροις.

[111] See Section 4.4.

[112] See Part 2.

did not write at all, we can have no reliable picture of their approximate number, their possible locations, their proximity to each other and to the Greek and Roman sites, and so on.

There is, however, one Greek author who does mention several local settlements, not indeed by name but still providing substantial information about them: Xenophon. This is an additional reason for considering the *Anabasis* one of the most valuable sources for any student of the southern Black Sea littoral. Let us see what knowledge we obtain from his – in this regard – unique work. Inevitably, it is the eastern part of the littoral that we learn most about, since Xenophon's army spent longer there, but there is nonetheless some information recorded for the rest of the littoral as well. Since Xenophon is our only source on this subject, we shall, just this once, follow his direction and move from east to west.

Just before reaching the sea at Trapezous Xenophon's army took up quarters in numerous villages, which contained provisions in abundance. It is there that they found the celebrated 'mad honey' (*Anabasis* 4. 8. 19–20).[113] Xenophon's formulation is 'ἐν πολλαῖς κώμαις', which indicates a multitude of local settlements, probably Colchian, judging by the ascription of the land around Trapezous to the Colchians.[114]

Of particular interest is Xenophon's reference to local coastal cities to the west of Trapezous and towards Kerasous: he persuaded the (local) inhabitants of the cities situated along the sea (Colchians and Drilae are mentioned around Trapezous) to repair the roads, which were difficult to travel, arguing that they would be rid of the Greeks the more quickly if the roads should be made serviceable (5. 1. 13–14; *cf.* 5. 3. 2). The term used here is πόλεις. Unfortunately, there is no description of these cities, but the mere mention of them, and indeed in the plural, is valuable confirmation that the southern Black Sea coast was not settled only by Greeks.

The settlements in mountainous country, on the other hand, are described in much more detail, since they became theatres of operations. With regard, first of all, to the Drilae (5. 2. 3–27), whose settlements have already been described,[115] we read of several settlements, both fortified and unfortified,[116] for which the term χωρίον is used. The most important of them is termed μητρόπολις; it was naturally well protected, but also had a citadel (ἄκρα). However, while it is usually called χωρίον, at the end of his account of the battle that took

[113] See Part 1.
[114] See Section 2.15.
[115] See Section 2.12.2.
[116] The unfortified ones were probably those that the Drilae themselves burned down, seeing the Greeks approaching, because they considered them easy to capture (5. 2. 3).

place there Xenophon suddenly uses the word πόλις (5. 2. 27).[117] 'Barbarian' χωρία are also mentioned in the mountainous area near Kerasous, in a speech delivered by Xenophon later, in Cotyora (5. 7. 13–17); these are reported to have actually been friendly towards the Greek inhabitants of Kerasous.

Quite a large part of Xenophon's narrative is dedicated to the land of the Mossynoikoi (*Anabasis* 5. 4), thus providing us with an opportunity to learn about the several types of settlements that they had established all over their land, from simple strongholds to villages, cities, and, again, a chief stronghold, described as the Metropolis, which had an acropolis. Here too, despite the predominantly mountainous landscape, coastal or near-coastal settlements must also have existed, even though not directly mentioned, since the Mossynoikoi constructed and used large numbers of wooden boats and, besides, practised fishing, as indicated by their varied use of dolphins.[118]

We also learn something about the density of the Mossynoikan settlements: the average distance between them was, according to Xenophon, 80 stadia (or 14-15 km), and 'the inhabitants could hear one another shouting from one town to the next' (*Anabasis* 5. 4. 31).[119] It is noteworthy that, when providing this information, Xenophon speaks of πόλεις, while in the preceding sentence he had referred to the several χωρία visited by the Greeks as they passed through the land of the Mossynoikoi (5. 4. 30). In addition, while he terms this Metropolis a πόλις he also uses the word χωρίον in close juxtaposition, much as in the case of the Drilae.

The next people to the west whose settlements are mentioned are the Tibarenoi. Here again Xenophon speaks clearly of coastal χωρία (5. 5. 2). This information, combined with what he reports of the Colchians and the Mossynoikoi, shapes an indisputable image of a coast dotted with not only Greek but also local settlements, some of which are actually called *poleis*. It would, therefore, not be unreasonable to wonder why there was never any kind of reference to these settlements in the *Periploi* that have survived. One possible explanation might be that the (Greek) users of these *Periploi* were more interested in Greek settlements, which played a role in trade, than local ones (for which no such activity may be inferred, at least from the sources), or in Greek installations,[120]

[117] For a possible explanation, see below.

[118] See Section 2.11.2.

[119] Doonan (2019, 96) suggests that the distance between the settlements might less, perhaps 5–8 km, if their inhabitants could hear one another.

[120] Not even in every single Greek settlement, as the cases of the χωρία that belonged to the Cotyorites (Xenophon *Anabasis* 5. 5. 6) or the coastal χωρία of Sinope (5. 5. 23) indicate. See above, n. 61.

where they would find better anchorages and repair facilities for their ships – but this is mere speculation.

Then comes a long spatial gap before Xenophon again mentions 'barbarian' settlements, this time in Bithynia, in the area around Calpe. Several κῶμαι are reported there, about 30 stadia (5.5 km) inland – that is to say, in an area of modest elevation. Some of them were larger and therefore two companies led by generals were needed against them, while others were smaller (6. 3. 1–3). Many captives were taken by the Greeks, which suggests quite a large number of villages. This is in fact later confirmed by Xenophon's own words (6. 4. 6) praising the natural benefits of the region, to which he ascribes the 'many inhabited κῶμαι'. It was around these villages that the Greeks would face the horsemen of Pharnabazus, who had come to the aid of the Bithynians (6. 4. 24). Xenophon also reports that some of these villages were linked by a central road, with smaller roads leading to it from the several villages (6. 5. 6), and consistently uses the term κώμη every time he refers to these places (*cf.* 6. 5. 7).

To sum up: we have three terms used by Xenophon to describe local settlements, *polis*, *kome* and *chorion*, except for some few occasions where he indirectly refers to local settlements or just posts without being more specific about them.[121] Let us try now to understand what the author might have meant by each of these terms. As has been noted, the *Anabasis* is the only Classical work in which the word *kome* is widely used[122] and it is applied exclusively to 'barbarian' settlements, since in our sources the term is not used for Greek ones until the Roman period (Table 4).[123] Judging by the information Xenophon provides for the southern Black Sea littoral, most 'barbarian' settlements there were *choria* or *poleis*, contrary to the view that 'most barbarians lived in *komai*'.[124] It is only in Bithynia and Colchis that *komai* appear, yet this does not exclude the existence of other types of settlement in those lands. However, there is no case where Greek *poleis* are mentioned close to 'barbarian' ones, while there are instances of 'barbarian' *komai* being near Greek *poleis*, as at Trapezous (4. 8. 22).[125]

[121] For example, in 5. 1. 9: 'Our enemies have leisure for plundering and they are plotting against us – quite properly, seeing that we have appropriated what was theirs; and they are posted up above us.' Also, in many cases where the Greeks are reported to have made forays in quest of booty, during their stay in foreign lands (e.g. 6. 1. 1). Xenophon also uses the term *polisma* twice, but outside our area of study. See Nielsen 2000, 137–38.

[122] In the whole work, not just the part that concerns us here. For the references of the ancient literature to 'barbarian' settlements in the whole Black Sea, see Kvirkvelia 2005.

[123] Hansen 1995, 51; Nielsen 2000, 133.

[124] Hansen 1995, 78.

[125] *Cf.* Hansen 1995, 79–80.

Furthermore, although the term *polis*, as opposed to *kome*, is widely used for Greek settlements in the literature, Xenophon also applies it to local settlements, usually in order to denote a settlement larger in size than a *kome*.[126] Based on his account, in the area of our study we would accord the distinction of having large cities – *poleis* – only to the Drilae, the Mossynoikoi and the Colchians.

Moreover, while in the Greek world particularly[127] several *komai* might be clustered around a *polis*, this is not necessarily the case with the 'barbarian' *komai*, or with the *choria*, as indicated for example by the case of the Tibarenian ones (5. 5. 2), which, like those that were friendly towards Kerasous (5. 7. 13–17), appear as autonomous urban units. Instances of dependency on *poleis*[128] should not be excluded, however, as the cases of Mossynoikan *choria* could imply. In reality, trying to clarify exact definitions for the terms and the differences between 'barbarian' *choria* and *komai*, is rather like looking for a needle in a haystack. Besides, Xenophon was under no obligation to be accurate and consistent when using all these terms, since his object was not a precise description of the urban landscape of Asia Minor. One may even doubt whether he always used a specific term consciously, at least as regards its meaning in terms of the settlement's size. In fact, *chorion* is used in more than one sense: apart from a town, it can also mean a stronghold or even a wider area – generally, a position or a place; in the *Anabasis*, there are passages where the term can be translated in any of these ways.

The best place to look more closely into this question is the land of the Mossynoikoi, where the terms *chorion* and *polis* are used together more frequently than anywhere else.[129] *Chorion* seems to be used here in several different senses, judging from its context in a sentence or its use in the singular or plural. A position or stronghold must be meant in 5. 4. 2, 14, 15, 23 and 26 (twice). But the next occurrence of the term (5. 4. 27) more likely describes a town, since houses are mentioned where the Greeks saw storerooms with food supplies, of a kind more to be expected in houses than strongholds. The next three references, which occur in the same passage (5. 4. 30–31), may be

[126] *Cf.* Nielsen 2000, 135–37. However, when the terms are used for Greek settlements, the difference may also concern political status. *Cf.* Hansen 1995, 61–62. For the use of the term *polis* for 'barbarian' settlements, see also Hansen 2000, 180–82; Hansen and Nielsen 2004, 36–37.

[127] The term *kome* frequently appears in the plural form, to denote villages scattered over a large territory. See Hansen 1995, in detail; Nielsen 2000.

[128] As in the Greek world: *cf.* the *choria* of the Cotyorites and the Sinopeans (5. 5. 6, 23).

[129] *Chorion*: ten times; *polis*: four times. When describing the land of the Drilae, Xenophon had used the term *chorion* 13 times, but *polis* only once.

considered rather enigmatic, and it is here that we encounter two of the four occurrences of the term *polis*. Here is what we read:

> The Greeks had their breakfast there, and then took up their onward march, after handing over the *chorion* to the Mossynoikoi who had fought alongside them. As for the other *choria* they passed, which belonged to those who sided with the enemy, their occupants were either abandoning them, if they were easily accessible, or voluntarily surrendering them. Most of the *choria* were like that. The *poleis* were 80 stadia distant from one another, some more, and some less; but the inhabitants could hear one another shouting from one *polis* to the next, such heights and valleys there were in the country [*chora*].

In the first instance *chorion* is probably a stronghold, since the specific place has already been mentioned. In the second, where the word is in the plural, it could equally be translated as 'towns'. It is the third one that gives us pause, with its crucial conundrum of how to interpret the words 'Most of the *choria* were like that': does 'like that' refer to what follows, namely the description of the cities and their distances, or to what precedes, namely the *choria* that were surrendered? Both translations have been proposed,[130] and the truth is that both seem correct. But in the former case, the term *chorion* would denote a wider area, to which the *poleis* belonged, while in the latter it could denote either a stronghold or a town, as presented above. In this case, we would have here Xenophon making a distinction between smaller towns and larger cities, which would be of interest as regards the urban environment of the Mossynoikan land. In any case, the Mossynoikoi certainly had *poleis*, as is suggested elsewhere in the text (5. 4. 15, 24), one of which Xenophon alls their Metropolis. Whether the *choria* belonged in any manner to these cities is not clear.

The account of the Drilae's land may be critical. The term *chorion* appears there a total of 13 times, but apart from the first (5. 2. 3), where it is used in the plural for the *choria* of the Drilae generally and may thus encompass positions of any kind, all the others refer to a specific settlement (τὸ χωρίον), this people's most important town, which happened to become the theatre of operations between them and the Greeks. Xenophon calls it Metropolis, as he does the chief settlement of the Mossynoikoi (5. 4. 15), but terms it a *chorion*, whereas he called the latter a *polis*. This in itself is an indication that *chorion* could have quite a broad meaning.

Xenophon's narrative about the battles with the Drilae is rather detailed and so, inevitably, is the description of the metropolis (5. 2. 3–27). This '*chorion*'

[130] See, indicatively, the translation of the *Anabasis* by C.L. Brownson (Loeb, London 1922) for the former; the translations by R. Warner (Harmondsworth 1949) and R. Waterfield (Oxford 2005) for the latter.

did indeed, apparently, merit the label of a city. First of all, it was big enough to shelter all the Drilae in case of war. Moreover, it had an acropolis, reached by a road, with many houses on both sides of it. Within it one could find all manner of supplies (5. 2. 7). Indeed, it seems that the whole city was also fortified, since gates are mentioned not only to the acropolis (5. 2. 23), but also to the city itself (5. 2. 16–17). In the last sentence of his account Xenophon is more explicit: the Greeks eventually managed to withdraw from the *chorion* after burning down the whole *polis* (πᾶσα ἡ πόλις) – houses, towers, palisades and everything else except the acropolis (5. 2. 27). The two terms are again used side by side, as in the account of the Mossynoikan land (5. 4. 23–24). It seems that *chorion* may be used to denote either a part of a *polis* or the *polis* itself, thus serving as a substitute for *settlement*.[131] As for the whole region around the *polis*, Xenophon uses the term χώρα (5. 4. 31).

Be that as it may, from the scant references in our sources to local settlements and populations[132] it seems that most of the latter based their economy more on exploitation of the natural resources of their land than on commercial interaction with other peoples. Any maritime activities mentioned, for example for the Mossynoikoi (*Anabasis* 5. 4. 28), concerned fishing and not trade.[133] And any inland relations with neighbouring peoples, such as between the Mossynoikoi and the Chalybes (*Anabasis* 5. 5. 1), seem to have been mostly of a political and military rather than a commercial nature – or at least this was the impression gained by the Greek authors who wrote about them.[134] Thus it seems that the local populations of the southern Black Sea littoral had a totally different perspective on the broad sea that spread out before them, and the opportunities it offered, than the Greek colonisers.

It would be truly thrilling to have some archaeological evidence to set alongside the information derived from the *Anabasis*. Despite the generally disappointing extent of the archaeological investigation of the littoral, recent surveys conducted in the broader promontory of Sinope[135] show how much more complete our image would be had we more archaeological evidence from the area. We are speaking of just a few years' surveys in only a very small part of the littoral, but the results are still impressive. Several Iron Age settlements have been located on the eastern part of the promontory, roughly

[131] See also Maksimova's remarks (1956, 128–30).

[132] *Cf.* Doonan 2019, 98, table 1, with a categorisation of the local settlements.

[133] *Cf.* Doonan 2019, 93, 96–97; Doonan *et al.* 2016.

[134] See below, for some possible evidence provided by archaeological investigation in the area around Sinope.

[135] Işın 1998; Doonan 2004; 2006; 2009; 2019; Doonan *et al.* 2015; 2017 – all with more bibliography.

between Sinope and Gerze (Caroussa), which have yielded significant amounts of local pottery.[136] The similarities with the indigenous settlement patterns described by Xenophon are notable,[137] especially as regards the existence of highland strongholds with palisades and ramparts straddling both sides of valleys.[138]

Two settlement patterns belonging to the 1st millennium BC have been distinguished in the region. One consists of dispersed networks of settlements (the upland ones up to 5–6 ha) in the foothills and highlands overlooking river-based communication routes. The second indicates an 'engagement of indigenous and local communities through exchange with outsiders at near-coastal sites, and continues via pre-existing indigenous networks up to the mountain communities'.[139] More specifically, the pattern of occupation in the wider Sinop promontory that has been traced across the millennium displays four distinct stages, with the following progression: sparse settlements based mainly on agriculture and only seasonal exploitation of maritime resources; gradual increase in maritime activity due to outsiders bringing in pottery; emergence of larger and sometimes fortified indigenous settlements after the colonisation of Sinope and limited economic connections between the colony and its hinterland; gradual increase in connections and establishment of transportation routes leading inland to the growing indigenous settlements in the highlands.[140]

One may wonder whether the results of a survey can actually provide such a multi-dimensional hypothesis and have reasonable reservations about how much of it might reflect reality, even with ample finds of pottery. Nonetheless, there are some points that deserve attention. First of all, the significance of Greek colonisation for the change in image: not only do the indigenous settlements expand, but their inhabitants are tempted to engage in transactions with the colonists, something they do not seem to have done with their indigenous neighbours (see above). At the same time, however, we begin to see some fortified settlements. This might indicate a concomitant reserve towards the outsiders' departure, which would scarcely encompass only trade opportunities.

136 For all the sites that have been documented in the surveys, see Işın 1998; Doonan 2007.

137 Although far away from the areas described by Xenophon, the wider region around Sinope provides general similarities with the former in terms of resources, climate and landforms (*cf.* Doonan 2019, 94, although with a probable overestimation of the similarities between the two areas, especially the social and economic ones).

138 Doonan 2019, 98–104.

139 Doonan 2019, 105.

140 Doonan 2019, 105.

Moreover, the initial groundwork for trade relations with the newcomers might have been largely laid by 'a population of specialist mobile fishermen'.[141]

Obviously, much needs to be done to help scholars shape a reliable image of the indigenous peoples' spatial organisation both before and after colonisation – systematic excavations, first of all. Be that as it may, the existence of a wide variety of local settlements parallel to the Greek colonial sprawl along the littoral cannot be denied;[142] neither can the Greek authors' obvious disregard of those settlements.

4. Houses and Other Architectural Structures

Knowledge of the names and numbers of settlements and installations may form a bare-bones image of the urban landscape, but how thrilling it would be to have some idea of what they looked like! What were the houses built of? How big were they? What did city streets and urban quarters look like? What sort of port facilities did they have? And what about the evolution of building technology and its products in the southern Black Sea littoral in antiquity? The bibliography on domestic architecture and urban design in the ancient world, especially the ancient Greek world, is huge, and our area of interest should not look much different from the areas that have been studied to date, at least as far as the Greek settlements from the late 6th century BC on are concerned.[143] Therefore, we shall not initiate here an attempt to fully reconstruct the urban environment of the littoral[144] – this would demand a separate monograph – but rather focus on some primary evidence relating to indigenous peoples' constructions.

In our scarcely excavated area, ancient literature proves once again to be of significant help, but recent excavation results from some sites add crucial evidence to the tale. As has already been noted in our surveys of the several indigenous peoples, wood was by far the commonest building material used in the eastern part of the littoral, doubtless due to the abundance of timber in that woody and mountainous region. Xenophon's narrative is quite revealing. The settlements of both the Drilae and the Mossynoikoi were full of wooden houses

[141] Doonan 2019, 93. *Cf.* Doonan *et al.* 2016.

[142] For evidence form the area of Cide, see Şerifoğlu 2015, 221–29.

[143] For the poor documentation of domestic architecture through archaeological evidence before the Classical period, see, indicatively, Tsiafakis 2010, 379.

[144] Unfortunately, not a single one of the settlements has been adequately excavated to provide even a rudimentary image of it. Moreover, of all the *poleis* it is only for Sinope that we have any description at all of its interior in the ancient literature, and that is a really brief one: Strabo (12. 3. 11) wrote merely that the city was beautifully walled, and splendidly adorned with a gymnasium, *agora* and colonnades.

and towers, the latter having after all taken their very name from the wooden constructions they so widely used.[145] These constructions, which have been responsible for some attempts to compare the settlement pattern of the area with that of the Caucasus region,[146] must have disappeared relatively early, together with their creators and inhabitants.

As far as the other parts of the littoral are concerned, there is nothing in the literature to suggest a preference for wooden houses, and only the conclusions drawn from surveys around Sinope could indicate the existence of some wooden constructions.[147] Nonetheless, excavations in Sinope, as well as in Tieion, further west, have recently yielded some evidence of subterranean architecture. Subterranean constructions were quite widespread until the end of the Archaic period over a large part of the world, the northern Black Sea region being among the areas with the greatest wealth of examples – settlements with entire quarters of pit-houses.[148] Subterranean and semi-subterranean constructions, known also as dugouts and semi-dugouts, or pit-houses and semi-pit-houses, were favoured, *inter alia*, by harsh climatic conditions and the absence of building-stone in many areas. Furthermore, not all of them were necessarily dwellings; some might be also have been used for production activities, for example as workshops.

While a semi-dugout is a construction set more than 0.3 m into the ground but with its supporting walls projecting above ground, a dugout burrows deeper: its walls are the sides of the pit, sometimes faced with wood, wattle-and-daub or stone, and the roof cornice is usually at or slightly above ground level. These subterranean constructions could be of various shapes (round, oval, rectangular, trapezoidal), were probably roofed with thatch held together with clay, and had clay-coated floors, sometimes with depressions for cooking vessels and amphorae. Their size might vary from 3 to 14 m^2, and they almost always consist of a single room. In some of them tables and benches have survived along the walls.[149]

The finds from Tieion and Sinope are very recent and the areas where they came to light are still being investigated by the excavation teams. At Tieion, four pit-houses (two rounded and two rectangular) have been found on the acropolis, with a lot of pottery, both local and East Greek. Based on the latter,

[145] See Sections 2.12.2 and 2.11.1, in detail, also about the possible interpretations of the name Strabo cites for the Mossynoikoi: 'Heptacomitae'.

[146] Maksimova 1956, 128–36.

[147] See previous section.

[148] See most recently Tsetskhladze 2021a, with the whole bibliography. For examples from north Greece, see Tsiafakis 2013, with bibliography.

[149] Tsetskhladze 2004; 2021a, with many illustrations.

the constructions are dated to the 7th–6th century BC. Their size varies from 3 to 7 m^2.[150] In Sinope, the evidence so far is as follows: a pit-house was revealed by the excavations of 2015–17 (Fig. 15). It was badly damaged by the construction trench for the Hellenistic city wall, and no pavement was preserved within it. The preserved corner suggests a rectangular shape. It was lined with rounded river cobbles and associated with a series of short-lived surfaces with ceramics primarily originating in the north and west Black Sea.[151] Other pit-houses were excavated in a scarp in the province of Sinop in 2000.[152]

Apart from the issue of their use, two main questions preoccupy the scholarship on subterranean constructions to date: were they autonomous constructions or just the basements of above-ground buildings, and, were they inhabited by indigenous peoples or by Greeks (as well)? From the archaeological standpoint, the answer to the first question rather tends away from the basement view, for in that case the remains from the collapse of the above-ground structures would surely have left some trace identifiable during the excavation, which is nowhere the case.[153] As regards the identity of their inhabitants, the subterranean constructions have always been attributed to the indigenous peoples, but it is widely accepted that the Greeks adopted them in the new areas they came to colonise, which would be natural for newcomers in any case, and especially since the Greek type of house had not yet emerged even in Greece proper. It is indicative that in some cases in the northern Black Sea region, where the vast majority of these constructions have been found, Greek pottery accounts for 89% of all pottery finds.[154]

According to another theory, however, the Greeks of the Black Sea colonies were for the most part Milesians, and since subterranean dwellings were already known to the Greeks in Miletus from their own neighbours there, there was no need for Milesian colonists to copy them from the local Black Sea populations.[155] This theory may sound interesting, but the problem is that the existence of subterranean dwellings in Miletus and its environs is not substantiated. The same holds true for the rest of central and south Greece

[150] Yıldırım 2022, figs. 3–5.

[151] Doonan 2016, 217–18. See also Doonan *et al.* 2016; 2017, 184 and 183, fig. 9.4. There is also a second structure, which was initially thought to be another pit-house but eventually turned out to be a stone-faced glacis built of water-smoothed cobbles topped with mud brick. I am most grateful to the excavator of Sinope, O. Doonan, for providing me with additional information to that given in his publications cited here.

[152] Doonan 2007, 615.

[153] Tsetskhladze 2021a.

[154] Tsetskhladze 2021a, referring to Berezan.

[155] Tsetskhladze 2004; 2021a.

proper.[156] That the post-holes in Miletus 'may indicate the presence of wooden houses but are just as likely to be evidence for pit-houses'[157] is speculation and cannot been proven, at least for the time being. On the contrary, we know for certain that modest above-ground cottages or houses were constructed in Miletus at least from the Late Geometric Period, like the well-known 'huts'.[158] Thus, if the Milesians wanted to take to the colonies the houses and technical know-how they used at home, why would it not be the huts, which, after all, were neither an expensive nor a difficult kind of construction to build, wherever they might find themselves.

Assembling the elements we have from the few subterranean constructions in Tieion and Sinope, as well as their excavators' remarks, the use there of this type of dwelling must have been initiated by the locals[159] and adopted by the Greeks. In Tieion both Greek and local pottery was found in the pit-houses, but the majority was local.[160] In the constructions of Sinope no Greek material has been located so far, and only a couple of Iron Age painted sherds were mixed with the fills adjacent to the construction excavated in 2015, which could just indicate that it might fall into the colonial phase of the site but not necessarily that it was constructed by colonists.[161] The pottery dates to the Iron Age,

> including fragments of an Iron Age painted bowl rim, and a burnished faceted handle with parallels in the Iron Age Bafra plain, which was found just outside the walls of this structure. The other ceramics from this house were burnished, low-fired red and black handmade wares with shell temper. One sherd was decorated with a horizontal band of finger impressions.[162]

According to the excavator, these pit-houses constitute 'the first clear, precolonial, settlement evidence in the urban area of Sinop'.[163] Moreover, 'stone

[156] For the subterranean dwellings dating from the 8th–7th centuries BC that have been excavated at Karabournaki, a settlement close to Thessaloniki, which must have been inhabited by both local Thracians and Greek colonists, it has been argued that they represented the local (Thracian) element of the population (Tiverios 2008, 28), but were probably soon employed by the Greeks as well (Tsiafakis 2010; 2013).

[157] Tsetskhladze 2021a.

[158] The relevant bibliography is huge. See, indicatively, Morris 1998; Mazarakis-Ainian 2001; Lang 2005. For Miletus in specific, see Senff 2000.

[159] Trying to identify the locals in these regions is a risky proposition. For which indigenous peoples inhabited the areas in question, see Part 2. For whether the Phrygians might have belonged to them (*cf.* Tsetskhladze 2021), see Section 3.4.

[160] Information provided in October 2020 by the excavator, Ş. Yıldırım, to whom I am most grateful. See also Yıldırım 2022.

[161] Information provided in July 2020 by the excavator, O. Doonan, to whom I am most grateful.

[162] Doonan *et al.* 2017, 184.

[163] Doonan 2016, 217–18. See also Doonan *et al.* 2017, 183–84.

architecture is not characteristic of non-Greek settlement in Sinop province, where wattle-and-daub pit-houses predominate'.[164]

The excavation team's working hypothesis is that the houses with the smoothed cobbles were built and rebuilt in many phases, before and maybe parallel to the Greek colonisation of the region.[165] The earliest finds from the site belong to the Early Bronze Age, and the ceramics are diverse. In addition to ceramics, an Early Bronze Age incised spindle whorl, a fish-hook, incised bone plaques, fish bones and sheep bones were documented in the pit-house excavated in 2015–17. The excavators suggest that the site of the pre-colonial village demonstrates a practice of fishing from a high lookout that traces all the way back to the Bronze Age and corresponds to later fishing practices recorded by Aelian in the 2nd century.[166] Thus, they are postulating a long-standing fishing station used on a seasonal basis by itinerant fisher folk, probably hailing from the north and west.[167]

5. Roads

One could argue that this section should have appeared earlier, before discussion of the settlements. For there persists the difficult question of which came first, the settlements or the roads. In his extremely influential book on the Mediterranean, F. Braudel implied the problem by originally entitling a corresponding chapter 'Routes et villes, villes et routes', and by writing that 'The Mediterranean city creates routes and at the same time is created by them.'[168] As I see it, it is a matter, among other things, of definition and size. A distinction should be made between natural routes, created by the landscape, and roads, constructed by people, as well as between big cities and small settlements, installations, etc. Moreover, each case should be considered on its own terms: the Mediterranean and the southern Black Sea littoral are as dissimilar as the 16th century AD and the 1st millennium BC.

164 Doonan 2007, 615.

165 Similar are the assumptions of the Tieion excavators. Again, I am much obliged to both O. Doonan and Ş. Yıldırım for being always willing to share with me in detail their views on the matter and more information about their excavation results.

166 Doonan 2016, 217–18.

167 Because 'the handmade ceramics with horizontal bands of finger-impressed decoration were made of non-local pastes that showed the closest parallels to examples from the north and west coasts of the Black Sea. Also, the structures resemble the dug-out and semi dug-out houses from the same region' (Doonan *et al.* 2016).

168 Braudel 1972, 276–77. It is interesting that in the English translation the word is 'roads', but the original word used by Braudel is 'routes', a word that the translator maintains on other occasions.

In any case, natural routes play the foremost role. They are crucial to the establishment of big cities. Once a city exists and has reached a degree of development, the construction of roads leading to it – on natural routes or otherwise – is open, for I assume that a community would not expend effort and resources on constructing roads were they not sure that the place they had chosen to establish was an appropriate one – let alone if they already had good sea routes to serve them. The existence of such roads would in turn certainly favour and cause the creation of smaller settlements along them, just as the sea routes would determine the location of port installations and facilities. The southern Black Sea littoral lacks good natural overland routes, for reasons that have already been pointed out, and many of the north–south ones that exist are determined by the routes of several rivers that flow into the Black Sea.[169] Be that as it may, it is mainly roads that are discussed here, as a sequel to discussion of the settlements.

A littoral fringed with a lacework of settlements, from cities that functioned as busy 'international' harbours to installations and small villages, both Greek and local, would be expected to have several adequate roads connecting them along the coast – all the more so since most of these settlements were closely interconnected, for example with some being founded by others. Stressing the crucial importance of roads for the existence of settlements, Braudel claimed that 'a map of the cities closely corresponds to a map of the roads'.[170] Even if he was referring primarily to significant cities, the southern Black Sea littoral had several of those, and besides, there are numerous specific narratives of overland movements between coastal settlements, the most celebrated of these being, of course, the march of the Ten Thousand. Nevertheless, the existence of good coastal roads does not seem to have been the case at any time in the whole 1st millennium BC.

First of all, in our coastal area of study, with the additional particularity of mountainous or rocky terrain often reaching the sea, tracing ancient roads is a difficult task, as difficult as tracing the ruins of the settlements and installations.[171] Many of the latter may be submerged now, others overbuilt, and still others destroyed by various agencies over the last two millennia. The same holds true for the roads. Apart from the assistance provided by the relief and by the spatial arrangement of the settlements, other ways to trace an ancient road are archaeological evidence, such as milestones,[172] pavements and

[169] See below and Part 1.
[170] Braudel 1972, 312–19.
[171] See above, in the introduction to Section 5.1.
[172] See most recently French 2012; 2013; 2016a, with bibliography.

bridges,[173] and literary descriptions, including itineraries.[174] But examples of the former only start further inland, and here it is only the latter that may shed some light on the coast.

To begin with, one should roughly differentiate between two – at least – main categories of overland road: trunk roads, serving 'international' commercial purposes, and minor, local roads, primarily connecting neighbouring settlements and other places of interest.[175] Attempts to define main transport axes in northern Anatolia in antiquity, and chiefly in the Roman period, began in the 19th century. This first phase culminated with the work of J.A.R. Munro, who wrote about the 'trunk road' from Niksar to Taşköprü and the 'commercial road' that intersected it, leading to the Black Sea, at Amisos.[176] After recent archaeological surveys, Munro's reconstruction has been slightly amended,[177] while other variants of this road, or even different east–west roads in northern Anatolia, starting from the Bosporus or the Propontis and extending as far as the Caucasus and central Asia, have also been proposed as being in existence from the Bronze Age.[178] However, none of these actually traversed the littoral, but rather ran parallel to it, to the south of the high mountain summits, at a distance varying roughly from 50 to 110 km from the coast as the crow flies.[179] In fact, it seems that these routes were more or less defined by the south slopes of the mountains, that is, the North Anatolian Fault, as a simple look at a physical map of Anatolia confirms.[180] Thus, stressing the significance of these roads to the fruition of Greek colonisation in the southern Black Sea region[181] might sound bizarre.

If this is so, then what kind of roads were used by Xenophon and his army to march from Trapezous to Cotyora (*Anabasis* 5), or by the several people of

[173] Since the rivers of our area of interest were mostly impassable. See, for example, Xenophon *Anabasis* 5. 6. 9. *Cf.* Bryer and Winfield 1985, 18.

[174] For a more detailed procedure in searching for ancient roads, see Starr 1962; French 1974.

[175] For a more specific and wider categorisation, see French 1974, 144.

[176] Munro 1901, with the previous bibliography, chief in which is Ramsay (1890, 27–82). *Cf.* Maksimova 1951, 79.

[177] See Bekker-Nielsen and Czichon 2015.

[178] Ramsay 1890, 35–62; Burney 1956; Birmingham 1961; Danoff 1962, 1046–47; Mellink 1974, 105; Winfield 1977, 152, 158–59; Özgüç 1980, 305; French 1982, 26–27; Bryer and Winfield 1985, 20–39, with all the variants; Marek 1993, 65; Boardman 1999, 240; Easton *et al.* 2002, 104; González Salazar and Lucia Castejón 2008, 11–17; van Dongen 2013.

[179] Which in that mountainous terrain of course means a much greater distance on the ground.

[180] Among the main points on them are Gerede, Vezirköprü, Amasya, Niksar, Koyulhisar, Sadak and Erzurum: Maksimova 1956, 14; Winfield 1977; Bryer and Winfield 1985, 20–39; Lendle 1995; Bekker-Nielsen and Czichon 2015. Many important cities would be founded on this route in the Hellenistic and Roman periods, like Nicomedia, Prusias, Bithynion, Pompeiopolis, Amaseia, Comana, Nicopolis, etc.

[181] Lenschau 1913; Miltner 1939; *cf.* Saprykin 1997, 26.

the littoral who occasionally came to meet them (for example, the Sinopean ambassadors who came to Cotyora to talk to them, as reported in *Anabasis* 5. 5. 7)? Xenophon's narrative provides some clues. Most telling is the fact that, at almost every station on their route along the southern Black Sea littoral (Trapezous, Kerasous, Cotyora, Cape Acherousias near Heraclea), the Greek soldiers wondered and debated whether they would do better to proceed by way of the inland roads or by sea. This is a strong indication of the unsuitability of the east–west roads near the coast. Sometimes this is made explicit, as in the case of the roads leading west from Trapezous: Xenophon reports that they were difficult to travel, and thus he had to persuade the inhabitants of the cities situated along the sea to repair them, arguing that they would be rid of the Greeks the more quickly if the roads should be made serviceable (5. 1. 13–14, *cf.* 3. 2). Furthermore, in Cotyora, Hecatonymus, the ambassador sent by the Sinopeans to negotiate with the Ten Thousand, clearly warns them that the journey to Sinope by land 'is not merely difficult, but a thing of utter impossibility' (5. 6. 10).[182] This is because the land to the west of Cotyora is full of lofty mountains with very narrow passes, and large impassable rivers. Hecatonymus therefore strongly recommends that they continue by sea (5. 6. 5–9).

In other cases, we read of the Greeks taking the road from Kerasous to Cotyora, through the lands of the Mossynoikoi, the Chalybes and the Tibarenoi (5. 4. 1; 5. 5. 1–3), with no information about the status of the road. However, the distances given by Xenophon seem excessive and suggest that the Greek army most probably turned away from the coast and followed a long inland curve before reaching Cotyora,[183] which means that there was no good coastal road.

Things seem to have been better in the western part of the littoral, in Bithynia, where indeed the land is much flatter. Thus Cheirisophus appears to have had no trouble in marching from Heraclea to Calpe by the coastal road, especially 'after entering Thrace',[184] nor does Xenophon in traversing the Thracian (Bithynian) countryside (6. 2. 18–19; 6. 3. 10). The same is indicated by the whole account of the Greeks' movements in Bithynia (6. 3; 6. 5. 6; 6. 6. 38[185]). This in fact is the land where we first hear of a main road linking some of the villages, with local roads leading to it from the others (6. 5. 6). Thus, the existence of a road from Calpe to the south may be considered very possible.

[182] *Cf.* Hamilton 1842, 244.

[183] See Section 2.9.1 in more detail.

[184] Probably after the Hyrius or the Sangarius: see Sections 2.2.1 and 2.4.1. For this road, see Lendle 1995, 377–78.

[185] For the inland road mentioned in this last passage of Book 6, between Calpe and Chrysopolis, see Lendle 1995, 410.

All in all, it appears that, for most of the 1st millennium BC at least, the greater part of the littoral did not have good coastal roads. 'Travel along much of the southern Black Sea coast cannot have been at all easy';[186] it has even been suggested that the coastal settlements between Heraclea and Amisos were to all intents and purposes cordoned off.[187] The preference for sea routes implied by the Argonautic myth was the natural consequence.[188] Travelling by sea from the western to the eastern end of northern Anatolia was, moreover, faster, cheaper and safer.[189] Indeed, to the commercial use of these sea routes 'the Romans added a serious naval and military purpose'.[190] This becomes apparent with the plethora of relevant settlements and installations created in the Roman period[191] and the interest of the Romans in inspecting them, which is clearly demonstrated in Arrian's *Periplus*. This in turn implies that until the Roman era no good road had been built along the coast, and it is only in the Late Imperial period that the several itineraries demonstrate regular communication between the coastal cities.[192] Therefore, when Apollodorus states, according to Strabo (12. 3. 26), that Homer – in the early 7th century BC – knew of the Paphlagonians from men who had passed through their country on foot, he was certainly referring to the main roads in the hinterland, south of the mountains.

This does not mean that no coastal roads existed. Xenophon's statement that the roads were rough and needed repair indicates that the roads were there, although evidently little used, at least not in 400 BC, long after the founding of the colonies. Such being the case, links to the east–west trunk roads would be vital to the prosperity of the larger coastal towns. Thus we shall be looking for north–south routes leading inland from the coast. As Braudel points out, 'all ports, by definition, stand where land and water meet. Every one stands at the end of a road or inland waterway.'[193] Notwithstanding the fact that the most significant ports on the southern Black Sea were products of Greek colonisation, which was accomplished via the sea, Braudel's axiom is something that one would expect to have been taken into account in the establishment of

[186] Burney 1956, 180; Bryer and Winfield 1985, 18–19, with reports about the specific difficulty dated from much later periods.

[187] Lendle 1995, 339. *Cf.* Marek 1993, 9–10.

[188] Winfield 1977, 153; González Salazar and Lucia Castejón 2008, 11–17. *Cf.* Lendle 1995, 296.

[189] Bryer and Winfield 1985, 18.

[190] Winfield 1977, 153.

[191] See Section 5.2.

[192] For the coastal roads and their stations presented in the Roman itineraries, see most recently French 2016b, especially the maps on pp. 8, 9, 83–86, with bibliography. See also French 1980 (e.g. p. 710 for a coastal road between Amastris and Sinope).

[193] Braudel 1972, 316–17.

the big port cities, even in cases where the landscape was not so favourable. In the wider area around Sinope, for example, the recent archaeological investigations suggest a network connecting the numerous local settlements, which became busier after the establishment of the Greek colony.[194] The status of these roads – whether most of them were just natural routes – is indeterminable, as is the degree of human intervention. But we can at least imagine a rudimentary network that was improved in the Roman period, as indicated in the itineraries.[195]

The truth is that there are very few cases of well-attested roads leading from significant port cities into the hinterland and the trunk road beyond the mountains. The most incontrovertible among them was favoured by the natural relief: constituting such a uniquely privileged natural access to the hinterland, the valley along the Halys, as well as the one along the Iris, would inevitably be developed as significant overland routes, all the more since these, as two of the very few navigable rivers on the littoral, were also potential river routes.[196] Consequently, the establishment of an important settlement on the coast at the northern end of this route (Amisos) was only a matter of time. This road (or roads?)[197] must have played a vital role in the interaction between several cultures and civilisations that flourished in the central part of northern Anatolia.[198]

With the road leading to Trapezous the case is quite different. Although the city was also established at the mouth of a river,[199] the latter's course through the steep mountains was much more difficult and formed no notable valley. The road nonetheless became famous, thanks to Xenophon and the accounts of many subsequent travellers. Frequent attempts have been made to reconstruct the route of the Ten Thousand north to Trapezous via Gymnias, Mt Theches and the land of the Macrones (*Anabasis* 4. 7–8): Bayburt, Torul, Zigana, Maçka, Trabzon.[200] It was probably on the roads of this mountainous region that the Mossynoikoi placed their famous crazing honey, to poison and subsequently attack Pompey's troops (Strabo 12. 3. 18; *cf.* Xenophon *Anabasis* 4. 8.

[194] Doonan 2019, 98–106.

[195] For roads leading to the south of Sinope, see Bryer and Winfield 1985, 39–40.

[196] Bryer and Winfield 1985, 18–19; French 2016b, 28–29. In contrast, Maksimova (1956, 14–15) supported the position that these rivers actually made communication with the hinterland more difficult rather than easier.

[197] See Bryer and Winfield 1985, 40–41, for other possible roads on the valley.

[198] Later, this road would benefit other significant cities, such as Amaseia, the capital of the Pontic kingdom. See, indicatively, Ramsay 1890, 27–28; Munro 1901, 53; Burney 1956, 193; Saprykin 1997, 24; Summerer 2005a, 147; 2007, 29; Tsetskhladze 2013, 77; French 2016b.

[199] The Değirmendere, which has erroneously been identified with the Pyxites. See Part 1, Table 1.

[200] E.g. Munro 1901; Winfield 1977, 155–57; Bryer and Winfield 1985, 48–53; Lendle 1995, 250–87; French 2016b, 28. See also Section 2.6.3.

20–21).[201] And it was probably one of these roads that led from the Metropolis of the Drilae to Trapezous and was taken by the Greeks after their battle with them. According to Xenophon, this road was steep and narrow (*Anabasis* 5. 2. 28), and it was certainly not the only one like that in the region. This is an indication that most of the roads leading to the sea could not have been especially commodious.

Other significant settlements on the coast also had access to the hinterland, but not necessarily by way of a river. A. Bryer and D. Winfield pointed out that ancient north-south routes did not necessarily follow the valley bottoms in their courses through the mountains, 'and a mountain crossing was often best made by keeping along a winding route on gradually ascending ridges', which could be considered safer.[202] As examples they mention the roads from Oenoe (Ünye) and Kerasous (Giresun),[203] both of which, however, lie on river estuaries. In such cases, it is usually the existence of known settlements or strongholds, mainly Hellenistic, to the south of the coastal sites that leads to the assumption of the existence of a road connecting them. Sometimes these routes are confirmed by the Roman itineraries, as is the case with the route from Artane to Nicomedia and Prusias ad Hypium, and the two different routes from Polemonion, or even Phadissa, to Neocaesaria and Nicopolis.[204]

The establishment of cities like Prusias ad Hypium, Bithynion, Neocaesaria and Nicopolis on or near the trunk road makes it difficult not to speculate about the existence of roads connecting them to the sea, and sites like Dia and Lillaion, Cotyora and Kerasous (Giresun), all built at river mouths, are perfect candidates for the north ends of these roads.[205] However, we know that Cotyora had lost its significance by the 2nd century BC at the latest, as Arrian (*Periplus* 16) notes. Conversely, the establishment of new cities on the coast (e.g. Amastris, Abonou Teichos) or the continued significance of other big cities (e.g. Tieion, Heraclea) in the Hellenistic and Roman periods would be good reasons for the creation of roads connecting them with the trunk road. Especially as regards Tieion, its location near the mouth of the Billaeus is highly advantageous, since the western edge of the Paphlagonian mountain range allows a natural pass from the Billaeus (Yenice) via the Devrek Çayı and the Bolu Stream to the plain of Bolu; this is considered the equivalent of the Halys valley marking the eastern edge of the range, but is of much greater

[201] For the crazing honey, see Part 1.
[202] Bryer and Winfield 1985, 39, 41–42, 45–46.
[203] See also Munro 1901; Winfield 1977, 155–57.
[204] Munro 1901; Winfield 1977, 155–57; Bryer and Winfield 1985, 42–44; French 2016b, 28–29.
[205] E.g. Bryer and Winfield 1985, 44–46.

accessibility.[206] However, there is no positive evidence for such roads earlier than the mediaeval period. In other cases, the existence of roads leading south from coastal settlements sited at the mouth of a usually homonymous river is merely hypothetical, even for later periods.[207]

We might thus deduce that most of the littoral's roads were not widely used or even created until the Hellenistic and Roman periods. If we look at the 35 settlements established earlier (the two installations did not necessarily need inland access),[208] we observe that very few had the privilege of a convenient road connection with the hinterland, and sometimes not even with other coastal settlements. In fact, the only one that certainly had such a feature was Amisos, while Tieion and Trapezous at least had the natural potentiality for it, as did Calpe and Themiscyra. The site of Sinope, the earliest and most important Greek settlement, was apparently selected for its outstanding possibilities regarding sea routes and safe docking, and any inland roads that may have existed at the time of its founding were of absolutely secondary significance.

It is therefore tempting to question whether every coastal settlement was in fact accessible by any road, inland or coastal. Even the establishment of the secondary colonies might have taken place primarily by way of the sea, and only in some cases of nearby settlements, like Harmene and Caroussa in relation to Sinope, might coastal roads have been created or repaired. Indeed, Xenophon does not specify how the ambassadors from Sinope came to Cotyora to meet the Ten Thousand (*Anabasis* 5. 5. 7), and it is highly likely that they arrived by ship. As for the installations, and perhaps even some small settlements, they might not have been created at the end of a road or inland waterway but just at a place that was considered a safe station on a sea route, and, for the rest, inaccessible by land. It would be no exaggeration to say that, before the Hellenistic period, the littoral, being naturally cut off from the hinterland, was consequently also to a great degree politically isolated from the rest of Anatolia;[209] and the absence of convenient roads must have played a crucial role in this.

[206] Marek 1993, 10.

[207] For example, from the areas of Zephyrios Limen, Tripolis, Kerasous (Vakfıkebir), Hermonassa, Hyssos, Ophious, Rhizeon, Athenai, Xyline: Munro 1901; Winfield 1977, 155–57; Bryer and Winfield 1985, 46–48, 53–57. Hypothetical is also the existence of a road leading to the south of Themiscyra (Munro 1901; Winfield 1977, 155–57; Bryer and Winfield 1985, 41). Despite the existence of the Thermodon, there is no evidence of such a road, at least as late as the Roman period.

[208] See Section 5.2.2 and Table 5.

[209] Marek 1993, 9–10.

BIBLIOGRAPHY

Adrimi-Sismani, V., Guerra, M.F. and Walter, P. 2009: 'La tombe mycénienne de Kazanaki (Volos) et le mythe de la Toison d'or'. *ArchéoSciences* 33, 135–41.

Aksoy, E. and Yıldırım, Ş. 2017: 'Rise and Fall of Tios-Tieion'. *IOP Conference Series: Materials Science and Engineering* 245, 1–10.

Akurgal, E. 1955a: *Phrygische Kunst* (Ankara).

—. 1955b: 'Summary of Archaeological Work in Turkey, 1954. Sinope'. *AS* 5, 23.

—. 1956: 'Die Ausgrabungen von Sinope'. *Türk Arkeoloji Dergisi* 6.1, 54–61.

—. 1967: 'Eine Silberschale aus dem Pontus'. *Antike Kunst* 10, 32–38.

Akurgal, E. and Budde, L. 1956: *Vorläufiger Bericht über die Ausgrabungen in Sinope* (Ankara).

Akyüz, U. 2013: 'Samsun-Kurupelit'de Ele Geçen Bir Grup Mezar Steli ve Çakalca-Karadoğan Höyük Üzerine Bir İnceleme'. *Kubaba* 22, 33–50.

Algan, O., Ergin, M., Keskin, Ş., Gökaşan, E., Alpar, B., Ongan, D. and Kırcı-Elmas, E. 2007: 'Sea-level changes during the late Pleistocene-Holocene on the southern shelves of the Black Sea'. In Yanko-Hombach *et al.* 2007, 603–31.

Alkım, U.B., Alkım, H. and Bilgi, Ö (eds.), *İkiztepe Üçüncü, Dördüncü, Beşinci, Altıncı ve Yedinci Dönem Kazıları 1976–1980* (Türk Tarih Kurumu yayınlarından. V. dizi 39a) (Ankara).

Allen, T.W. 1921: *The Homeric Catalogue of Ships* (Oxford).

—. 1924: *Homer: The Origins and the Transmission* (Oxford).

Antonaccio, C.M. 2001: 'Ethnicity and Colonization'. In Malkin, I. (ed.), *Ancient Perceptions of Greek Ethnicity* (Cambridge, MA/London), 113–57.

Arnaud, P. 2015: 'Marcus Vipsanius Agrippa and his Geographical Work'. In Bianchetti *et al.* 2015, 205–22.

Asheri, D. 1972: 'Über die Frühgeschichte von Herakleia Pontike'. In Asheri, D., Hoepfner, W. and Erichson, A., *Forschungen an der Nordküste Kleinasiens* 1 (Denkschriften, Österreichische Akademie der Wissenschaften. Philosophisch-Historische Klasse 106; Ergänzungsbände zu den Tituli Asiae Minoris 5) (Vienna), 11–34.

—. 1998: 'The Achaeans and the Heniochi. Reflections on the Origins and History of a Greek Rhetorical Topos'. In Tsetskhladze 1998b, 265–85.

—. 2007: 'General Introduction'. In Asheri, D., Lloyd, A. and Corcella, A., *A Commentary on Herodotus Books I–IV*, ed. O. Murray and A. Moreno (Oxford), 1–56.

Aslan, E. 2014: 'Port of Kalpe in the Bithynian Region'. *Olba* 22, 129–53.

Atasoy, S. 2003: 'Amisos'. In Grammenos, D.V. and Petropoulos, E.K. (eds.), *Ancient Greek Colonies in the Black Sea [1]*, vol. 2 (Publications of the Archaeological Institute of Northern Greece 4) (Thessaloniki), 1331–77.

—. 2016: 'Excavations at Tios: 2006–2015'. In Manoledakis, M. (ed.), *The Black Sea, in the Light of New archaeological Data and Theoretical Approaches* (Proceedings of the 2nd International Workshop on the Black Sea in Antiquity, Thessaloniki, 18–20 September 2015) (Oxford), 207–16.

—. 2018: 'Surveys and excavations on the southern Black Sea coast'. In Manoledakis *et al.* 2018, 109–46.

—. 2019: 'The Baruthane Tumuli at Amisos/Samsun'. In Tsetskhladze and Atasoy 2019, 167–76.
Atasoy, S. and Erpehlivan, H. 2012: 'Filyos-Tios Kenti: İlk yerleşmeye ait keramik buluntular'. In Şahin, I (ed.), *İ.Fazlıoğlu Anı Kitabı* (Edirne), 1–11.
Atasoy, S. and Ertuğrul, Ö. 1998: '1996 Amisos Kazısı'. In *XIX. Kazı Sonuçları Toplantısı. 26–30 Mayıs 1997 Ankara* (Ankara), 523–30.
Atasoy, S. and Yıldırım, Ş. (eds.) 2015a: *Tios: Zonguldak'ta Bir Antik Kent. 2006–2012 Arkeolojik .alışmaları ve Genel Değerlendirme/ Tios: An Ancient City in Zonguldak. General Assessment of Works Between 2006 and 2012* (Ankara).
—. 2015b: 'Recent discoveries at Tios and its territory'. In Tsetskhladze *et al.* 2015, 441–44.
Ateşoğulları, A. and Şimşek, Ö. 1995: 'Bartın İli Amasra İlçesi, Küçüksanayi Sitesi İnşaatında Bulunan Roma Dönemi Heykellerini Kurtarma Kazısı'. In *V. Müze Kurtarma Kazıları Semineri. 25–28 Nisan 1994* (Didim/Ankara), 93–111.
Aujac, G. 2015: 'The "Revolution" of Ptolemy'. In Bianchetti *et al.* 2015, 313–34.
Avi-Yonah, M. 1962: 'Scythopolis'. *Israel Exploration Journal* 12.2, 123–34.
Avram, A. 2007: 'Some Thoughts about the Black Sea and the Slave Trade before the Roman Domination (6th–1st Centuries BC)'. In Gabrielsen and Lund 2007, 239–52.
—. 2018: 'Paphlagonian Notes'. *AWE* 17, 65–81.
—. 2021: 'Studies in Pontic epigraphy 1997–2017. Reviews and prospects'. In Tsetskhladze *et al.* 2021, 10–18.
Avram, A., Hind, J. and Tsetskhladze, G. 2004: 'The Black Sea Area'. In Hansen, M.H. and Nielsen, T.H. (eds.), *An Inventory of Archaic and Classical Poleis* (Oxford), 924–74.
Bäbler, B. 1998: *Fleissige Thrakerinnen und Wehrhafte Skythen: Nichtgriechen im klassischen Athen und ihre Archäologische Hinterlassenschaft* (Beiträge zur Altertumskunde 108) (Stuttgart).
Ballard, R.D., Coleman, D.F. and Rosenburg, G. 2000: 'Further evidence of abrupt Holocene drowning of the Black Sea shelf'. *Marine Geology* 173, 253–61.
Ballard, R.D., Hiebert, F.T., Coleman, D.F., Ward, C., Smith, J.S., Willis, K., Foley, B., Croff, K., Major, C. and Torre, F. 2001: 'Deepwater Archaeology of the Black Sea: The 2000 Season at Sinop, Turkey'. *AJA* 105.4, 607–23.
Baralis, A., Bivolaru, A., Marriner, N., Morhange, C., Porotov, A. and Zin'ko, V. 2016: 'Introduction, Geomorphology and geoarchaeology of Black Sea coasts'. *Méditerranée* 126, 3–14.
Barat, C. 2012: 'La ville de Sinope, réflexions historiques et archéologiques'. In Kassab Tezgör 2012, 25–64.
Barnett, R.D. 1975: 'Phrygia and the Peoples of Anatolia in the Iron Age'. *CAH* II.2^3, 417–42.
—. 1982: 'Urartu'. *CAH* III.1^2, 314–71.
Baş, A. 2018: 'Finds from the Excavations at the Foundation of the Sinop Teacher Training High School'. In Yiğitpaşa *et al.* 2018, 93–101.
Baschmakoff, A. 1948: *La synthèse des périples pontiques* (Paris).
Bauer, A.A. 2006: 'Between the Steppe and the Sown: Prehistoric Sinop and Inter-Regional Interaction along the Black Sea Coast'. In Peterson, D.L., Popova, L.M. and Smith, A.T. (eds.), *Beyond the Steppe and the Sown: Integrating Local and Global Visions* (Colloquia Pontica 13) (Leiden/Boston), 227–46.

—. 2014: 'The "Sea Peoples" as an emergent phenomenon'. In Galanakis, Y., Wilkinson, T.C. and Bennet, J. (eds.), *ΑΘΥΡΜΑΤΑ. Critical Essays on the Archaeology of the Eastern Mediterranean in Honour of E. Susan Sherratt* (Oxford), 31–39.

Bauer, A.A. and Doonan, O.P. 2012: 'Fluid Histories: Culture, Community, and the *Longue Durée* of the Black Sea World'. In Ivan, R. (ed.), *New Regionalism or No Regionalism? Emerging Regionalism in the Black Sea Area* (Burlington, VT), 13–30.

Baumann, H. 2011: *Flora mythological: Griechische Pflanzenwelt in der Antike* (Akanthus Crescens 8), 2nd ed. (Kilchberg).

Baynham, E. 1998: *Alexander the Great: The Unique History of Quintus Curtius* (Ann Arbor).

Beal, R.H. 2011: 'Hittite Anatolia: A Political History'. In Steadman and McMahon 2011, 579–603.

Beaumont, R.L. 1936: Greek Influence in the Adriatic Sea before the Fourth Century BC'. *JHS* 56.2, 159–204.

Beckman, G. 2011: 'The Hittite Language: Recovery and Grammatical Sketch'. In Steadman and McMahon 2011, 517–33.

Bekker-Nielsen, T. (ed.) 2005: *Ancient Fishing and Fish Processing in the Black Sea Region* (Black Sea Studies 2) (Aarhus).

—. 2008: *Urban Life and Local Politics in Roman Bithynia: The Small World of Dion Chrysostomos* (Black Sea Studies 7) (Aarhus).

—. 2014: 'To be or not to be Paphlagonian? A question of identity'. In Bekker-Nielsen, T. (ed.), *Space, Place and Identity in Northern Anatolia* (Geographica Historica 29) (Stuttgart), 63–74.

Bekker-Nielsen, T. and Czichon, R.M. 2015: 'Ancient Roads and Bridges of the Vezirköprü District'. In Winther-Jacobsen, K. and Summerer, L. (eds.), *Landscape Dynamics and Settlement Patterns in Northern Anatolia during the Roman and Byzantine Period* (Geographica Historica 32) (Stuttgart), 295–305.

Bianchetti, S., Cataudella, M.R. and Gehrke, H.-J. (eds.) 2015: *Brill's Companion to Ancient Geography: The Inhabited World in Greek and Roman Tradition* (Leiden/Boston).

Bilabel, F. 1920: *Die ionische Kolonisation* (Leipzig).

Bilgi, Ö. 1999: 'Ikiztepe in the Late Iron Age'. *AS* 49, 27–54.

—. 2001: *Metallurgists of the Central Black Sea Region: A New Perspective on the Question of the Indo-Europeans' Original Homeland* (Istanbul).

Birmingham, J.M. 1961: 'The Overland Routes across Anatolia in the Eighth and Seventh Centuries B.C.'. *AS* 11, 185–95.

Bittel, K. 1976: *Die Hethiter: Die Kunst Anatoliens vom Ende des 3. bis zum Anfang des 1. Jahrtausends vor Christus* (Munich).

Bittel-Heidenheim, K. 1976: 'Die zweite vorchristliche Jahrtausend im östlichen Mittelmeer und im Vorderen Orient: Anatolien und Aegaeis'. *Gymnasium* 83, 513–33.

Boardman, J. 1963: 'Greek Archaeology on the Shores of the Black Sea'. *Archaeological Reports* 9 (for 1962–63), 34–51.

—. 1991: 'Early Greek Pottery on Black Sea Sites?'. *The Oxford Journal of Archaeology* 10.3, 387–90.

—. 1999: *The Greeks Overseas: Their Early Colonies and Trade*, 4th ed. (London).

Bolling, G.M. 1925: *The External Evidence for Interpolation in Homer* (Oxford).

Bonfante, L. 2011: 'Classical and Barbarian'. In Bonfante, L. (ed.), *The Barbarians of Ancient Europe* (Cambridge), 1–36.

Boshnakov, K. 2004: *Pseudo-Skymnos (Semos von Delos?): ta aristera tou Pontou. Zeugnisse griechischer Schriftsteller über den westlichen Pontosraum* (Palingenesia 82) (Stuttgart).

Bossert, E.-M. 2000: *Die Keramik phrygischer Zeit von Boğazköy: Funde aus den Grabungskampagnen 1906, 1907, 1911, 1912, 1931–1939 und 1952–1960* (Boğazköy-Hattusa. Ergebnisse der Ausgrabungen 18) (Mainz).

Boulay, T. 2015: 'Wine Appreciation in Ancient Greece'. In Wilkins and Nadeau 2015, 273–82.

Bouzek, J. 1983: 'Les Cimmériens en Anatolie'. In *Modes de contacts et processus de transformation dans les sociétés anciennes* (Actes du colloque de Cortone, 24–30 mai 1981) (Rome), 145–61.

—. 1990: *Studies of Greek Pottery in the Black Sea Area* (Prague).

—. 2005: 'Early Greek Fine Pottery in the Black Sea Area'. In Kacharava *et al.* 2005, 63–75.

—. 2008: 'Archaic Greek pottery in the Black Sea region'. *Encyclopaedia of the Hellenic World, Black Sea* <http://www.ehw.gr/l.aspx?id=12222>.

—. 2013: 'Pre-Colonisation in the Black Sea: The First Greek Pottery Finds in the Black Sea'. In Tsetskhladze *et al.* 2013, 19–21.

Boysal, Y. 1959: 'Über die älteren Funde von Sinope und die Kolonisationsfrage'. *Archäologischer Anzeiger*, 8–20.

Branting, S., Özarslan, Y., Lehner, J., Marston, J.M and Graff, S.R. 2019: 'Kerkenes and Phrygia: Old and New Directions of Research'. In Tsetskhladze 2019b, 539–59.

Bratianu, G.I. 1969: *La Mer Noire: Des origines à la conquête ottomane* (Acta Historica 9) (Munich).

Braudel, F. 1972: *The Mediterranean and the Mediterranean World in the Age of Philip II*, vol. 1 (London/Berkeley/Los Angeles).

Braund, D. 1994: *Georgia in Antiquity. A History of Colchis and Transcaucasian Iberia 550 BC–AD 562* (Oxford).

—. 2005a: 'Reflections on Eumelus' Black Sea Region'. In Kacharava *et al.* 2005, 99–112.

—. 2005b: 'Pericles, Cleon and the Pontus. The Black Sea in Athens c. 440–421'. In Braund, D. (ed.), *Scythians and Greeks: Cultural Interactions in Scythia, Athens and the Early Roman Empire (Sixth Century BC–First Century AD)* (Exeter), 80–99.

—. 2005c: 'Across the Black Sea. Patterns of maritime exchange on the northern periphery of Roman Asia Minor'. In Mitchell and Katsari 2005, 115–38.

—. 2015: 'Food among Greeks of the Black Sea: the Challenging Diet of Olbia'. In Wilkins and Nadeau 2015, 296–308.

—. 2021: '"Colchians did not like to write": reflections on Greek epigraphy in the eastern Black Sea region and its hinterland'. In Manoledakis 2021b, 131–40.

Braund, D.C. and Tsetskhladze, G.R. 1989: 'The Export of Slaves from Colchis'. *Classical Quarterly* 39.1, 114–25.

Brennan, M.L., Davis, D., Roman, C., Buynevich, I., Catsambis, A., Kofahl, M., Ürkmez, D., Vaughn, J.I., Merrigan, M. and Duman, M. 2013: 'Ocean dynamics and anthropogenic impacts along the southern Black Sea shelf examined

through the preservation of pre-modern shipwrecks'. *Continental Shelf Research* 53, 89–101.
Brennan, S. 2015: 'The Nobble Kardouchoi and the Barbarous Mossynoikoi: Remembering and Forgetting Ancient Anatolian Peoples'. *Pamukkale Üniversitesi Sosyal Bilimler Enstitüsü Dergisi* 20, 100–07.
Bresson, A. and Rouillard, P. (eds.) 1993: *L'Emporion* (Publications du Centre Pierre Paris 26) (Paris).
Briant, P. 1995 (ed.): *Dans les pas des Dix-Mille: peuples et pays du Proche-Orient vus par un Grec* (= *Pallas* 43) (Toulouse).
—. 2002: *From Cyrus to Alexander: A History of the Persian Empire* (Winona Lake).
Brixhe, C. and Lejeune, M. 1984: *Corpus des inscriptions paléo-phrygiennes* (Mémoire 45) (Paris).
Brodersen, K. 2015: 'The Geographies of Pliny and his "Ape" Solinus'. In Bianchetti *et al.* 2015, 298–310.
Brosius, M. 2010: '*Pax Persica* and the Peoples of the Black Sea Region: Extent and Limits of Achaemenid Imperial Ideology'. In Nieling and Rehm 2010, 29–40.
Bryce, T. 2006: *The Trojans and their Neighbours* (London/New York).
—. 2009: *The Routledge Handbook of the Peoples and Places of Ancient Western Asia: From the Early Bronze Age to the Fall of the Persian Empire* (London/New York).
Bryer, A. 1982: 'The Question of Byzantine Mines in the Pontos: Chalybian Iron, Chaldian Silver, Koloneian Alum and the Mummy of Cheriana'. *AS* 32, 133–50.
Bryer, A. and Winfield, D. 1985: *The Byzantine Monuments and Topography of the Pontos*, vol. 1 (Washington, DC).
Budde, L. 1956: 'Kurzer vorläufiger Bericht über die Grabungen in Sinope der Kampagnen 1951–1953'. *Türk Arkeoloji Dergisi* 6.2, 5–10.
Burney, C.A. 1956: 'Northern Anatolia before Classical Times'. *AS* 6, 179–203.
Burstein, S.M. 1976a: *Outpost of Hellenism: The Emergence of Heraclea on the Black Sea* (University of California Classical Studies 14) (Berkeley/Los Angeles/London).
—. 1976b: 'Fragment 53 of Callisthenes and the Text of Iliad 2. 850-55'. *Classical Philology* 71.4, 339–41.
Çalışkan Akgül, H. 2019: 'The Traces of the Chalcolithic Culture of Alaca Höyük in the Coastal Settlements of the Central Black Sea Region of Turkey'. In Tsetskhladze and Atasoy 2019, 157–66.
Çam, F.B., Bora, A. and Altunkayalıer, H.B. 2019: 'New Archaeological Expeditions in the Ancient City of Amastris'. In Tsetskhladze and Atasoy 2019, 190–207.
Cameron, G.G. 1973: 'The Persian Satrapies and Related Matters'. *JNES* 32.1–2, 47–56.
Carpenter, R. 1948: 'The Greek Penetration of the Black Sea'. *AJA* 52, 1–10.
Cartledge, P. 2002: *The Greeks: A Portrait of Self and Others* (Oxford).
Cawkwell, G. 2004: 'When, How and Why did Xenophon Write the Anabasis?'. In Lane Fox 2004b, 47–67.
Chadwick, J. and Baumbach, L. 1963: 'The Mycenaean Greek Vocabulary'. *Glotta* 41.3/4, 157–271.
Chamoux, F. 2002: *Hellenistic Civilization* (Oxford).
Chantraine, P. 1968: *Dictionnaire étymologique de la langue grecque. Histoire des mots* (Paris).

Chepalyga, A.L. 2007: 'The late glacial great flood in the Ponto-Caspian basin'. In Yanko-Hombach *et al.* 2007, 119–48.

Cline, E.H. 1991: 'A Possible Hittite Embargo Against the Mycenaeans'. *Historia* 40, 1–9.

—. 1996: 'Assuwa and the Achaeans: the 'Mycenaean' Sword at Hattusas and its Possible Implications'. *The Annual of the British School at Athens* 91, 137–51.

Coleman, J.E. 2000: 'An Archaeological Scenario for the Coming of the Greeks'. *Journal of Indo-European Studies* 28.1/2, 101–53.

Coleman, D.F. and Ballard, R.D. 2007: 'Submerged paleoshorelines in the southern and western Black Sea – Implications for inundated prehistoric archaeological sites'. In Yanko-Hombach *et al.* 2007, 671–96.

Conovici, N. 2005: 'The Dynamics of Trade in Transport Amphoras from Sinope, Thasos and Rhodos on the Western Black Sea Coast: a Comparative Approach'. In Stolba, V.F. and Hannestad, L. (eds.), *Chronologies of the Black Sea Area in the Period c. 400–100 BC* (Black Sea Studies 3) (Aarhus), 101–04.

Cook, J.M. 1960: 'Greek Archaeology in Western Asia Minor'. *Archaeological Reports* 6 (for 1959–60), 27–57.

—. 1985: 'The Rise of the Achaemenids and Establishment of their Empire'. In *The Cambridge History of Iran*, vol. 2 (Cambridge), 200–91.

Cook, R.M. 1946: 'Ionia and Greece in the Eighth and Seventh Centuries BC'. *JHS* 66, 67–98.

Corcella, A. 2007: 'Book IV'. In Asheri, D., Lloyd, A. and Corcella, A., *A Commentary on Herodotus Books I–IV*, ed. O. Murray and A. Moreno (Oxford), 543–721.

Counillon, P. 1993: '*L'emporion* des géographes grecs'. In Bresson and Rouillard 1993, 47–57.

—. 2004: *Pseudo-Skylax: le Périple du Pont-Euxin* (Scripta Antiqua 8) (Bordeaux).

Cramer, J.A. 1832: *A Geographical and Historical Description of Asia Minor*, vol. 1 (Oxford).

Cummer, W.W. 1976: 'Iron Age Pottery from Akalan'. *IstMitt* 26, 31–39.

Curtis, R.I. 2005: 'Sources for Production and Trade of Greek and Roman Processed Fish'. In Bekker-Nielsen 2005, 31–46.

Dan, A. 2012: 'Les Leukosyriens: quelques notes d'ethnographie sinopéenne'. In Kassab Tezgör 2012, 73–102.

—. 2012–13: 'From Imagined Ethnographies to Invented Identities. The Homeric Halizones'. *Orbis Terrarum* 11, 33–69.

—. 2013: 'Achaemenid World Representations in Herodotus' Histories: some geographic examples of cultural translation'. In Geus, K., Irwin, E. and Poiss, T. (eds.), *Herodots Wege des Erzählens: Logos und Topos in den Historien* (Zivilisationen und Geschichte 22) (Frankfurt), 83–121.

Dana, M. 2011: *Culture et mobilité dans le Pont-Euxin: Approche régionale de la vie culturelle des cités grecques* (Scripta Antiqua 37) (Bordeaux).

—. 2016: 'Histoire et historiens de Propontide et de Bithynie: mythes, récits et identités'. In Dana, M. and Prêteux, F. (eds.), *Identités civiques, identité régionale: autour des Détroits des Dardanelles et du Bosphore (Ve siècle av. J.-C.–IIe siècle après J.-C.)* (Actes de la journée d'études du centre ANHIMA, Paris, 23 mars 2013) (*Dialogues d'histoire ancienne* Suppl. 15) (Clermont-Ferrand), 171–240.

Dana, D. and Dana, M. 2014: 'Arrien avant Arrien: une jeunesse entre Bithynie, Grèce et Rome'. *Ktèma* 39, 19–35.

Danoff, C.M. 1962: 'Pontos Euxeinos'. *RE* Suppl. IX, 866–1175.
de Boer, J.G. 2005: 'The foundation of Agathopolis/Alaeouteichos and the Athenian Black Sea policy in the 5th century BC'. In Kacharava *et al.* 2005, 167–80.
—. 2006: 'The Cimmerian Invasions in Anatolia and the Earliest Greek Colonies in the Black Sea Area'. *Eirene* 42, 43–55.
—. 2006–07: 'Phantom-Mycenaeans in the Black Sea'. *Talanta* 38–39, 277–302.
—. 2013: 'Stamped Amphorae from the Greek Black Sea Colony of Sinope in the Mediterranean During the Hellenistic Period'. In Tsetskhladze *et al.* 2013, 109–14.
—. 2015: 'Sinope and Colchis: colonisation, or a Greek population in "poleis barbaron"?'. In Tsetskhladze *et al.* 2015, 73–80.
De Graauw, A. 2016: 'Catalogue of potential ancient ports in the Black Sea'. *Méditerranée* 126, 129–44.
De Jesus, P.S. 1978: 'Metal Resources in Ancient Anatolia'. *AS* 28, 97–102.
De Planhol, X. 1963: 'Geographica Pontica'. *Journal Asiatique* 251, 293–310.
Debord, P. 1999: *L'Asie Mineure au IVe siècle (412-323 a.C.): Pouvoirs et jeux politiques* (Études 3) (Bordeaux).
Delage, É. 1930: *La géographie dans les Argonautiques d'Apollonios de Rhodes* (Bordeaux/Paris).
Demetriou, D. 2012: *Negotiating Identity in the Ancient Mediterranean: The Archaic and Classical Greek Multiethnic Emporia* (Cambridge).
Demirel, S. 2019: 'Some Observations about the Problem of Late Bronze Age in the Settlements of Sinop Region'. In Kaba *et al.* 2019, 39–51.
Descœudres, J.-P. 2008: 'Central Greece on the Eve of the Colonisation Movement'. In Tsetskhladze 2008b, 289–382.
Devereux, G. 1964: 'The Enetian Horses of Hippolytos (Euripides, *Hippolytos*, 231, 1131)'. *L'antiquité classique* 33.2, 375–83.
Dickey, E. 2007: *Ancient Greek Scholarship* (Oxford).
Dihle, A. 1998: *Οι Έλληνες και οι ξένοι*, transl. into Greek by T. Siete (Athens).
Diller, A. 1952: *The Tradition of the Minor Greek Geographers* (Philological Monographs 14) (Lancaster, PA).
—. 1975: *The Textual Tradition of Strabo's Geography* (Amsterdam).
Doksanaltı, E.M. 2012: 'Open air cult sites in eastern Pontus region'. *Anodos* 12, 61–78.
Doksanaltı, E.M. and Ekici, M. 2017: 'Giresun-Aretias/Khalkeritis Adası Kazıları Seramik Buluntuları'. *Pamukkale University Journal of Social Sciences Institute* 27, 389–403.
Doksanaltı, E.M. and Mimiroğlu, I.M. 2010: 'Giresun/Aretias – Khalkeritis Island.' *Anodos* 10, 85–101.
Domínguez, A.J. 2006: 'Greeks in Sicily'. In Tsetskhladze 2006b, 253–357.
Dönmez, Ş. 2002: 'The 2nd Millennium BC Settlements in Samsun and Amasya Provinces, Central Black Sea Region, Turkey'. *AWE* 1.2, 243–83.
—. 2006: 'Some Observations of the Socio–Economic Structure and Ethnic Make-up of the Central Black Sea Region of Turkey during the Iron Age in the Light of New Evidence'. *AWE* 5, 13–43.
—. 2007a: 'Sinop Province during the Iron Age in the Light of New Research'. *Anatolia Antiqua* 15, 59–65.
—. 2007b: 'The Central Black Sea Region, Turkey, during the Iron Age: the Local Cultures and the Eurasian Horse-Riding Nomads'. In Grammenos, D.V. and

Petropoulos, E.K. (eds.), *Ancient Greek Colonies in the Black Sea 2*, vol. 2 (BAR International Series 1675) (Oxford), 1207–16.
—. 2007c: 'The Achaemenid Impact on the Central Black Sea Region'. In Delemen, I., Casabonne, O., Karagöz, Ş. and Tekin, O. (eds.), *The Achaemenid Impact on Local Populations and Culture in Anatolia (Sixth–Fourth Centuries B.C.)* (Papers presented at the International Workshop, Istanbul, 20-21 May 2005) (Istanbul), 107–16.
—. 2010: 'Sacral Monuments of the North-Central Anatolia'. In Petropoulos, E.K. and Maslennikov, A.A. (eds.), *Ancient Sacral Monuments in the Black Sea* (Thessaloniki), 515–62.
—. 2011: 'Avrasyalı Savaşçı-Atlı Göçebeler'in Karadeniz Bölgesi'ndeki Varlıklarına Ait Yeni Kanıtlar/ New Evidences on the Existence of Eurasian Horse-Riding Nomads in the Black Sea Region, Turkey'. In Öztan, A. and Dönmez, S. (eds.), *Karadeniz'den Fırat'a bilgi üretimleri: Önder Bilgi'ye armağan yazılar/ Knowledge Production from the Black Sea to the Euphrates. Studies Presented in Honour of Önder Bilgi* (Ankara), 135–46.
—. 2012: 'An Overview of the 2nd Millennium BC and Iron Age Cultures of the Province of Sinop in the Light of New Research'. In Kassab Tezgör 2012, 153–74.
—. 2015: 'Achaemenid Presence at Oluz Höyük, north-central Anatolia'. In Tsetskhladze *et al.* 2015, 467–73.
—. 2019a: 'The Ethnic Composition of the Halys Basin and its Surrounding Area in the Ptrotohistorical Times and the Iron Age'. In Tsetskhladze 2019b, 429–78.
—. 2019b: 'The Land of Sacred Fire: Amasya - Oluz Höyük'. In Tsetskhladze and Atasoy 2019, 244–57.
Dönmez, Ş. and Ulugergerli, E.U. 2008: 'Geophysical Research at Akalan'. *AWE* 7, 249–64.
Doonan, O.P. 2002: 'Production in a Pontic Landscape: The Hinterland of Greek and Roman Sinope'. In Faudot, M., Fraysse, A. and Geny, É. (eds.), *Pont Euxin et commerce: La genése de la 'route de la soie'* (Actes du IXe Symposium de Vani, Colchide, 1999) (Publications de l'Institut des Sciences et Techniques de l'Antiquite 853) (Besançon), 185–98.
—. 2003: 'Sinope'. In Grammenos, D.V. and Petropoulos, E.K. (eds.), *Ancient Greek Colonies in the Black Sea [1]*, vol. 2 (Publications of the Archaeological Institute of Northern Greece 4) (Thessaloniki), 1379–1402.
—. 2004: *Sinop Landscapes: Exploring Connection in a Black Sea Hinterland* (Philadelphia).
—. 2006: 'Exploring Community in the Hinterland of a Black Sea Port'. In Bilde, P.G. and Stolba, V.F. (eds.), *Surveying the Greek Chora: The Black Sea Region in a Comparative Perspective* (Black Sea Studies 4) (Aarhus), 47–58.
—. 2007: 'Colony and Conjuncture: The Early Greek Colony at Sinope'. In Cobet, J., von Graeve, V., Niemeier, W.-D. and Zimmermann, K. (eds.), *Frühes Ionien: Eine Bestandsaufnahme* (Panionion-Symposion, Guzelcamlı, 26. September–1. Oktober 1999) (Milesische Forschungen 5) (Mainz), 613–20.
—. 2009: 'Sacred Landscapes and the Colonization of the Sinop Promontory'. In Gates, C., Morin, J. and Zimmermann, T. (eds.), *Sacred Landscapes in Anatolia and Neighboring Regions* (BAR International Series 2034) (Oxford), 69–78.
—. 2012: 'Sinop Landscapes: Towards an Archaeology of Community in the Hinterland of a Black Sea Port'. In Kassab Tezgör 2012, 175–87.

—. 2016: 'Sinope, new understandings of the early colony based on recent research at Sinop Kale'. In Manoledakis, M. (ed.), *The Black Sea in the Light of New Archaeological Data and Theoretical Approaches* (Proceedings of the 2nd International Workshop on the Black Sea in Antiquity held in Thessaloniki, 18–20 September 2015) (Oxford), 217–23.

—. 2019: 'Xenophon in the Black Sea Landscape: Settlement models from the Iron Age on the Sinop Promontory (Turkey)'. *European Journal of Archaeology* 22.1, 91–110.

Doonan, O.P., Bauer, A., Sherratt, S.E., Casson, A., Feathers, J., Conrad, M., Besonen, M., Evren, E., Domzalski, K. and Smokotina, A. 2015: 'Sinop Regional Archaeological Project: Report on the 2010–2012 Field Seasons'. In Steadman, S.R. and McMahon, G. (eds.), *The Archaeology of Anatolia I: Recent Discoveries (2011–2014)* (Newcastle-upon-Tyne), 298–327.

Doonan, O.P., Vural, H., Goldman, A., Bauer, A., Rempel, J., Sherratt, S., Krotschek, U., Maranzana, P. and Sökmen, E. 2017: 'Sinop Kalesi Archaeological Excavations, 2015–2016 Field Seasons'. In Steadman, S.R. and McMahon, G. (eds.), *The Archaeology of Anatolia II: Recent Discoveries (2015–2016)* (Newcastle-upon-Tyne), 178–99.

Doonan, O.P., Vural, H., Goldman, A., Bauer, A., Sheratt, S., Rempel, J., Domzallski, K. and Smokotina, A. 2016: 'Sinope ancient Kale excavations 2015: towards a new model of mobile fishing communities and incipient trade in the Black Sea'. *Antiquity Project Gallery* 90 (351) <http://www.antiquity.ac.uk/projgall/doonan351>.

Dorati, M. 1999-2000: 'Le testimonianze relative alla περίοδος τῆς γῆς di Ecateo'. *Geographia Antiqua* 8–9, 120–27.

Dougherty, C. 1993a: *The Poetics of Colonization* (New York/Oxford).

—. 1993b: 'It's Murder to Found a Colony'. In Dougherty, C. and Kurke, L. (eds.), *Cultural Poetics in Archaic Greece: Cult, Performance, Politics* (Cambridge), 178–98.

Draycott, C.M. and Summers, G.D. 2008: *Sculpture and Inscriptions from the Monumental Entrance to the Palatial Complex at Kerkenes Dağ, Turkey* (Kerkenes Special Studies 1) (Chicago).

Drews, R. 1976: 'The Earliest Greek Settlements on the Black Sea'. *JHS* 96, 18–31.

Dueck, D. 2000: *Strabo of Amasia: A Greek Man of Letters in Augustan Rome* (New York).

Dunham, A.G. 1915: *The History of Miletus: Down to the Anabasis of Alexander* (London).

Düring, B.S. and Glatz, C. (eds.) 2015a: *Kinetic Landscapes: The Cide Archaeological Project: Surveying the Turkish Western Black Sea Region* (Warsaw/Berlin).

—. 2015b: 'Remote Landscapes in Flux – The Cide and Şenpazar Region'. In Düring and Glatz 2015a, 9–35.

Dusinberre, E.R.M. 2013: *Empire, Authority, and Autonomy in Achaemenid Anatolia* (Cambridge).

Easton, D.F, Hawkins, J.D., Sherratt, A.G. and Sherratt, E.S. 2002: 'Troy in Recent Perspective'. *AS* 52, 75–109.

Edwards, R.W. 1988: 'The Vale of Kola: A Final Preliminary Report on the Marchlands of Northeast Turkey'. *Dumbarton Oaks Papers* 42, 119–41.

Ehrhardt, N. 1988: *Milet und seine Kolonien: Vergleichende Untersuchung der kultischen und politischen Einrichtungen* (Europäische Hochschulschriften, Reihe III: Geschichte und ihre Hilfswissenschaften Bd 206) (Frankfurt/Bern/New York).

Elderkin, G.W. 1935: 'Oitosyros and Oiorpata'. *American Journal of Philology* 56.4, 342–46.
Erciyas, D.B. 2006: *Wealth, Aristocracy and Royal Propaganda under the Hellenistic Kingdom of the Mithradatids in the Central Black Sea Region of Turkey* (Colloquia Pontica 12) (Leiden/Boston).
—. 2007: 'Cotyora, Kerasous and Trapezous: The Three Colonies of Sinope'. In Grammenos, D.V. and Petropoulos, E.K. (eds.), *Ancient Greek Colonies in the Black Sea 2*, vol. 2 (BAR International Series 1675) (Oxford), 1195–1206.
Erdem, Z.K. 2021: 'Terracotta production of the Black Sea coast of Turkey: the example of Sinope'. In Tsetskhladze *et al.* 2021, 429–34.
Erol, A.F. 2015: 'New Findings on the History and Archaeology of the Eastern Black Sea Region of Turkey: The Excavation of Cıngırt Kayası'. In Tsetskhladze *et al.* 2015, 453–62.
—. 2018: 'An Archaeological Assessment of the Turkish Eastern Black Sea Region in the Light of Fatsa Surveys and Cıngırt Kayası Excavations'. In Yiğitpaşa *et al.* 2018, 102–09.
Erol, A.F. and Tamer, D. 2021: 'Recent finds of Hellenistic mould-made relief bowls from the Ordu/Fatsa Cingirt Kayasi excavations in the southern Black Sea region'. In Tsetskhladze *et al.* 2021, 561–70.
Erol, A.F. and Yıldırım, E. 2019: 'Evaluations of Iron Finds from the Fatsa Cıngırt Kayası Excavations'. In Tsetskhladze and Atasoy 2019, 177–89.
Étienne, R. 1993: '*L'emporion* chez Strabon'. In Bresson and Rouillard 1993, 23–34.
Faulkner-Gentry, I. 2021: 'Colonisation and foundation myth in the Pontic regions'. In Tsetskhladze *et al.* 2021, 27–32.
Ferrarese, P. 1974: 'La spedizione di Pericle nel Ponto Eusino'. *Contributi dell'Istituto di storia antica* 2, 7–19.
Figueira, T. 2008: 'Colonisation in the Classical Period'. In Tsetskhladze 2008b, 427–523.
Finkelberg, M. 1997: 'Anatolian Languages and Indo-European Migrations to Greece'. *The Classical World* 91.1, 3–20.
Flensted-Jensen, P. and Hansen, M.H. 1996: 'Pseudo-Skylax' Use of the Term Polis'. In Hansen, M.H. and Raaflaub, K. (eds.), *More Studies in the Ancient Greek Polis* (*Historia* Einzelschriften 108) (Stuttgart), 137–67.
Flower, M.A. 2012: *Xenophon's Anabasis, or The Expedition of Cyrus* (Oxford).
Fol, A. and Hammond, N.G.L. 1988: 'Persia in Europe, apart from Greece'. *CAH* IV^2, 234–53.
Forrer, E. 1921: 'Ausbeute aus den Bogazkoi-Inschriften'. *Mitteilungen der Deutschen Orient-Gesellschaft* 61, 20–39.
Francis, E.D. 1980: 'Greeks and Persians: The Art of Hazard and Triumph'. In Schmandt-Besserat, D. (ed.), *Ancient Persia: The Art of an Empire* (Invited Lectures on the Middle East at the University of Texas at Austin 4) (Malibu), 53–86.
Frazer, J.G. 1894: *The Golden Bough* (New York/London).
French, D.H. 1974: 'A Study of Roman Roads in Anatolia: Principles and methods'. *AS* 24, 143–49.
—. 1980: 'The Roman Road System of Asia Minor'. In Temporini, H. and Haase, W. (eds.), *Aufstieg und Niedergang der Römischen Welt* 2.7.2 (Berlin), 698–729.
—. 1982: 'Mycenaeans in the Black Sea?'. *Thracia Pontica* 1, 19–30.

—. 2004: *The Inscriptions of Sinope* (Inschriften Griechischer Städte aus Kleinasien 64) (Bonn).
—. 2012: *Roman Roads and Milestones of Asia Minor 3: Milestones, Fasc. 3.3: Cappadocia* (BIAA Electronic Monograph 3) (London).
—. 2013: *Roman Roads and Milestones of Asia Minor 3: Milestones, Fasc. 3.4: Pontus et Bithynia (with Northern Galatia)* (BIAA Electronic Monograph 4) (London).
—. 2016a: *Roman Roads and Milestones of Asia Minor 3: Milestones, Fasc. 3.9, An Album of Maps* (BIAA Electronic Monograph 9) (London).
—. 2016b: *Roman Roads and Milestones of Asia Minor 4: The Roads, Fasc. 4.1: Notes on the Itineraria* (BIAA Electronic Monograph 10) (London).
Frye, R.N. 1992: 'Assyria and Syria: Synomyns'. *JNES* 51.4, 281–85.
Gabrielsen, V. and Lund, J. (eds.) 2007: *The Black Sea in Antiquity: Regional and Interregional Economic Exchanges* (Black Sea Studies 6) (Aarhus).
von Gall, H. 1966: *Die paphlagonischen Felsgräber: Eine Studie zur kleinasiatischen Kunstgeschichte* (*IstMitt* Beih. 1) (Tübingen).
Garelli, P. 1963: *Les Assyriens en Cappadoce* (Bibliothèque archéologique et historique de l'Institut français d'archéologie d'Istanbul 19) (Paris).
Garlan, Y. 2007: 'Échanges d'amphores timbrées entre Sinope et la Méditerranée aux époques classique et hellénistique'. In Gabrielsen and Lund 2007, 143–48.
Garnsey, P. 1996: *Ideas of Slavery from Aristotle to Augustine* (Cambridge).
Genz, H. 2003: 'The Early Iron Age in Central Anatolia'. In Fischer, B., Genz, H., Jean, É. and Köroğlu, K. (eds.), *Identifying Changes: The Transition from Bronze to Iron Ages in Anatolia and its Neighbouring Regions* (Proceedings of the International Workshop, Istanbul, November 8–9, 2002) (Istanbul), 179–91.
—. 2011: 'The Iron Age in Central Anatolia'. In Tsetskhladze, G.R. (ed.), *The Black Sea, Greece, Anatolia and Europe in the First Millennium BC* (Colloquia Antiqua 1) (Leuven/Paris/Walpole, MA), 331–68.
Gisinger, F. 1927: 'Skylax 2'. *RE* 2. III.A.1, 619–46.
Glatz, C. and Matthews, R. 2005: 'Anthropology of a Frontier Zone: Hittite-Kaska Relations in Late Bronze Age Central Anatolia'. *Bulletin of the American Schools of Oriental Research* 339, 47–65.
Glatz, C., Matthews, R. and Schachner, A. 2009: 'A Landscape of Conflict and Control: Paphlagonia during the Second Millennium BC'. In Matthews and Glatz 2009, 107–47.
Goetze, A. 1957: *Kulturgeschichte des Alten Orients III.1: Kleinasien*, 2nd ed. (Munich).
González Salazar, J.M. and Lucia Castejón, R. 2008: 'Itineraria Anatolica Antiqua. The Road Network of the Northern mountainous sectors of Anatolia: the North-eastern routes and the Hittite Empire'. *Gerión* 26.2, 7–22.
Graham, A.J. 1958: 'The Date of the Greek Penetration of the Black Sea'. *Bulletin of the Institute of Classical Studies* 5, 25–42.
—. 1971: 'Patterns in Early Greek Colonization'. *JHS* 91, 35–47.
—. 1982: 'The Colonial Expansion of Greece'. *CAH* III.3^2, 83–162.
—. 1989: 'Pre-Colonial Contacts: Questions and Problems'. In Descœudres, J.-P. (ed.), *Greek Colonists and Native Populations* (Proceedings of the First Australian Congress of Classical Archaeology held in Honour of Emeritus Professor A.D. Trendall, Sydney, 9–14 July 1985) (Oxford), 45–60.

—. 1994: 'Greek and Roman Settlements on the Black Sea Coasts: Historical Background'. *Colloquenda Pontica* 1, 4–10.
—. 2001: *Collected Papers on Greek Colonization* (*Mnemosyne* Suppl. 214) (Leiden/Boston).
—. 2005: 'The Colonial Expansion of Greece'. *CAH* III.3^2, 83–162.
Greaves, A. 2007: 'Milesians in the Black Sea: Trade, Settlement and Religion'. In Gabrielsen, V. and Lund, J. (eds.), *The Black Sea in Antiquity: Regional and Interregional Economic Exchanges* (Black Sea Studies 6) (Aarhus), 9–21.
Greco, E. 2006: 'Greek Colonisation in Southern Italy: A Methodological Essay'. In Tsetskhladze 2006b, 169–200.
Gruen, E.S. 2011: *Rethinking the Other in Antiquity* (Princeton/Woodstock).
Haas, O. 1966: *Die Phrygischen Sprachdenkmäler* (Balkansko Ezikoznanie 10) (Sofia).
Hall, E. 1989: *Inventing the Barbarian: Greek Self-definition through Tragedy* (Oxford).
Hall, J.M. 1997: *Ethnic Identity in Greek Antiquity* (Cambridge).
—. 2002: *Hellenicity: Between Ethnicity and Culture* (Chicago/London).
—. 2008: 'Foundation Stories'. In Tsetskhladze 2008b, 383–426.
Hamilton, W.J. 1842: *Researches in Asia Minor, Pontus and Armenia*, vol. 1 (London).
Handberg, S. 2013: 'Milesian Ktiseis and Aeolian Potters in the Black Sea Region'. In Manoledakis 2013b, 1–18.
Hanell, K. 1934: *Megarische Studien* (Lund).
Hannestad, L. 1996: '"This Contributes in No Small Way to One's Reputation": The Bithynian Kings and Greek Culture'. In Bilde, P. (ed.), *Aspects of Hellenistic Kingship* (Studies in Hellenistic Civilization 7) (Aarhus), 67–98.
Hansen, M.H. 1995: '*Kome*. A Study in how the Greeks Designated and Classified Settlements which were not *Poleis*'. In Hansen, M.H. and Raaflaub, K. (eds.), *Studies in the Ancient Greek Polis* (*Historia* Einzelschriften 95) (Stuttgart), 47–81.
—. 1997: 'Hekataios' Use of the word Polis in his Periegesis'. In Nielsen, T.H. (ed.), *Yet More Studies in the Ancient Greek Polis* (*Historia* Einzelschriften 117) (Stuttgart), 17–27.
—. 2000: 'A Survey of the Use of the Word Polis in Archaic and Classical Sources'. In Flensted-Jensen, P. (ed.), *Further Studies in the Ancient Greek Polis* (*Historia* Einzelschriften 138) (Stuttgart), 173–206.
—. 2006: '*Emporion*. A Study of the Use and Meaning of the Term in the Archaic and Classical Periods'. In Tsetskhladze 2006b, 1–39.
Hansen, M.H. and Nielsen, T.H. (eds.) 2004: *An Inventory of Archaic and Classical Poleis* (Oxford).
Harding, A.F. 2005: 'Horse-harness and the origins of the Mycenaean Civilisation'. In Dakouri-Hild, A. and Sherratt, S (eds.), *Autochthon: Papers Presented to O.T.P.K. Dickinson on the Occasion of his Retirement* (Oxford), 296–300.
Harrison, C.M. 1982: 'Persian Names on Coins of Northern Anatolia'. *JNES* 41.3, 181–94.
Hartog, F. 1988: *The Mirror of Herodotus: The Representation of the Other in the Writing of History* (Berkeley/Los Angeles/London).
Haywood, J. 2016: 'Divine Narratives in Xenophon's *Anabasis*'. *Histos* 10, 85–110.
Hellmuth, A. 2008: 'The Chronological Setting of the so-called Cimmerian and Early Scythian Material from Anatolia'. *ANES* 45, 102–22.

Herrmann, A. 1931: *Die Erdkarte der Urbibel, mit einem Anhang über Tartessos und die Etruskerfrage* (Brunswick).
Herzfeld, E. 1968: *The Persian Empire: Studies in Geography and Ethnography of the Ancient Near East*, ed. G. Walser (Wiesbaden).
Heskel, J. 1997: *The North Aegean Wars, 371–360 BC* (*Historia* Einzelschriften 102) (Stuttgart).
Hewsen, R.H. 2009: 'Armenians on the Black Sea? The Province of Trebizond'. In R.G. Hovannisian (ed.), *Armenian Pontus* (Costa Mesa), 37–65.
Hiller, S. 1991: 'The Mycenaeans and the Black Sea'. In Laffineur, R and Basch, L. (eds.), *Thalassa: L'Egée préhistorique et la mer* (Actes de la troisième Rencontre égéenne internationale de l'Université de Liège, Station de recherches sous-marines et océanographiques (StaReSO), Calvi, Corse, 23–25 avril 1990) (Aegaeum 7) (Liège), 207–16.
Hind, J.G.F. 1964: 'Pamyantiki Antichnykh Gorodov Yuzhnogo Prichernomor'ya (Po dannym poezdki v 1962)'. *Sovetskaya Arkheologiya* 3, 172–87.
—. 1984: 'Greek and Barbarian Peoples on the Shores of the Black Sea'. *Archaeological Reports* 30 (for 1983–84), 71–97.
—. 1988: 'The Colonisation of Sinope and the South-East Black Sea Area'. In Lordkipanidze, O.D. (ed.), *Myestnye Ethno-politicheskiye Obyedinenia Prichernomor'ya v VII-IV vyekakh do n.e.* (Maleriali IV Vsesojuznogo simpoziuma po drevnei istorii Prichernomor'ya, Tskhaltubo-Vani, 1985) (Tbilisi), 207–23.
—. 1993: 'Archaeology of the Greeks and Barbarian Peoples around the Black Sea (1982–1992)'. *Archaeological Reports* 39 (for 1992–93), 82–112.
—. 1995–96: 'Traders and Ports-of-Trade (emporoi and emporia) in the Black Sea in Antiquity'. *Il Mar Nero* 2, 113–26.
—. 1999: 'The Dates and Mother Cities of the Black Sea Colonies (Pseudo-Scymnus and the Pontic Contact Zone)'. In Fraysse, A. and Geny, É. (eds.), *La Mer Noire zone de contacts* (Actes du VIIe Symposium de Vani, 26–30 IX 1994) (Publications de l'Institut des Sciences et Techniques de l'Antiquité 715) (Besançon), 25–34.
Hirsch, S.V. 1985: *The Friendship of the Barbarians: Xenophon and the Persian Empire* (Hanover, NH/London).
Hobden, F. and Tuplin, C.J. (eds.) 2012: *Xenophon: Ethical Principles and Historical Enquiry* (*Mnemosyne* Suppl. 348) (Leiden/Boston).
Hodkinson, S. 2009: *Property and Wealth in Classical Sparta* (London).
Hoefer, U. 1904: 'Pontosvoelker, Ephoros and Apollonios von Rhodos'. *Rheinisches Museum für Philologie, Geschichte und griechische Philosophie* 59, 542–64.
Hoepfner, W. 1972: 'Topographische Forschungen'. In Asheri, D., Hoepfner, W. and Erichson, A., *Forschungen an der Nordküste Kleinasiens* 1 (Denkschriften, Österreichische Akademie der Wissenschaften. Philosophisch-Historische Klasse 106; Ergänzungsbände zu den Tituli Asiae Minoris 5) (Vienna), 37–46.
Højte, J.M. 2005: 'The Archaeological Evidence for Fish Processing in the Black Sea Region'. In Bekker-Nielsen 2005, 133–60.
—. 2008: 'The Cities that Never Were. Failed Attempts at Colonization in the Black Sea'. In Bilde, P.G. and Petersen, J.H. (eds.), *Meetings of Cultures in the Black Sea Region: Between Conflict and Coexistence* (Black Sea Studies 8) (Aarhus), 149–62.
—. (ed.) 2009: *Mithridates VI and the Pontic Kingdom* (Black Sea Studies 9) (Aarhus).
Hornblower, S. 1991: *The Greek World 479–323 BC*, 2nd ed. (London).

—. 2008: 'Greek Identity in the Archaic and Classical Periods'. In Zacharia, K. (ed.), *Hellenisms. Culture, Identity, and Ethnicity from Antiquity to Modernity* (Aldershot), 37–58.
Huxley, G.L. 1960: 'Homerica'. *Greek, Roman, and Byzantine Studies* 3, 17–30.
—. 1990: 'Eusebius on the Foundation of Trapezous'. In Lordkipanidze, O.D. (ed.), *Prichernomor'e v VII–V vv. do n.e.: pis'mennye istochniki i arkheologiya* (Materialy V Mezhdunarodnogo simpoziuma po drevneĭ istorii Prichernomor'ya) (Tbilisi), 198–201.
Işın, M.A. 1998: 'Sinop Region Field Survey'. *Anatolia Antiqua* 6, 95–139.
Ivantchik, A.I. 1993: *Les Cimmeriens au Proche-Orient* (Orbis biblicus et orientalis 127) (Fribourg/Göttingen).
—. 1997: 'Les légendes de fondation de Sinope du Pont'. *Revue des Études Anciennes* 99.1–2, 33–45.
—. 1998: 'Die Gründung von Sinope und die Probleme der Anfangsphase der griechischen Kolonisation des Schwarzmeergebietes'. In Tsetskhladze 1998b, 297–330.
—. 1999: 'The Scythian "Rule over Asia": The Classical Tradition and the Historical Reality'. In Tsetskhladze, G.R. (ed.), *Ancient Greeks West and East* (*Mnemosyne* Suppl. 196) (Leiden/Boston/Cologne), 497–520.
—. 2001a: *Kimmerier und Skythen* (Moscow).
—. 2001b: 'The Current State of the Cimmerian Problem'. *ACSS* 7.3, 307–39.
—. 2001c: 'Die archäologischen Spuren der Kimmerier im Vorderen Orient und das Problem der Datierung der vor- und frühskythischen Kulturen'. In Eichmann, R. and Parxinger, H. (eds.), *Migration und Kulturtransfer. Der Wandel vorder- und zentralasiatischer Kulturen im Umbruch vom 2. zum 1. vorchristlichen Jahrtausend* (Akten des Internationalen Kolloquiums, Berlin, 23. bis 26. November 1999) (Bonn), 329–42.
—. 2005a: *Am Vorabend der Kolonisation. Das nördliche Schwarzmeergebiet und die Steppennomaden des 8.–7. Jhs. v. Chr. in der klassischen Literaturtradition: Mündliche Überlieferung, Literatur und Geschichte* (Pontus Septentrionalis 3) (Berlin/Moscow).
—. 2005b: 'Le *dêmos* et la *polis* des Cimmériens dans l'*Odyssée* (XI, 14): le contenu de l'image épique'. in Kacharava *et al.* 2005, 83–98.
—. 2010: 'Sinope et les Cimmériens'. *ACSS* 16, 65–72.
—. 2012: 'Cimmerians'. *Encyclopedia of the Bible and Its Reception* 5, 323–24.
—. 2018: 'Scythians'. *Encyclopaedia Iranica*, online edition <https://www.iranicaonline.org/articles/scythians>.
Jablonka, P. 2003: 'The Link Between the Black Sea and the Mediterranean since the End of the Last Ice Age: Archaeology and Geology'. In Wagner, G., Pernicka, E. and Uerpmann, H.-P. (eds.), *Troia and the Troad: Scientific Approaches* (Berlin/Heidelberg), 77–94.
Jablonka, P. and Rose, C.B. 2004: 'Forum Response: Late Bronze Age Troy: A Response to Frank Kolb'. *AJA* 108, 615–30.
Jacoby, F. 1912: 'Hekataios 3'. *RE* VII.2, 2667–2750.
—. 1957: *Die Fragmente der griechischen Historiker. Teil 1, Genealogie und Mythographie. Kommentar, Nachträge* (Leiden).
Janko, R. 2012: 'Πρῶτόν τε καὶ ὕστατον αἰὲν ἀείδειν: Relative Chronology and the Literary History of the Early Greek Epos'. In Andersen, Ø. and Haug, D.T.T. (eds.), *Relative Chronology in Early Greek Epic Poetry* (Cambridge), 20–43.

Janse, M. 2002: 'Aspects of Bilingualism in the History of the Greek Language'. In Adams, J.N., Janse, M. and Swain, S. (eds.), *Bilingualism in Ancient Society. Language Contact and the Written World* (Oxford), 346–59.
Jones, A.H.M. 1971: *The Cities of the Eastern Roman Provinces* (Oxford).
Jones, S. 1997: *The Archaeology of Ethnicity: Constructing Identities in the Past and Present* (London/New York).
Kaba, H., Şahin, K., Akarsu, B.M and Bozoğlan, O. 2019 (eds.): *Proceedings of the International Symposium on Sinope and Black Sea Archaeology 'Ancient Sinope and the Black Sea'* (Sinop).
Kac, V.I. 2003. 'A New Chronology for the Ceramic Stamps of Herakleia Pontike'. In Bilde, P.G., Højte, J.M. and Stolba, V. (eds.), *The Cauldron of Ariantas: Studies Presented to A.N. Ščeglov on the Occasion of His 70th Birthday* (Black Sea Studies 1) (Aarhus), 261–78.
Kacarava, D., Kvirkvélja, G. and Lordkipanidzé, O.D. 1999: 'Les contacts entre les Grecs et les populations locales de la Mer noire. Chronologie et typologie'. In Fraysse, A. and Geny, É. (eds.) 1999, *La Mer Noire zone de contacts* (Actes du VIIe Symposium de Vani, 26–30 IX 1994) (Publications de l'Institut des Sciences et Techniques de l'Antiquité 715) (Besançon), 65–100.
Kacharava, D. 2005: '*Polis hellenis* in the Black Sea area'. In Kacharava *et al.* 2005, 9–31.
Kacharava, D., Faudot, M. and Geny, É. (eds.) 2005: *Pont-Euxin et Polis. Polis Hellenis et Polis Barbaron* (Actes du Xe Symposium de Vani, 23–26 Septembre 2002: hommage à Otar Lordkipanidzé et Pierre Lévêque) (Publications de l'Institut des Sciences et Techniques de l'Antiquité 979) (Besançon).
Karasalihoğlu, M. and Düring, B.S. 2015: 'Traveller's Notes on the Cide Region through the Ages'. In Düring and Glatz 2015a, 36–52.
Kassab Tezgör, D. (ed.) 2012: *Sinope: The Results of Fifteen Years of Research* (Leiden/Boston).
Kaya, E. 1999: 'Trabzon Il Merkezindeki Kurtarma Kazısında Ortaya Çıkan Buluntular'. In *IX. Müze Kurtarma Kazıları Semineri: 27–29 Nisan 1998, Türkiye Cumhuriyeti Kültür BakanlığıYayınları, Antalya* (Ankara), 321–35.
Kaymakçi, S. 2021: 'Giresun and its vicinity in the Greek colonization period'. In Tsetskhladze *et al.* 2021, 94–98.
Kavtaradze, G.L. 2002: 'An attempt to interpret some Anatolian and Caucasian ethnonyms of the classical sources'. In *Sprache und Kultur 3: Gregor Giorgadze von Kollegen und ehemaligen Studenten zum 75. Geburtstag gewidmet* (Tbilisi), 68–83.
Kealhofer, L. and Grave, P. 2011: 'The Iron Age on the Central Anatolian Plateau'. In Steadman and McMahon 2011, 415–42.
Keleş, V. 2006: 'Sikkeler isiginda Sinope'de Pers etkisi (Persian Influence at Sinope based on the Evidence of Coins)'. In Erciyas, D.B. and Koparal, E. (eds.), *Karadeniz Araştırmaları Sempozyum Bildirileri/ Black Sea Studies Symposium Proceedings 16–17 April 2004* (Ankara), 99–110.
Kerestecioğlu, M., Borotav, Z. and Ercan, H. 1998: *Black Sea Environmental Priorities Study: National Report of Turkey* (Istanbul).
Kerschner, M. and Stoltzhauer, U. 2005: 'A New Classification System form East Greek Pottery'. *AWE* 4.1, 1–56.
Kidd, I.G. 1988: *Posidonius, vol. 2: The Commentary* (Cambridge).

Kiessling, E. 1912: 'Ἡνίοχοι'. *RE* VIII.1, 259–80.

Kimball Armayor, O. 1978a: 'Did Herodotus ever go to the Black Sea ?'. *Harvard Studies in Classical Philology* 82, 45–62.

—. 1978b: 'Herodotus' Catalogues of the Persian Empire in the Light of the Monuments and the Greek Literary Tradition'. *TAPA* 108, 1–9.

Kirk, G.S. 1985: *The Iliad: A Commentary*, vol. 1 (Cambridge).

Klingenberg, A. 2014: 'Die "Iranische Diaspora". Kontext, Charakter und Auswirkung persischer Einwanderung nach Kleinasien'. In Olshausen, E. and Sauer, V. (eds.), *Mobilität in den Kulturen der antiken Mittelmeerwelt* (Stuttgarter Kolloquium zur Historischen Geographie des Altertums II, 2011) (Geographica Historica 31) (Stuttgart), 309–24.

Klinkott, H. 2000: *Die Satrapienregister der Alexander- und Diadochenzeit* (*Historia* Einzelschriften 145) (Stuttgart).

Kõiv, M. 2007: 'Cimmerians in the Western Anatolia: A Chronological Note'. In Kämmerer, T.R. (ed.), *Studien zu Ritual und Sozialgeschichte in Alten Orient/ Studies on Ritual and Society in the Ancient Near East* (Tartuer Symposien 1998–2004) (Beihefte zur Zeitschrift für die alttestamentliche Wissenschaft 374) (Berlin/New York), 153–70.

Kolb, F. 2004: 'Troy VI: A Trading Center and Commercial City?' *AJA* 108.4, 577–613.

Korfmann, M. 1997: 'Das homerische Troia war größer – Ergebnisse der Grabungen 1988–96'. In Galter, H.D. (ed.), *Troia. Mythen und Archäologie* (Grazer morgenländische Studien 4) (Graz), 67–95.

—. 2001: 'Troia als Drehscheibe des Handels im 2. und 3. vorchristlichen Jahrtausend'. In Theune-Grosskopf, B. *et al.* (eds.), *Troia – Traum und Wirklichkeit* (Exhibition Catalogue) (Stuttgart), 355–68.

Krapivina, V.V. 2012: 'Ceramics from Sinope in Olbia Pontica'. In Kassab Tezgör 2012, 467–77.

Krappe, A.H. 1946: 'The Leucosyrians'. *Armenian Quarterly* 1, 101–07.

Kretschmer, P. 1896: *Einleitung in die Geschichte der griechischen Sprache* (Göttingen).

Kuhrt, A. 2007: *The Persian Empire: A Corpus of Sources from the Achaemenid Period* (London/New York).

Kullmann, W. 2011: 'Ilias'. In Rengakos, A. and Zimmermann, B. (eds.), *Homer-Handbuch. Leben – Werk – Wirkung* (Stuttgart), 78–119.

Kvirkvelia, G. 2005: '*Polis barbaron* in the Black Sea area in archaic and classical periods'. In Kacharava *et al.* 2005, 33–40.

—. 2010: 'Colchis – North Black Sea Area – Sinope: The System of Interregional Trade'. In Kassab Tezgör, D. and Inaishvili, N. (eds.), *PATABS I: Production and Trade of Amphorae in the Black Sea* (Actes de la Table Ronde internationale de Batoumi et Trabzon, 27–29 avril 2006) (Varia Anatolica XXI) (Paris), 127–31.

Labaree, B.W. 1957: 'How the Greeks Sailed into the Black Sea'. *AJA* 61, 29–33.

Lane Fox, R. 2004a: 'Introduction'. In Lane Fox 2004b, 1–46.

—. (ed.) 2004b: *The Long March. Xenophon and the Ten Thousand* (New Haven, CT/ London).

Lang, F. 2005: 'The Structural Change in Archaic Greek Housing'. In Ault, B.A. and Nevett, L.C. (eds.), *Ancient Greek Houses and Households: Chronological, Regional, and Social Diversity* (Philadelphia), 12–35.

Langella, A. 1989: 'Sinope, Datame e la Persia'. *Dialoghi di archeologia* 7, 93–104.

—. 1997: *Sulle origini di Sinope: Analisi della tradizione preconiale e coloniale* (Naples).
Lasserre, F. 1978: 'Strabon 12,3,14: deux mots inédits sur l'histoire d'Amisos'. *Museum Helveticum* 35, 212–20.
—. (ed.) 1981: *Strabon, Géographie (livre XII)*, vol. 9 (Paris).
Lauffer, S. 1979: *Die Bergwerkssklaven von Laureion*, 2nd ed. (Forschungen zur antiken Sklaverei 11) (Stuttgart).
Lauriola, R. 2006: 'Athena and the Paphlagonian in Aristophanes' Knights. Re-Considering Equites 1090–5, 1172–81'. *Mnemosyne* 59, 75–94.
Lawall, M.L. 2010: 'Pontic, Aegean and Levantine Amphorae at Gordion'. In Kassab Tezgör, D. and Inaishvili, N. (eds.), *PATABS I: Production and Trade of Amphorae in the Black Sea* (Actes de la Table Ronde internationale de Batoumi et Trabzon, 27–29 avril 2006) (Varia Anatolica XXI) (Paris), 159–65.
Leaf, W. 1923: *Strabo on the Troad: Book XIII, Cap. I* (Cambridge).
Lehmann-Haupt C.F. 1931: *Armenien, einst und jetzt*, vol. 2 (Berlin).
Lendle, O. 1995: *Kommentar zu Xenophons Anabasis: Bücher 1–7* (Darmstadt).
Lenschau, T. 1913: 'Zur Geschichte Ioniens'. *Klio* 13, 175–83.
Leonhard, R. 1903: *Paphlagonische Denkmäler (Tumuli, Felsengräber, Befestigungen): Ergebnisse einer Reise* (Breslau).
—. 1915: *Paphlagonia: Reisen und Forschungen im Nördlichen Kleinasien* (Berlin).
Lewis, D. 2011: 'Near Eastern slaves in Classical Attica and the slave trade with Persian territories'. *The Classical Quarterly* 61.1, 91–113.
Lightfoot, J.L. 2014: *Dionysius Periegetes. Description of the Known World, with Introduction, Text, Translation, and Commentary* (Oxford).
Lipka, M. 1995: 'Anmerkungen zu geographischen, wirtschaftlichen und sozialen Verhältnissen an der südöstlichen Schwarzmeerküste Ende des fünften/Anfang des vierten Jhs. v. Chr.'. *Klio* 77, 65–74.
Long, T. 1986: *Barbarians in Greek Comedy* (Carbondale, IL).
Lotze, D. 1959: *Metaxy eleutherōn kai doulōn: Studien zur Rechtsstellung unfreier Landbevölkerungen in Griechenland bis zum 4. Jahrhundert v. Chr.* (Schriften der Sektion für Altertumswissenschaft 17) (Berlin).
Lund, J. and Gabrielsen, V. 2005: 'A Fishy Business: Transport Amphorae of the Black Sea Region as a Source for the Trade in Fish and Fish Products in the Classical and Hellenistic Periods'. In Bekker-Nielsen 2005, 161–69.
Luraghi, S. 1998: 'The Anatolian Languages'. In Ramat, A.G. and Ramat, P. (eds.), *The Indo-European Languages* (London/New York), 169–96.
Ma, J. 2004: 'You can't go home again: Displacement and Identity in Xenophon's Anabasis'. In Lane Fox 2004b, 330–45.
McInerney, J. 2014: 'Pelasgians and Leleges: Using the Past to Understand the Present'. In Ker, J. and Pieper, C. (eds.), *Valuing the Past in the Greco-Roman World* (Proceedings from the Penn-Leiden Colloquia on Ancient Values 7) (*Mnemosyne* Suppl. 369) (Leiden/Boston), 25–55.
Magdalino, P. 1998: 'Paphlagonians in Byzantine Society'. In *Η Βυζαντινή Μικρά Ασία: 6ος–12ος αι./ Byzantine Asia Minor (6th–12th cent.)* (Speros Basil Vryonis Center for the Study of Hellenism 27) (Athens), 141–50.
Magie, D. 1950: *Roman Rule in Asia Minor to the End of the Third Century after Christ*, 2 vols. (Princeton).
Maksimova, M.I. 1951: 'Hittites in the Black Sea Region'. *JNES* 10.2, 74–81.

—. 1956: *Antichnye goroda Yugo-Vostochnogo Prichernomor'ya: Sinopa, Amis, Trapezunt* (Moscow/Leningrad).
Malkin, I. 1987: *Religion and Colonization in Ancient Greece* (Studies in Greek and Roman Religion 3) (Leiden).
Malloy, A.G. 1970: *The Coinage of Amisus* (New York).
Manoledakis, M. 2005: *Δίκη Πολύποινος. Η απονομή δικαιοσύνης στην αρχαία ελληνική μυθολογία και τέχνη* (Thessaloniki).
—. 2010a: 'Choirades, Kerasous, Pharnakeia. Observations on three ancient placenames in the southern Black Sea'. *AWE* 9, 137–55.
—. 2010b: 'On the cults of Sinope and the founders of the city'. In Petropoulos, E.K. and Maslennikov, A.A. (eds.), *Ancient Sacral Monuments in the Black Sea* (Thessaloniki), 563–76.
—. 2013a: 'The Southern Black Sea in the Homeric Iliad: Some Geographical, Philological and Historical Remarks'. In Manoledakis 2013b, 19–37.
—. (ed.) 2013b: *Exploring the Hospitable Sea* (Proceedings of the International Workshop on the Black Sea in Antiquity, Thessaloniki, 21–23 September 2012) (BAR International Series 2498) (Oxford).
—. 2015a: 'Greek colonies and the southern Black Sea hinterland: looking closer into a long, complex and multidimensional relationship'. In Tsetskhladze *et al.* 2015, 81–89.
—. 2015b: 'The *Odyssey*, the Black Sea, and an endless voyage to a utopian destination'. *AWE* 14, 301–12.
—. 2015c: 'The Greek colonisation in the southern Black Sea from the viewpoint of the local populations'. In Adam-Veleni, P. and Tsangari, D. (eds.), *Greek Colonisation: New Data, Current Approaches* (Proceedings of the Scientific Meeting held in Thessaloniki, 6 February 2015) (Athens), 59–71.
—. 2016: 'From Macedonia to Anatolia. Some comments on the Phrygians and their migration'. In Giannopoulou, M. and Kallini, C. (eds.), *Volume in Honour of Prof. Stella Drougou* (Athens), 48–72.
—. 2018a: 'The local peoples of the southern Black Sea coast'. In Manoledakis *et al.* 2018, 147–71.
—. 2018b: 'Early Greek Presence in the Southern Black Sea'. In Manoledakis *et al.* 2018, 173–241.
—. 2019: 'Politics and Diplomacy in Paphlagonia'. In Tsetskhladze and Atasoy 2019, 214–25.
—. 2021a: '"Barbarians" in the southern Black Sea: the extreme case of the Mossynoikoi'. In Manoledakis 2021b, 53–63.
—. (ed.) 2021b: *Peoples in the Black Sea from the Archaic to the Roman Period* (Proceedings of the 3rd International Workshop on the Black Sea in Antiquity, held in Thessaloniki, 21–23 September 2018) (Oxford).
—. 2021c: 'The Paphlagonians and the Greeks' Perception of them'. In Tsetskhladze *et al.* 2021, 167–78.
—. 2021d: 'Paphlagonians and Phrygians'. In Boardman, J., Hargrave, J., Avram, A. and Podossinov, A. (eds.), *Connecting the Ancient West and East: Studies Presented to Prof. Gocha R. Tsetskhladze* (Monographs on Antiquity 8) (Leuven/Paris/Bristol, CT), 575–91 [published in 2022].
—. 2021e: 'Ionians in the Southern Black Sea Littoral'. In Tsetskhladze, G.R. (ed.), *Proceedings of the International Congress 'Ionians in the East and West', Empurias,*

26-29 October 2015 (Colloquia Antiqua 27) (Leuven/Paris/Bristol, CT), 895–914 [published in 2022].

—. 2021f: 'The districts of the southern Black Sea littoral in the era of Alexander the Great'. In Akamatis, I., Manoledakis, M., Nigdelis, P. and Xydopoulos, I. (eds.), *Ancient Macedonia* 8. *Macedonia from the Death of Philip II to Augustus' Rise to Power. Papers Read at the Eighth International Symposiun Held in Thessaloniki, November 21–24, 2017* (Thessaloniki), 629–49.

—. 2022a: 'The southern Black Sea in the Roman geographic texts'. In Braund, D., Chaniotis, A. and Petropoulos, E. (eds.), *The Black Sea Region in the Context of the Roman Empire* (International Symposium Dedicated in Memory of Victor I. Sarianidi, Athens 5–8 May 2016) (Athens), 33–58.

—. 2022b: 'The Identification of Xenophon's Kerasous: A New Solution to an Old Problem'. *AWE* 21, 109–25.

Manoledakis, M. and Livieratos, E. 2006: 'On the digital placement of Aegae, the first capital of ancient Macedonia, according to Ptolemy's Geographia'. In *Proceedings of the First International Workshop 'Digital Approaches to Cartographic Heritage'* (Thessaloniki), 262–70. (Also in *e-Perimetron* 2.1 [2007], 31–41: e-journal: *www.e-perimetron.org*).

Manoledakis, M., Tsetskhladze, G.R. and Xydopoulos, I. (eds.) 2018: *Essays on the Archaeology and Ancient History of the Black Sea Littoral* (Colloquia Antiqua 18) (Leuven/Paris/Bristol, CT).

Marcotte, D. (ed.) 2004: *Les géographes grecs I: Introduction générale. Ps.-Scymnos. Circuit de la terre* (Collection des universités de France 403) (Paris).

Marek, C. 1993: *Stadt, Ära und Territorium in Pontus-Bithynien und Nord-Galatien* (= *Istanbuler Forschungen* 39) (Tübingen).

—. 2003: *Pontus et Bithynia: die römischen Provinzen im Norden Kleinasiens* (Mainz).

—. 2010: *Geschichte Kleinasiens in der Antike* (Munich).

Marsh, B., Roberts, N., Toprak, V., Matthews, R., Eastwood, W., Carolan, J., Arcasoy, A. and Süzen, M.L. 2009: 'Contexts of Human Interaction: Geology, Geography, Geomorphology and Environment'. In Matthews and Glatz 2009, 27–73.

Matthews, R. 2009a: 'Project Paphlagonia: Research Issues, Approaches and Methods'. In Matthews and Glatz 2009, 1–25.

—. 2009b: 'A Dark Age, Grey Ware and Elusive Empires: Paphlagonia through the Iron Age, 1200–330 BC'. In Matthews and Glatz 2009, 149–71.

Matthews, R. and Glatz, C. (eds.) 2009: *At Empire's Edge: Project Paphlagonia. Regional Survey in North-Central Turkey* (British Institute at Ankara Monograph 44) (London).

Matsumura, K. 2008: 'The Early Iron Age in Kaman Kalehöyük: The Search for its Roots'. In Bonatz, B., Czichon, R.M. and Kreppner, F.J. (eds.), *Fundstellen: Gesammelte Schriften zur Archäologie und Geschichte Altvorderasiens ad honorem Hartmut Kühne* (Wiesbaden), 41–50.

Matsumura, K. and Omori, T. 2010: 'The Iron Age Chronology in Anatolia Reconsidered: The Results of the Excavations at Kaman Kalehöyük'. In Matthiae, P., Pinnock, F., Nigro, L. and Marchetti, N. (eds.), *ICAANE 6: Proceedings of the 6th International Congress on the Archaeology of the Ancient Near East (5 May–10 May 2009, "Sapienza" Università di Roma)*, vol. 1 (Wiesbaden), 443–55.

Mayor, A. 1995: 'Mad Honey!'. *Archaeology* 48.6, 32–40.

Mazarakis-Ainian, A. 2001: 'From Huts to Houses in Early Iron Age Greece'. In Brandt, J.R. and Karlsson, L. (eds.), *From Huts to Houses: Transformations of*

Ancient Societies (Proceedings of an International Seminar organised by the Norwegian and Swedish Institutes in Rome, 21–24 September 1997) (Stockholm), 139–61.

Mee, C. 1978: 'Aegean trade and settlement in Anatolia in the 2nd millennium BC'. *AS* 28, 121–56.

—. 1997: 'Anatolia and the Aegean in the Late Bronze Age'. In Cline, E.H. and Harris-Cline, D. (eds.), *The Aegean and the Orient in the Second Millennium* (Proceedings of the 50th Anniversary Symposium, University of Cincinnati, 18–20 April 1997) (Aegaeum 18) (Liège/Austin, TX), 137–46.

Mehl, A. 1987: 'Der Überseehandel von Pontos'. In Olshausen, E (ed.), *Stuttgarter Kolloquium zur historischen Geographie des Altertums* [1, 1980] (Geographica Historica 4) (Bonn), 103–86.

Melchert, H.C. 1994: *Anatolian Historical Phonology* (Leiden Studies in Indo-European 3) (Amsterdam/Atlanta).

Mellaart, J. 1978: *The Archaeology of Ancient Turkey* (London/Sydney/Toronto).

Mellink, M.J. 1974: 'Archaeology in Asia Minor'. *AJA* 78, 105–30.

—. 1975: 'Archaeology in Asia Minor'. *AJA* 79, 201–22.

—. 1976: 'Archaeology in Asia Minor'. *AJA* 80, 261–89.

—. 1984: 'Archaeology in Asia Minor'. *AJA* 88, 441–59.

—. 1985: 'Archaeology in Anatolia'. *AJA* 89, 547–67.

Meritt, B.D., Wade-Gery, H.T. and McGregor, M.F. 1950: *The Athenian Tribute Lists*, vol. 3 (Princeton).

Meyer, E. 1893: *Geschichte des Altertums*, vol. 3 (Darmstadt).

—. 1913: *Geschichte des Altertums*, vol. 1.2, 3rd ed. (Stuttgart/Berlin).

—. 1925: *Die Grenzen der hellenistischen Staaten in Kleinasien* (Zürich).

Michels, C. 2017: 'The Persian Impact on Bithynia, Commagene, Pontus, and Cappadocia'. In Müller, S., Howe, T., Bowden, H. and Rollinger, R. (eds.), *The History of the Argeads: New Perspectives* (Classica et Orientalia 19) (Wiesbaden), 41–55.

Miller, J.A. 1995: 'Gordium'. In Ring, T. (ed.), *International Dictionary of Historical Places 3: Southern Europe* (Chicago/London), 288–91.

Miltner, F. 1939: 'Die erste milesische Kolonisation im Südpontus'. In Calder, W.M. and Keil, J. (eds.), *Anatolian Studies Presented to William Hepburn Buckler* (Publications of the University of Manchester 265) (Manchester), 191–95.

Minns, E.H. 1913: *Scythians and Greeks: A Survey of Ancient History and Archaeology on the North Coast of the Euxine from the Danube to the Caucasus* (Cambridge).

Mitchell, L. 2007: *Panhellenism and the Barbarian in Archaic and Classical Greece* (Swansea).

Mitchell, S. 2005: 'Olive Cultivation in the Economy of Roman Asia Minor'. In Mitchell and Katsari 2005, 83–113.

—. 2010: 'The Ionians of Paphlagonia'. In Whitmarsh, T. (ed.), *Local Knowledge and Microidentities in the Imperial Greek World* (Cambridge), 86–110.

Mitchell, S. and Katsari, C. (eds.) 2005: *Patterns in the Economy of Roman Asia Minor* (Swansea).

Monachov, S.Y. 2010: 'New Series of Amphorae from Southern Pontic *Poleis* 4th – First Third of the 3rd Centuries BC'. In Kassab Tezgör, D. and Inaishvili, N. (eds.), *PATABS I: Production and Trade of Amphorae in the Black Sea* (Actes de la Table Ronde internationale de Batoumi et Trabzon, 27–29 avril 2006) (Varia Anatolica XXI) (Paris), 23–27.

Morel, J.-P. 2006: 'Phocaean Colonisation'. In Tsetskhladze 2006b, 359–428.
Morris, I. 1998: 'Archaeology and archaic Greek history'. In Fisher, N. and van Wees, H. (eds.), *Archaic Greece: New Approaches and New Evidence* (London), 1–92.
Moreu, C.J. 2003: 'The Sea Peoples and the Historical Background of the Trojan War'. *Mediterranean Archaeology* 16, 107–24.
Morrison, A.D. 2020: *Apollonius Rhodius, Herodotus and Historiography* (Cambridge).
Müller, C. 1855: *Geographi Graeci Minores*, vol. 1 (Paris, repr. New York 1990).
—. 1861: *Geographi Graeci Minores*, vol. 2 (Paris).
—. 1878: *Fragmenta Historicorum Graecorum*, vol. 3 (Paris).
Munn, M. 2008: 'Kybele as Kubaba in a Lydo-Phrygian context'. In Collins, B.J., Bachvarova, M.R. and Rutherford, I.C. (eds.), *Anatolian Interfaces: Hittites, Greeks and their Neighbours* (Proceedings of an International Conference on Cross-cultural Interaction, September 17–19, 2004) (Oxford), 159–64.
Munro, J.A.R. 1901: 'Roads in Pontus, Royal and Roman'. *JHS* 21, 52–66.
Neumann, G. 1961: *Untersuchungen zum Weiterleben hethitischen und luwischen Sprachgutes in hellenistischer und römischer Zeit* (Wiesbaden).
Nieling, J. and Rehm, E. (eds.) 2010: *Achaemenid Impact in the Black Sea: Communication of Powers* (Black Sea Studies 11) (Aarhus).
Nielsen, T.H. 2000: 'Xenophon's Use of the Word Polis in the Anabasis'. In Flensted-Jensen, P. (ed.), *Further Studies in the Ancient Greek Polis* (*Historia* Einzelschriften 138) (Stuttgart), 133–39.
Nilsson, M.P. 1932: *The Mycenaean Origin of Greek Mythology* (Sather Classical Lectures 8) (Cambridge).
Nöldeke, T. 1871: 'ΑΣΣΥΡΙΟΣ ΣΥΡΙΟΣ ΣΥΡΟΣ'. *Hermes* 5, 443–68.
Nollé, J. 2005: 'Boars, Bears, and Bugs. Farming in Asia Minor and the protection of men, animals, and crops'. In Mitchell and Katsari 2005, 53–82.
Ökse, A.T. 2019: 'The Eastern Border of Phrygia'. In Tsetskhladze 2019b, 161–87.
Olbrycht, J. 2000a: 'The Cimmerian Problem Re-Examined: The Evidence of the Classical Sources'. In Ptsrusinska, J. and Fear, A. (eds.), *Collectanea Celto-Asiatica Cracoviensia* (Cracow), 71–99.
—. 2000b: 'Notes on the Presence of Iranian Peoples in Europe and their Asiatic Relations'. In Ptsrusinska, J. and Fear, A. (eds.), *Collectanea Celto-Asiatica Cracoviensia* (Cracow), 101–40.
Olshausen, E. 1994: 'Amisos'. In Zahariade, M. (ed.), *Lexicon of the Greek and Roman Cities and Place Names in Antiquity, c. 1500 BC–c. AD 500*, fasc. 3 (Amsterdam), 476–77.
—. 2012: 'Chalyben: Autonym oder Xenonym?'. In Olshausen, E. and Sauer, V. (eds.), *Die Schätze der Erde – Natürliche Ressourcen in der antiken Welt* (Geographica Historica 28) (Stuttgart), 337–44.
—. 2015: 'News from the East? Roman-Age Geographers and the Pontus Euxinus'. In Bianchetti *et al.* 2015, 259–73.
Omura, S. 2011: 'Kaman-Kalehöyük Excavations in Central Anatolia'. In Steadman and McMahon 2011, 1095–1111.
Oren, E.D. (ed.) 2000: *The Sea Peoples and their World: A Reassessment* (University Museum Monograh 108) (Philadelphia).
Özdoğan, M. 2003: 'The Black Sea, the Sea of Marmara and Bronze Age Archaeology: An Archaeological Predicament'. In Wagner, G., Pernicka, E. and

Uerpmann, H.-P. (eds.), *Troia and the Troad: Scientific Approaches* (Berlin/ Heidelberg), 105–20.
—. 2007: 'Coastal changes of the Black Sea and Sea of Marmara in archaeological perspective'. In Yanko-Hombach *et al.* 2007, 651–69.
Özgüç, T. 1978: *Maşat Höyük kazıları ve çevresindeki araştırmalar/Excavations at Maşat Höyük and Investigations in its Vicinity* (Türk Tarih Kurumu yayınlarından, V. dizi, 38) (Ankara).
—. 1980: 'Excavations at the Hittite Site, Maşat Höyük: Palace, Archives, Mycenaean Pottery'. *AJA* 84, 305–09.
—. 1982: *Maşat Höyük II: Boğazköy'ün kuzeydoğusunda bir Hitit merkezi/ A Hittite Center Northeast of Boğazköy* (Türk Tarih Kurumu yayınlarından, V. dizi, 38a) (Ankara).
Özlem-Aytaçlar, P. 2010: 'An Onomastic Survey of the Indigenous Population of North-Western Asia Minor'. In Catling, R.W.V. and Marchand, F. (eds.), *Onomatologos: Studies in Greek Personal Names presented to Elaine Matthews* (Oxford), 506–29.
Öztürk, B. 2013: 'The History of Tieion/Tios (Eastern Bithynia) in the Light of Inscriptions'. In Manoledakis 2013b, 147–64.
Özveren, Y.E. 1997: 'A Framework for the Study of the Black Sea World, 1789–1915'. *Review (Fernand Braudel Center)* 20.1, 77–113.
—. 2001: 'The Black Sea as a Unit of Analysis'. In Aybak, T. (ed.), *Politics of the Black Sea: Dynamics of Cooperation and Conflict* (London), 61–84.
Page, D.L. 1959: *History and the Homeric Iliad* (Berkeley).
Parker, G. 2008: *The Making of Roman India* (Cambridge).
Parker, V. 1998: 'O chem umalchivaet Gerodot. Zametki o peredache svedenii o Kimmeriitsakh u Grecheskikh avtorov pomimo Gerodota'. *Vestnik drevnei istorii* 4 (227), 93–102.
Peev, P. 2016: 'Using archaeological data to infer past relative sea-level positions along the Bulgarian coast of the Black Sea'. *Méditerranée* 126, 17–24.
Peretti, A. 1961: 'Eforo e Pseuso-Scilace'. *Studi Classici e Orientali* 10, 5–43.
—. 1979: *Il Periplo di Scilace. Studio sul primo portolano del Mediterraneo* (Biblioteca di Studi antichi 23) (Pisa).
Perez, A. 2010: 'Les Bébryces d'Occident ont-ils existé?' *Pallas* 84, 37–58.
Petropoulos, E.K. 2005a: *Hellenic Colonization in Euxeinos Pontos: Penetration, Early Establishment and the Problem of the 'Emporion' Revisited* (BAR International Series 1394) (Oxford).
—. 2005b: 'Emporion and apoikia-polis in the Northeast Black Sea Area during the 6th and 5th c. BC: an Urbanistic View'. In Kacharava *et al.* 2005, 207–32.
Pogrebova, M.N, Raevskii, D.S. and Yatsenko, I.V. 1998: 'Kimmeriiskaya problema (Po povodu knigi: A.Y. Alexeev, N.K. Kachalova, S.R. Tokhtasev, Kimmeriitsy: etnokulturnaya prinadlezhnost', Saint Petersburg, 1993)'. *Vestnik drevnei istorii* 3 (226), 69–87.
Popko, M. 2008: *Völker und Sprachen Altanatoliens* (Wiesbaden).
Porotov, A. 2007: 'Relative sea-level changes and submersion of archaeological sites along the northern shoreline of the Black Sea'. *Méditerranée* 108, 29–36.
Prêteux, F. 2005: 'Priapos Bébrykès dans le Propontide et les Détroits: succès d'un mythe local'. *Revue des Études Grecques* 118, 246–65.
Price, M. 1993: *Syllogae Nummorum Graecorum IX: The British Museum. Part 1: The Black Sea* (London).

Prontera, F. 2015: 'Strabo's *Geography*'. In Bianchetti *et al.* 2015, 239–58.
Rabe, H. 1906: *Scholia in Lucianum* (Leipzig).
Ramsay, W.M. 1890: *The Historical Geography of Asia Minor* (London).
Rapp, S.H. and Crego, P. (eds.) 2012: *Languages and Cultures of Eastern Christianity: Georgian* (The Worlds of Eastern Christianity 5) (New York).
Rathmann, M. 2018: *Tabula Peutingeriana: Die einzige Weltkarte aus der Antike* (Darmstadt).
Rayfield, D. 2012: *Edge of Empires: A History of Georgia* (London).
Redfield, J. 1985: 'Herodotus the Tourist'. *Classical Philology* 80.2, 97–118.
Rehm, E. 2010: 'The Classification of Objects from the Black Sea Region Made or Influenced by the Achaemenids'. In Nieling and Rehm 2010, 161–94.
Reinach, T. 1975: *Mithridates Eupator, König von Pontos* (Hildesheim/New York; first published Paris 1890).
Rempel, J. and Doonan, O.P. 2020: 'Rural hinterlands in the Black Sea during the fourth century BCE: Expansion, intensification and new connections'. *AS* 70, 127–51.
Renfrew, A.C. and Cherry, J.F. (eds.) 1986: *Peer Polity Interaction and Socio-Political Change* (Cambridge).
Rengakos, A. 1993: *Der Homertext und die hellenistischen Dichter* (Hermes 64) (Stuttgart).
Rigsby, K.J. 1980: 'Seleucid Notes'. *TAPA* 110, 233–54.
Robert, L. 1980: *À travers l'Asie Mineure: poètes et prosateurs, monnaies grecques, voyageurs et géographie* (Bibliothèque des écoles françaises d'Athènes et de Rome 239) (Paris).
Robinson, D.M. 1906: 'Ancient Sinope'. *American Journal of Philology* 27, 125–53, 245–69.
Robinson, E.S.G. 1920: 'A Find of Coins of Sinope'. *The Numismatic Chronicle and Journal of the Royal Numismatic Society* 20, 1–16.
Roebuck, C. 1959: *Ionian Trade and Colonization* (New York).
Roller, D.W. 2018: *A Historical and Topographical Guide to the Geography of Strabo* (Cambridge).
Roller, L.E. 1999: *In Search of God the Mother: The Cult of Anatolian Cybele* (Berkeley/London).
—. 2011, 'Phrygian and the Phrygians'. In Steadman and McMahon 2011, 560–78.
Rollinger, R. 2006: 'The Terms "Assyria" and "Syria" Again'. *JNES* 65.4, 283–87.
Romer, F.E. 2001: *Pomponius Mela's Description of the World* (Ann Arbor).
Rood, T. 2005: *Xenophon, The Expedition of Cyrus* (Oxford).
—. 2010: 'Xenophon's Parasangs'. *JHS* 130, 51–66.
Root, M.C. 1991: 'From the Heart: Powerful Persianisms in the Art of the Western Empire'. In Sancisi-Weerdenburg, H. and Kuhrt, A. (eds.), *Achaemenid History VI: Asia Minor and Egypt: Old Cultures in a New Empire* (Proceedings of the Groningen 1998 Achaemenid History Workshop) (Leiden), 1–29.
Rostovtzeff, M. and Ormerod, H.A. 1932: 'Pontus and its Neighbours: The First Mithridatic War'. *CAH* IX, 211–60.
Rouillard, P. 1993: '*L'emporion* chez Strabon'. In Bresson and Rouillard 1993, 35–46.
Ruge, W. 1941: 'Phrygia 1 (Topographie)'. *RE* XX.1, 781–868.
—. 1949: 'Paphlagonia'. *RE* XVIII.4, 2486–2537 (rev. by K. Bittel).
Ruscu, L. 2008: 'Sinopeans abroad and foreigners at Sinope'. *AWE* 7, 81–106.
Russell, T. 2016: *Byzantium and the Bosporus: A Historical Study, from the Seventh Century BC until the Foundation of Constantinople* (Oxford).

Ryan, W.B.F. 2007: 'Status of the Black Sea flood hypothesis'. In Yanko-Hombach *et al.* 2007, 63–88.

Ryan, W.B.F., Major, C.O., Lericolais, G. and Goldstein, S.L. 2003: 'Catastrophic Flooding of the Black Sea'. *Annual Review of Earth and Planetary Sciences* 31, 525–54.

Ryan, W.B.F. and Pitman, W.C. 1998: *Noah's Flood: The New Scientific Discoveries about the Event that Changed History* (New York).

Sağlan, S. and Korkmaz, Z. 2021: 'Attic red-figure pottery from Sinope'. In Tsetskhladze *et al.* 2021, 616–27.

Sakellariou, M.B. 1958: *La migration grecque en Ionie* (Athens).

Saprykin, S.J. 1991: 'ΠΑΦΛΑΓΟΝΙΚΑ'. *Thracia Pontica* 4, 241–54.

—. 1997: *Heracleia Pontica and Tauric Chersonesus before Roman Domination (VI–I Centuries BC)* (Amsterdam).

Sartre, M. 1991: *L'Orient romain: Provinces et sociétés provinciales en Méditerranée orientale d'Auguste aux Sévères (31 avant J.-C.–235 après J.-C.)* (Paris).

—. 2003: *L'Anatolie hellénistique de l'Égée au Caucase (334–31 av. J.-C.)* (Paris).

Šašel Kos, M. 2004: 'Mythological stories concerning Illyria and its name'. In *L'Illyrie méridionale et l'Épire dans l'Antiquité IV* (Actes du IVe colloque international de Grenoble, 10–12 octobre 2002) (Paris), 493–504.

Schachner, A. 2011: *Hattuscha: Auf der Suche nach dem sagenhaften Großreich der Hethiter* (Munich).

Schmitt Pantel, P. 2015: 'Dining in Ancient Greece'. In Wilkins and Nadeau 2015, 224–33.

Scholten, J. 2007: 'Building Hellenistic Bithynia'. In Elton, H. and Reger, G. (eds.), *Regionalism in Hellenistic and Roman Asia Minor* (Acts of the Conference, Hartford, Connecticut, August 22–24, 1997) (Études 20) (Paris), 17–24.

Schoville, K.N. 2005: 'Scythopolis, City of the Scyths?'. In Carnagey, G.A. (ed.), *Beyond the Jordan: Studies in Honor of W. Harold Mare* (Eugene, OR), 61–75.

Schwartz, E. 1907: 'Ephoros 1'. *RE* VI.1, 1–16.

Seeher, J. 2011: 'The Plateau: The Hittites'. In Steadman and McMahon 2011, 376–92.

Selden, J. 1617: *De Dis Syris* (Leipzig).

Senff, R. 2000. 'Die archaische Wohnebaung am Kalabaktepe in Milet'. In Krizinger, F. (ed.), *Die Ägäis und das westliche Mittelmeer. Beziehungen und Wechselwirkungen 8. bis 5. Jh. v. Chr.* (Akten des Symposions, Wien, 24. bis 27. März 1999) (Denkschriften, Österreichische Akademie der Wissenschaften. Philosophisch–Historische Klasse 288; Archäologische Forschungen 4) (Vienna), 29–38.

Şenyurt, S.Y. 2015: 'Kurul Kalesi Excavations: Archaeological Contributions to the Pontic Kingdom of Mithridates VI Eupator'. In Frecer, R. (ed.), *Contacts, Migrations and Climate Change. Conference in Honour of Jan Bouzek's illustrious 80th Birthday, May 19th–23rd, 2015* (Conference Abstracts) (Prague), 12.

Şerifoğlu, T.E. 2015: 'The Cide–Şenpazar Region in the Iron Age (ca. 1200 to 325/300 BC)'. In Düring and Glatz 2015a, 212–45.

Şerifoğlu, T.E. and Bakan, C. 2015: 'The Cide–Şenpazar Region during the Hellenistic Period (325/300 – 1 BC)'. In Düring and Glatz 2015a, 246–59.

Shipley, G. 2011: *Pseudo-Skylax's Periplous* (Exeter).

Simpson, R.H. and Lazenby, J.F. 1970: *The Catalogue of the Ships in Homer's Iliad* (Oxford).

Şirin, O.A. and Kolağasioğlu, M. 2016: *Çakalca–Karadoğan Höyüğü. Arkaik Dönemde Amisos ve Kybele Kültü* (Samsun).

Sivas, T.T. and Sivas, H. (eds.) 2012: *Frigler. Midas'in Ülkesinde, Anitlarin Gölgesinde/ Phrygians. in the Land of Midas, in the Shadow of Monuments* (Istanbul).
Six, M.J.P. 1885: 'Sinope'. *The Numismatic Chronicle and Journal of the Numismatic Society* 5, 15–65.
—. 1895: 'Monnaies grecques, inédites et incertaines (suite)'. *The Numismatic Chronicle and Journal of the Numismatic Society* 15, 169–210.
Stählin, F. 1967: *Das Hellenistische Thessalien: Landeskundliche und geschichtliche Beschreibung Thessaliens in der hellenistischen und römischen Zeit* (Amsterdam).
Starr, S.F. 1962: *The Ancient Roads of Asia Minor* (Yale).
Steadman, S.R. 2011: 'The Early Bronze Age on the Plateau'. In Steadman and McMahon 2011, 229–59.
Steadman, S.R. and McMahon, G. (eds.) 2011: *The Oxford Handbook of Ancient Anatolia (10,000–323 BCE)* (Oxford).
Stolba, V.F. 2005: 'Fish and Money: Numismatic Evidence for Black Sea Fishing'. In Bekker-Nielsen 2005, 115–32.
Stoop, M.W. 1977–78: 'Ancient Armene and its Harbour'. *Anatolica* 6, 117–28.
Stückelberger, A. and Grasshoff, G. (eds.) 2006: *Klaudios Ptolemaios – Handbuch der Geographie* (Basel).
Stylianou, P.J. 2004: 'One Anabasis or Two?'. In Lane Fox 2004b, 68–96.
Sulimani, I. 2011: *Diodorus' Mythistory and the Pagan Mission: Historiography and Culture-Heroes in the First Pentad of the Bibliotheke* (Leiden/Boston).
Sulimirski, T. and Taylor, T. 1991: 'The Scythians'. *CAH* III.2², 547–90.
Summerer, L. 1999: *Hellenistische Terrakotten aus Amisos: ein Beitrag zur Kunstgeschichte des Pontosgebietes* (Geographica Historica 13) (Stuttgart).
—. 2003: 'Achämenidische Silberfunde aus der Umgebung von Sinope'. *ACSS* 9.1–2, 17–42.
—. 2005a: 'Amisos – eine Griechische Polis im Land der Leukosyrer'. In Kacharava *et al.* 2005, 129–65.
—. 2005b: 'Griechische Tondächer im kappadokischen Kontext: Die Architekturterrakotten aus Akalan'. In Fless, F. and Treister, M. (eds.), *Bilder und Objekte als Träger kultureller Identität und interkultureller Kommunikation im Schwarzmeergebiet* (Kolloquium in Zschortau/Sachsen vom 13.2.–15.2.2003) (Internationale Archäologie. Arbeitsgemeinschaft, Symposium, Tagung, Kongress 6) (Rahden), 125–39.
—. 2005c: 'Achämeniden am Schwarzen Meer: Bemerkungen zum spätarchaischen Marmorkopf aus Herakleia Pontike'. *ANES* 42, 231–52.
—. 2007: 'Greeks and Natives on the Southern Black Sea Coast in Antiquity'. In Erkut, G. and Mitchell, S. (eds.), *The Black Sea: Past, Present and Future* (Proceedings of the International Interdisciplinary Conference, Istanbul, 14–16 October 2004) (British Institute of Archaeology at Ankara Monograph 42) (London/ Istanbul), 27–36.
—. 2008: 'Indigenous Responses to Encounters with the Greeks in Northern Anatolia: The Reception of Architectural Terracottas in the Iron Age Settlements of the Halys Basin'. In Bilde, P.G. and Petersen, J.H. (eds.), *Meetings of Cultures in the Black Sea Region: Between Conflicts and Coexistence* (Black Sea Studies 8) (Aarhus), 263–86.
—. 2009: 'Herakles in Paphlagonien'. In Einicke, R., Lehmann, S., Lohr, H., Mehnert, G., Mehnert, A. and Slawisch, A. (eds.), *Zurück zum Gegenstand. Festschrift für*

Andreas E. Furtwängler (Schriften des Zentrums für Archäologie und Kulturgeschichte des Schwarzmeerraumes 16) (Langenweißbach), 15–24.
—. 2014: 'Topographies of worship in northern Anatolia'. In Bekker-Nielsen, T. (ed.), *Space, Place and Identity in Northern Anatolia* (Geographica Historica 29) (Stuttgart), 189–213.
Summerer, L. and von Kienlin, A. 2010: 'Achaemenid Impact in Paphlagonia: Rupestral Tombs in the Amnias Valley'. In Nieling and Rehm 2010, 195–221.
Summers, G.D. 1994: 'Grey Ware and the Eastern Limits if Phrygia'. In Çilingiroğlu, A. and French, D.H. (eds.), *Anatolian Iron Ages 3: The Proceedings of the Third Anatolian Iron Ages Colloquium held at Van, 6–12 August 1990* (British Institute of Archaeology at Ankara Monograph 16) (London), 241–52.
—. 1997: 'The identification of the Iron Age City on Kerkenes Dağ in Central Anatolia'. *JNES* 56.2, 81–94.
—. 2006: 'Architectural Terracottas in Greater Phrygia: Problems of Chronology and Distribution'. In Erkanal-Öktü, A. (ed.), *Studies in Honor of Hayat Erkanal* (Istanbul), 684–88.
Szamalek, J.K. 2014: 'Greeks and the Peoples of the Black Sea Region. Beyond Ethnicity and Identity: an Archaeology of Commonalities'. *Dialogues d'histoire ancienne* Suppl. 10, 53–80.
Talbert, R.J.A. 2000 (ed.): *The Barrington Atlas of the Greek and Roman World*, 3 vols. (Princeton).
Taracha, P. 2009: *Religions of Second Millennium Anatolia* (Dresdner Beiträge zur Hethitologie 27) (Wiesbaden).
Temür, A. and Iltar, G. 2018: 'Giresun, the Excavation of the Island of Aretias/Khalkeritis'. In Yiğitpaşa *et al.* 2018, 1–11.
Thallon, I.C. 1924: 'The tradition of Antenor and its Historical Possibility'. *AJA* 28.1, 47–65.
Thalmann, W.G. 2011: *Apollonius of Rhodes and the Spaces of Hellenism* (Oxford).
Thomas, H. and Stubbings, F.H. 1963: *A Companion to Homer* (London).
Thompson, D.J. 2011: 'Slavery in the Hellenistic World'. In Bradley, K. and Cartledge, P. (eds.), *The Cambridge World History of Slavery 1: The Ancient Mediterranean World* (Cambridge), 194–213.
Thonemann, P. 2013: 'Phrygia: an anarchist history, 950 BC–AD 100'. In Thonemann, P. (ed.), *Roman Phrygia: Culture and Society* (Cambridge), 1–40.
Tiverios, M. 2008: 'Greek Colonisation of the Northern Aegean'. In Tsetskhladze 2008b, 1–154.
Tokhtas'ev, S.R. 1991: 'Cimmerians'. *Encyclopaedia Iranica* V.6, 563–67.
—. 1996: 'Die Kimmerier in der antiken Überlieferung'. *Hyperboreus* 2, 1–46.
Tozer, H.F. 1971: *A History of Ancient Geography* (New York).
Tsetskhladze, G.R. 1994a: 'Greek Penetration of the Black Sea'. In Tsetskhladze, G.R. and De Angelis, F. (eds.), *The Archaeology of Greek Colonisation: Essays Dedicated to Sir John Boardman* (Oxford), 111–35.
—. 1994b: 'Argonautica, Colchis and the Black Sea: Myth, Reality and Modern Scholarship'. *Thracia Pontica* 6.1, 337–42.
—. 1995: 'Did the Greeks go to Colchis for Metals?'. *Oxford Journal of Archaeology* 14.3, 307–31.
—. 1997: 'Plutarch, Pericles and Pontus: some thoughts'. In Schrader, C., Ramón, V. and Vela, J. (eds.), *Plutarco y la Historia* (Zaragoza), 461–66.

—. 1998a: 'Greek Colonisation of the Black Sea Area: Stages, Models, Native Population'. In Tsetskhladze 1998b, 9–68.
—. (ed.) 1998b: *The Greek Colonisation of the Black Sea Area: Historical Interpretation of Archaeology* (*Historia* Einzelschriften 121) (Stuttgart).
—. 1999: 'Between West and East: Anatolian Roots of Local Cultures of the Pontus'. In Tsetskhladze, G.R. (ed.), *Ancient Greeks West and East* (*Mnemosyne* Suppl. 196) (Leiden/Boston/Cologne), 469–96.
—. 2002a: 'Black Sea Piracy'. In Tsetskhladze, G.R. and de Boer, J.G. (eds.), *The Black Sea Region in the Greek, Roman and Byzantine Periods* (Amsterdam) (= *Talanta* 31–32 [for 2000–01]), 11–15.
—. 2002b: 'Ionians Abroad'. In Tsetskhladze, G.R. and Snodgrass, A.M. (eds.), *Greek Settlements in the Eastern Mediterranean and the Black Sea* (BAR International Series 1062) (Oxford), 81–96.
—. 2004: 'On the Earliest Greek Colonial Architecture in the Pontus'. In Tuplin, C.J. (ed.), *Pontus and the Outside World. Studies in Black Sea History, Historiography and Archaeology* (Colloquia Pontica 9) (Leiden/Boston), 225–78.
—. 2006a: 'Revisiting Ancient Greek Colonisation'. In Tsetskhladze 2006b, xxiii–lxxxiii.
—. (ed.) 2006b: *Greek Colonisation: An Account of Greek Colonies and Other Settlements Overseas*, vol. 1 (*Mnemosyne* Suppl. 193/1) (Leiden/Boston).
—. 2007: 'Greeks and Locals in the Southern Black Sea Littoral: A Re-Examination'. In Herman, G. and Shatzman, I. (eds.), *Greeks between East and West: Essays in Greek Literature and History in Memory of David Asheri* (Jerusalem), 160–95.
—. 2008a: 'Pontic Slaves in Athens: Orthodoxy and Reality'. In Mauritsch, P., Petermandl, W., Rollinger, R. and Ulf, C. (eds.), *Antike Lebenswelten. Konstanz – Wandel – Wirkungsmacht. Festschrift für Ingomar Weiler zum 70. Geburtstag* (Philippika 25) (Wiesbaden), 309–19.
—. (ed.) 2008b: *Greek Colonisation: An Account of Greek Colonies and Other Settlements Overseas*, vol. 2 (*Mnemosyne* Suppl. 193/2) (Leiden/Boston).
—. 2009: 'Secondary Colonisers in the Black Sea: Sinope and Panticapaeum'. In Lombardo, M. and Frisone, F. (eds.), *Colonie di colonie: le fondazioni subcoloniali greche tra colonizzazione e colonialismo* (Atti del Convegno Lecce, 22–24 Giugno 2006) (Università del Salento, Dipartimento di beni culturali 16) (Galantina), 229–54.
—. 2010: 'Beware of Greeks Bearing Gifts'. In Rollinger, R., Gufler, B., Lang, M. and Madreiter, I. (eds.), *Interkulturalität in der Alten Welt: Vorderasien, Hellas, Ägypten und die vielfältigen Ebenen des Kontakts* (Philippika 34) (Wiesbaden), 41–61.
—. 2011: 'The Scythians: Three Essays'. In Tsetskhladze, G.R. (ed.), *The Black Sea, Greece, Anatolia and Europe in the First Millennium BC* (Colloquia Antiqua 1) (Leuven/Paris/Walpole, MA), 95–139 (essay 2 = 'Scythians in the Central Black Sea Region of Turkey?'. In Öztan, A. and Dönmez, S. [eds.], *Knowledge Production for the Black Sea to the Euphrates. Studies Presented in Honour of Önder Bilgi* [Ankara 2011], 429–35).
—. 2012a: 'Pots versus People: Further Consideration of the Earliest Examples of East Greek Pottery in Native Settlements of the Northern Pontus'. In Hermary, A. and Tsetskhladze, G.R. (eds.), *From the Pillars of Hercules to the Footsteps of the Argonauts* (Colloquia Antiqua 4) (Leuven/Paris/Walpole, MA), 315–74.

—. 2012b: 'The Southern Black Sea Coast and its Hinterland: An Ethno-Cultural Perspective'. In Tsetskhladze, G.R. (ed.), *The Black Sea, Paphlagonia, Pontus and Phrygia in Antiquity: Aspects of Archaeology and Ancient History* (BAR International Series 2432) (Oxford), 235–41.

—. 2013: 'Not all Things Greek came through Greek Trade or at the Hands of Greeks'. In Tsetskhladze *et al.* 2013, 69–82.

—. 2014: 'Black Sea Ethnicities'. In McInerney, J. (ed.), *A Companion to Ethnicity in the Ancient Mediterranean* (Malden, MA/Oxford), 312–26.

—. 2015: 'Greeks, locals and others around the Black Sea and its hinterland: recent developments'. In Tsetskhladze *et al.* 2015, 11–42.

—. 2017: 'Bridging East and West: Locals, Greeks and Achaemenids around the Black Sea'. In Gallo, L. and Genito, B. (eds.), *Grecità di frontiera: Frontiere geografiche e culturali nell'evidenza storica e archeologica* (Atti del convegno internazionale, Napoli, 5–6 giugno 2014) (Studi di Storia greca e romana 14) (Alessandria), 27–44.

—. 2018: '"Greek Penetration of the Black Sea": Twenty Years on'. In Manoledakis *et al.* 2018, 1–64.

—. 2019a: 'Introduction. Pre-Hellenistic Phrygia: Some Matters in Debate'. In Tsetskhladze 2019b, 1–23.

—. (ed.) 2019b: *Phrygia in Antiquity: From the Bronze Age to the Byzantine Period* (Colloquia Antiqua 24) (Leuven/Paris/Bristol, CT).

—. 2019c: 'Once again abut the Establishment Date of Some Greek Colonies around the Black Sea'. In Tsetskhladze and Atasoy 2019, 1–41.

—. 2019d: 'Some Recent Developments in the Study of the Greeks Overseas'. In Moralis, R., Leão, D., Rodriguez Pérez, D. and Ferreira, D. (eds.), *Greek Art in Motion. Studies in Honour of Sir John Boardman on the Occasion of his 90th Birthday* (Oxford), 59–65.

—. 2019e: 'An Achaemenid Inscription from Phanagoria: Extending the Boundaries of Empire'. *AWE* 18, 113–51.

—. 2021a: 'Pontic Greeks and Locals: Subterranean Dwellings Once Again'. In Manoledakis 2021b, 149–84.

—. 2021b: 'Pontic studies twenty years on: terra incognita?'. In Tsetskhladze *et al.* 2021, 2–9.

Tsetskhladze, G.R. and Atasoy, S. (eds.) 2019: *Settlements and Necropoleis of the Black Sea and its Hinterland in Antiquity* (Select papers from the Third International Conference 'The Black Sea in Antiquity and Tekkeköy: An Ancient Settlement on the Southern Black Sea Coast', 27–29 October 2017, Tekkeköy, Samsun) (Oxford).

Tsetskhladze, G.R., Atasoy, S., Avram, A., Dönmez, Ş. and Hargrave, J.F. (eds.) 2013: *The Bosporus: Gateway between the Ancient West and East (1st Millennium BC–5th Century AD)* (Proceedings of the Fourth International Congress on Black Sea Antiquities, Istanbul, 14th–18th September 2009) (BAR International Series 2517) (Oxford).

Tsetskhladze, G.R., Avram, A. and Hargrave, J.F. (eds.) 2015: *The Danubian Lands between the Black, Aegean and Adriatic Seas (7th Century BC–10th Century AD)* (Proceedings of the Fifth International Congress on Black Sea Antiquities, Belgrade, 17–21 September 2013) (Oxford).

—. (eds.) 2021: *The Greeks and Romans in the Black Sea and the Importance of the Pontic Region for the Graeco-Roman World (7th Century BC–5th Century AD):*

20 Years On (1997–2017) (Proceedings of the Sixth International Congress on Black Sea Antiquities, Constanţa, 18–22 September 2017) (Oxford).
Tsetskhladze, G.R. and Hargrave, J.F. 2011: 'Colonisation from Antiquity to Modern Times: Comparison and Contrasts'. *AWE* 10, 161–82.
Tsiafakis, D. 2010: 'Domestic Architecture in the Northern Aegean: the Evidence from the ancient settlement of Karabournaki'. In Tréziny, H. (ed.), *Grecs et indigènes de la catalogne à la Mer Noire* (Paris), 379–87.
—. 2013: 'Architectural similarities(?) between the Black Sea and North Aegean settlements'. In Tsetskhladze *et al.* 2013, 61–68.
—. 2015: 'Thracian Tattoos'. In Boschung, D., Shapiro, A. and Waschek, F. (eds.), *Bodies in Transition: Dissolving the Boundaries of Embodied Knowledge* (Paderborn), 89–117.
Tülek, F. 2018: 'A Small Island of Great Importance in Maritime Trade of the South Black Sea Coast: Daphnousia Island/Kefken'. In Yiğitpaşa *et al.* 2018, 155–59.
Tuplin, C.J. 1987: 'The Administration of the Achaemenid Empire'. In Carradice, I. (ed.), *Coinage and Administration in the Athenian and Persian Empires* (The Ninth Oxford Symposium on Coinage and Monetary History) (BAR International Series 343) (Oxford), 109–66.
—. 1999: 'On the Track of the Ten Thousand'. *Revue des Études Anciennes* 101, 331–66.
—. 2004: 'The Persian Empire'. In Lane Fox 2004b, 154–83.
—. 2007a: 'A Foreigner's Perspective: Xenophon in Anatolia'. In Delemen, I., Casabonne, O., Karagöz, Ş. and Tekin, O. (eds.), *The Achaemenid Impact on Local Populations and Culture in Anatolia* (Istanbul), 7–32.
—. 2007b: 'Fear of Slavery and the Failure of the Polis'. In *Fear of Slaves – Fear of Enslavement in the Ancient Mediterranean (Discourse, Representations, Practices)* (Actes du XXIXe colloque du groupe international de recherche sur l'esclavage dans l'antiquité [GIREA], Rethymnon, 4–7 novembre 2004) (Besançon), 57–74.
Türker, A. and Özdemir, M.A. 2019: 'Iron Age Pottery from Suluca Karahöyük and its Possibility of Being "Phrygian"'. In Tsetskhladze 2019b, 411–28.
Tvedtnes, J.A. 1981: 'The Origin of the Name Syria'. *JNES* 40, 139–40.
Vachtina, M.Ju. 2007: 'Greek Archaic Orientalising Pottery from the Barbarian Sites of the Forest-steppe Zone of the Northern Black Sea Coastal Region'. In Gabrielsen and Lund 2007, 23–37.
Vaggione, R.P. 1973: 'Over All Asia? The Extent of the Scythian Domination in Herodotus'. *Journal of Biblical Literature* 92.4, 523–30.
van Dongen, E. 2013: 'The Overland Route: Intra-Anatolia Interaction *ca.* 1000–540 BC and the Transmission of the Alphabet'. *AWE* 12, 47–70.
Vanschoonwinkel, J. 2006: 'Mycenaean Expansion'. In Tsetskhladze 2006b, 41–113.
Vian, F. and Delage, É. 1974: *Apollonios de Rhodes Argonautiques* (Paris).
Vitucci, G. 1953: *Il regno di Bitinia* (Roma).
Vlassopoulos, K. 2013: *Greeks and Barbarians* (Cambridge).
Voigt, M.M. 2009: 'The Chronology of Phrygian Gordion'. In Manning, S.W. and Bruce, M.J. (eds.), *Tree-Rings, Kings, and Old World Archaeology and Environment: Papers Presented in Honor of P.I. Kuniholm* (Oxford/Oakville, CT), 219–37.
—. 2011: 'Gordion: The Changing Political and Economic Roles of a First Millennium B.C.E. City'. In Steadman and McMahon 2011, 1069–94.
Vryonis, S. 1971: *The Decline of Medieval Hellenism in Asia Minor and the Process of Islamization from the Eleventh through the Fifteenth Century* (Berkeley/Los Angeles/London).

Wainwright, G.A. 1936: 'Tabal, Tibareni, Tebarani'. *Orientalistische Literaturzeitung* 39, 479–81.

Weber, G. 2011: 'Der Troianische Krieg: Historische Realität oder poetische Fiktion'. In Rengakos, A. and Zimmermann, B. (eds.), *Homer-Handbuch. Leben – Werk – Wirkung* (Stuttgart), 228–56.

Weber, M. 1968: *Economy and Society* (Berkeley/Los Angeles/London).

West, M.L. 2002: '"Eumelos": A Corinthian Epic Cycle?'. *JHS* 122, 109–33.

West, S. 2003: '"The Most Marvellous of All Seas"; The Greek Encounter with the Euxine'. *Greece and Rome* 50.2, 151–67.

Westlake, H.D. 1987: 'Diodorus and the Expedition of Cyrus'. *Phoenix* 41.3, 241–54.

Whitley, J. 2001: *The Archaeology of Ancient Greece* (Cambridge).

Wilkins, J. 2005: 'Fish as a Source of Food in Antiquity'. In Bekker-Nielsen 2005, 21–30.

Wilkins, J. and Nadeau, R. (eds.) 2015: *A Companion to Food in the Ancient World* (Malden, MA/Oxford).

Winckler, H. 1903: *Die Völker Vorderasiens* (Leipzig).

Winfield, D. 1977: 'The Northern Routes across Anatolia'. *AS* 27, 151–66.

Winter, E. and Zimmermann, K. (eds.) 2013: *Neue Funde und Forschungen in Bithynien: Friedrich Karl Dörner zum 100. Geburtstag gewidmet* (Asia Minor Studien 69) (Bonn).

Wojan, F. 2003: 'Kérasonte du Pont sous l'Empire romain: étude historique et corpus monétaire'. *Revue numismatique*, ser. 6, 159, 257–90.

Woudhuizen, F.C. 2006: *The Ethnicity of the Sea Peoples* (Dissertation, Erasmus University of Rotterdam).

Wright, J.C. 1998: 'The Place of Troy among the Civilizations of the Bronze Age'. *The Classical World* 91.5, 356–68.

Xydopoulos, I. 2010: 'Defining Identities in the Black Sea Region: The Case of the Massagetae in Greek Literary Sources'. *Egnatia* 14, 19–28.

—. 2018: 'The Cimmerians and their Perception in Ancient Greek Sources'. In Manoledakis *et al.* 2018, 65–81.

Yalçın, Ü. 2008: 'Ancient Metallurgy in Anatolia'. In Yalçın, Ü., Özbal, H. and Paşamehmetoğlu, G. (eds.), *Ancient Mining in Turkey and Eastern Mediterranean* (Tagungsband International Symposium AMITEM 2008) (Ankara), 15–42.

Yanko-Hombach, V. 2007a: 'Late Quaternary history of the Black Sea: an overview with respect to the Noah's Flood Hypothesis'. In Erkut, G. and Mitchell, S. (eds.), *The Black Sea: Past, Present and Future* (Proceedings of the International Interdisciplinary Conference, Istanbul, 14–16 October 2004) (British Institute of Archaeology at Ankara Monograph 42) (London/Istanbul), 5–20.

—. 2007b: 'Controversy over Noah's Flood in the Black Sea: Geological and foraminiferal evidence from the shelf'. In Yanko-Hombach *et al.* 2007, 149–203.

Yanko-Hombach, V., Gilbert, A.S., Panin, N. and Dolukhanov, P.M. (eds.) 2007: *The Black Sea Flood Question. Changes in Coastline, Climate and Human Settlement* (Dordrecht).

Yanko-Hombach, V., Mudie, P. and Gilbert, A.S. 2011: 'Was the Black Sea Catastrophically Flooded during the Holocene? – Geological evidence and archaeological impacts'. In Benjamin, J., Bonsall, C., Pickard, C. and Fischer, A. (eds.), *Submerged Prehistory* (Oxford/Oakville, CT), 245–62.

Yavuz, V., Akcar, N. and Schlüchter, C. 2007: 'The frozen Bosphorus and its paleoclimatic implications based on a summary of the historical data'. In Yanko-Hombach *et al.* 2007, 633–49.

Yiğitpaşa, D. 2018: 'An Overview of the Iron Age Settlements in Samsun'. In Yiğitpaşa *et al.* 2018, 110–26.

Yiğitpaşa, D., Öniz, H. and Temür, A. (eds.) 2018: *Black Sea Archaeology Studies: Recent Developments* (Istanbul).

Yıldırım, Ş. 2022: 'Results of Recent Archaeological Investigations of Archaic Tieion'. In Boardman, J., Hargrave, J., Avram, A. and Podossinov, A. (eds.), *Connecting the Ancient West and East: Studies Presented to Prof. Gocha R. Tsetskhladze* (Monographs on Antiquity 8) (Leuven/Paris/Bristol, CT), 855–75.

Yılmaz, Y. 2007: 'Morphotectonic development of the southern Black Sea region and the Bosporus channel'. In Yanko-Hombach *et al.* 2007, 537–69.

Zacharia, K. 2008: 'Herodotus' Four Markers of Greek Identity'. In Zacharia, K. (ed.), *Hellenisms: Culture, Identity, and Ethnicity from Antiquity to Modernity* (Aldershot), 21–36.

Zournatzi, A. 2000: 'The Processing of Gold and Silver Tax in the Achaemenid Empire: Herodotus 3.96.2 and the Archaeological Realities'. *Studia Iranica* 29, 241–71.

INDEX

Fig. 1: Geophysical map of the Black Sea (by the European Environment Agency – https://www.eea.europa.eu/legal/copyright).

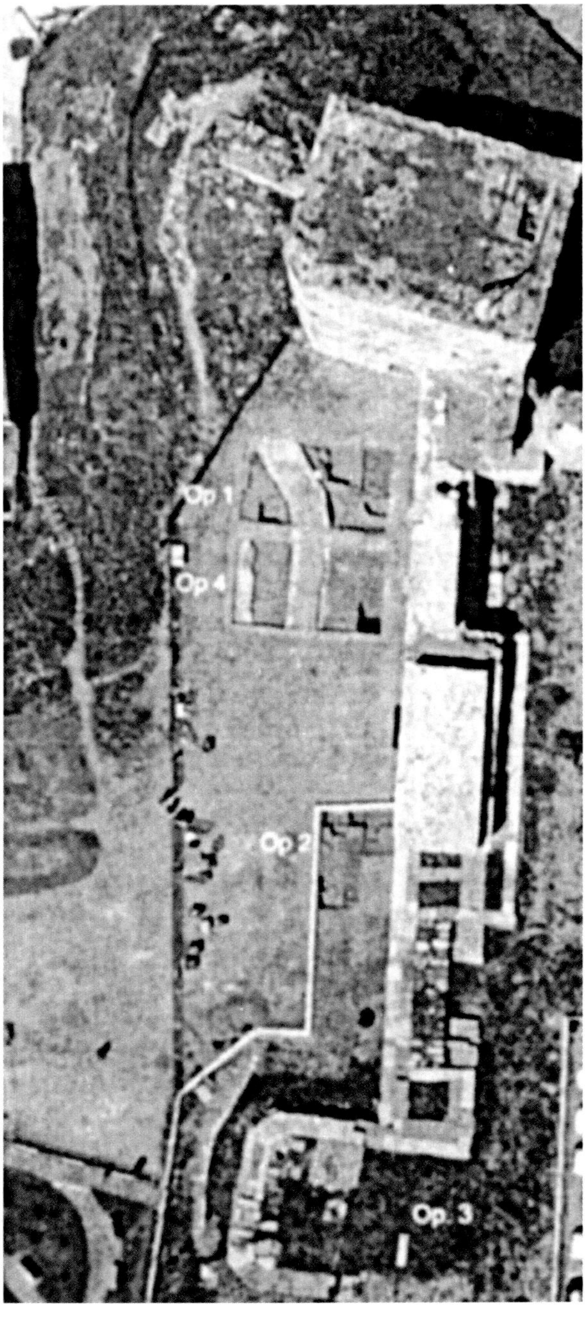

Fig. 2: Overview of the trenches opened close to the castle of Sinop (after Doonan *et al.* 2017, fig. 9-1).

Fig. 3: General view (from the north) of the site of Tieion and the modern village of Filyos (photograph by Ş. Yıldırım).

Fig. 4: General view (from the south) of the acropolis of Tieion and the estuary of the Billaeus (after Atasoy 2016, fig. 1).

Fig. 5: Aerial photograph of the acropolis of Tieion
(photograph by Ş. Yıldırım).

Fig. 6: The ancient theatre of Tieion
(after Atasoy 2016, fig. 8).

Fig. 7: The submerged breakwater(s) off the coast of Filyos
(photograph by Ş. Yıldırım).

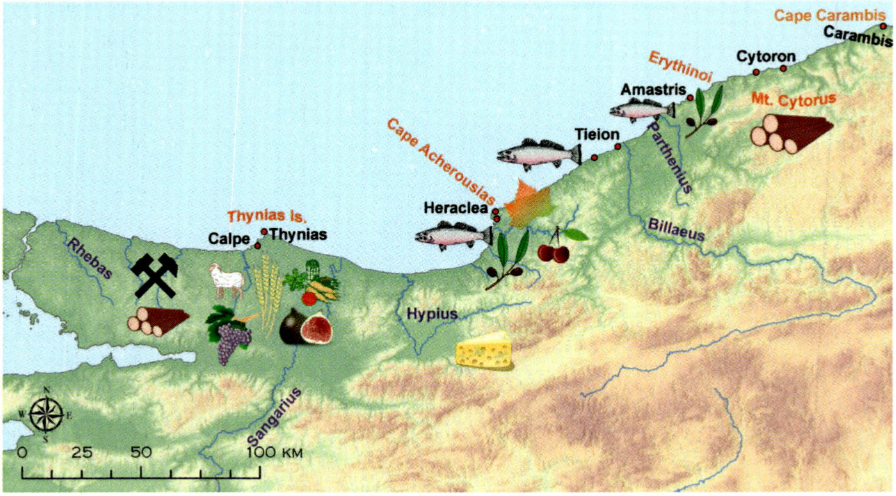

Fig. 8: Map of the southern Black Sea littoral's natural resources.
1. The western part (by J. Donati and M. Manoledakis).

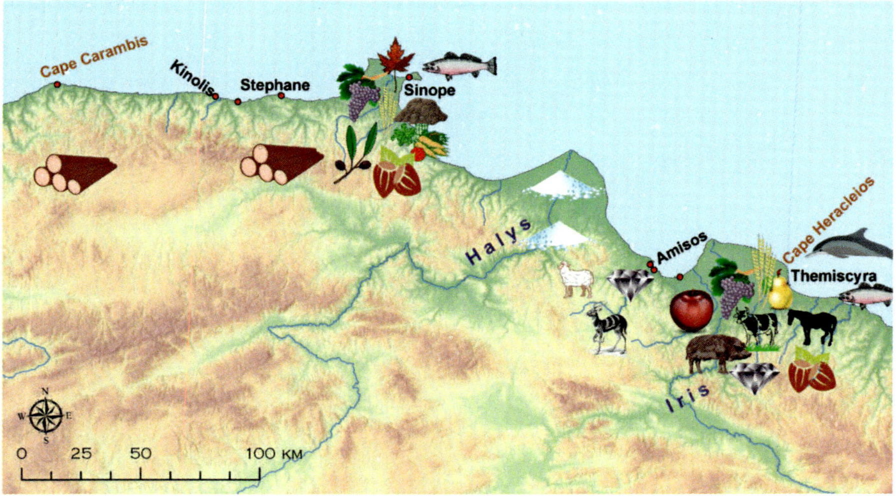

Fig. 8: Map of the southern Black Sea littoral's natural resources.
2. The central part (by J. Donati and M. Manoledakis).

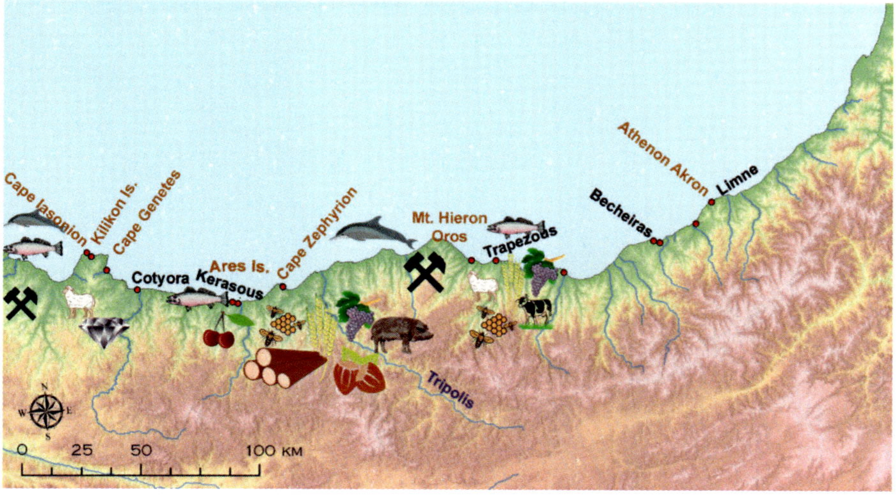

Fig. 8: Map of the southern Black Sea littoral's natural resources.
3. The eastern part (by J. Donati and M. Manoledakis).

Fig. 9: The natural environment around the region of Tieion
(after Atasoy 2018, fig. 5).

Fig. 10: Map of the southern Black Sea littoral with the estimated territories of the indigenous peoples. 1. The western part (by J. Donati and M. Manoledakis).

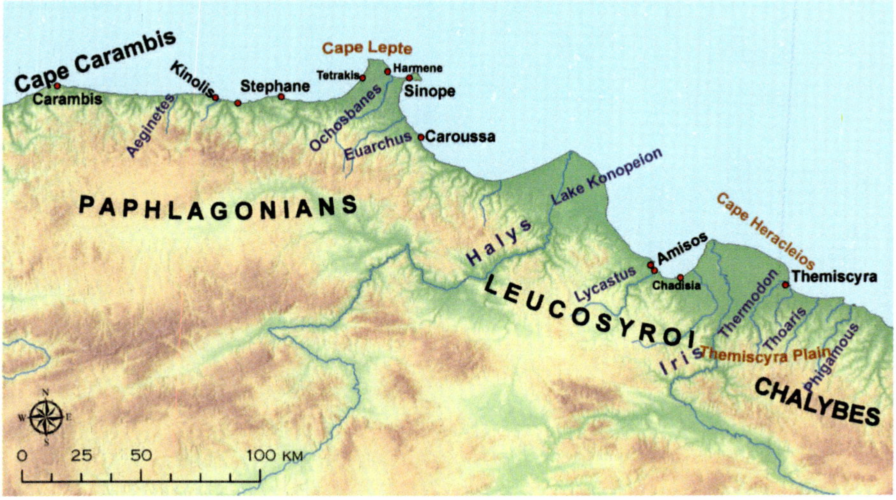

Fig. 10: Map of the southern Black Sea littoral with the estimated territories of the indigenous peoples. 2. The central part (by J. Donati and M. Manoledakis).

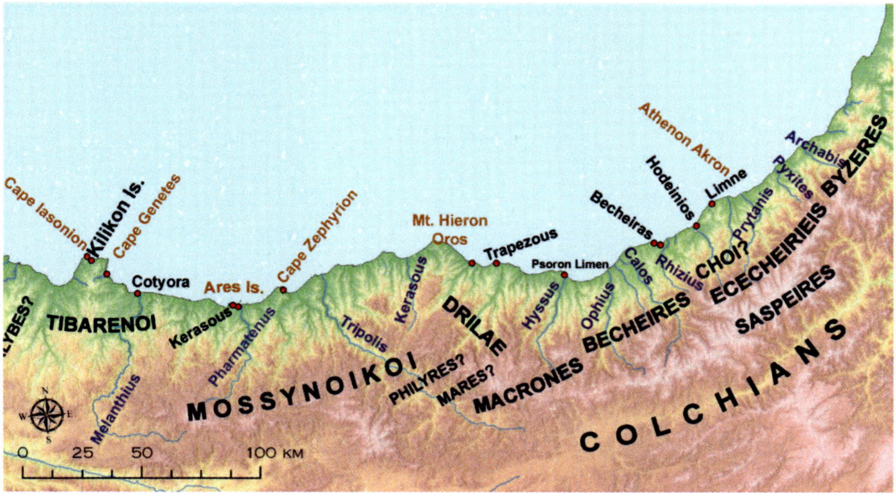

Fig. 10: Map of the southern Black Sea littoral with the estimated territories of the indigenous peoples. 3. The eastern part (by J. Donati and M. Manoledakis).

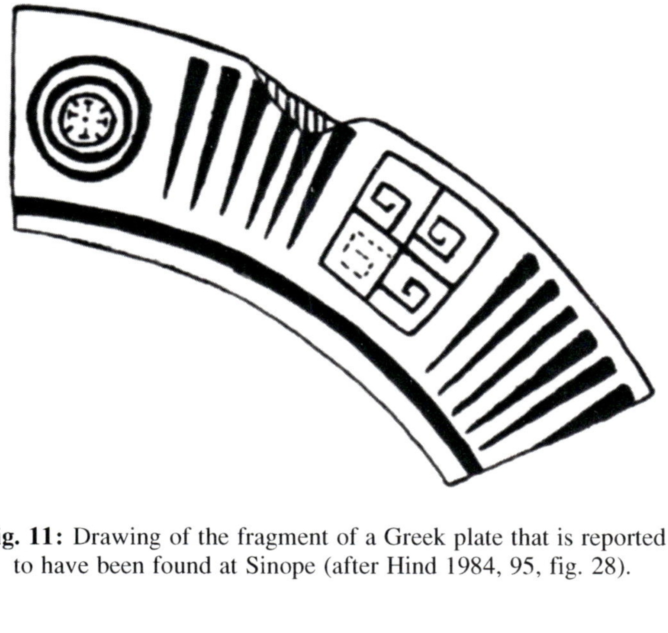

Fig. 11: Drawing of the fragment of a Greek plate that is reported to have been found at Sinope (after Hind 1984, 95, fig. 28).

Fig. 12: Greek pottery fragments from Tieion
(after Atasoy and Erpehlivan 2012, fig. 2).

Fig. 13.1–8: Map of the southern Black Sea littoral (divided into eight segments) with all the ancient settlements and installations, as well as the geographical features that are mentioned in the ancient literature (by J. Donati and M. Manoledakis).

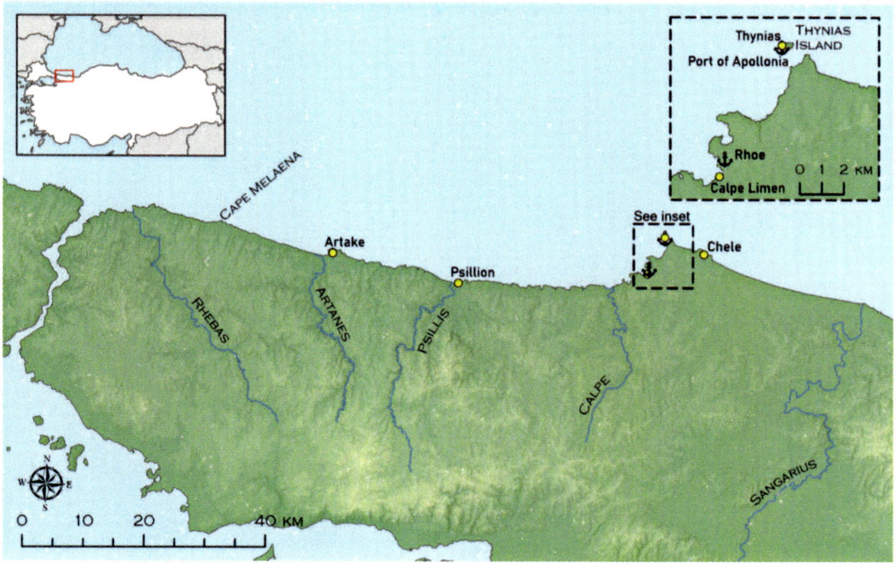

Fig. 13.1

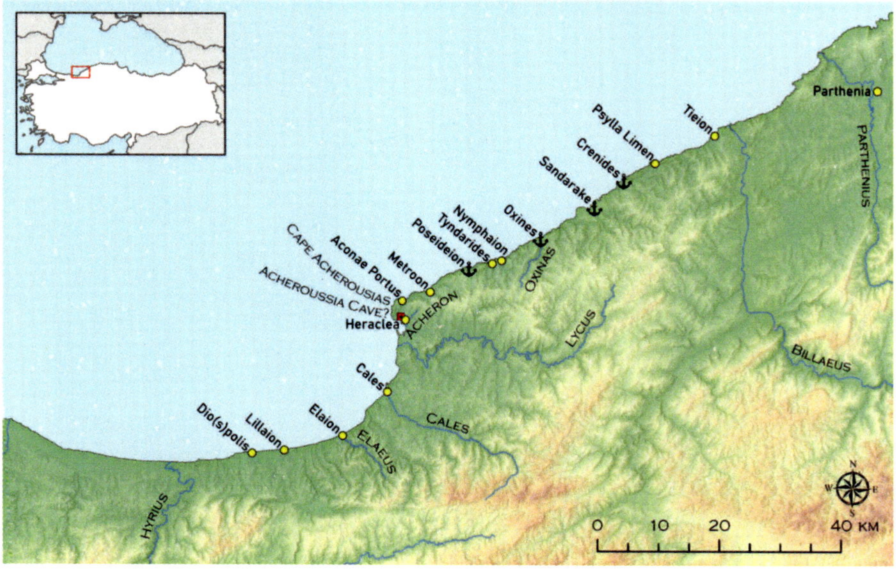

Fig. 13.2

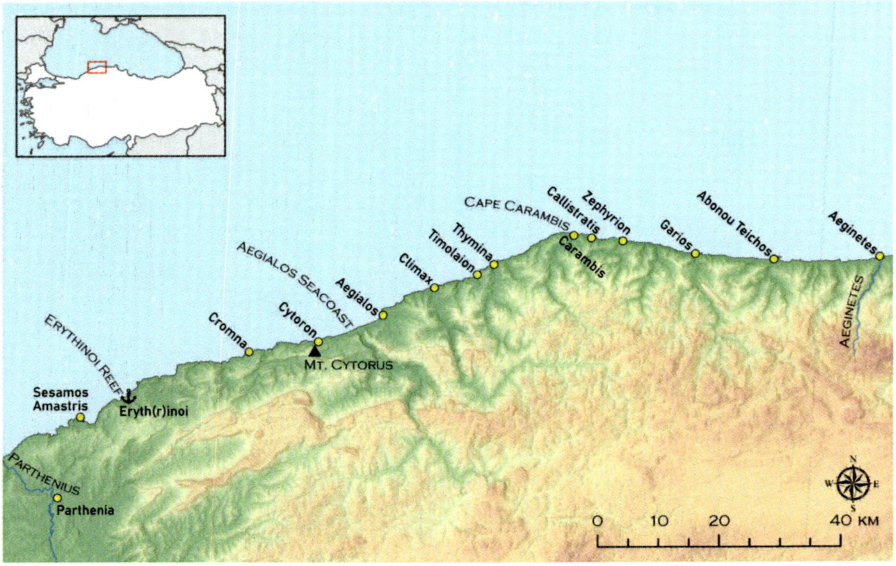

Fig. 13.3

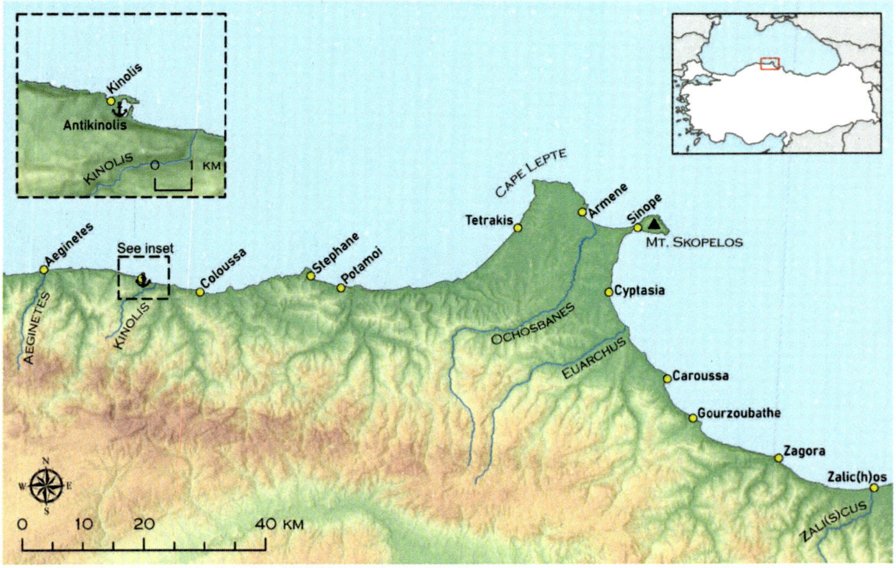

Fig. 13.4

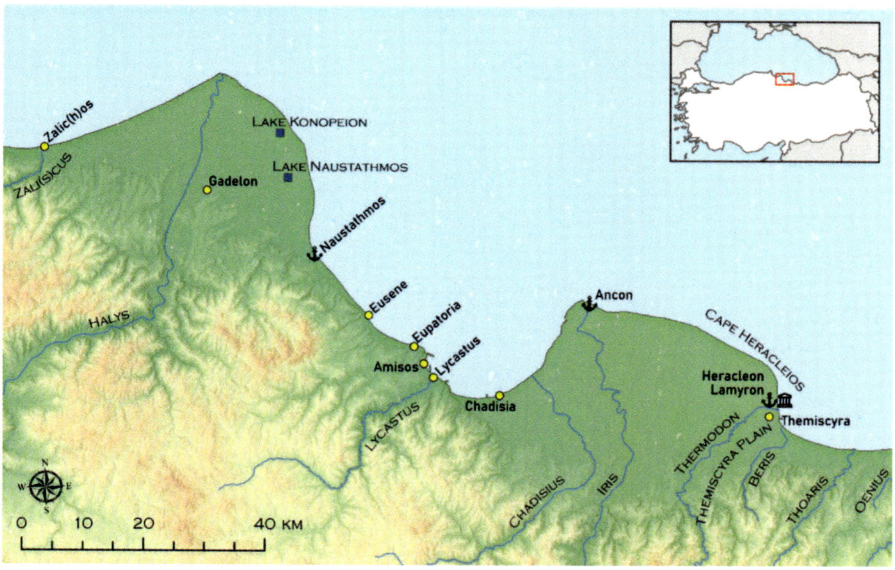

Fig. 13.5

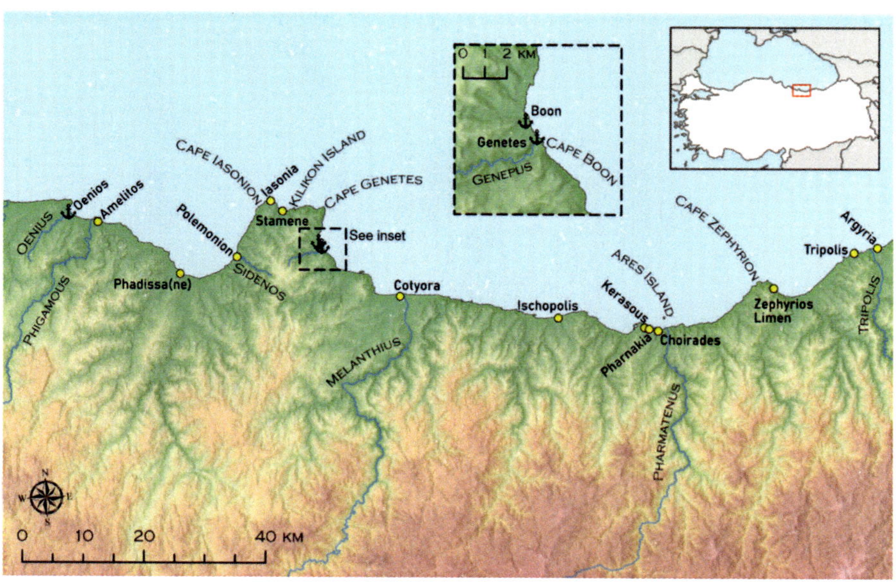

Fig. 13.6

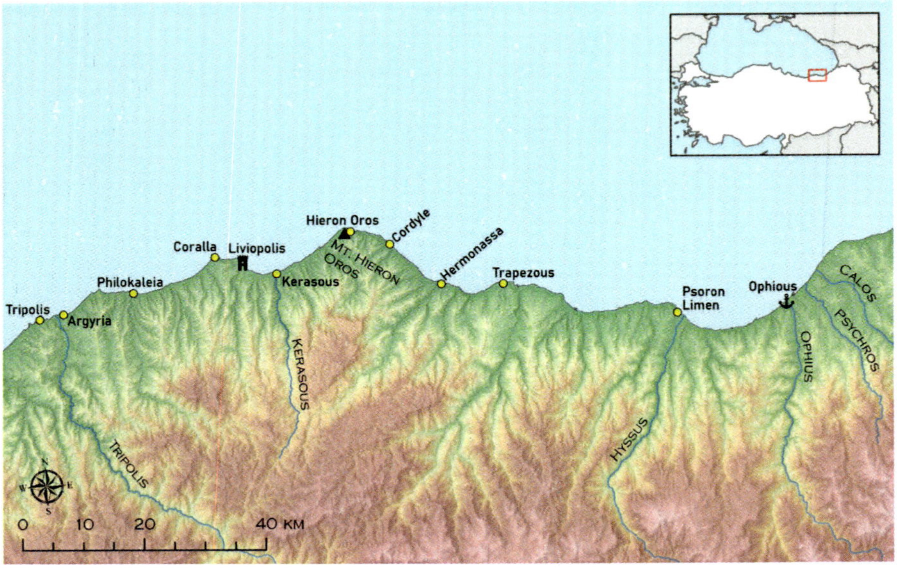

Fig. 13.7

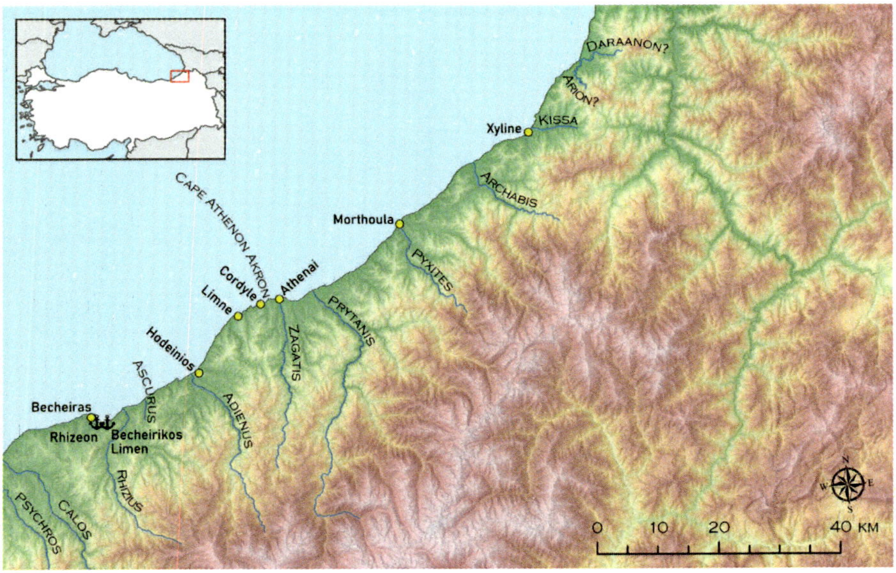

Fig. 13.8

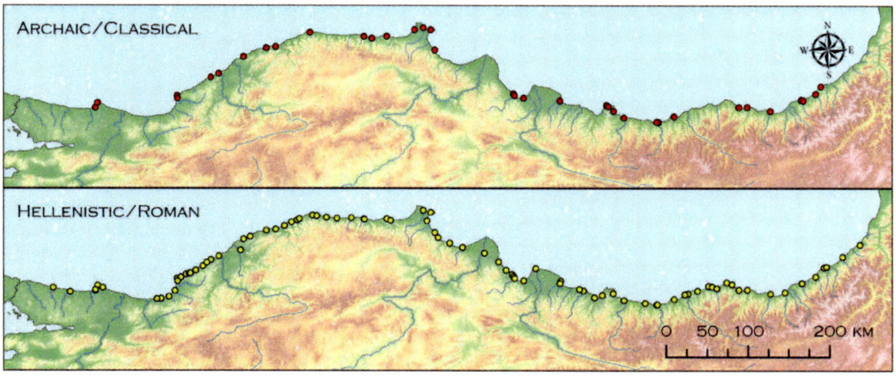

Fig. 14: Maps of the southern Black Sea littoral showing the number and density of settlements and installations in the Archaic/Classical and in the Hellenistic/Roman periods.
The increase during the last two periods is obvious (by J. Donati and M. Manoledakis).

Fig. 15: Part of a pit-house revealed in Sinope (after Doonan *et al.* 2017, fig. 9-4).

COLLOQUIA ANTIQUA

1. G.R. Tsetskhladze (ed.), The Black Sea, Greece, Anatolia and Europe in the First Millennium BC.
2. H. Genz and D.P. Mielke (eds.), Insights into Hittite History and Archaeology.
3. S.A. Kovalenko, Sylloge Nummorum Graecorum. State Pushkin Museum of Fine Arts. Coins of the Black Sea Region. Part I: Ancient Coins from the Northern Black Sea Littoral.
4. A. Hermary and G.R. Tsetskhladze (eds.), From the Pillars of Hercules to the Footsteps of the Argonauts.
5. L. Mihailescu-Bîrliba, *Ex Toto Orbe Romano*: Immigration into Roman Dacia. With Prosopographical Observations on the Population of Dacia.
6. P.-A. Kreuz, Die Grabreliefs aus dem Bosporanischen Reich.
7. F. De Angelis (ed.), Regionalism and Globalism in Antiquity: Exploring Their Limits.
8. A. Avram, Prosopographia Ponti Euxini Externa.
9. Y.N. Youssef and S. Moawad (eds.), From Old Cairo to the New World. Coptic Studies Presented to Gawdat Gabra on the Occasion of his Sixty-Fifth Birthday.
10. R. Rollinger and K. Schnegg (eds.), Kulturkontakte in antiken Welten: vom Denkmodell zum Fallbeispiel.
11. S.A. Kovalenko, Sylloge Nummorum Graecorum. State Pushkin Museum of Fine Arts. Coins of the Black Sea Region. Part II: Ancient Coins of the Black Sea Littoral.
12. A.V. Podossinov (ed.), The Periphery of the Classical World in Ancient Geography and Cartography.
13. A.M. Madden, Corpus of Byzantine Church Mosaic Pavements from Israel and the Palestinian Territories.
14. A. Petrova, Funerary Reliefs from the West Pontic Area (6th–1st Centuries BC).
15. A. Fantalkin and O. Tal, Tell Qudadi: An Iron Age IIB Fortress on the Central Mediterranean Coast of Israel (with References to Earlier and Later Periods).
16. C.M. Draycott and M. Stamatopoulou (eds.), Dining and Death: Interdisciplinary Perspectives on the 'Funerary Banquet' in Ancient Art, Burial and Belief.
17. M.-P. de Hoz, J.P. Sánchez Hernández and C. Molina Valero (eds.), Between Tarhuntas and Zeus Polieus: Cultural Crossroads in the Temples and Cults of Graeco-Roman Anatolia.

18. M. MANOLEDAKIS, G.R. TSETSKHLADZE and I. XYDOPOULOS (eds.), Essays on the Archaeology and Ancient History of the Black Sea Littoral.
19. R.G. GÜRTEKİN DEMİR, H. CEVİZOĞLU, Y. POLAT and G. POLAT (eds.), Archaic and Classical Western Anatolia: New Perspectives in Ceramic Studies.
20. C. KÖRNER, Die zyprischen Königtümer im Schatten der Großreiche des Vorderen Orients. Studien zu den zyprischen Monarchien vom 8. bis zum 4. Jh. v. Chr.
21. G.R. TSETSKHLADZE (ed.), Pessinus and Its Regional Setting. Volume 1.
22. G.R. TSETSKHLADZE (ed.), Pessinus and Its Regional Setting. Volume 2: Work in 2009–2013.
23. I. MOGA, Religious Excitement in Ancient Anatolia. Cult and Devotional Forms for Solar and Lunar Gods.
24. G.R. TSETSKHLADZE (ed.), Phrygia in Antiquity: From the Bronze Age to the Byzantine Period.
25. L. MIHAILESCU-BÎRLIBA (ed.), *Limes*, Economy and Society in the Lower Danubian Roman Provinces.
26. M. COSTANZI and M. DANA (eds.), Une autre façon d'être grec: interactions et productions des Grecs en milieu colonial/Another Way of Being Greek: Interactions and Cultural Innovations of the Greeks in a Colonial Milieu.
27. G.R. TSETSKHLADZE (ed.), Ionians in the West and East.
28. G.R. TSETSKHLADZE (ed.), Archaeology and History of Urartu (Biainili).
29. M.-P. DE HOZ, J.L. GARCÍA ALONSO and L.A. GUICHARD ROMERO (eds.), Greek *Paideia* and Local Tradition in the Graeco-Roman East.
30. A.V. BELOUSOV, Defixiones Olbiae Ponticae (*DefOlb*).
31. J. PORUCZNIK, Cultural Identity within the Northern Black Sea Region in Antiquity. (De)constructing Past Identities.
32. M.G. ABRAMZON and V.D. KUZNETSOV, Coin Hoards Volume XI: Greek Hoards, The Cimmerian Bosporus.
33. A. COŞKUN (ed.), Galatian Victories and Other Studies into the Agency and Identity of the Galatians in the Hellenistic and Early Roman Periods.
34. V.D. KUZNETSOV and M.G. ABRAMZON, The Beginning of Coinage in the Cimmerian Bosporus (a Hoard from Phanagoria).
35. A.-I. PÁZSINT, Private Associations in the Pontic Greek Cities (6th century BC–3rd century AD).

PRINTED ON PERMANENT PAPER • IMPRIME SUR PAPIER PERMANENT • GEDRUKT OP DUURZAAM PAPIER - ISO 9706
N.V. PEETERS S.A., WAROTSTRAAT 50, B-3020 HERENT